Bioreaction Engineering Principles

Second Edition

Bioreaction Engineering Principles

Second Edition

Jens Nielsen and John Villadsen
Technical University of Denmark
Lyngby, Denmark

and

Gunnar Lidén
Lund University
Lund, Sweden

Kluwer Academic / Plenum Publishers
New York, Boston, Dordrecht, London, Moscow

Library of Congress Cataloging-in-Publication Data

Nielsen, Jens Høiriis.
    Bioreaction engineering principles.—2nd ed./Jens Nielsen and John Villadsen and Gunnar Lidén.
      p.   cm.
    Includes bibliographical references and index.
    ISBN 0-306-47349-6
    1. Bioreactors.   I. Villadsen, John.   II. Lidén, Gunnar.   III. Title.

TP248.25.B55 N53 2002
660.6—dc21

                                                                                                                                          2002034178

The First Edition of *Bioreaction Engineering Principles* by Jens Nielsen and John Villadsen was published in 1994 by Plenum Press, New York

ISBN 0-306-47349-6

©2003 Kluwer Academic/Plenum Publishers, New York
233 Spring Street, New York, N.Y. 10013

http://www.wkap.nl/

10 9 8 7 6 5 4 3 2 1

A C.I.P. record for this book is available from the Library of Congress

All rights reserved

No part of this book may be reproduced, stored in a retrieval system, or transmitted in any form or by any means, electronic, mechanical, photocopying, microfilming, recording, or otherwise, without written permission from the Publisher, with the exception of any material supplied specifically for the purpose of being entered and executed on a computer system, for exclusive use by the purchaser of the work

Printed in the United States of America

Preface

This is the second edition of the text "Bioreaction Engineering Principles" by Jens Nielsen and John Villadsen, originally published in 1994 by Plenum Press (now part of Kluwer).

Time runs fast in Biotechnology, and when Kluwer Plenum stopped reprinting the first edition and asked us to make a second, revised edition we happily accepted. A text on bioreactions written in the early 1990's will not reflect the enormous development of experimental as well as theoretical aspects of cellular reactions during the past decade.

In the preface to the first edition we admitted to be newcomers in the field. One of us (JV) has had 10 more years of job training in biotechnology, and the younger author (JN) has now received international recognition for his work with the hottest topics of "modern" biotechnology. Furthermore we are happy to have induced Gunnar Lidén, professor of chemical reaction engineering at our sister university in Lund, Sweden to join us as co-author of the second edition. His contribution, especially on the chemical engineering aspects of "real" bioreactors has been of the greatest value.

Chapter 8 of the present edition is largely unchanged from the first edition. We wish to thank professor Martin Hjortso from LSU for his substantial help with this chapter.

As was the case for the first edition numerous people helped us by carefully reviewing individual chapters. Professor Lars K Nielsen of University of Queensland was a constant sparring partner, both in Australia and lately as a visiting professor at DTU. The help of Dr. Mats Åkesson and of our PhD students, in particular Mikkel Nordkvist, Thomas Grotkjær, Jochen Förster and Morten Skov Hansen is also gratefully acknowledged. MSc student Rebecca Munk Vejborg was of great help in her careful editing of the final version of the manuscript.

All three authors are chemical engineers by education, and we followed in the footsteps of other chemical engineers who "converted" to biotechnology, but retained their passion for a quantitative treatment of problems from the physical world. One of the greatest innovators of biochemical engineering, professor James E. Bailey was also a chemical engineer by education. We wish to dedicate this book to the memory of this eminent scientist, who was a close colleague and a friend (of the senior author for more than 35 years), and whose work is admired by all three of us. If the pages of this book could inspire some students in the way Jay Bailey inspired hundreds of chemical engineering and biochemical engineering students we could hope for no better reward.

John Villadsen and Jens Nielsen	Gunnar Lidén.
BioCentrum-DTU	Kemicentrum, Lund University

Contents

List of Symbols	xi
Chapter 1. Bioreaction Engineering: From Bioprocess Design to Systems Biology	1
1.1 The Structure of the Book	3
1.2 Some Comments on Nomenclature used in the Book	7
1.3 A Final Note	8
References	8
Chapter 2. From Cellular Function to Industrial Products	9
2.1 Cellular Growth	10
2.1.1 From Genotype to Phenotype	13
2.1.2 Transport Processes	15
2.1.2.1 Free Diffusion	16
2.1.2.2 Facilitated Diffusion	20
2.1.2.3 Active Transport	22
2.1.3 Catabolism	24
2.1.3.1 Glycolysis	25
2.1.3.2 TCA Cycle and Oxidative Phosphorylation	27
2.1.3.3 Fermentative Pathways	29
2.1.4 Anabolism	30
2.1.5 Secondary Metabolism	35
2.1.6 Secreted Proteins	37
2.2 Biotech Processes – An Overview	37
2.2.1 Strain Design and Selection	38
2.2.2 Fermentation Media	40
2.2.3 Criteria for Design and Optimization	41
2.2.4 Strain Improvement	42
References	45
Chapter 3. Biochemical Reactions – A First Look	47
3.1 The Continuous Stirred Tank Reactor	47
3.2 Yield Coefficients	53
3.3 Black Box Stoichiometries	57
3.4 Degree of Reduction Balances	60
3.5 Systematic Analysis of Black Box Stoichiometries	73
3.6 Identification of Gross Measurement Errors	77
Problems	88
References	92
Chapter 4. Thermodynamics of Biochemical Reactions	95
4.1 Chemical Equilibrium and Thermodynamic State Functions	95
4.1.1 Changes in Free Energy and Enthalpy	97
4.1.2 Combustion – A Change in Reference State	102
4.2 Heat of Reaction	103
4.3 Non-equilibrium Thermodynamics	109
Problems	115
References	118

Chapter 5. Biochemical Reaction Networks — 119
    5.1 Basic Concepts — 119
    5.2 Growth Energetics — 124
        5.2.1 Consumption of ATP for Cellular Maintenance — 125
        5.2.2 Energetics of Anaerobic Processes — 128
        5.2.3 Energetics of Aerobic Processes — 132
    5.3 Simple Metabolic Networks — 142
    5.4 Flux Analysis in Large Metabolic Networks — 151
        5.4.1 Use of Measurable Rates — 153
        5.4.2 Use of Labeled Substrates — 163
        5.4.3 Use of Linear Programming — 171
    Problems — 179
    References — 186

Chapter 6. Enzyme Kinetics and Metabolic Control Analysis — 189
    6.1 Michaelis-Menten and Analogous Enzyme Kinetics — 190
    6.2 More Complicated Enzyme Kinetics — 195
        6.2.1 Variants of Michaelis-Menten Kinetics — 195
        6.2.2 Cooperativity and Allosteric Enzymes — 201
    6.3 Metabolic Control Analysis — 207
    Problems — 233
    References — 234

Chapter 7. Modeling of Growth Kinetics — 235
    7.1 Model Structure and Complexity — 237
    7.2 A General Structure for Kinetic Models — 240
        7.2.1 Specification of Reaction Stoichiometries — 240
        7.2.2 Reaction Rates — 242
        7.2.3 Dynamic Mass Balances — 244
    7.3 Unstructured Growth Kinetics — 245
        7.3.1 The Black Box Model — 245
        7.3.2 Multiple Reaction Models — 253
        7.3.3 The Influence of Temperature and pH — 261
    7.4 Simple Structured Models — 265
        7.4.1 Compartment Models — 265
        7.4.2 Cybernetic Models — 274
    7.5 Mechanistic Models — 278
        7.5.1 Genetically Structured Models — 279
        7.5.2 Single Cell Models — 289
    7.6 Morphologically Structured Models — 290
        7.6.1 Oscillating Yeast Cultures — 295
        7.6.2 Growth of Filamentous Microorganisms — 300
    Problems — 306
    References — 311

Chapter 8. Population Balance Equations — 315
    Problems — 335
    References — 338

Chapter 9. Design of Fermentation Processes — 339
    9.1 The Stirred Tank Bioreactor — 340
        9.1.1 Batch Operation — 342
        9.1.2 The Continuous Stirred Tank Reactor — 352
        9.1.3 Biomass Recirculation — 359
        9.1.4 The Stirred Tank with Substrate Extracted from a Gas Phase — 364

		9.1.5 Fed-batch Operation	367
	9.2	The Plug Flow Reactor	372
	9.3	Dynamic Analysis of Continuous Stirred Tank Bioreactors	380
		9.3.1 Dynamic Response of the Reactor for Simple, Unstructured Kinetic Models	380
		9.3.2 Stability Analysis of a Steady State Solution	388
		9.3.3 Dynamics of the Continuous Stirred Tank for a Mixed Microbial Population	397
	Problems		409
	References		420

Chapter 10. Mass Transfer — 423

10.1	Gas-Liquid Mass Transfer	425
	10.1.1 Models for k_l	428
	10.1.2 Interfacial Area and Bubble Behavior	430
	10.1.3 Empirical Correlations for $k_l a$	438
	10.1.4 Mass Transfer Correlations Based on Dimensionless Groups	442
	10.1.5 Gas-Liquid Oxygen Transfer	448
	10.1.6 Gas-Liquid Mass Transfer of Components Other than Oxygen	453
10.2	Mass Transfer to and into Solid Particles	456
	10.2.1 External Mass Transfer	456
	10.2.2 Intraparticle Diffusion	460
Problems		469
References		474

Chapter 11. Scale-Up of Bioprocesses — 477

11.1	Scale-up Phenomena	477
11.2	Bioreactors	478
	11.2.1 Basic Requirements and Reactor Types	478
	11.2.2 The Stirred Tank Bioreactor	480
11.3	Physical Processes of Importance for Scale-Up	482
	11.3.1 Mixing	482
	11.3.2 Power Consumption	486
	11.3.3 Heat Transfer	491
	11.3.4 Scale-Up Related Effects on Mass Transfer	495
	11.3.5 Rheology of Fermentation Broths	496
	11.3.6 Flow in Stirred Tank Reactors	501
11.4	Metabolic Processes Affected by Scale-up	508
11.5	Scale-up in Practice	510
Problems		514
References		517

Index — 519

List of Symbols

Symbols that are defined and used only within a particular Example, Note, or Problem are not listed. It should be noted that a few symbols are used for different purposes in different chapters. For this reason more than one definition may apply for a given symbol.

a	Cell age (h)
a	Specific interfacial area (m^2 per m^3 of medium)
a_d	Specific interfacial area (m^2 per m^3 of gas-liquid dispersion)
a_{cell}	Specific cell surface area (m^2 per gram dry weight)
$\mathbf{A}$	Matrix of stoichiometric coefficients for substrates, introduced in Eq. 7.2
$b(y)$	Breakage frequency (h^{-1})
Bi	Biot number, given by Eq. (10.59)
$\mathbf{B}$	Matrix of stoichiometric coefficients for metabolic products, introduced in Eq. 7.2
c_i	Concentration of the ith chemical compound (kg m^{-3})
c_i^*	Saturation concentration of the ith chemical compound (kg m^{-3})
c	Vector of concentrations (kg m^{-3})
C_{ij}	Concentration control coefficient of the jth intermediate with respect to the activity of the ith enzyme
C_i^J	Flux control coefficient with respect to the activity of the ith enzyme
$\mathbf{C}^*$	Matrix containing the control coefficients [defined in Eq. (6.44)]
d_b	Bubble diameter (m)
δ_f	Thickness of liquid film (m)
d_{mean}	Mean bubble diameter (m)
d_{mem}	Lipid membrane thickness (m)
d_s	Stirrer diameter (m)
d_{Sauter}	Mean Sauter bubble diameter (m), given by Eq. (10.18)
D	Dilution rate (h^{-1}), given by Eq. (3.1)
D_{max}	Maximum dilution rate (h^{-1})
D_{mem}	Diffusion coefficient in a lipid membrane ($m^2\ s^{-1}$)
D_{eff}	Effective diffusion coefficient ($m^2\ s^{-1}$)
D_i	Diffusion coefficient of the ith chemical compound ($m^2\ s^{-1}$)
Da	Damköhler number, given by Eq. (10.37)
e_0	Enzyme concentration (g enzyme L^{-1})
E_g	Activation energy of the growth process in Eq. (7.27)
$\mathbf{E}$	Elemental matrix for all compounds
$\mathbf{E_c}$	Elemental matrix for calculated compounds
$\mathbf{E_m}$	Elemental matrix for measured compounds
$f(y,t)$	Distribution function for cells with property $\mathbf{y}$ in the population, Eq. (8.1)
$\mathbf{F}$	Variance-covariance matrix
g	Gravity (m s^{-2})
G	Gibbs free energy (kJ $mole^{-1}$)
G^0	Gibbs free energy at standard conditions (kJ $mole^{-1}$)
ΔG_{ci}	Gibbs free energy of combustion of the ith reaction component (kJ $mole^{-1}$)

ΔG_d	Gibbs free energy of denaturation (kJ mole^{-1}), Eq. (7.28)
ΔG^0_{ci}	Gibbs free energy of combustion of the ith reaction component at standard conditions (kJ mole^{-1})
ΔG^0_f	Gibbs free energy of formation at standard conditions (kJ mole^{-1})
Gr	Grashof number, defined in Table 10.6
h	Test function, given by Eq. (3.52)
$h(y)$	Net rate of formation of cells with property y upon cell division (cells h^{-1})
$h^+(y)$	Rate of formation of cells with property y upon cell division (cells h^{-1})
$h^-(y)$	Rate of disappearance of cells with property y upon cell division (cells h^{-1})
H_A	Henry's constant for compound A (atm L mole^{-1})
ΔH_{ci}	Enthalpy of combustion of the ith reaction component (kJ mole^{-1})
ΔH^0_f	Enthalpy of formation (kJ mole^{-1})
$\mathbf{I}$	Identity matrix (diagonal matrix with 1 in the diagonal)
$\mathbf{J}$	Jacobian matrix, Eq. (9.102)
k_0	Enzyme activity (g substrate [g enzyme]$^{-1}$ h^{-1})
k_i	Rate constant (e.g. kg kg^{-1} h^{-1})
k_g	Mass transfer coefficient for gas film (e.g. mole atm^{-1} s^{-1} m^{-2})
k_l	Mass transfer coefficient for a liquid film surrounding a gas bubble (m s^{-1})
$k_l a$	Volumetric mass transfer coefficient (s^{-1})
k_s	Mass transfer coefficient for a liquid film surrounding a solid particle (m s^{-1})
K_a	Acid dissociation constant (moles L^{-1})
K_l	Overall mass transfer coefficient for gas-liquid mass transfer (m s^{-1})
K	Partition coefficient
K_{eq}	Equilibrium constant
K_m	Michaelis constant (g L^{-1}), Eq. (6.1)
m	Amount of biomass (kg)
m	Degree of mixing, defined in Eq. (11.1)
m_{ATP}	Maintenance-associated ATP consumption (moles ATP [kg DW]$^{-1}$ h^{-1})
m_s	Maintenance-associated specific substrate consumption (kg [kg DW]$^{-1}$ h^{-1})
$M_n(t)$	The nth moment of a one-dimensional distribution function, given by Eq. (8.9)
n	Number of cells per unit volume (cells m^{-3}), Eq. (8.1)
N	Stirring speed (s^{-1})
N_A	Aeration number, defined in Eq. (11.9)
N_f	Flow number
N_p	Power number, defined in Eq. (11.5)
p	Extracellular metabolic product concentration (kg m^{-3})
p_A	Partial pressure of compound A (e.g. atm)
$p(y, y^*, t)$	Partitioning function, Eq. (8.5)
P_i	Productivity of species i in a chemostat (e.g. kg m^{-3} h^{-1})
P	Dimensionless metabolic product concentration
P	Permeability coefficient (m s^{-1})
P	Power input to a bioreactor (W)
P_g	Power input to a bioreactor at gassed conditions (W)
$\mathbf{P}$	Variance-covariance matrix for the residuals, given by Eq. (3.46)
Pe	Peclet number, defined in Table 10.6
q^t_A	Volumetric rate of transfer of A from gas to liquid (moles L^{-1} h^{-1})
$q_{A,Obs}$	Observed volumetric formation rate of A (kg m^{-3} h^{-1}), Eq. (10.45)
q_x	Volumetric rate of formation of biomass (kg DW m^{-3} h^{-1})
$\mathbf{q}$	Volumetric rate vector (kg m^{-3} h^{-1})
$\mathbf{q^t}$	Vector of volumetric mass transfer rates (kg m^{-3} h^{-1})

List of Symbols

Q	Number of morphological forms
Q	Heat of reaction (kJ mole^{-1})
Q_1	Fraction of repressor-free operators, given by Eq. (7.52)
Q_2	Fraction of promotors being activated, given by Eq. (7.58)
Q_3	Fraction of promoters, which form complexes with RNA polymerase, in Eq. (7.60)
r	Specific reaction rate (kg [kg DW]$^{-1}$ h^{-1})
r	Enzymatic reaction rate (Chapter 6) (g substrate L^{-1} h^{-1})
r_{ATP}	Specific ATP synthesis rate (moles of ATP [kg DW]$^{-1}$ h^{-1})
r	Specific reaction rate vector (kg [kg DW]$^{-1}$ h^{-1})
r$_s$	Specific substrate formation rate vector (kg [kg DW]$^{-1}$ h^{-1})
r$_p$	Specific product formation rate vector (kg [kg DW]$^{-1}$ h^{-1})
r$_x$	Specific formation rate vector of biomass constituents (kg [kg DW]$^{-1}$ h^{-1})
r(y,t)	Vector containing the rates of change of properties, in Eq. (8.2)
R	Gas constant (=8.314 J K^{-1} mole^{-1})
R	Recirculation factor
R	Redundancy matrix, given by Eq. (3.39)
R$_r$	Reduced redundancy matrix
Re	Reynolds number, defined in Table 10.6
s	Extracellular substrate concentration (kg m^{-3})
s	Extracellular substrate concentration vector (kg m^{-3})
s_f	Substrate concentration in the feed to the bioreactor (kg m^{-3})
S	Dimensionless substrate concentration
ΔS	Entropy change (kJ mole^{-1} K^{-1})
Sc	Schmidt number, defined in Table 10.6
Sh	Sherwood number, defined in Table 10.6
t	Time (h)
t_c	Circulation time (s)
t_m	Mixing time (s)
T	Temperature (K)
T	Total stoichiometric matrix
T$_1$	Stoichiometric matrix corresponding to non-measured rates in rows of **T**^{T}
T$_2$	Stoichiometric matrix corresponding to known rates of **T**^{T}
u_b	Bubble rise velocity (m s^{-1})
u_i	Cybernetic variable, given by Eq. (7.41)
u_s	Superficial gas velocity (m s^{-1})
u	Vector containing the specific rates of the metamorphosis reaction (kg kg^{-1} h^{-1})
v	Liquid flow (m^3 h^{-1})
v_e	Liquid effluent flow from the reactor (m^3 h^{-1})
v_f	Liquid feed to the reactor (m^3 h^{-1})
v_g	Gas flow (m^3 h^{-1})
v_i	Flux of reaction i (kg [kg DW]$^{-1}$ h^{-1})
v_{pump}	Impeller induced flow (m^3 s^{-1})
v	Flux vector, i.e. vector of specific intracellular reaction rates (kg [kg DW]$^{-1}$ h^{-1})
V	Volume (m^3)
V_d	Total volume of gas-liquid dispersion (m^3)
V_g	Dispersed gas volume (m^3)
V_l	Liquid volume (m^3)
V_y	Total property space, Eq. (8.2)
w_i	Cybernetic variable, given by Eq. (7.42)
x	Biomass concentration (kg m^{-3})
X	Dimensionless biomass concentration

X_i	Concentration of the ith intracellular component (kg [kg DW]$^{-1}$)
X	Vector of concentrations of intracellular biomass components (kg [kg DW]$^{-1}$)
y	Property state vector
Y_{ij}	Yield coefficient of j from i (kg j per kg of i or C-mole of j per kg of i)
Y_{xATP}	ATP consumption for biomass formation (moles of ATP [kg DW]$^{-1}$)
Z_i	Concentration of the ith morphological form (kg [kg DW]$^{-1}$)

Greek Letters

α_{ji}	Stoichiometric coefficients for substrate i in intracellular reaction j
β_{ji}	Stoichiometric coefficient for metabolic product i in intracellular reaction i
$\dot{\gamma}$	Shear rate (s^{-1})
γ_{ji}	Stoichiometric coefficient for intracellular component i in intracellular reaction j
Γ	Matrices containing the stoichiometric coefficients for intracellular biomass components
δ	Vector of measurement errors in Eq. (3.41)
Δ	Matrix for stoichiometric coefficients for morphological forms
ε	Gas holdup (m^3 of gas per m^3 of gas-liquid dispersion)
ε	Porosity of a pellet
ε_{ji}	Elasticity coefficients, defined in Eq. (6.37)
ε	Vector of residuals in Eq. (3.44)
E	Matrix containing the elasticity coefficients
η	Dynamic viscosity (kg m^{-1} s^{-1})
η_{eff}	Internal effectiveness factor, defined in Eq. (10.46)
π_i	Partial pressure of compound i (atm)
θ	Dimensionless time
κ_i	Degree of reduction of the ith compound
μ	The specific growth rate of biomass (h^{-1})
μ_{max}	The maximum specific growth rate (h^{-1})
μ_q	The specific growth rate for the qth morphological form (kg DW [kg DW]$^{-1}$ h^{-1})
ρ_{cell}	Cell density (kg wet biomass [m^{-3} cell])
ρ_l	Liquid density (kg m^{-3})
σ	Surface tension (N m^{-1})
σ^2	Variance
τ	Space time in reactor (h)
τ_s	Shear stress (N m^{-2})
τ_t	Tortuosity factor, used in Eq. (10.43)
Φ	Thiele modulus, given by Eq. (10.49)
Φ_{gen}	Generalized Thiele modulus, given by Eq. (10.55)
$\psi(X)$	Distribution function of cells, Eq (8.8)

Abbreviations

ADP	Adenosine diphosphate
AMP	Adenosine monophosphate
ATP	Adenosine triphosphate
CoA	Coenzyme A
DNA	Deoxyribonucleic acid
E_c	Energy charge

List of Symbols

EMP	Embden-Meyerhof-Parnas
FAD	Flavin adenine dinucleotide (oxidized form)
$FADH_2$	Flavin adenine dinucleotide (reduced form)
FDA	Food and Drug Administration
F6P	Fructose-6-phosphate
GTP	Guanosine triphosphate
G6P	Glucose-6-phosphate
MCA	Metabolic control analysis
NAD^+	Nicotinamide adenine dinucleotide (oxidized form)
NADH	Nicotinamide adenine dinucleotide (reduced form)
$NADP^+$	Nicotinamide adenine dinucleotide phosphate (oxidized form)
NADPH	Nicotinamide adenine dinucleotide phosphate (reduced form)
PEP	Phosphoenol pyruvate
PP	Pentose phosphate
PSS	Protein synthesizing system
PTS	Phosphotransferase system
PYR	Pyruvate
P/O ratio	Number of molecules of ATP formed per atom of oxygen used in the oxidative phosphorylation
RNA	Ribonucleic acid
mRNA	Messenger RNA
rRNA	Ribosomal RNA
tRNA	Transfer RNA
RQ	Respiratory quotient
R5P	Ribose-5-phosphate
TCA	Tricarboxylic acid
UQ	Ubiquinone

1

Bioreaction Engineering:
From Bioprocess Design to Systems Biology

Biotechnology is a key factor in the development and implementation of processes for the manufacture of new food products, animal feedstuffs, pharmaceuticals, and a number of speciality products through the application of microbiology, enzyme technology, and engineering disciplines such as reaction engineering and separation technology. With the introduction of the so-called "new" biotechnologies since 1970, directed manipulation of the cell's genetic machinery through recombinant DNA techniques and cell fusion became possible. This has fundamentally expanded the potential for biological systems to make important biological molecules that cannot be produced by other means. Existing industrial organisms can be systematically altered to produce useful products in cost-efficient and environmentally acceptable ways. Thus, progress in genetic engineering has led to directed genetic changes through recombinant DNA technology, which allows a far more rational approach to strain improvement than by classical methods. This is referred to as *metabolic engineering* (Bailey, 1991), and in recent years, metabolic engineering has been applied for improvement of many different microbial fermentation processes (Ostergaard *et al.*, 2000; Nielsen, 2001). Initially, metabolic engineering was simply the technological manifestation of molecular biology, but with the rapid development in new analytical techniques, in cloning techniques, and in theoretical tools for analysis of biological data, it has become possible to rapidly introduce directed genetic changes and subsequently analyze the consequences of the introduced changes at the cellular level. Often the analysis will point towards an additional genetic change that may be required to further improve the cellular performance, and metabolic engineering therefore involves a cyclic operation with a close integration between analysis of the cellular function and genetic engineering.

The pervasive influence that biotechnology is bound to have on everyday life in the 21st century is recognized by scientists, industrialists, and politicians in industrialized countries and certainly also in the less industrially developed countries of the world, where biotechnology will lead to revolutionary changes in traditional agricultural economies. In order to reap the benefits of development in biology there is, however, an urgent need for industrial microbiologists with experience in solving quantitative problems, particularly as applied to industrial bioreactors. Such persons have traditionally been referred to as biochemical engineers or bioprocess engineers. They should ideally combine a generalist's knowledge of the major topics in molecular biology, microbial physiology, and process engineering with an expert's insight into one particular field.

Traditionally biochemical engineers had an important function in the design and scale up of bioprocesses. Today they are heavily involved also in the very early design phase of a new process, as it has become of utmost importance to apply an integrated process design wherein the prospective production organism is made fit for large scale operation even at the early stages of laboratory strain development. Thus, biochemical engineers have been very active in the rapid progress of metabolic engineering. Teams of engineers and biologists will be responsible for the implementation of an integrated approach to process design. It is therefore important that main stream biologists obtain some insight into quantitative analysis of cellular function and bioreactor operation, and that biochemical engineers continue to learn more about fundamental biological processes.

Besides their role in process design and in metabolic engineering, biochemical engineers must also play an increasing part in fundamental biological research. The genome of a large number of organisms has been completely sequenced, and it has become a major research goal both to assign function to all genes in the genome, referred to as *functional genomics*, and to understand how all the components within the cellular system interact. This can only be done through the use of complex mathematical models, and this field is referred to as *systems biology* (see Fig. 1.1).

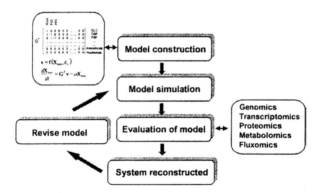

Figure 1.1 Schematic representation of systems biology.
Based on empirical data and knowledge of cellular function a mathematical model is proposed. The model is used to simulate the overall cell function, and model simulations are compared with experimental data. Experimental data may be obtained from: 1) Genomics; information about the genomic sequence; 2) Transcriptomics; data on the expression of all genes obtained by measurement of the complete mRNA pool using DNA arrays; 3) Proteomics; data on all available proteins in the cell obtained by 2D-gel electrophoresis or protein chips; 4) Metabolomics; data on the metabolite profiles inside the cells are obtained using different analytical techniques; and 5) Fluxomics; fluxes through all the cellular reactions are quantified. If there is a good fit between experimental data and model simulations the model is likely to be a good representation of the biological system, which can therefore be reconstructed from its essential parts. A poor fit shows that the model needs to be revised, and often the discrepancy between model simulations and the experimental data will point to where the model needs to be revised [Adapted from Nielsen and Olsson (2002)].

Table 1.1 Definition of research areas where biochemical engineers play an important role

Term	Definition
Bioprocess design	The overall design of a bioprocess. This involves both design of the equipment to be used in the process and quantitative evaluation of how the process is to be operated most efficiently. A key element in scale up of processes from laboratory scale to industrial scale.
Metabolic engineering	The use of directed genetic modification to improve the properties of a given cell, *e.g.* improved yield or productivity, expanded substrate range, and production of novel products. Quantitative analysis of cellular function plays an important role in this field.
Functional genomics	The qualitative assignment of function to open reading frames (ORFs). This includes assignment of function to ORFs that have been identified but have no known function as well as assignment of additional functions to ORFs with already assigned functions. With the interaction of many different processes it is necessary to consider interactions between the many different components, and this may require quantitative analysis
Systems biology	Description of overall cell function through a quantitative study of the interaction between all the individual components in the system (the cell), *e.g.* gene transcription, translation, protein-protein interaction, enzyme catalysis of biochemical reactions, and receptor-metabolite interaction. With a detailed description of the individual molecular events it is also possible to consider cell-cell interactions, and hereby whole cultures can be quantitatively described.

In the future it is expected that the distance will be very short between fundamental discoveries and process design, and biochemical engineers will play an important role in the different research fields mentioned above. Table 1.1 gives our definition of these different areas.

1.1 The Structure of the Book

The present text has been named *Bioreaction Engineering Principles*, and it is the second edition of a textbook that was first published in 1994. The text has been extensively rewritten and many new topics are included. The goal is the same as in the original text: To provide students and industrial researchers with some of the tools needed to analyze, and by analysis to improve the outcome of a bioreaction process. The book can by no means claim to present the desired integrated view of the whole bioprocess from selection of the strain to the downstream processing and further to the final marketable product (separation processes are entirely absent from the text). Our focus is on the central unit of the bioprocess, the bioreactor and the processes that occur in the reactor. Basically a bioreaction can be divided into two parts: operation of the cell factory and the interaction of the cell factories with each other and the environment imposed via operation of the bioreactor. With the above mentioned developments in metabolic engineering and systems biology a fundamental understanding of the cell factory, i.e. how the cells function at different environmental conditions, has become even more important, not only for design of bioreactions but also to gain detailed insight into cellular function. Whether one wants to improve a bioprocess or to understand cellular function at a fundamental level the tools are to a large extent the same. However, as will be discussed in Chapter 7 the structure of the model used to describe cellular function depends on the purpose of the study.

What the text does – hopefully in a useful manner – is to integrate the concepts of mathematical modeling on reasonably general systems with some of the fundamental aspects of microbial physiology. The cell is the ultimate reactor, and everything that is going to come out of this reactor has to pass the boundary between the cell and the environment. But what happens inside the cell, in the *biotic phase*, is intimately coupled with the conditions in the environment, the *abiotic phase*. Therefore the coupling between cell and environment must be given a very serious treatment, although much idealization is necessary in order to obtain a model of reasonable complexity that can still be used to study certain general features of bioreactions. The real bioreaction system is an immensely complicated agglomerate of three phases – gas, liquid, and solid – with concentration gradients and time constants of greatly different magnitudes. This system is beyond the scope of any textbook; it is in fact hardly touched upon in front-line research papers. But the individual steps of a bioreaction, transport to or from the cells, and mixing in a vessel can be treated and will be illustrated with numerous examples, most of which are simple enough to be solved without recourse to a computer (and therefore perhaps better suited to impart the understanding of the underlying mechanisms).

The intended target group for this textbook is students who have studied both natural sciences and engineering sciences. This includes most students following a chemical engineering curriculum. Some knowledge of biology will be advantageous, but not mandatory for reading the book. The book divides the topic into several different themes, as illustrated in Fig. 1.2. It is of little use to investigate the kinetics of bioreactions without a certain appreciation of the biochemistry of living organisms. The ingestion of substrate components from the abiotic medium and the fate of a substrate as it is being converted through metabolic pathways must be known, and the widely different product distribution under varying environmental conditions must be recognized. Most chemical engineering students and all microbiologists and biochemists have a working knowledge of the major pathways of microorganisms. Still, a brief summary of the subject is given in Chapter 2, which at the same time gives an introduction to design of biotech processes. A cursory study of the many examples dispersed throughout the book may give the impression that *Escherichia coli*, *Saccharomyces cerevisiae*, lactic acid bacteria, and certain filamentous fungi are our favored microbial species, but it is important to emphasize that the concepts described in this textbook are equally well suited to analyze also other cellular systems, i.e. other microbes, cell cultures, plants, animal cells and even human cells.

It is often painful to analyze kinetic data from industrial (or, indeed, academic) research where the mass balances do not even approximately close. A microorganism grows and produces metabolites from substrates. Since all the input carbon and nitrogen must be found in one of the effluents from the bioreactor, the biomass, the remaining substrates or the metabolic products, it appears to be fairly elementary to check whether the essential mass balances close. It may be inferred from the opening remark of the paragraph that this is rarely the case. Lack of instrumentation, the inherent difficulties of making consistent measurements in biological systems (a fact not readily recognized by researchers of less complex systems), or – less easily forgivable – a lack of insight into the biochemistry of the process, that leads to omission of a significant metabolic product, may all contribute to make the raw experimental data unsuitable for analysis. In Chapter 3 we describe methods to check the consistency of experimental data on the overall conversion of substrates to metabolic products.

There is no way in which the myriad reactions occurring inside a microorganism can be described in a consistent set of equations. There is not nearly enough data to do so – many reaction steps are unknown even qualitatively – and the result would anyhow be useless for practical purposes. Thus we shall leave out many reaction steps for either one of two reasons: The rate may be so low that it does not influence the process during the time of observation, or the rate may be so high compared to the frequency of our observations of the system that the step can be regarded as being in equilibrium. The rate of mutation of a microorganism is hopefully much smaller than the specific growth rate of the biomass. Thus mutation can usually be neglected when calculating the result of a batch experiment (an assumption of a *frozen state*). Similarly many steps of a metabolic pathway can safely be assumed to be in a *pseudosteady state* because other steps are orders of magnitude slower and represent the *bottlenecks* of the metabolism. To pinpoint fast and slow steps, the concept of a *time constant* or *characteristic time* for a certain step is useful. We are usually not interested in processes with time constants on the order of milliseconds (although these may be the key objects of spectroscopic studies in fundamental biochemistry), nor are we interested in time constants of several months. In between these very wide limits there is, however, plenty of scope for the modeling of bioreactions.

In Chapter 5 we describe methods for analyzing the cell factory in detail – in particular we focus on the pathways that operate at different growth conditions. This involves both application of simple models where all the reactions are lumped into a few overall pathways and very detailed models that consider a large number of reactions in the metabolic network. Concepts for quantification of the fluxes through the different branches of the metabolic network are presented. These concepts turn out to be very useful to gain further insight into how the cell operates and hereby one may design strategies for improving the cell factory. This is often referred to as metabolic flux analysis, and we present several different methods that can be applied for flux analysis.

Concepts from metabolic flux analysis are clearly useful to gain insight into cell function, but it does not supply any information about how the fluxes are controlled. Here it is necessary to include information about the enzyme kinetics in the analysis. In Chapter 6 we give a short review of enzyme kinetics, and then move on to metabolic control analysis (MCA), a concept that enables quantification of flux control within a given pathway. MCA is extensively described in other textbooks and in research publications, but the short introduction is illustrative for teaching the concepts of flux control in biochemical pathways and may be helpful as an introduction to the subject.

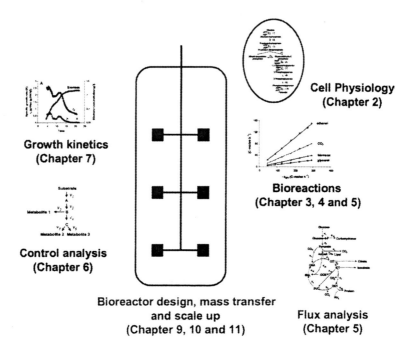

Figure 1.2 An overview of the topics in this textbook. See the text for details.

A requirement for any reaction to proceed is that the laws of thermodynamics are fulfilled, and in Chapter 4 we shortly review some fundamental aspects of classical thermodynamics. The emphasis is on evaluating the feasibility of specific biochemical reactions, but we also give simple engineering approaches for calculation of the heat generation – a topic of importance for the proper design of heat exchangers in bioreactors.

Chapter 7 deals with *unstructured* and *structured microbial kinetics*. One may think of an unstructured model as a primitive data-fitter, but it always tells something about the influence of major substrate concentrations, and surprisingly often it can serve a useful pedagogical purpose – for example in the discussion of ideal reactor performance in Chapter 9 – and it can be used as a control model with adjustable parameters. Broadly speaking, most steady-state microbial reactions can be fitted to fairly simple unstructured models, but only when the internal composition of the cell is in balance with the medium composition – the so called *balanced growth* situation. In the case of rapid transients, where this assumption does not hold, an unstructured model may be of little use for extrapolation purposes. Here, structured models with much more biochemical detail concerning the influence of cell composition on the bioreaction must be used. Structured models may be extended to describe very detailed cellular mechanisms, but often the aim of such models is to understand the interaction between the different processes in the cell, i.e. as specified in the field of systems biology, rather than to simulate a bioprocess. Another type of structure in the biomass is imposed by the changing morphology of the whole culture, either as an ageing phenomenon or as a response to a changing environment. The concept of a homogeneous culture of identical reactors (the individual cells) breaks down in a number of situations, and population balances based on biochemical diversity of the culture have to be introduced. Here the mathematical complexity becomes substantial, but in Chapter 8 we introduce the basic concepts and provide some examples to illustrate how heterogeneity in a population can be quantitatively described.

The *bioreactor* is the subject of Chapters 9, 10 and 11. In a sense, the whole treatment up to that point could be said to lead up to these chapters, where stoichiometry and reaction kinetics together with transport phenomena come together in an engineering design problem. Much has been said and written on the analysis of an *ideal bioreactor*, a reactor with no spatial variation in the medium or the biomass. We have chosen to discuss *steady-state* and *transient operation* of the bioreactor as equally important subjects. Much material on the application of more or less complicated empirical kinetics in reactor design has been left out in order to highlight the basic aspects of operating the *tank reactor* at a steady state and in a dynamic situation caused by changes in the environment. The *plug flow reactor* is given much less space than the tank reactor. We do not wish to get involved with the complexity of modeling tubular reactors (a major subject of most textbooks on chemical reaction engineering), and the stirred tank is by far the most important bioreactor.

Transport processes of a physical nature are well-known complements to kinetics in classical reaction engineering. Since oxygen is a substrate in countless bioreactions and has to be transferred from a gas phase through the liquid phase to the cell – the ultimate reactor – it becomes necessary to treat some concepts of *mass transfer*. These concepts are examined in Chapter 10, but mostly on a general basis; we refrain from citing the many correlations that exist for particular pieces of equipment but rather concentrate on a few fundamental aspects, illustrated with some practical applications from laboratory and pilot plant experimental design.

One might have hoped that in a text as long as the present one it would be possible to give precise design advice for *industrial bioreactors*. Unfortunately it is not possible to give simple design advice on a general basis, but in the final chapter of the book, Chapter 11, we present some general concepts related to design of industrial bioreactors. We do, however, believe that a proper understanding of the topics discussed in this text will be of substantial help to the designer of new industrial scale bioprocesses.

1.2 Some Comments on Nomenclature used in the Book

Biochemical engineering is a multidisciplinary subject, and a unified nomenclature has not yet been developed. As far as possible we have followed the nomenclature used in the standard biotechnology journals (such as D for dilution rate, μ for specific growth rate and so on), but in one respect the nomenclature may differ from that used in other textbooks. As already argued in the introduction we insist on treating the cell as the real bioreactor and the rate of reaction based on cell reactor volume or weight is consequently called r in accordance with the usual practice. The rates **r** are normally referred to as specific rates, and in any physiological study these are the important rates. That the cells live and interact in a vessel, the bioreactor, is incidental to the cellular reaction, but we need another term for the rate of reaction based on reactor volume. Here we choose to use q as in the first edition of the text. We know that in many other textbooks q is used for specific rates. However, we trust that the reader will accept our choice – even a cursory study of the text will soon make the nomenclature acceptable, also if the reader happens to be familiar with another nomenclature.

Also in order to make cross reference to texts in chemical engineering easier we have used a slightly different nomenclature for concentrations in Chapter 10 and 11 than in the remainder of the book.

1.3 A Final Note

As educators and academic researchers, we wish to promote an understanding of the subject-- perhaps sometimes to search for the mechanism of a physical or biological process – and in our effort to contribute to development in general, we wish to assist in the improvement of industrial processes in the broadest possible meaning of the word *improvement*.

On the whole, the present text hopefully illustrates the fundamental engineering cyclic approach to problem solution:
> *Ideas breed experiments, which are systematically described by models, which further lead to new experiments and new model structures – a dictúm valid both in bioprocess design and in systems biology.*

REFERENCES

Bailey, J. E. (1991) Toward a science of metabolic engineering. *Science* **252**:1668-1674
Nielsen, J. (2001) Metabolic engineering. *Appl Microbiol Biotechnol* **55**:263-283
Nielsen, J., Olsson, L. (2002) An Expanded Role for Microbial Physiology in Metabolic Engineering and Functional Genomics: Moving towards Systems Biology. *FEMS Yeast Research*, in press
Ostergaard, S., Olsson, L., Nielsen, J. (2000) Metabolic engineering of *Saccharomyces cerevisiae*. *Microbiol. Mol. Biol. Rev.* **64**:34-50

2

From Cellular Function to Industrial Products

The Biotech industry is one of the fastest growing industrial sectors, and in recent years many new products have been launched. Thus, many new pharmaceuticals are currently produced based on growth of microbial and cell cultures, but also - to the benefit of the environment - many classical chemical products are today produced through recruitment of cell factories. The exploitation of cell cultures for production of industrial products involves growth of the cells in so-called bioreactors, and this is often referred to as a fermentation process. The term fermentation is derived from Latin *fervere*, to boil, and it has been used to describe the metabolism of sugars by microorganisms since ancient times. Thus fermentation of fruits is so old that ancient Greeks attributed its discovery to one of their gods, Dionysos. Among the classical fermentation processes are: beer brewing, which is documented to have been widely known 3000 years B.C. in Babylonia, soy sauce production in Japan and China, and fermented milk beverages in the Balkans and in Central Asia. Before WWII fermentation processes, however, mainly found their application in the production of food and beverages, and it was only after the introduction of the penicillin production in the late 1940's that large-scale fermentation found a use in the production of pharmaceuticals. Today fermentation processes are used in the production of a broad spectrum of products and these processes can be divided into seven categories according to the product made (Table 2.1). Besides the use of fermentation processes for the production of specific products, these processes are also used extensively for specific biotransformations, e.g., in the conversion of sorbose to sorbitol (an important step in the chemical synthesis of vitamin C), as they offer the possibility to perform site-specific chemical modifications. Thus, in the production of complex molecules with one or more chiral centers the use of whole cells to carry out specific biotransformations offers unique possibilities, and today many fine chemicals are produced using specific biotransformations. Another area where microbial cultures are extensively used is in the environmental sector, to clean wastewater and contaminated soil. Here the product is clean water and decontaminated soil.

A key issue in the design of novel fermentation processes, and in the optimization of existing fermentation processes is a quantitative description of the cellular reactions that play an important role in each particular cell factory and of the processes by which the cells interact with the environment imposed on them through the bioreactor operation. These aspects are the essence of this whole textbook, but in this chapter we give a short overview of the processes underlying cellular function and some general aspects concerning design of fermentation processes.

Table 2.1 List of some fermentation products and some market values in year 2000

Category of Product	Product	Typical organism	Market value
Whole cells	Baker's yeast	*Saccharomyces cerevisiae*	
	Lactic acid bacteria	Lactic acid bacteria	
	Single cell protein	Methylotrophic bacteria	
Primary metabolites	Beer, wine	*Saccharomyces cerevisiae*	
	Ethanol	*Saccharomyces carlsbergensis*	12 billion US$
		Saccharomyces cerevisiae	
	Lactic acid	*Zymomonas mobilis*	200 million US$
		Lactic acid bacteria	
	Citric acid	*Rhizopus oryzae*	1.5 billion US$
	Glutamate	*Aspergillus niger*	1 billion US$
	Lysine	*Corynebacterium glutamicum*	500 million US$
	Phenylalanine	*Corynebacterium glutamicum*	200 million US$
		Escherichia coli	
Secondary metabolites	Penicillins	*Penicillium chrysogenum*	4 billion US$
	Cephalosporins	*Acremonium chrysogenum*	11 billion US$
		Streptomyces clavuligerus	
	Statins	*Aspergillus terreus*	9 billion US$
	Taxol	Plant cells	1 billion US$
Recombinant proteins	Insulin	*Saccharomyces cerevisiae*	3 billion US$
		Escherichia coli	
	tPA	Chinese Hamster Ovary cells	1 billion US$
	Erythropoietin	Chinese Hamster Ovary cells	3.6 billion US$
	Human growth hormone	*Escherichia coli*	1 billion US$
	Interferons	*Escherichia coli*	2 billion US$
	Vaccines	Bacteria and yeast	
	Monoclonal antibodies	Hybridoma cells	700 million US$
Enzymes	Detergent enzymes	*Bacilli, Aspergilli*	600 million US$
	Starch industry	*Bacilli, Aspergilli*	200 million US$
	Chymosin	*Aspergilli*	
Polymers	Xanthan gum	*Xanthomonas campestris*	400 million US$
	Polyhydroxyalkanoates	*Alcaligenes erytrophus*	
DNA	Vaccines	*Escherichia coli*	
	Gene therapy	*Escherichia coli*	

2.1 Cellular Growth

Cellular growth is the net result of the uptake and conversion of nutrients – often called substrates, into new cell material – often called biomass. Fig. 2.1 is a schematic view of cellular growth and some of the associated reactions. Cellular growth is an autocatalytic process, as one cell is responsible for the synthesis of more cells, and at conditions with excess nutrients growth is therefore exponential with a certain doubling time for the number of cells. Besides the formation of new biomass (or new cells), the break down (or catabolism) of substrates leads to the formation of several by-products, with carbon dioxide being a typical by-product formed, especially in respiring cells. Here we will simply refer to by-products formed as a result of the catabolism as *metabolic products*. They include many different types of compounds, some of significant industrial interest, e.g., ethanol or lactic acid. The cells may also secrete other metabolic products that have other functions, e.g., acting as hormones or toxins for other cells (antibiotics), and even though the biosynthesis of these may not be a direct result of cellular growth we will also refer to them as metabolic products. Finally, cells may secrete macromolecules that have specialized functions, e.g. insulin by pancreatic cells or hydrolytic enzymes that can degrade complex polymers like starch and xylanases (some of which represent industrial products of significant value).

In order to ensure cell growth the necessary nutrients (or substrates) must be available. Nutrients can roughly be divided into: 1) carbon source, 2) energy source, 3) nitrogen source, 4) minerals, and 5) vitamins. The energy source ensures supply of the necessary Gibbs free energy for cell growth, and often the carbon and energy source are identical. The most typical carbon and energy source is glucose, but carbohydrates like maltose, sucrose, dextrins or starch are also frequently used. Many cells may also use organic acids, e.g., acetic acid or lactic acid, polyols, e.g. glycerol, or alcohols, e.g. ethanol, as carbon and energy source. A few microorganisms grow on hydrocarbons and on CO_2 (as do the plant cells), but using H_2 as energy source. In section 2.2.2 we discuss typical media used in industrial fermentation processes and will there return to the cellular requirements for different nutrients.

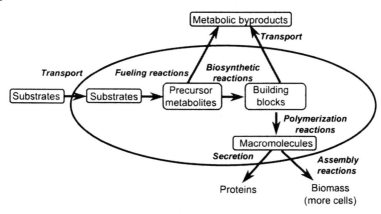

Figure 2.1 Overview of reactions and processes involved in cellular growth and product formation. Substrates are taken up by the cells by transport processes and converted into precursor metabolites via fueling reactions. The precursor metabolites are converted to building blocks that are polymerized to macromolecules. Finally macromolecules are assembled into cellular structures like membranes, organelles etc. that make up the functioning cell. Precursor metabolites and building blocks may be secreted to the extracellular medium as metabolites, or they may serve as precursors for metabolites that are secreted. The cell may also secrete certain macromolecules – primarily proteins that can act as hydrolytic enzymes, but some cells may also secrete polysaccharides.

The conversion of substrates to biomass is clearly the result of a very large number of chemical reactions and events like gene expression, translation of mRNA into functional proteins, further processing of proteins into functional enzymes or structural proteins, and long sequences of biochemical reactions leading to building blocks needed for synthesis of cellular components. The many reactions and processes have traditionally been categorized into four types (see Fig. 2.1): 1) fueling reactions; 2) biosynthesis reactions; 3) polymerization reactions; and 4) assembly reactions (Neidhardt *et al.*, 1990). Reactions involved in transport of compounds across the cytoplasmic membrane do, however, also play a very important role, and in our discussion of processes underlying cellular growth we therefore consider five categories of reactions:

- **Transport reactions**, which are involved in the uptake of substrates and export of metabolites. Transport reactions may also ensure secretion of macromolecules needed outside the cells, e.g., enzymes that are involved in degradation of complex carbohydrates to monomers that can serve as carbon and energy sources for the cells.
- **Fueling reactions**, which convert the substrates into 12 so-called *precursor metabolites* (see Table 2.4) which forms the basis for biosynthesis of cell mass. Additionally, the fueling reactions generate Gibbs free energy in the form of ATP that is used for biosynthesis, polymerization, and assembling reactions (see below). Finally, the fueling reactions produce reducing power needed for biosynthesis. The fueling reactions include all biochemical pathways generally referred to as the *catabolism* that are involved in degrading and oxidizing substrates. As indicated in Fig. 2.1 some of the precursor metabolites may be secreted as metabolic products or they may be converted via a few reaction steps to other metabolites that are secreted. The reactions leading to these metabolic products, e.g., ethanol, lactic acid or acetic acid, are included in the group of fueling reactions.
- **Biosynthetic reactions**, which convert the precursor metabolites into so-called *building blocks* used in the synthesis of macromolecules (in the polymerization reactions, see below). These reactions also produce co-enzymes and related metabolic factors, including signal molecules. There is a large number of biosynthetic reactions, which occur in functional units called biosynthetic pathways, each consisting of one to a dozen sequential reactions leading to the synthesis of one or more building blocks. Pathways are easily recognized and are often controlled *en bloc*. In bacteria and other prokaryotes their reactions are often catalyzed by enzymes made from a single piece of mRNA transcribed from a set of contiguous genes forming an operon, whereas in eukaryotes genes encoding enzymes of a given pathway are often collected in so-called gene clusters where the genes are expressed from similar promoters. All biosynthetic pathways begin with one of 12 precursor metabolites. Some pathways begin directly with such a precursor metabolite, others indirectly by branching from an intermediate or an end product of a related pathway. As indicated in Fig. 2.1 some of the building blocks may be secreted as metabolic products, e.g. amino acids, or they may serve as precursors for more complex metabolites – often referred to as secondary metabolites (see Section 2.1.5).
- **Polymerization reactions**, which represent directed, sequential linkage of the building blocks into long (branched or unbranched) polymeric chains. The macromolecules formed by these reactions makes up almost the whole cell mass. The sum of the biosynthetic reactions and the polymerization reactions is generally referred to as the *anabolism*.
- The last group of reactions is the **assembly reactions**, which include chemical modifications

of macromolecules, their transport to pre-specified locations in the cell, and, finally their association to form cellular structures like cell wall, membranes, the nucleus, etc. Thus, the assembly reaction results in the formation of a functional cell from the individual macromolecules synthesized by the polymerization reactions.

Besides the above listed categories of reactions there are other processes that play a very important role. In particular expression of genes determines the overall function of the cell, and by an altered gene expression the cell can vary its active metabolic network (or its catabolic and anabolic activity). As gene expression to a large extent determines which enzymes are active within the cell, one can argue that gene expression represents an overall control of cell function. However, there is a close interaction between metabolism and gene expression, and it is therefore not possible to divide the control into different levels that are separated from each other.

The distinction made here between substrates and metabolic products is in most cases obvious, but in a few cases the cells may reuse a metabolic product, i.e., the product may serve as a second substrate, and here it is more difficult to categorize the compound as either a substrate or a product. An example is the diauxic growth of *Saccharomyces cerevisiae* at aerobic conditions. Here ethanol is formed together with biomass when the yeast grows on glucose present in excess, and when the glucose is exhausted, the growth continues with ethanol as a substrate. Due to glucose repression on the genes encoding enzymes involved in ethanol metabolism there is a lag-phase between exhaustion of glucose and initiation of ethanol consumption. Throughout this text we will only consider substances originally present in the medium as substrates, i.e. in the example stated above, glucose is a substrate and ethanol is a metabolic product. Depending on the process to be described, a given compound may therefore appear both as a substrate and as a metabolic product.

In the following we give a separate treatment of the key processes involved in cellular growth and product formation: 1) gene transcription and translation; 2) transport processes (substrate uptake and product excretion); 3) catabolism (fueling reactions); 4) anabolism (biosynthetic and polymerization reactions); and 5) secondary metabolism. Only those aspects of metabolic processes that appear to be important to a biochemical engineer who wishes to understand how fermentation processes work under various operating conditions will be discussed. In a text of reasonable length and with the above-mentioned aim, it will be impossible to give even a reasonably correct summary of the vast repository of knowledge concerning the biochemistry of cellular function, and we therefore refer to standard textbooks in microbiology and biochemistry. One may also consult the many different reaction databases available on-line, e.g., www.genome.ad.jp.

2.1.1 From Genotype to Phenotype

The genes in living cells are located in the chromosome(s). In prokaryotes (bacteria) there is normally only one chromosome whereas in eukaryotes (fungi, plants, animals, human) there may be several chromosomes of varying size. The number of genes in the chromosome(s) depends significantly on the complexity of the cell. A few bacteria have only a few hundred genes, e.g. *Mycoplasma genitalium* has only about 500 genes, other bacteria have thousands of genes, e.g.,

Escherichia coli has about 5000 genes. Fungi normally have several thousand genes, e.g., baker's yeast *Saccharomyces cerevisiae* has about 6000 genes and filamentous fungi typically more than 8000 genes. Higher eukaryotes such as plants, animals and humans clearly have a much larger number of genes. In these organisms many functions may also be derived from the individual genes through alternative splicing of the mRNA, i.e., the specific mRNA formed by gene transcription can be modified and hereby encode the formation of several different proteins.

The collection of genes (normally referred to as the genome) represents the genotype of a given organism, and through genomic sequencing programs all the genes of many different organisms have become known (see e.g. www.genome.ad.jp for a list of sequenced organisms). The genome represents all the possible functions a cell can express – often referred to as the *genotype*. The actual functions expressed by the cell do, however, depend on which genes are expressed and further translated into active proteins, which are the key factors for determining the cellular function. The term *phenotype* is often used to express the actual functions expressed by a given cell, and clearly the functions may depend on the environmental conditions experienced by the cells. A given cell may therefore express several different phenotypes.

The phenotype of a given cell is controlled at many different levels (see Fig. 2.2):
   a) transcriptional control
   b) control of mRNA degradation
   c) translational control
   d) protein activation/inactivation
   e) allosteric regulation of enzymes.

There is a certain degree of hierarchy in the overall control of cell function. The genes have an overall control since they define all possible phenotypes, and the proteins have a control of the actual phenotype. Signal transduction pathways that convey signals from the environment to specific proteins controlling e.g. gene transcription are placed relatively high in the hierarchy. There are, however, many feed-back loops in the hierarchical control structure, and it is therefore difficult to predict the exact function of the individual genes and to assign function to genes identified via sequencing of whole genomes (this is referred to as functional genomics). In this textbook we are not going to consider in further detail the functional relationship between the genotype of the cell and the phenotype. In most cases we shall take a given phenotype for granted and then analyze it at the quantitative level to see what the consequences of this phenotype are for industrial exploitation of the cell factory.

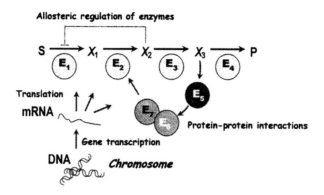

Figure 2.2 Overview of process involved in expressing a specific phenotype in a cell. The chromosome contains a large number of genes, which represents the genotype of the cell. The genes are transcribed into mRNA, which is further translated into proteins. Some of these proteins are enzymes involved in metabolism, e.g., the catabolism of substrates to metabolic products. Some enzymes are allosteric enzymes, which means that their *in vivo* activity is regulated by other metabolites, e.g., there may be feed-back inhibition at high concentration of a certain metabolite. Other proteins are involved in overall regulation of cellular function. Thus, protein kinases may phosphorylate other proteins and hereby activate or deactivate them. Some proteins are involved in regulation of gene expression, and a feed-back from the metabolite levels to gene expression via these regulatory proteins is clearly possible.

2.1.2 Transport Processes

Two structures surround the cytoplasm, of most microbial cells, the cell wall and the cytoplasmic membrane. These structures are normally referred to as the *cell envelope*, and their chemical composition is a determining factor for the transport of species between the abiotic phase, i.e., the exterior of the cell, and the cytoplasm. The cell wall has a rigid structure of cross-linked dissacharides and peptides (peptidoglycans). Its major function is to prevent the cell from bursting due to a high intracellular osmotic pressure. The cytoplasmic membrane, which mainly consists of phospholipids, has a fluid structure with properties that may change dynamically during growth. Most small molecules can easily pass through the cell wall, and therefore the cytoplasmic membrane is determining the transport processes. Larger molecules (e.g., proteins) can pass through the cell wall and the cytoplasmic membrane only if the cell is equipped with special mechanisms for their excretion. We will not describe these mechanisms further, and in the following only focus on the transport of small molecules. Furthermore, as the cytoplasmic membrane is the important barrier between the intracellular and extracellular environment, we restrict the discussion to transport across this membrane.

Table 2.2. Summary of transport processes for different compounds.

Compound	Bacteria	Fungi
Amino acids	Active transport	Active transport
Glucose	Active transport (PTS[#] and permease)	Facilitated diffusion and active transport
Lactose	Active transport (PTS and permease)	Facilitated diffusion and active transport
Glycerol	Free and facilitated diffusion	Free and facilitated diffusion
Ethanol	Free diffusion	Free diffusion
Lactic acid	Active transport and free diffusion	Free diffusion
Acetic acid	Free diffusion	Free diffusion
Carbon dioxide	Free diffusion	Free diffusion
Oxygen	Free diffusion	Free diffusion
Water	Free diffusion	Free diffusion

[#] PTS – phosphotransferase system

There are three different transport mechanisms across the cytoplasmic membrane:
1. free diffusion
2. facilitated diffusion
3. active transport

By the first two mechanisms species are transported down a concentration gradient, i.e. they are passive processes that in principle require no extraneous energy supply to run properly, whereas species transported against a concentration gradient by an active process require a considerable input of Gibbs free energy. Table 2.2 summarizes the type of transport process for a few substrates and metabolic products in bacteria and fungi, respectively. It is observed that most of the substances are transported by the same type of process in the two groups of microorganisms, an important exception being sugars, which are transported actively in bacteria, whereas they may also be transported passively in fungi.

2.1.2.1. Free Diffusion

Transport of a substance across a lipid membrane by free diffusion involves three steps: (1) transfer of the substance from the extracellular medium to the membrane phase, (2) molecular diffusion through the lipid membrane, and (3) transfer from the lipid phase to the cytoplasm. Normally the physical and chemical properties of the cytoplasm are similar to the properties of the extracellular medium, and steps 1 and 3 are therefore similar. Furthermore, the interphase processes can be assumed to be in equilibrium; i.e., the characteristic time for these processes to equilibrate is much smaller than the characteristic time for the molecular diffusion through the lipid layer. The concentration in the lipid layer at the interface can therefore be described as the product of the concentration in the water phase and the so-called partitioning coefficient K_{par}, the ratio of the solubility of the compound in the lipid layer to its solubility in water. Mass flux due to molecular diffusion follows Fick's first law, and the rate of mass transport of a compound into the cell (J, unit: mass per membrane area per time, e.g., g m^{-2} s^{-1}) through a lipid membrane of thickness d_{mem} can therefore be described by

$$J = \frac{D_{mem}}{d_{mem}} K_{par}(c_a - c_b) \quad (2.1)$$

D_{mem} is the diffusion coefficient for the compound under consideration in the lipid membrane, and c_a and c_b are the concentrations of the compound in, respectively, the abiotic phase (extracellular medium) and the biotic phase (the cytoplasm). The ratio $D_{mem}K_{par}/d_{mem}$ is called the permeability coefficient P, and it is frequently used for calculation of the mass transport (Stein, 1990). A collection of permeability coefficients for a few compounds in the cytoplasmic membrane of the plant cell *Chara ceratophylla* is given in Table 2.3. In the absence of experimental data for a particular permeability coefficient, one may use the following to get a rough estimate:

$$P\sqrt{M_w} = 0.028 K_{par}^{oil} \quad (2.2)$$

Table 2.3 Permeability coefficients for compounds in membranes of the plant cell *Chara ceratophylla* and the olive oil-water partitioning coefficient. To evaluate the permeability of other compounds one may use certain rough measures of how chemical groupings on a permeant can be expected to affect the membrane permeability (Stein, 1990): An extra hydroxyl group on the molecule decreases the permeability 100- or 1000-fold. A carboxyl group has an even larger effect. An extra amide group is more or less equivalent to two extra hydroxyl groups. Conversely, an extra methyl group in the compound is likely to increase the permeability five-fold, while a doubling of molecular volume decreases the permeability 30-fold.

Compound	Permeability coefficient (cm s^{-1})	Partitioning coefficient
Carbon dioxide	4.5 10^{-1}	
Bicarbonate	5.0 10^{-7}	
Water	6.6 10^{-4}	
Urea	2.8 10^{-7}	1.5 10^{-4}
Methanol	2.5 10^{-4}	
Ethanol	1.4 10^{-4}	
Ethanediol	1.7 10^{-5}	4.9 10^{-4}
1,2 Propanediol		1.7 10^{-3}
1,3 Propanediol		2.1 10^{-3}
Formic acid		1.5 10^{-2}
Acetic acid		3.0 10^{-2}
Propionic acid		1.5 10^{-1}
Butyric acid		4.4 10^{-1}
Acetamide	1.4 10^{-5}	8.3 10^{-4}
Formamide	2.0 10^{-5}	7.6 10^{-4}
Lactamide	1.5 10^{-6}	
Butyramide	5.0 10^{-5}	
Glucose	5.0 10^{-8}	
Glycerol	2.2 10^{-7}	7.0 10^{-5}

M_w is the molecular weight of the compound, and K_{par}^{oil} is the olive oil water-partitioning coefficient for the compound; P has the unit cm s^{-1}. The correlation has been obtained from measurements on a large number of different compounds. When using the correlation one should, however, recognize that for some compounds P might deviate by a factor of 10 from the value predicted by the correlation; see, e.g., Stein (1990). When calculating a value for the permeability coefficient it is important to consider an appropriate lipid membrane system, since by its definition the permeability coefficient is inversely proportional to the membrane thickness. Strictly speaking, Eq. (2.2) therefore holds only for the system for which it was derived, namely the plant cell *Chara ceratophylla*, but lacking better information it may be used also for other cell types.

In order to calculate the net rate of transport the flux given by Eq. (2.1) is multiplied by the specific surface area of the cell a_{cell} (unit: area per cell dry weight, e.g., m^2 (g DW)$^{-1}$):

$$r = Ja_{cell} = Pa_{cell}(c_a - c_b) \tag{2.3}$$

For a spherical cell with a water content w (g (g cell)$^{-1}$) and cell density ρ_{cell} (g m^{-3}), the specific surface area is $a_{cell} = 6/(d_{cell}(1-w)\rho_{cell})$. Thus, with a few parameters the rate of free diffusion can easily be calculated for a given compound.

The most important compounds transported by free diffusion are oxygen, carbon dioxide, water, organic acids, and alcohols. In their dissociated form, small organic acids are practically insoluble in the lipid membrane, and one should therefore replace their total concentrations in Eq. (2.1) with the concentrations of the undissociated acid on each side of the membrane. These can be calculated from

$$c_{i,undiss} = \frac{c_i}{K_a 10^{pH_i} + 1} \tag{2.4}$$

where K_a is the acid dissociation constant. It is seen that the pH in the aqueous phase at the membrane surface has an influence on $c_{i,undiss}$, and since extra- and intracellular pH are often different, it is therefore possible to have a flux of the acid across the membrane even when $c_a = c_b$.

Example 2.1 Free diffusion of organic acids
Maintenance of intracellular pH is very important for overall cell function, and the cell is equipped with specific enzymes located in the cytoplasmic membrane that ensures that the intracellular pH is kept constant by pumping protons out of the cell. This proton pumping is an active process, and involves consumption of Gibbs free energy in the form of ATP (hereby the name ATPases for these proton pumping enzymes). In the presence of high concentrations of organic acids it has been found that there is a substantial drain of ATP to maintain the intracellular pH, and this is due to an increased flux of protons into the cell via rapid diffusion of the undissociated acids. This effect has been illustrated in a study of Verduyn *et al.* (1992), who analyzed the influence of benzoic acid on the respiration of *Saccharomyces cerevisiae*. They found that the biomass yield on glucose decreased with increasing concentration of the acid. At the same time the specific uptake rates of glucose and oxygen increased. Thus, there is a less

efficient utilization of glucose for biomass synthesis, and the drain of ATP due to the increased proton influx with the presence of benzoic acid explains this. In another study, Schulze (1995) analyzed the influence of benzoic acid on the ATP costs for biomass synthesis in anaerobic cultures of *S. cerevisiae*. He found that the ATP costs for biomass synthesis increased linearly with the benzoic acid concentration, also a consequence of the increased proton influx when this acid is present in the medium.

Henriksen et al. (1998) derived a set of equations that allows quantification of the ATP costs resulting from uncoupling of the proton gradient by organic acids. The aim of the study was to quantify the uncoupling effect of phenoxyacetic acid, a precursor for penicillin V production, on the proton gradient in *Penicillium chrysogenum*. Both forms of this acid may diffuse passively across the plasma membrane, but the undissociated acid has a much larger solubility, *i.e.*, a larger partition coefficient, and is therefore transported much faster. To describe the mass flux of the two forms across the plasma membrane, Henriksen et al. (1998) applied Eq. (2.3). The specific cell area is about 2.5 m^2 (g DW)$^{-1}$ for *P. chrysogenum*, and the permeability coefficients for the undissociated and dissociated forms of phenoxyacetic acid have been estimated to be 3.2 x 10^{-6} and 2.6 x 10^{-10} m s^{-1}, respectively (Nielsen, 1997).

Because the undissociated and dissociated forms of the acid are in equilibrium on each side of the cytoplasmic membrane (HA ↔ H$^+$ + A$^-$) with equilibrium constant K_a, Eq. (2.4) which correlates the two forms with the total acid concentration can be written as

$$c_{undiss} = c_{diss} 10^{pK_a - pH} = \frac{c_{total}}{1 + 10^{pH - pK_a}} \tag{1}$$

where the pK_a for phenoxyacetic acid is 3.1. At pseudo-steady state conditions, the net influx of undissociated acid will equal the net outflux of the dissociated form of the acid:

$$r_{undiss,in} = r_{diss,out}, \text{ or } P_{undiss} a_{cell}(c_{undiss,a} - c_{undiss,b}) = P_{diss} a_{cell}(c_{diss,b} - c_{diss,a}) \tag{2}$$

where subscript *a* and *b* indicate the abiotic and biotic (cytosolic) side of the membrane, respectively. By substituting for the undissociated and dissociated acid concentrations on the abiotic and cytosolic sides of the membrane in terms of the total concentrations on the abiotic and cytosolic sides from eq. (1) and rearranging, we obtain the following equation for the ratio of the total concentrations on the two sides of the membrane:

$$\frac{c_{b,tot}}{c_{a,tot}} = \frac{P_{undiss}\frac{1+10^{pH_b-pK_a}}{1+10^{pH_a-pK_a}} + P_{diss}\frac{1+10^{pH_b-pK_a}}{1+10^{pK_a-pH_a}}}{P_{diss}10^{pH_b-pK_a} + P_{undiss}} \tag{3}$$

Because the permeability coefficient for the undissociated form of the acid is orders of magnitude greater than that of the dissociated form, this equation can be reduced to

$$\frac{c_{b,tot}}{c_{a,tot}} = \frac{1+10^{pH_b-pK_a}}{1+10^{pH_a-pK_a}} \tag{4}$$

Now, the intracellular pH is usually greater than the typical pH of the medium in penicillin cultivations. Equation (4) then indicates that there is a higher total concentration of the acid inside the cells than in the extracellular medium. Using this equation Henriksen *et al.* (1998) calculated the concentration ratio at different extracellular pH values and an intracellular pH of 7.2. For an extracellular pH of 6.5 the accumulation is low (about 2.3-fold), whereas at an extracellular pH of 5.0 the accumulation is high (about 100-fold).

For a given total extracellular acid concentration, the concentrations of both forms of the acid on each side of the cytoplasmic membrane can be calculated using Eq. (1), from which the mass flux of acid across the membrane can be calculated using Eq. (2.3). Because the net outflux of dissociated acid equals the net influx of undissociated acid, the result of acid transport is a net influx of protons, which have to be re-exported by the cytoplasma membrane bound ATPase in order to maintain a constant intracellular pH. If it is assumed that the export of each proton requires the expenditure of one ATP by the ATPase reaction, Henriksen *et al.* (1998) calculated that the ATP consumption resulting from this futile cycle amounts to 0.15 mmol of ATP (g DW)$^{-1}$ h^{-1} at an extracellular pH of 6.5 and an intracellular pH of 7.2. This is a low value compared with other non-growth-associated processes that also consume ATP (see Section 5.2). However, at an extracellular pH of 5.0 the ATP loss is about 7 mmol of ATP (g DW)$^{-1}$ h^{-1} (again with an intracellular pH of 7.2), which is a significant drain of cellular free energy. It is thus seen how the maintenance of acid concentration gradients across the plasma membrane contributes to the decoupling of ATP generation and ATP consumption used strictly for biosynthetic demands.

2.1.2.2 Facilitated Diffusion

In the cytoplasmic membrane there are a number of carrier proteins that allow specific compounds to be transported passively, but considerably faster than by free diffusion across the membrane. This process is referred to as facilitated diffusion, and this transport mechanism is typical for fungi, but much rarer for bacteria – thus glycerol is the only substrate which is known to enter *E. coli* by facilitated diffusion (Ingraham *et al.*, 1983). Facilitated diffusion resembles free diffusion since transport occurs only in the downhill direction of a concentration gradient. The compound can enter the membrane only if there is an available free carrier, and the rate of the transport process therefore follows typical saturation-type kinetics; i.e., at low concentrations the rate is first order with respect to the substrate concentration, whereas it is zero order at high concentrations (see Example 2.2). The most important substances transported by facilitated diffusion are glucose and other sugars in fungi.

Example 2.2. Facilitated diffusion
A substrate is transported across a (lipid) membrane by a carrier molecule, which is present in the membrane either in free form (concentration e) or bound to the substrate (concentration $c_m e$). The free substrate concentration just inside the membrane at $z = 0$ is $c_{ma} = K c_a$, and at the other face of the membrane, at $z = d$ the concentration is $c_{mb} = K_b c_b$ (see figure below).

From Cellular Function to Industrial Products

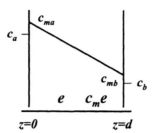

Binding between substrate and carrier is reversible, and the net rate of complex formation is r_m. Mass balances for each of the membrane components are:

$$D_{c_m}\frac{d^2c_m}{dz^2} - r_m = 0 \tag{1}$$

$$D_e\frac{d^2e}{dz^2} - r_m = 0 \tag{2}$$

$$D_{c_me}\frac{d^2c_me}{dz^2} + r_m = 0 \tag{3}$$

The total carrier concentration is

$$e_t = \frac{1}{d}\int_0^d (e + c_me)dz \tag{4}$$

To simplify the discussion we shall assume that $K_a = K_b = K$. This is not necessarily true: the partitioning coefficient may be different at the outside face of the membrane and at the inside face since the environment inside the cell is different from that in the abiotic phase. In derivation of Eq. (2.1) we also implicitly assumed that the two partitioning coefficients were identical. We shall further take the three diffusivities to be constant and equal to D. Finally we shall assume that both the forward and backward reactions are much faster than the diffusion process. Effectively, this means that the chemical reaction is at equilibrium everywhere:

$$c_m + e \leftrightarrow c_me \quad \text{with equilibrium constant} \quad K_e = \frac{c_me}{e \cdot c_m} \tag{5}$$

Adding Eqs. (2) and (3) and integrating one obtains:

$$D\left(\frac{de}{dz} + \frac{dc_me}{dz}\right) = k_1 \tag{6}$$

At the two surfaces

$$\frac{de}{dz} = \frac{dc_m e}{dz} = 0 \qquad (7)$$

since neither carrier nor carrier-substrate complex can leave the membrane. Hence the arbitrary constant k_1 in Eq. (6) is zero, and on further integration one obtains

$$e + c_m e = k_2 \qquad (8)$$

Comparison with Eq. (4) shows that the arbitrary constant $k_2 = e_t$ and hence that the sum of the two carrier species is e_t at all positions in the membrane. Adding Eqs. (1) and (3) and integrating yields

$$-J = D\left(\frac{dc_m}{dz} + \frac{dc_m e}{dz}\right) \qquad (9)$$

where J is the desired total flux of substrate from $z = 0$ to $z = d$ through each unit area of the membrane. Since $c_m e = e_t - e = K_e \cdot e \cdot c_m$ one obtains

$$-J = D\left(\frac{dc_m}{dz} + \frac{d}{dz}\left(\frac{e_t K_e}{1 + K_e c_m} c_m\right)\right) \qquad (10)$$

Separation of variables and integration from $z = 0$ to $z = d$ yields

$$J = \frac{D}{d} K(c_a - c_b) + \frac{D}{d} K \frac{e_t K_e (c_a - c_b)}{(1 + K_e K c_a)(1 + K_e K c_b)} \qquad (11)$$

The first term on the right-hand side of Eq. (11) is obviously the free diffusion term corresponding to Eq. (2.1). If the partition coefficient K is so small that $K_e K c_a$ and $K_e K c_b$ are small compared with 1, it is seen that the last term is larger by a factor $e_t K_e$ than the free diffusion term. This is the effect of the facilitated diffusion. It is seen that the facilitated diffusion, just like free diffusion is dependent on the concentration gradient, but also it is a function of the concentration of the carrier protein.

2.1.2.3 Active Transport

Active transport resembles facilitated diffusion since specific membrane-located proteins mediate the transport process. In contrast to facilitated diffusion, the transport can be in the uphill direction of a concentration gradient, and active transport is therefore a free energy consuming process. The free energy required for the transport process may be supported by consumption of high-energy phosphate bonds in ATP (primary active transport), or the process may be coupled to another transport process with a downhill concentration gradient (secondary active transport). Finally, in a special type of active transport process for some substrates – the so-called group translocation – the substrate is converted to an impermeable derivative as soon as it crosses the cell membrane. The different types of active transport are illustrated in Fig. 2.3.

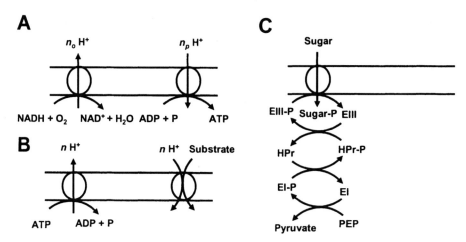

Figure 2.3 Different active transport processes.
A. Primary active transport illustrated by the oxidative phosphorylation. In the oxidation of NADH to NAD^+ electrons are donated to oxygen via the electron transport chain (not shown in full detail). When the electrons are transported through the electron transport chain protons are pumped across the membrane (the cytosolic membrane in prokaryotes and the inner mitochondrial membrane in eukaryotes). The protons are pumped against a proton and electrochemical gradient and the Gibbs free energy gained in the oxidation of NADH is hereby transferred to a concentration and electrochemical gradient. The proton gradient can be used to generate high-energy phosphate bonds in the form of ATP when the protons slide down the concentration gradient via the enzyme F_0F_1-ATPase, i.e., the Gibbs free energy is reconverted from a concentration and electrochemical gradient to high-energy chemical bonds. The stoichiometry in the process, i.e., the number of protons pumped in the electron transport chain and the number of protons involved in the F_0F_1-ATPase depends on the cell type. There is also not a fixed stoichiometric relationship between oxidation of NADH and ATP synthesis since the proton gradient may be used for other purposes, e.g., proton symport.
B. Secondary active transport, where the transport of one compound, e.g. a substrate that is to be transported into the cells, is associated with the transport of another compound, typically protons, that are transported down a concentration gradient. In order to maintain the proton gradient primary active transport of protons is required by an ATPase, which is a primary active transporter.
C. Transport by group translocation where the transport is associated with a conversion of the substrate to another compound. The most well-known group translocation process is the phosphotransferase system in bacteria. Here sugars are phosphorylated upon transport, and the phosphate group is donated by phosphoenolpyruvate (PEP), which is converted to pyruvate. The phosphate group is transferred from PEP to the sugar via several proteins, some of which are specific for the individual sugar and some of which are used in the translocation of different sugars.

In secondary active transport the compound is transported across the cytoplasmic membrane at the expense of a previously established gradient of another substance. If the two substances are transported in the same direction, the transport process is called *symport* (as illustrated in Fig. 2.3); if they are transported in opposite directions, it is called *antiport*; and if an electrochemical potential drives the flow of ions, it is called *uniport*. Often, secondary active transport is coupled to the pH gradient across the cytoplasmic membrane, and in order to keep the intracellular pH constant it is necessary to pump protons out of the cells by means of ATPase. In eukaryotes there are specific ATPases – different from the F_0F_1-ATPase (see Fig. 2.3), which are located in the cytosolic membrane. Since Gibbs free energy is used in this process, the overall effect is that a secondary active transport process requires free energy generated inside the cell. Examples of secondary active transport are the uptake of sugars in microorganisms by so-called permeases, where the sugar is transported into the cytoplasm together with a proton – i.e., there is proton symport. A well-studied system is lactose permease in *E. coli*, where a stoichiometric ratio of 1:1 in the lactose proton transport has been found. A similar simple stoichiometric ratio is not necessarily found for other transport processes.

In group translocation the transport process is coupled to a concomitant conversion of the transported substance. The best-known example of group translocation is the *phosphotransferase system* (PTS), which is used by many bacteria for uptake of different sugars. In this system the sugar is phosphorylated upon uptake and the phosphate group is donated from phosphoenolpyruvate (PEP), which is an intermediate in the Embden-Meyerhof-Parnas pathway (see Section 2.1.2.1). The transfer of the phosphate group involves at least four separate proteins (see Fig. 2.2), of which the last member of the chain also serves as the carrier protein that transports the sugar across the cytoplasmic membrane. The last two proteins in the chain are specific to the particular sugar, whereas the first two are identical in different PTSs. When glucose is transported to the cell by means of a PTS, it is directly converted to glucose-6-phosphate (G6P). The high-energy phosphate bond originally present in PEP is therefore conserved, and the uptake process is more economical from an energy point of view than glucose uptake by a permease. Furthermore, the PTSs may operate at very high rates of sugar uptake compared with other uptake systems. This may explain why the PTSs are predominant in fermentative bacteria, where the ATP generation resulting from sugar metabolism is less than in respirative bacteria; i.e., strict aerobes such as *Azotobacter* do not possess PTSs, whereas anaerobes and facultative anaerobes such as *Lactococcus* and *Escherichia* possess PTSs for several different sugars. One may finally speculate why PTS systems use an apparently complicated transfer of the free energy to the membrane bound protein by a chain of 3 cytosolic proteins. But this multi step transfer process serves as a fine-tuned regulator for the energy status of the cell-see Lengeler *et al.* (1999) for more details of this fascinating biological system.

2.1.3 Catabolism

When the substrates have been transported into the cytoplasm, they are converted to metabolic products and biomass components in a large number of biochemical reactions. As mentioned earlier the first reactions are the fueling reactions where the substrate (carbon and energy source) is converted to precursor metabolites (Table 2.4) and Gibbs free energy. The Gibbs free energy is captured in the form of high-energy phosphate bonds in ATP (anhydride-bound phosphate groups) or as reduction equivalents typically harbored in the reduced form of two related, but distinct compounds NADPH/NADP$^+$ and NADH/NAD$^+$.

The hydrolysis of a high-energy phosphate bond in ATP results in release of a large quantity of Gibbs free energy:

$$ADP + \sim P - ATP - H_2O = 0 \quad ; \quad \Delta G^0 = -30.5 \text{ kJ/mole} \quad (2.5)$$

where ~P is a phosphate group. The Gibbs free energy released by hydrolysis of ATP can be used to drive biosynthetic reactions with an otherwise positive Gibbs free energy, and ATP hereby represents a key link between the fueling reactions and the anabolism. When NAD(P)H is oxidized to NAD(P)$^+$ two electrons are released[1], and these are transferred to other compounds inside the cells, which consequently becomes reduced. In the following we consider the key pathways of the catabolism.

2.1.3.1 Glycolysis

The most frequently applied energy source for cellular growth is sugars, which are converted to metabolic products (e.g., carbon dioxide, lactic acid, acetic acid, and ethanol) with concurrent formation of ATP, NADH, and NADPH. NADH is produced together with NADPH in the catabolic reactions, but whereas NADPH is consumed mainly in the anabolic reactions (see Table

Table 2.4 List of precursor metabolites and some of the building blocks derived from these. The total requirements for the precursor metabolites to synthesize *E. coli* cell mass is also specified.

Precursor metabolite	Building blocks	Amount required (μmoles (g DW)$^{-1}$)
Glucose-6-phosphate	UDP-glucose, UDP-galactose	205
Fructose-6-phosphate	UDP-*N*-acetylglucosamine	71
Ribose-5-phosphate	Histidine, Tryptophane, Nucleotides	898
Erythrose-4-phosphate	Phenylalanine, Tryptophane, Tyrosine	361
Glyceraldehyde-3-phosphate	Backbone of phospholipids	129
3-Phosphoglycerate	Cysteine, Glycine, Serine, Choline, Nucleotides	1,496
Phosphoenolpyruvate	Phenylalanine, Tryptophane, Tyrosine	519
Pyruvate	Alanine, Isoleucine, Valine	2,833
Acetyl-CoA	Lipids	3,747
2-Oxoglutarate	Arginine, Glutamate, Glutamine, Proline	1,079
Succinyl-CoA	Hemes	-
Oxaloacetate	Aspartate, Asparagine, Isoleucine, Methionine, Threonine, Lysine, Nucleotides	1,787

---

[1] The nomenclature NAD(P)H with a parenthesis around P is used to designate either of the two co-factors.

2.5), NADH is consumed mainly within the catabolic reaction pathways, e.g., by oxidation with free oxygen in respiration (see Section 2.1.3.2). Most sugars are converted to glucose-6-phosphate (G6P) or fructose-6-phosphate (F6P) before being metabolized. The intracellular isomerization of G6P to F6P is normally in equilibrium, and G6P can therefore be considered a common starting point in many catabolic pathways. In some microorganisms formation of G6P from glucose occurs in the transport process (see Section 2.1.1.3), but in others this compound is formed from intracellular glucose in a reaction coupled with the hydrolysis of ATP. The catabolism of sugars from G6P is traditionally divided into glycolysis and pyruvate metabolism. Glycolysis is defined as the sum of all pathways by which glucose (or G6P) is converted to pyruvate. The two major pathways are the Embden-Meyerhof-Parnas (EMP) pathway and the pentose phosphate (PP) pathway (see Fig. 2.4).

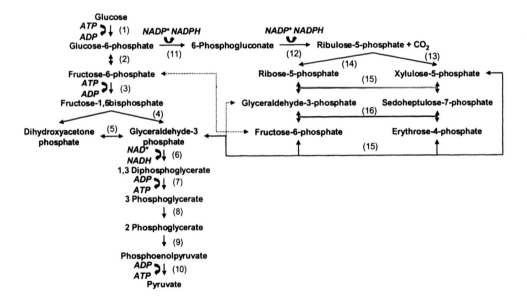

Figure 2.4 Embden-Meyerhof-Parnas (EMP) pathway and the pentose phosphate (PP) pathway. The scheme is without a PTS transport system. The enzymes are: (1) hexokinase; (2) phosphohexoseisomerase; (3) phosphofructokinase; (4) aldolase; (5) triosephosphate isomerase); (6) 3-phosphoglyceraldehyde dehydrogenase; (7) 3-phosphoglycerate kinase; (8) phosphoglycerate mutase; (9) enolase; (10) pyruvate kinase; (11) glucose-6-phosphate dehydrogenase; (12) 6-phosphogluconate dehydrogenase; (13) ribulosephosphate-3-epimerase; (14) ribosephosphate isomerase; (15) transketolase; (16) transaldolase. Fructose-6-phosphate and glyceraldehyde-3-phosphate takes part in both pathways and hereby allows recycling of carbon from the PP pathway back to the EMP pathway. Transaldolase can catalyze two reactions: 1) the interconversion of xylulose-5-phosphate and ribose-5-phosphate to glyceraldehydes-3-phosphate and sedoheptulose-7-phosphate, and 2) the interconversion of xylulose-5-phosphate and erythrose-4-phosphate to glyceraldehyde-3-phosphate and fructose-6-phosphate.

In the EMP pathway G6P is converted to pyruvate, and the overall stoichiometry from glucose is given by Eq. (2.6). One reaction (the conversion of F6P to fructose-1,6-diphosphate) requires a concomitant hydrolysis of ATP to proceed, but two other reactions which run twice for every molecule of G6P produce enough Gibbs free energy to give a net production of ATP in the pathway (see also Section 4.1). Since ATP (or PEP) is used for formation of G6P from glucose, the net yield of ATP is 2 moles per mole of glucose converted to pyruvate. The four electrons liberated by the partial oxidation of 1 mole glucose to 2 moles pyruvate are captured by 2 moles of NAD^+ leading to formation of 2 moles NADH.

$$2\text{ PYR} + 2\text{ ATP} + 2\text{ H}_2\text{O} + 2\text{ NADH} + 2\text{ H}^+ - \text{Glucose} - 2\text{ ADP} - 2 \sim \text{P} - 2\text{ NAD}^+ = 0 \qquad (2.6)$$

The major function of the PP pathway is to supply the anabolic reactions with reducing equivalents in the form of NADPH and to produce the precursor metabolites ribose-5-phosphate (R5P) and erythrose-4-phosphate. Due to the branch points present in the PP pathway (see Fig. 2.4), it is possible to adjust the fate of G6P in this pathway exactly to the cellular need for R5P and NADPH. If necessary a glucose molecule can be ground up to form 12 NADPH and 6 CO_2 by six "passages" in the loop G6P → PP pathway → F6P → G6P.

2.1.3.2 TCA Cycle and Oxidative Phosphorylation

The pyruvate formed in the glycolysis can be oxidized completely to carbon dioxide and water in the tricarboxylic acid (TCA) cycle, which is entered via acetyl-CoA (see Fig. 2.5). Here one mole of GTP, an "energy package" equivalent to ATP, and five reduced cofactor molecules are formed for each pyruvate molecule. Four of these are NADH, the fifth is $FADH_2$. A prerequisite for the complete conversion of pyruvate in the TCA cycle is that NAD^+ and FAD can be regenerated from NADH and $FADH_2$. This is done in the respiratory chain (see Fig. 4.1 for details), an oxidative process involving free oxygen and therefore operable only in aerobic organisms. In the respiratory chain, electrons are passed from NADH to a co-enzyme called ubiquinone (UQ) by NADH dehydrogenase. They are carried on from UQ through a sequence of cytochromes (proteins containing a heme group), and are finally donated to oxygen, resulting in the formation of water. The cytochromes and the co-enzyme UQ are positioned at or near the cytoplasmic membrane (or the inner mitochondrial membrane in eucaryotes), and when electrons pass through the respiratory chain protons are pumped across the membrane (in prokaryotes it is the cytosolic membrane and in eukaryotes it is the inner mitochondrial membrane). When the protons re-enter the cell (or the mitochondria) through the action of the enzyme F_0F_1-ATPase, as shown in Fig. 2.3A, ADP may be phosphorylated to form ATP, and the respiratory chain is therefore often referred to as *oxidative phosphorylation*. The number of sites where protons are pumped across the membrane in the respiratory chain depends on the organism. In many organisms there are three sites, and here ideally three moles of ATP can be formed by the oxidation of NADH. $FADH_2$ enters the respiratory chain at UQ. The electrons therefore do not pass the NADH dehydrogenase and the oxidation of $FADH_2$ therefore results only in the pumping of protons across the membrane at two sites. The number of moles of ATP formed for each oxygen atom used in the oxidative phosphorylation is normally referred to as the P/O ratio, and the value of this stoichiometric coefficient indicates the overall thermodynamic efficiency of the process. If NADH was the only co-enzyme formed in the catabolic reactions, the theoretical P/O ratio would be exactly 3, but since some $FADH_2$ is also formed the P/O ratio is always less than 3. Furthermore, the proton and electrochemical gradient is also used for solute transport and the overall stoichiometry in the process is therefore substantially smaller than the upper value of 3 (see Section 4.3). As the different reactions in the oxidative phosphorylation are not directly coupled the P/O-ratio varies with the growth conditions, and the overall stoichiometry is therefore written as:

$$NAD^+ + (1 + P/O) H_2O + P/O\ ATP - NADH - 0.5\ O_2 - P/O\ ADP - H^+ - P/O \sim P = 0 \quad (2.8)$$

In many microorganisms one or more of the sites of proton pumping are lacking, and this of course results in a substantially lower P/O-ratio.

Since the electron transport chain is located in the inner mitochondrial membrane in eukaryotes, and since NADH cannot be transported from the cytosol into the mitochondrial matrix NADH formed in the cytosol needs to be oxidized by another route. Strain specific NADH dehydrogenases face the cytosol, and these proteins donate the electrons to the electron transport chain at a later stage than the mitochondrial NADH dehydrogenase. The theoretical P/O ratio for oxidation of cytoplasmic NADH is therefore lower than that for mitochondrial NADH. To calculate the overall P/O ratio it is therefore necessary to distinguish between reactions in the cytoplasm and reactions in the mitochondria.

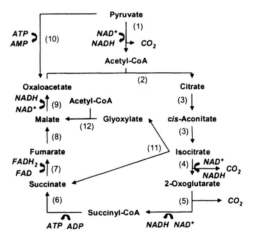

Figure 2.5 TCA cycle, pyruvate carboxylation and the glyoxylate cycle. To simplify the diagram CoA is not shown. The reaction from succinyl-CoA to succinate does in reality involve GTP, an analogue of ATP, but again to simplify the diagram the two species are lumped together. The enzymes are: (1) pyruvate dehydrogenase; (2) citrate synthase; (3) aconitase; (4) isocitrate dehydrogenase; (5) 2-oxoglutarate dehydrogenase; (6) succinate thiokinase; (7) succinate dehydrogenase; (8) fumarase; (9) malate dehydrogenase; (10) pyruvate carboxylase; (11) isocitrate lyase; (12) malate synthase.

From Cellular Function to Industrial Products

2.1.3.3 Fermentative Pathways

When the oxidative phosphorylation is inactive (due to the absence of oxygen or lack of some of the necessary proteins), pyruvate is not oxidized in the TCA cycle since that would lead to an accumulation of NADH inside the cells. In this situation NADH is oxidized with simultaneous reduction of pyruvate to acetate, lactic acid, or ethanol. These processes are collectively called *fermentative metabolism*. Fermentative metabolism is not the same in all microorganisms, but there are many similarities.

Bacteria can regenerate all NAD^+ by reduction of pyruvate to lactic acid (reaction (1) in Fig. 2.6A and B). They can also regenerate all NAD^+ by formation of ethanol in the so-called mixed acid fermentation pathway for which the entry point is the compound Acetyl-CoA. CoA is a cofactor

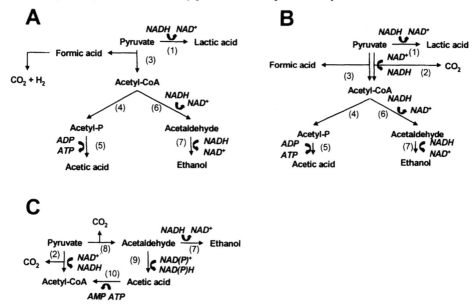

Figure 2.6 Different fermentative pathways for reduction of pyruvate. Only the main fermentative pathways are shown. The enzymes are: (1) lactic acid dehydrogenase; (2) pyruvate dehydrogenase; (3) pyruvate-formate lyase; (4) phosphate acetyltransferase; (5) acetate kinase; (6) and (9) acetaldehyde dehydrogenase; (7) alcohol dehydrogenase; (8) pyruvate decarboxylase; (10) acetyl-CoA synthase.
A. The fermentative (or mixed acid) metabolism of *Escherichia coli*.
B. The fermentative metabolism of lactic acid bacteria.
C. The fermentative metabolism in the yeast *Saccharomyces cerevisiae*. Not all reactions occur in the same compartment, i.e. the pyruvate dehydrogenase catalyzed conversion of pyruvate to acetyl-CoA occurs in the mitochondrion whereas the other reactions occur in the cytosol.

with a free –SH group that can be acetylated to CH_3CO-S-, either directly from acetate or by capturing two of the carbon atoms of pyruvate with the last carbon atom liberated as carbon dioxide or formic acid, HCOOH. Lactic acid bacteria (Fig. 2.6B) have both pathways for conversion of pyruvate to acetyl-CoA, whereas *E. coli* only has the pyruvate formate lyase catalyzed reaction. In yeast the fermentative pathway does not proceed via acetyl-CoA, but instead by decarboxylation of pyruvate to acetaldehyde. From acetate cytosolic acetyl-CoA may be synthesized, and this serves as precursor for fatty acid biosynthesis, whereas the mitocondrial acetyl-CoA that is formed directly from pyruvate serves as an entry point to the TCA cycle. In yeast the primary metabolic product is ethanol, but even with respiratory growth, where complete re-oxidation of NADH is possible by oxidative phosphorylation the pyruvate dehydrogenase complex (reaction (2) in Fig. 2.6C), which catalyzes the direct conversion of pyruvate to acetyl-CoA, may be by-passed as indicated. Above a certain glucose uptake rate the respiratory capacity becomes limiting and this leads to overflow in the by-pass and consequently ethanol is formed. This over-flow metabolism is traditionally referred to as the *Crabtree effect*[2].

Acetyl-CoA can be regarded as an activated form of acetic acid as it can be converted to acetic acid via reactions (4) and (5) in Fig. 2.6A and B. As seen in reaction (5) an ATP is released, hereby doubling the ATP yield by catabolism of glucose from 2 to 4 ATP per glucose molecule. This is the reason why bacteria use the mixed acid pathways at very low glucose fluxes. To obtain a complete regeneration of NAD^+ the flow of carbon to the metabolic end products formic acid, ethanol and acetic acid must, however, be balanced as will be discussed in Chapter 5.

Finally it should be noted that the pathways shown in Fig. 2.6 are of necessity quite simplified. Thus, in *E. coli* succinate may be an end product. Furthermore, in some bacteria alternative pathways from pyruvate to other end products such as butanol (together with butyric acid and acetone) or to 2,3 butanediol (together with acetoin) may be active.

2.1.4 Anabolism

Formation of macromolecules which constitute the major part of the cell mass requires production of the necessary building blocks followed by polymerization of the building blocks. In Table 2.5 the composition of an *E. coli* cell is shown together with the energy requirement for synthesis of the individual macromolecules, i.e., requirements for both biosynthesis and polymerization. It is observed that approximately 70%, of the total requirements for energy and reduction equivalents are used for synthesis of proteins. The precise values should therefore be used with caution for other microbial species since the protein content may vary considerably, not only among microbial species, but also with the operating conditions. It is furthermore observed that the requirements for Gibbs free energy and reduction equivalents are strongly dependent on whether the building blocks are present in the medium. It is therefore difficult to make detailed physiological studies when

---

[2] The Crabtree effect is a term often misused to describe over-flow metabolism. In *S. cerevisiae* the mechanisms behind over-flow metabolism are quite complex as discussed in Example 7.3, and it involves both redirection of carbon fluxes and repression of respiration. The term is named after Herbert G. Crabtree who studied sugar metabolism in tumor cells.

using a complex medium that contains some but not all of the building blocks. In physiological characterization of cells it is always preferable to use well-defined growth media, or even minimal media containing only glucose, ammonia, salts and a few vitamins.

Since protein synthesis requires a large input of free energy, it is important for the cell to adjust the protein synthesis precisely to its needs. Proteins are synthesized by the so-called *protein synthesizing system* (PSS) that primarily consists of ribosomes. In adjusting the size and activity of the PSS, the cell can control the protein synthesis and thereby the energy expenditure. When the cells experience a change from an energy-sufficient to an energy-deficient environment, they will fairly rapidly (time constant of 1-2 h) adjust the size of the PSS to the new conditions.

With the large costs of ATP for biomass synthesis it is important to have a continuous supply of this compound. The total pool of adenylate phosphates is, however, small in the cell, and the turnover time of ATP in living cells is therefore on the order of 1-10 s. Thus, if the continuous generation of high-energy phosphate bonds in the form of ATP is halted for just a few seconds free energy to drive biosynthesis will rapidly be used up, and consequently energy utilizing reactions must be tightly coupled to energy producing reactions inside the cell. The regulatory mechanisms in different pathways are very complex and involve many different components, and since ATP formation and utilization involve a cyclic flow through ADP and/or AMP[3], it is not surprising that all three adenylates play a regulatory role in the cellular reactions. Some enzymes are regulated by the concentration of one of the three components, whereas others are regulated by the concentration ratio of two of the components, e.g., [ATP]/[ADP]. Atkinson (1977) collected the control action of the adenylates in a single variable, the *energy charge* E_c.

$$E_c = \frac{[ATP] + \frac{1}{2}[ADP]}{[ATP] + [ADP] + [AMP]} \qquad (2.9)$$

Since ADP contains only one energized phosphate bond while ATP has two, a factor of ½ is used for ADP in the numerator. In theory the energy charge could vary between 0 and 1, but in living cells at balanced growth conditions E_c varies only between 0.65 and 0.9. The pool of adenylates is usually found to increase with the specific growth rate of the cells, whereas the energy charge is almost invariant with the specific growth rate (Atkinson, 1977). Table 2.6 collects steady state measurements of AMP, ADP and ATP at different specific growth rates of *Lactococcus lactis*. With both glucose and maltose as the limiting substrate the total adenylate pool is found to increase with the specific growth rate, but the energy charge is almost constant at 0.68 for all dilution rates and independent of the applied sugar.

---

[3] AMP (adenosine monophosphate) is formed by cleavage of the anhydride-bound phosphate from ADP. The amount of Gibbs free energy produced in the decomposition is slightly larger, i.e., 32.0 kJ/mole, than that of ATP hydrolysis.

Table 2.5 Composition and ATP and NADPH requirements of *E. coli* cell mass.[#]

Species	Content (g (g DW)$^{-1}$)	ATP (µmoles (g DW)$^{-1}$)		NADPH (µmoles (g DW)$^{-1}$)	
Protein	0.55	29,257	(21,970)	11,523	(0)
RNA	0.20	6,796	(2,146)	427	(0)
rRNA	0.16				
tRNA	0.03				
mRNA	0.01				
DNA	0.03	1,240	(450)	200	(0)
Lipid	0.09	2,836	(387)	5,270	(0)
Lipopolysaccharide	0.03	470	(125)	564	(0)
Peptidoglycan	0.03	386	(193)	193	(0)
Glycogen	0.03	154	(154)	0	(0)
Building blocks	0.04				
Total	1.00	41,139	(25,425)	18,177	(0)

[#] The data are for balanced growth at 37 °C on a glucose minimal medium and a specific growth rate of 1.04 h^{-1}. The content of species is given as mass fraction of total cell weight. The data in parenthesis for ATP and NADPH requirements are for growth on a rich medium containing all the necessary building blocks (amino acids, nucleotides, fatty acids, etc.) The data are taken from Ingraham *et al.* (1983).

Table 2.6 Measurements of the concentrations of AMP, ADP, and ATP in *Lactococcus* lacks at different specific growth rates obtained in a chemostat at steady state and calculation of the total adenylate pool and the energy charge.[#]

Specific growth rate (h^{-1})	Substrate	[AMP]	[ADP]	[ATP]	Total adenylate pool	Energy charge
0.03	Glucose	12	17	25	54	0.62
0.48	"	17	23	52	92	0.69
0.69	"	15	27	50	92	0.69
0.15	Maltose	5	8	20	33	0.73
0.32	"	12	23	41	76	0.69
0.58	"	17	26	44	87	0.66

[#] All concentrations are in µmole per gram dry weight. The data are taken from Sjöberg and Hahn-Hägerdal (1989).

The results of Table 2.6 are only true for balanced growth where the anabolic processes immediately utilize the ATP generated by the catabolic processes. Dramatic, and very fast changes in the energetic state of the cell will accompany abrupt changes in the environment – for example when a pulse of glucose is added to a steady state cultivation growing a low glucose concentrations. Reuss and coworkers (Theobald *et al.*, 1993; Theobald *et al.*, 1997; Vaseghi *et al.*, 1999) have made some – both from a scientific and from an experimental methodology point of view – very interesting studies of the rapid changes of metabolite levels that accompany an abrupt change in the environment. The experiments were made with an aerobic yeast culture grown on glucose, and results have been obtained in several time windows (depending on the experimental technique) ranging from 0 to 100 seconds or from 0 to 1 second. Figure 2.7 shows results in the 0 to 100 seconds time window for the changes in adenylate concentrations.

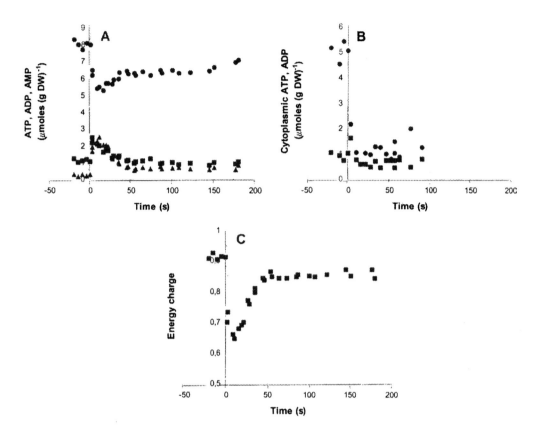

Figure 2.7 Rapid dynamics of adenylate concentrations. When a pulse of glucose is applied to an energy-starved continuous culture, the levels of all intermediate metabolites in the EMP pathway experience rapid changes. The data shown are obtained from a steady-state culture of *S. cerevisiae* pulsed with 1 g L^{-1} glucose. The rapid sampling technique and the analytical methods are described in Theobald, *et al.* (1993). Within the first 5 s of the transient, the ATP concentration of the whole cell has decreased by 30%. An even more pronounced drop in cytoplasmic ATP (different from the overall cell concentration due to the ATP involving reactions in the mitochondria) is observed. This precipitous drain of ATP must be due to the rapid formation of glucose-6-phosphate and the subsequent formation of fructose-1,6-diphosphate. ADP peaks after less than 5 s and is slowly consumed in the last reactions of the EMP pathway or by mitochondrial processes. Note that the ATP concentration has not reached the steady-state level even after 3 min, while the energy charge is back in the 0.85-0.90 range after less than 1 min. On the time scale of the growth process (i.e. from 1 hour and up) a pseudo steady-state assumption (see also Eq. (5.2)) for all adenylate concentrations is certainly satisfied. The data are from Theobald *et al.* (1997).
A. Total concentrations of ATP (●), ADP (■) and AMP (▲)
B. Cytoplasmic concentrations of ATP (●) and ADP (■)
C. Energy charge

ATP is not the only compound involved in coupling catabolism and anabolism tightly together. Also NADH and NADPH have small turnover times, and the cell must also exercise a strict control of the level of these compounds. The tight control of these co-factors may be represented by the following variables that are analogous with the energy charge (Ingraham *et al.*, 1983):

$$\text{Catabolic reduction charge} = \frac{[NADH]}{[NADH]+[NAD^+]} \tag{2.10}$$

$$\text{Anabolic reduction charge} = \frac{[NADPH]}{[NADPH]+[NADP^+]} \tag{2.11}$$

In growing cells the catabolic reduction charge is maintained at a low level of 0.03-0.07 and the anabolic reduction charge at the higher level of 0.4-0.5. In the catabolism, NAD^+ is a substrate (and NADH is a product), and this explains why catabolic reduction charge is controlled at a low level when these reactions are to proceed. NADPH, on the other hand, is a substrate for the anabolic reactions, and the anabolic reduction charge therefore has to be controlled at a higher level. The different levels of the two reduction charges explain the necessity for two different co-enzymes in the cell – they have different functions, and therefore have to be controlled at different levels in order to ensure proper function of the cell during growth.

The number of building blocks necessary for cellular synthesis varies between 75 and 100, and these are all synthesized from 12 precursor metabolites (see Table 2.4), which are intermediates in the fueling reactions (or the catabolism). Thus, the supply of these precursor metabolites further links the catabolism and the anabolism. When intermediates in the TCA cycle are used for biosynthesis, it is necessary to replenish these compounds since the TCA cycle activity would otherwise decline. The reactions involved are called the *anaplerotic pathways*, and the most important pathways are:

- Carboxylation reactions, where either pyruvate or phosphoenolpyruvate are carboxylated leading to formation of the TCA-cycle intermediate oxaloacetate (which at the same time is a precursor metabolite). Carboxylation of pyruvate is illustrated in Fig. 2.5.
- The glyoxylate cycle, which involves several steps of the TCA cycle and two additional reaction steps (see Fig. 2.5). In these steps the TCA cycle intermediate isocitrate is cleaved to succinate (another TCA cycle intermediate) and glyoxylate and subsequently glyoxylate reacts with acetyl-CoA to form malate (also a TCA cycle intermediate). The glyoxylate cycle allows for net synthesis of 4 carbon containing compounds from acetyl-CoA, and it plays an important role when cells are growing on C2 carbon compounds such as acetate and ethanol.

Clearly the formation of all the building blocks needed to synthesize a cell involves a large number of reactions and often very long pathways are involved in the synthesis of a single building block. To get an overview of biosynthetic pathways we therefore refer to standard biochemistry textbooks or reaction databases available on the World Wide Web, e.g., www.genome.adj.jp.

2.1.5 Secondary Metabolism

We have now treated the part of cellular metabolism that is associated with the growth process. This is referred to as the primary metabolism. A large number of industrially important products are, however, formed in cellular reactions that are not directly associated with the growth process. Secondary metabolites may be derived from precursor metabolites or building blocks – but typically only through a large number of reaction steps. This part of cellular metabolism is referred to as secondary metabolism, and the products formed in these reactions are called *secondary metabolites* just as metabolites formed in the primary metabolism are called primary metabolites. Secondary metabolites serve many different functions, and in many cases their exact role in the overall cell function is unknown. It is believed that many secondary metabolites serve as a defense system for the producing microorganisms against other microorganisms (this is particularly the case for many secondary metabolites produced by plant cells).

The most important group of secondary metabolites produced by microorganisms is antibiotics, and today about 12,000 antibiotics have been discovered. Of these about 150 have been approved for human use, and Table 2.7 lists some of the more common antibiotics. The world market for antibiotics has increased from 18 bio US$ in 1994 to 23 bio US$ in 1996, and it currently growing with about 10% per year, even though only a few new antibiotics are approved for human use every year. With the increasing demand for antibiotics it becomes important to improve the yield and productivity of existing processes as well as developing synthesis routes for novel compounds. The possibilities offered by the secondary metabolism of microorganisms are of such magnitude that the topic must fall outside the scope of the present text, but we will treat biosynthesis of β-lactams shortly.

β-Lactams constitute by far the largest group of antibiotics, with about 120 compounds approved for human use. The β-lactams can be divided into penicillins and cephalosporins, with penicillins having a five member thiazolidine ring associated with the four member β-lactam ring and cephalosporins having a six member dihydrothiazine associated with the four member β-lactam (Fig. 2.8). Different microorganisms can form both types of compounds naturally, and the pathways for their synthesis are shown in Fig. 2.8. Cephalosporins can also be derived chemically from penicillins by ring expansion, and most of the clinically applied β-lactams are chemically synthesized from penicillin V or G, which are produced by fermentation with the filamentous fungus *Penicillium chrysogenum*. Much effort has been spent to improve the yield and productivity of this classical fermentation process. As seen in Fig. 2.8 the two first steps are identical in the biosynthesis of penicillins and cephalosporins, and the two enzymes catalyzing these two reactions are therefore present in all β-lactam producing organisms. In the production of penicillin the side-chain of isopenicillin N is exchanged with a phenoxyacetic acid or phenylacetic acid, which are added to the medium, leading to the formation of penicillin V or penicillin G, respectively. In the production of cephalosporins isopenicillin is converted to penicillin N by an epimerase. The next step is ring expansion of the thiazolidine ring resulting in deacetoxycephalosporin C, which is a common precursor in the synthesis of cephalosporins, *e.g.* cephalosporin C by *A. chrysogenum* and cephamycin by *S. clavuligerus*.

Table 2.7 Some commonly used antibiotics, their use and their production host.

Antibiotic	Type of compound[1]	Typical use	Production organism
Penicillins[2]	β-lactam	Bacterial infections	*Penicillium chrysogenum*
Cephalosporins	β-lactam	Gram positive and Gram negative	*Acremonium chrysogenum* *Streptomyces*
Clavulanic acid	β-lactam	β-Lactam resistant infections[3]	
Nystatin	polyketide	Topical, fungal infections	*Streptomyces noursei*
Erythromycin	polyketide		
Tetracyclin			
Bacitracin	polypeptide	Animal treatment	*Bacillus licheniformis*
Vancomycin	glycopeptide	MRSA infections[4]	
Fusidic acid		Topical, Gram positive	

[1] Classification of natural products is very complicated, and the type of compound is only an indication of whether the compounds are of similar nature.

[2] Penicillin is a common name for a very broad group of compounds. Most penicillins are produced semi-synthetically from penicillin V or penicillin G (see text), and the most commonly used penicillins are amoxicillin and ampicillin.

[3] Clavulanic acid is not an antibiotic in itself, but it is a β-lactamase inhibitor and may together with other β-lactams be used to treat β-lactam resistant infections.

[4] MRSA – Methicillin Resistant *Staphylococcus aureus*, a Gram positive bacteria often encountered in hospital infections.

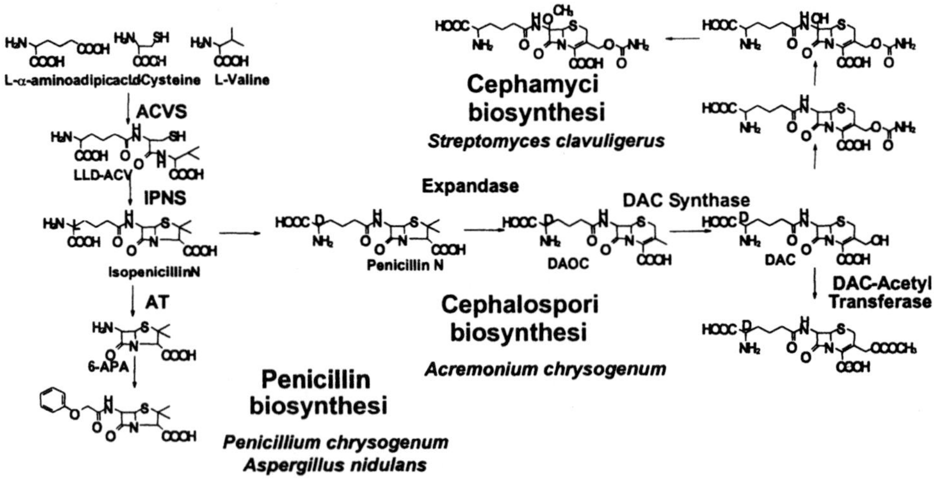

Figure 2.8 The biosynthetic pathway of β-lactams. For the pathway leading towards penicillin the enzymes are: δ-(L-α.aminoadipyl)-L-cysteinyl-D-valine synthetase (ACVS); isopenicillin N synthase (IPNS); acyl-CoA:isopenicillin N acyltransferase (AT). The step catalyzed by acyl-CoA:isopenicillin N acyltransferase may also proceed in a two step reaction via 6-APA, but both these reactions are also catalyzed by AT. In *Acremonium chrysogenum* the same enzyme carries out expansion and hydroxylation, whereas in *Streptomyces clavuligerus* two independent enzymes carry out these reactions.

2.1.6 Secreted Proteins

Many microorganisms secrete hydrolytic enzymes that degrade macromolecules to monomers that may serve as carbon, energy and nitrogen sources. Among the most frequently secreted enzymes are proteases (degradation of proteins), peptidases (degradation of peptides), amylases (degradation of starch), xylanases (degradation of xylans), cellulases (degradation of cellulose). Through the secretion of enzymes some microorganisms may grow on very complex nutrients, and the ability for microorganisms to decompose leaves and other plant materials plays a very important role in the overall carbon cycle. The ability of microorganisms to secrete enzymes has been exploited for many generations, particularly in the food and feed industry. Thus, the use of *Aspergillus oryzae*, which is an efficient producer of the starch degrading enzymes α-amylase and glucoamylase, in the koji-sauce production has been practiced for more than a thousand years. Furthermore, the secretion of proteases and peptidases by lactic acid bacteria plays a very important role in many dairy processes, both because the hydrolysis of proteins and peptides ensures supply of carbon and energy sources that may be converted to acids and other flavors and because hydrolysis of many peptides are important for proper flavor development. Today, enzymes are also used in detergents to improve the washing process and in many different industrial processes, e.g., in the treatment of cotton (see e.g. www.novozymes.com for more details on the industrial application of enzymes). Many of these industrial enzymes are today produced using a few host cells where genes encoding the enzymes are introduced by genetic engineering.

The possibility to introduce foreign genes into a microbial host by genetic engineering and hereby produce a specific protein in high amounts also paved the way for a completely new route for production of pharmaceutical proteins like human growth hormone (hGH) and human insulin. The first products (human insulin and hGH) were produced in recombinant *E. coli*, but soon followed the exploitation of other expression systems like *S. cerevisiae* (introduced for production of human insulin), insect cells, and mammalian cells (Chinese hamster ovary cells and hybridoma cells). Today there are more than 55 protein drugs, largely recombinant proteins and monoclonal antibodies that are often referred to as biotech drugs, and the top selling drugs produce revenues of billions of US$. The biochemical processes underlying the synthesis of a heterologous protein in a given host cell may be quite complex as many post-translational modifications may take place and the secretory pathway may involve many individual steps. Many of these processes are specific for the host system and it is therefore not possible to give a short overview of the subject.

2.2 Biotech Processes – An Overview

When a microorganism or cell type has been identified to produce an interesting compound there are a number of considerations to be made before an economically viable, industrial process can be realized. In companies with a solid experience in design and scale-up of fermentation processes there can be a relatively fast implementation of a new process, but still there are many hurdles on the road to an efficient industrial process. Development of a fermentation process can roughly be divided into four phases (see Fig. 2.9). First the product is identified. In the case of a pharmaceutical this may be a result of random screening for different therapeutic effects by microbial metabolites, *e.g.* by high throughput screening of secondary metabolites from *Actinomycetes*, or it may be the result of a targeted identification of a novel product, *e.g.* a peptide hormone with known function. Outside the pharmaceutical sector the product may also be chosen after a random screening procedure, *e.g.* screening for a novel enzyme to be used in detergents, or it may be chosen in a more rational fashion. With the rapid progress in genomic sequencing programs, it is now possible to search for new target proteins directly in the sequenced genomes by using bioinformatics.

When the product has been identified the next step is to choose or construct the strain to be used for production, and thereafter follows design of the process. This involves choosing an appropriate fermentation medium and the optimal process conditions. In parallel to the design of the production process further research, e.g. clinical tests, is done in order to have the product approved. When the strain has been constructed one of the first aims is therefore to produce sufficient cell material for further research, and this is typically done in pilot plant facilities. For a pharmaceutical compound sufficient material must be produced for clinical trials. For other products it may be necessary to carry out tests of the product and examine any possible toxic effects. The final steps in the development are product approval by the proper authorities and construction of the production facility (which in some cases may be through retrofitting of an existing plant). In the following we will consider some of the different aspects on process development as an introduction to the more detailed quantitative analysis that will be introduced in the later chapters.

2.2.1 Strain Design and Selection

A key step in the development phase of a fermentation process is to choose an appropriate strain. In the past this choice was normally obvious after the product had been identified, *e.g. P. chrysogenum* was chosen for penicillin production since it was this organism that was first

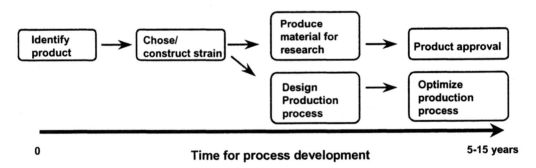

Figure 2.9 Different phases in the development of a fermentation process.

identified to produce penicillin. With the introduction of recombinant DNA technology it is, however, now possible to chose almost any production host for the production. Thus, a strain of *E. coli* has been constructed that can produce ethanol at a high yield, and a recombinant strain of *P. chrysogenum* can now be used to produce 7-ADCA (a precursor used for synthesis of cephalosporins) directly by fermentation. The choice of strain does, however, often depend much on tradition within the company, and most of the fermentation industries have a set of favorite organisms that are used in the production of many different products. For production of a heterologous protein, *i.e.* expression of a foreign gene in a given organism, it is also necessary to consider many other aspects, *e.g.* is the protein correctly folded and glycosylated, and Table 2.8 gives an overview of the advantages/disadvantages of different cellular systems for the production of recombinant proteins. Although optimization of the process continues even after large scale production has started it is important to choose a good host system from the beginning, particularly in the production of pharmaceuticals, since the introduction of new strains requires a new approval of the process, and the associated costs may in some cases prevent the introduction of a new process even though it gives a better process economy.

As indicated in Table 2.8 the choice of expression system depends on many factors, but the main factors are: 1) the desirability of post-translational modification and secretion, 2) the stability of the protein in question, and 3) the projected dose of protein per patient (which determines whether the cost of the drug becomes critical). Thus for proteins used in large doses, such as

Table 2.8 Advantages and disadvantages of different hosts for production of recombinant proteins.

Host	Advantages	Disadvantages
Bacteria (*E. coli*)	Wide choice of cloning vectors	Post-translational modifications lacking
	Gene expression easy to control	
	Large yields possible	High endotoxin content
	Good protein secretion	Protein aggregation (inclusion bodies)
Yeast (*S. cerevisiae*)	Generally regarded as safe (GRAS)	Less cloning vectors available
	No pathogens for humans	Glycosylation not identical to mammalian glycosylation
	Large scale production established	
	Some post-translational modifications possible	Genetics less understood
Filamentous fungi	Experience with large scale production	High level of heterologous protein expression has not been achieved
	Source of many industrial enzymes	
	Excellent protein secretion	
		Genetics not well characterized
Mammalian cells	Same biological activity as natural protein	Cells difficult to grow in bioreactors
	Expression vectors available	Expensive
		Slow growth
		Low productivity
Cultured insect cell	High level of gene expression possible	Not always 100% active proteins
	Post-translational modification possible	Mechanisms largely unknown

human insulin, it is important that the production costs are kept low, which requires an expression system with a high productivity, *i.e. E. coli* or *S. cerevisiae*. For very complex molecules like tissue plasminogen activator (tPA) and erythropoietin (EPO) it is, however, not possible to obtain sufficiently active compounds in microbial systems, and here a higher eukaryotic expression system is required.

2.2.2 Fermentation Media

An important aspect in design of a fermentation process is the choice of fermentation medium, which represents the raw material for the process. The fermentation medium should fulfill the following criteria:
- It should contain a carbon, nitrogen and energy source.
- It should contain all essential minerals required for growth.
- It should contain all necessary growth factors to ensure rapid growth and high yield of the desired product.
- It should be of a consistent quality and be readily available throughout the year.
- It causes a minimum of problems in the downstream processing.
- It causes a minimum of problems in other aspects related to the fermentation process, i.e., it has no negative effect on the gas-liquid mass transfer.

Traditionally complex media from the agricultural sector were used, as these were present in large quantities and were relatively cheap. Table 2.9 lists some typical complex media components used in the fermentation industry together with some typical products produced using the different media.

The advantages of applying complex media are that they often contain an organic nitrogen source, essential minerals and different growth factors. Thus, two frequently applied complex media, corn steep liquor and pharmamedia; both contain a large variety of amino acids, many different minerals and many different vitamins. The disadvantages of complex media are that: 1) there may be a seasonal variation in the composition, 2) the composition changes with storage,

Table 2.9 Complex fermentation media often applied in the fermentation industry.

Medium	Contents	Origin	Typical application
Corn steep liquor	Lactate, amino acids, minerals, vitamins	Starch processing from corn	Antibiotics
Corn starch	Starch, glucose	Corn	Ethanol, industrial enzymes
Barley malt	Starch, sucrose	Barley	Beer, whiskey
Molasses	Sucrose, raffinose, glucose, fructose, betain	Sugar cane or sugar beet	Bakers yeast, ethanol
Pharmamedia	Carbohydrates, minerals, amino acids, vitamins, fats	Cotton seed	Antibiotics
Serum	Amino acids, growth factors	Serum	Recombinant proteins
Whey	Lactose, proteins	Milk	Lactic acid
Yeast extract	Peptides, amino acids, vitamins	Yeast	Enzymes

and 3) there may be compounds present that are undesirable. With a requirement for increased documentation and reproducibility in the fermentation industry there is trend towards application of more defined media, and often minimal media with a carbon and energy source (glucose, sucrose or starch), an inorganic nitrogen source (ammonia or urea), a mixture of minerals and a perhaps a few vitamins may replace complex media to the benefit to the producer. In this process it is important to consider also the benefits of using a defined medium in the subsequent downstream processing. Also in the pharmaceutical sector there is a desire to use defined media, and particularly the use of serum free medium is today considered a standard requirement in connection with production of heterologous proteins using mammalian cell cultures. In the future it is expected that lignocellulose containing material may serve as a cheap and efficient carbon and energy source. Lignocellulose basically consists of three components: lignin, cellulose, and hemicellulose (a complex pentose rich polymer), and it is present in many different plant fibers like corn cob, bagasse, straw and wood. The exploitation of lignocellulose as raw material may allow many commodity products like fuels, polymers, and chemicals to be produced through fermentation processes in a cost efficient manner.

2.2.3 Criteria for Design and Optimization

The criteria used for design and optimization of a fermentation processes depends on the product. Thus, the criteria used for a high volume/low value added product are normally completely different than the criteria used for a low volume/high value added product. For products belonging to the first category (which includes most whole cell products, most primary metabolites, many secondary metabolites, most industrial enzymes, and most polysaccharides) the three most important design parameters are:
- Yield of product on the substrate (typical unit: g product per g substrate)
- Productivity (typical unit: g product per L reactor volume per hour)
- Final titer (typical unit: g product per L reactor volume)

Yield of product on the substrate is here very important since the raw materials often account for a significant part of the total costs. Thus in penicillin production the costs of glucose alone may account for up to 15% of the total production costs. Productivity is important since this ensures an efficient utilization of the production capacity, *i.e.* the bioreactors. Especially in an increasing market it is important to increase the productivity since this may prevent new capital investments. Final titer is of importance for the further treatment of the fermentation medium, *e.g.* purification of the product. Thus if the product is present in a very low concentration at the end of the fermentation it may be very expensive even to extract it from the medium with a satisfactory yield.

In Chapter 3 we are going to introduce reaction rates and yield coefficients that enables a quantification of productivity and yield, but the concept can be illustrated if we return to Fig. 2.1, which is a representation of the overall conversion of substrates into metabolic products and biomass components (or total biomass). As we will discuss in Section 3.1 the rates of substrate consumption can be determined during a fermentation process by measuring the concentration of these substrates in the medium. Similarly the formation rates of metabolic products and biomass are determined from measurements of the corresponding concentrations. It is therefore possible to determine what flows into the total pool of cells and what flows out of this pool. The inflow of a substrate is normally referred to as the substrate uptake rate and the outflow of a metabolic product is normally referred to as the product formation rate. Clearly the product formation rate directly gives a measure for the productivity of the culture. Furthermore, the yield can be derived from the ratio of product formation rate and the substrate uptake rate, and this quantifies the efficiency in the overall conversion of the substrate to the product of interest.

In the production of novel pharmaceuticals, which typically belong to the category of low volume/high value added products the three above mentioned design parameters are normally not that important. For these processes time-to-market and product quality is generally much more important, and change of the process after implementation is often complicated due to a requirement of FDA approval. In the initial design phase it is, however, still important to keep these three design criteria in mind, and especially the requirements for high final titer is important since the cost for purification (or down stream processing) often accounts for more than 90% of the total production costs.

2.2.4 Strain Improvement

A key issue in process optimization is to improve the properties of the applied strain, as this is clearly responsible for the overall conversion of substrates to the product of interest. One may see the cell as a small factory, and through engineering of the pathways it may be possible to redirect the carbon fluxes such that the productivity or yield is improved. Engineering of the cellular pathways can be obtained through altering the genome, as this may lead to different expressions of the enzymes that catalyze the individual biochemical reactions or processes. Traditionally strain improvement through introduction of mutations was done through random mutagenesis and selection of strains with improved properties. This is well illustrated by the industrial penicillin production, where introduction of new strains has resulted in a significant increase in the final titer of penicillin (see Fig. 2.10). In this process there have been several strain improvement programs, some carried out by the major penicillin producing companies and some by companies designated to carry out strain improvement. Similarly there has been made improvements in the properties of baker's yeast *S. cerevisiae* and in microorganisms applied for the production of enzymes and many different metabolites.

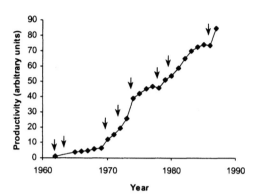

Figure 2.10 Increase in productivity, in the period between 1962 and 1987, (output rate/unit volume, arbitrary units) of penicillin G production by Gist-brocades, Delft (now DSM). The introduction of new strains is marked with arrows.

With the rapid development in recombinant DNA technology it has, however, become possible to apply a more directed approach to strain improvement. This is often referred to as *metabolic engineering*, but terms like molecular breeding or cellular engineering have also been applied. Several different definitions have been given for metabolic engineering, but they all convey similar meanings that are captured in the definition: *the directed improvement of product formation or cellular properties through the modifications of specific biochemical or introduction of new one(s) with the use of recombinant DNA technology.* Among the applications of metabolic engineering are:

- **Heterologous protein production**. Examples are found in the production of pharmaceutical proteins (hormones, antibodies, vaccines *etc.*) and in the production of novel enzymes. Initially the heterologous gene needs to be inserted in the production host. Subsequently it is often necessary to engineer the protein synthesis pathway, *e.g,* to have an efficient glycosylation or secretion of the protein from the cell. In some cases it may also be necessary to engineer the strain to obtain an improved productivity.

- **Extension of substrate range.** In many industrial processes it is interesting to extend the substrate range for the applied microorganism in order to have a more efficient utilization of the raw material. Initially, it is necessary to insert the necessary pathway (or enzyme) for utilization of the substrate of interest. Subsequently, it is important to ensure that the substrate is metabolized at a reasonable rate, and that the metabolism of the new substrate does not result in the formation of undesirable by-products. This may in some cases involve extensive pathway engineering.
- **Pathways leading to new products.** It is often of interest to use a certain host for production of several different products, especially if a good host system is available. This can be achieved by extending existing pathways by recruiting heterologous enzymes. Another approach is to generate completely new pathways through gene shuffling or other methods of directed evolution. In both cases it is often necessary to further engineer the organism to improve the rate of production and eliminate by-product formation.
- **Pathways for degradation of xenobiotics.** Many organisms naturally degrade xenobiotics, but there are few organisms that degrade several different xenobiotics. In bioremediation it is attractive to have a few organisms that can degrade several different compounds. This may be achieved either by inserting pathways from other organisms or through engineering of the existing pathways.
- **Engineering of cellular physiology for process improvement.** In the industrial exploitation of microorganisms or higher organisms it may be of significant interest to engineer the cellular physiology for process improvement, *e.g.* make the cells tolerant to low oxygen concentration, less sensitive to high glucose concentrations, improve their morphology, or increase flocculation. In cases where the underlying mechanisms are known this can be achieved by metabolic engineering. This may involve expression of heterologous genes, disruption of genes or over-expression of homologous genes.
- **Elimination or reduction of by-product formation.** In many industrial processes by-products are formed. This constitutes a problem due to the loss of carbon to the by-products, due to toxicity of these compounds, or due to interference of the by-products with the purification of the product. In some cases the by-products can be eliminated through simple gene disruption, but in other cases the formation of the by-product is essential for the overall cellular function and disruption of the pathway leading to the by-product may be lethal for the cell. In the last case it is necessary to analyze the complete metabolic network, and based on this analysis design a strategy for reduction of the by-product formation.
- **Improvement of yield or productivity.** In many industrial processes, especially in the production of low-value added products, it is important to continuously improve the yield and/or productivity. In some cases this can be achieved simply be increasing the activity of the biosynthetic pathway, *e.g.* by inserting additional gene copies. In other cases the pathways leading to the product of interest involves many steps, and it is therefore not possible to increase the activity of all the enzymes. Analysis of the flux regulation in the pathway is therefore required. Finally, in some cases the limitation is not in the actual pathway, and it may therefore be necessary to engineer the central carbon metabolism, which is generally very difficult due to the many control structures present here.

Besides application of these tools from the biosciences the design and proper operation of an industrial plant for low-value added products requires a deep technical insight into many of the disciplines that together makes up the science of chemical engineering: stoichiometric analysis, modeling of reaction kinetics, transport phenomena and bioreactor design, and unit operations. In this textbook we will highlight the chemical engineering disciplines that are associated with bioreactions, we will give a quantitative description of a given phenotype of a cell, and will illustrate how it operates in and interacts with technical equipment. A comprehensive treatment of metabolic engineering is given in Stephanopoulos et al. (1998) or Nielsen (2001), while a short (mostly provocative) account of the needs for engineering tools is given in Leib et al. (2001).

REFERENCES

Atkinson, D. E. (1977) Cellular Energy Metabolism and its Regulation. Academic Press, New York

Henriksen, C. M., Nielsen, J. & Villadsen, J. (1998). Modelling of the protonophoric uncoupling by phenoxyacetic acid of the plasma membrane of *Penicillium chrysogenum*. *Biotechnol. Bioeng.* **60**:761-767

Ingraham, J. L., Maaløe, O. & Neidhardt, F. C. (1983). Growth of the Bacterial Cell. Sinnauer Associates, Sunderland

Leib, T. M., Pereira, C. J. & Villadsen, J. (2001) Bioreactors: a chemical engineering perspective. *Chem. Eng. Sci.* **56**:5485-5497

Lengeler, J. W., Drews, G. & Schlegel, H. G. (1999) Biology of the Prokaryotes. Thieme, Stuttgart

Neidhardt, F. C., Ingraham, J. L. & Schaechter, M. (1990). Physiology of the Bacterial Cell. A   Molecular Approach. Sinauer Associates, Sunderland

Nielsen, J. (1997). Physiological engineering aspects of *Penicillium chrysogenum*. World Scientific Publishing Co., Singapore

Nielsen, J. (2001). Metabolic Engineering. *Appl. Microbiol. Biotechnol.* **55**:263-283

Schulze, U. (1995). Anaerobic physiology of *Saccharomyces cerevisiae*. Ph.D. Thesis, Technical University of Denmark

Sjöberg, A. & Hahn-Hägerdal, B. (1989) β-glucose-1-phosphate, a possible mediator for polysaccharide formation in maltose-assimilating *Lactococcus lactis*. *Appl. Environ. Microbiol.* **55**:1549-1554

Stein, W. D. (1990). Channels, Carriers and Pumps. An Introduction to Membrane Transport. Academic Press, San Diego

Stephanopoulos, G., Aristodou, A. & Nielsen, J. (1998). Metabolic Engineering. Principles and Methodologies. Academic Press, San Diego

Theobald, U., Mailinger, W. Reuss, M. & Rizzi, M. (1993). *In vivo* analysis of glucose-induced   fast changes on yeast adenine nucleotide pool applying a rapid sampling technique. *Analytical Biochemistry* **214**, 31-37.

Theobald, U., Mailinger, W., Baltes, M., Rizzi, M. & Reuss, M. (1997). In vivo analysis of metabolic dynamics in *Saccharomyces cerevisiae*: I. Experimental observations. *Biotechnol. Bioeng.* **55**:305-316

Vaseghi, S., Baumeister, A., Rizzi, M. & Reuss, M. (1999) *In vivo* dynamics of the pentose phosphate pathway in *Saccharomyces cerevisiae*. *Metabol. Eng.* **1**:128-140

Verduyn, C., Postma, E., Scheffers, W. A. & van Dijken, J. P. (1992) Effect of benzoic acid on metabolic fluxes in yeast. A continuous culture study on the regulation of respiration and alcoholic fermentation. *Yeast* **8**, 501-517

3

Biochemical Reactions – A First Look

Having had a brief overlook of the processes, which determine the life processes in a microbial cell we shall now turn to a quantitative treatment of these processes. The goal of all the remaining chapters of the book is, indeed, to understand at least at a superficial level how the cell functions, and by calculations predict what we may expect when living cells are exploited for the production of valuable products.

The first step in this progress towards a quantitative understanding of cell processes is to make experiments with cultivation of cells in a reactor, and to extract the rates of the biological reactions from these experiments. This is the subject of the present chapter. No biological interpretation of the data is offered – this has to wait until more fundamental knowledge has been obtained using methodologies described in Chapters 4 to 7, but the basic tools of quantitative analysis of fermentation data will be presented, and it will be shown how the detrimental effect of experimental errors and of a less than perfect experimental procedure may be minimized.

3.1 The Continuous Stirred Tank Reactor

A very efficient bioreactor set-up for analysis of overall conversion rates in bioprocesses is the continuous stirred tank reactor, which is schematically shown in Fig. 3.1. There are two feed ports, one for liquid feed and one for gaseous feed. In the liquid reaction medium the reactants, called *substrates*, are converted to *biomass* and *metabolic products*. Cells and metabolic products dissolved in the aqueous medium leave the reactor through the liquid effluent, while gaseous products such as carbon dioxide leave through the headspace of the reactor (exhaust gas). When the liquid feed rate v (L h^{-1}), the gas feed rate v_g (L h^{-1}), the reactor volume V (L) and the concentrations of substrates in the liquid and gas feed streams are independent of time, i.e., when all the input variables to the system have constant values, we will expect all the output variables, i.e., biomass concentration and activity of the cells as well as the concentration of metabolic products in liquid and gas effluents to have constant values. The bioreactor is operating in *steady state continuous mode*. In a large majority of cases all the output variables have time independent values when the feed variables are constant, but there are exceptions, and these exceptions can have considerable scientific and practical interest as will be qualitatively discussed in Note 3.1 and considered again in Chapters 8 and 9.

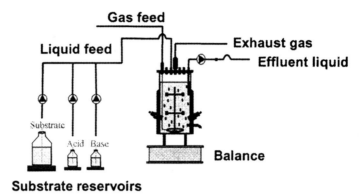

Figure 3.1 A continuous, stirred tank bioreactor with feed of a liquid medium from a substrate reservoir and a feed of gaseous medium through a sparger. pH is typically kept constant by adding acid or base from separate reservoirs.

Note 3.1 Time dependent output with constant input variables.
Sometimes a time independent set of input variables does not lead to time independent output variables.
- Some microbial cultures seem unable to reach steady state in a certain range of (time independent) operating conditions. All outputs oscillate with an oscillation time that depends on the input variables. This is further discussed in Chapter 8 and 9.
- The cells may suddenly, and for no apparent reason change their behavior. A product – it could be penicillin – is suddenly not being produced at the same constant rate, which has been measured for many hours of the apparently steady state operation. After a relatively short period the cells may totally have lost their ability to produce a particular metabolic product. In other cases the cells may become "sick" and die. A cell count of viable cells will show a gradually diminishing fraction of viable cells, i.e. cells that are able to grow. All these cases are typically a consequence of occurrence of natural mutations, which may be followed by selection of the mutated cells in the bioreactor.

When we use the presumably steady state, continuous reactor to measure what should be steady state rates of production of biomass and metabolic products it is, of course a surprise not to have time independent outputs. Still, the unexpected results might offer opportunities for challenging research, or the observation of the phenomena in well-controlled laboratory reactors may prevent later disasters when the laboratory results are to be used in an industrial scale process. Thus the observation of oscillating reactions indicates that the kinetics of the overall reaction is of a kind, which does not admit to a steady state output, even when the input is constant – a situation also encountered in chemical reaction engineering. The gradual cessation of synthesis of a desired product indicates that the environment has changed in a subtle fashion which it will take much experimental effort to explore – a slight difference in medium sterilization or in inoculation of the reactor, a different batch of yeast extract used as nitrogen source in the feed medium. Events observed in the steady state laboratory culture could also have their origin in the genome of the cell – a slight change in the expression of certain genes could lead to a cascade of events observed on the macroscopic level in the reactor.

Biochemical Reactions – A First Look

These observations – which as stated initially are not commonplace – do, however, show that experimental studies of bioreactions in principle try to extract useful information from a system, which is not at all observable. It is vastly more complex than the gas-phase reaction catalyzed by a solid catalyst, which is a standard topic in textbooks on chemical reaction engineering. An almost infinite possibility of reaction paths with a hierarchal control structure which fine-tunes the active paths in response to a changing environment, to the age of the culture, or to signals that we have not even begun to explore is the standard scenario of cell reaction studies.

Besides setting the substrate concentrations to certain values in the liquid and the gas feed streams the attainment of a steady state continuous culture requires that v_g, and especially v, the feed rate of the liquid feed, are set to constant values relative to the medium volume V in the reactor. Using gas flow meters v_g/V is controlled at a given set point to obtain a certain gas flow rate – often specified in terms of v.v.m. – volume gas per volume liquid per minute. The ratio between v and V is one of the most important input variables in bioreactors. The ratio is called the *dilution rate* and is measured in units of reciprocal time, usually h^{-1}.

$$\text{Dilution rate} = D = \frac{v}{V} \tag{3.1}$$

D is the space-time or the reciprocal of the holding time, the usual term in chemical reaction engineering.

To attain a constant D several different control strategies may be used:

- The volume of the reaction medium or the weight of reactor and medium is measured with a frequency of e.g. 10 $\min^{-1}$ and the liquid feed rate v is controlled to give a certain set-point for V. When D is fixed in this way, i.e. by controlling one input variable v by means of measuring another input variable V the reactor is said to operate as a *chemostat*. This is the mode of operation for a vast majority of laboratory continuous stirred tank reactors – and for many industrial reactors.

Control of D can also be achieved by measurement of one of the output variables:

- In the *turbidostat* v is manipulated at a constant V to obtain a constant biomass concentration x (g L^{-1}) in the effluent. In this way a certain value of D is obtained which corresponds to the set point value of x.
- The feed of a nutrient (e.g. glucose) can be manipulated to obtain a certain pH in the effluent. Many bioreactions produce or consume protons and by separating the nutrient feed from the alkali/acid feed used to neutralize the proton production the rate of the bioreaction can likewise be controlled. This is the *pH-auxostat*.
- Measurement of the effluent concentration of one of the metabolic products, e.g. ethanol in fermentations with *Saccharomyces cerevisiae*, can also be used to obtain the D value, which corresponds to a given set point for the effluent concentration (or the rate of production) of one of the products. This is called a *productostat* (Andersen et al., 1997).

These three strategies (which are called "closed loop control" strategies by control engineers) are used together with a more or less detailed mathematical model of the process, which occurs in the bioreactor. As in any textbook treatment of stirred tank reactors it is assumed that the effluent concentrations of biomass and metabolic products are identical to those found at any point in the reactor. The reactor is then called an *ideal bioreactor*. Unless the steady state to be explored is unstable all the four strategies discussed above can be used to reach a desired steady state. The choice of strategy should ideally depend on the sensitivity of the steady state to changes in the variable used to establish the control policy. The strategy, which gives the highest sensitivity, should be chosen – although for practical reasons most investigations are carried out with only one control policy to obtain the steady state at all different dilution rates investigated.

As seen in Fig. 3.2 the biomass concentration in the effluent from a steady state aerobic fermentation with the yeast *S. cerevisiae* on glucose is virtually independent of D at small values of D. Consequently a steady state at a small D cannot be accurately fixed in a turbidostat. The substrate concentration s in the effluent may well vary significantly on a relative scale (we shall see in chapter 7 that s is proportional to D at small D) but it is difficult to measure the small substrate concentration accurately enough to fix v or D at the desired value. Operation of the bioreactor as a chemostat is therefore the preferred strategy. At a steady state close to the so-called "wash-out" the substrate concentration has increased significantly and the biomass concentration has decreased from its high and almost constant value at small D. Here the turbidostat is working very well since even small variations in D give rise to large changes in x, and the steady state is pinpointed by basing the control on a given set point for x. The chemostat is totally unsuited near wash out, but the pH-stat is also very satisfactory since the rate of proton production is strongly coupled (perhaps even proportional) to the value of biomass production. Around the so called "critical dilution rate" D_{crit} where ethanol production sets in (see Example 7.3) the ethanol concentration p (or the ethanol production rate) depends strongly on D. Hence a productostat is the ideal control strategy whereas both control of x or of the respiratory quotient, the ratio between oxygen consumption and carbon dioxide production, are less sensitive, and the chemostat is unsuited since one cannot control v accurately enough to obtain a steady state with a desired p.

It always takes patience to reach a steady state in a stirred tank continuous reactor and the time constant for the transient between one steady state and the next varies with the steady state. It takes on the order of five holding times, i.e., $5 \cdot D^{-1}$, to attain a new steady state and the measured rates can be far off their true steady state values if the approach to the new steady state is not within 95-99%. The time between steady states is not wasted since the transient itself contains much information on the physiology of the organism. The transient time is definitely a function of the control strategy used to fix the next steady state. Postma *et al.* (1989) found that around D_{crit} of an aerobic yeast fermentation on glucose the chemostat strategy had a transient time close to 50 holding times whereas Lei (2001) obtained a steady state within 5 holding times using the productostat control strategy.

Biochemical Reactions – A First Look

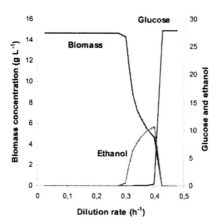

Figure 3.2 Schematic view of the concentration profiles of biomass, ethanol and glucose at different dilution rates in a continuous, stirred tank bioreactor containing the yeast *S. cerevisiae*.

The despair of experimenters who have waited many hours to obtain a steady state in continuous stirred tank reactors when the pH control for no apparent reason fails or a rubber tube breaks is all too well known. Likewise the loss of plasmids leading to gradual loss of productivity of the culture is a well-known source of frustration. Still, the steady state continuous reactor is the ideal equipment for physiological studies – and also to obtain trustworthy data for design of an industrial production. The set of steady state data is the foundation of a quantitative treatment of bioreactions, and deeper layers of metabolic response are revealed in the transients from one steady state to the next. These transient experiments give the necessary input for the modeling of non-ideal reactors in which the effects of spatial in-homogeneities of e.g. glucose or oxygen on the performance of industrial bioreactors are investigated (see Section 11.3.6). Without at least a semi quantitative knowledge of how rapidly a change in e.g. a vitamin concentration changes the productivity of a desired metabolite the detailed calculations of flow patterns in the bioreactor obtained by computational fluid dynamics are of little value.

Based on the measured feed and effluent concentrations in the steady state continuous bioreactor the reaction rates are easily calculated from steady state mass balances for the bioreactor. Thus all three equations (3.2) to (3.4) express that the mass of the compound produced by the reaction is equal to the difference in mass between a liquid feed and the outlet from the reactor.

$$q_{s_i} V + v(s_{i,f} - s_i) = 0 \qquad (3.2)$$

$$q_{p_i} V + v(p_{i,f} - p_i) = 0 \qquad (3.3)$$

$$q_x V + v(x_f - x) = 0 \qquad (3.4)$$

q_i is the volumetric production rate, i.e. the mass of compound i produced per volume reactor and per unit time. s_i and p_i are the concentrations of the ith substrate and metabolic product, respectively. x is the concentration of the biomass. Subscript f indicates the concentration of the variable in the liquid feed to the bioreactor.

For a gaseous substrate or product Eqs. (3.2) and (3.3) are slightly modified

$$\left(q^t_{s_i} + q_{s_i}\right)V + v\left(s_{i,f} - s_i\right) = 0 \tag{3.5}$$

$$\left(q^t_{p_i} + q_{p_i}\right)V + v\left(p_{i,f} - p_i\right) = 0 \tag{3.6}$$

In Eqs. (3.5) and (3.6) q_i is again the rate of production of the component by the biochemical reaction, whereas q^t_i is the rate at which the component is transferred from the gas phase to the liquid medium, which is given by:

$$q^t_{s_i} = k_l a\left(s^*_i - s_i\right) \tag{3.7}$$

$$q^t_{p_i} = k_l a\left(p^*_i - p_i\right) \tag{3.8}$$

$k_l a$ is a mass transfer coefficient which may depend on the fermentation medium and on the component being transferred, but certainly depends on the agitation of the liquid medium or the construction of the sparger used to introduce the gas. s^*_i and p^*_i are the concentrations of the component in the liquid which is in equilibrium with the gas phase. Both depend on the partial pressure of the component in the gas phase, on the medium temperature, pH etc. as will be further discussed in Chapter 10.

$$s^*_i = H_{s_i} \pi_{s_i} \quad ; \quad p^*_i = H_{p_i} \pi_{p_i} \tag{3.9}$$

In Eq. (3.9) π_{s_i} and π_{p_i} are the partial pressures in the gas phase and H_i is a so-called Henry's law coefficient, which as described above is dependent on the medium and on the operating conditions. A typical unit for H is mol (L atm)$^{-1}$.

Finally the gas phase partial pressures are given from mass balances for the gas phase over the whole system:

$$\frac{1}{RT}\left(v_{gf} \pi_{s_i,f} - v_g \pi_{s_i}\right) = k_l a\left(s^*_i - s_i\right)V \tag{3.10}$$

where v_{gf} and v_g are the inlet and head space volumetric gas flow rates, respectively. v_{gf} is different from v_g if the volumetric production rate of gaseous products is different from the consumption rate of gaseous substrates or if liquid phase components, typically water are stripped off from the

Biochemical Reactions – A First Look

medium.

From measurements of the concentrations of the variables at a fixed value of the input variable D the reaction rates can easily be determined from the mass balances (3.2)-(3.4) or (3.5)-(3.10) if there is an exchange with a gas phase.

In an experiment used to determine the reaction rates a number of the variables in the mass balances are usually zero.
- Typically there is neither biomass nor product in the feed streams, i.e.,

$$x_f = p_{i,f} = \pi_{if} = 0 \qquad (3.11)$$

- Components transferred from the gas phase are typically only sparingly soluble. In particular this is true for oxygen and also for natural gas, the carbon and energy source used in modern single cell protein production. Typically $s^*_{O_2} \cong 1\,mM$ and $s_{O_2} = 10-20\,\mu M$ while the corresponding values for s_{CH_4} are a little lower for SCP production. Thus in (3.7) and (3.10) s_i can be set to zero. In other aerobic bioreactions s_{O_2} has to be a sizeable fraction of $s^*_{O_2}$ - perhaps 10% for penicillin fermentation and even 25-35% for baker's yeast production.

In equations (3.2) to (3.4) q_i is the volumetric production rate, i.e., the mass of compound i produced per volume reactor and per unit time. The rates of the bioreactions, i.e., of the reactions within the cell, conventionally measured in mass of the ith component produced per unit weight of cell (rather than per unit volume of cell) and per unit time are:

$$r_i = \frac{q_i}{x} \qquad (3.12)$$

and specifically for the biomass:

$$\mu = r_x = \frac{q_x}{x} \equiv \text{the specific growth rate} \qquad (3.13)$$

As mentioned in Section 1.2 r_i are the rates associated with the "real" reactions in the "real" reactor – the cell, whereas q_i is used in mass balances set up for the reactor vessel.

3.2 Yield Coefficients

When any particular rate, whether q_i or r_i is scaled with another rate q_j or r_j one obtains the *yield coefficient* Y_{ji}.

$$Y_{ji} \equiv \left|\frac{r_i}{r_j}\right| = \left|\frac{q_i}{q_j}\right| \quad \text{and} \quad Y_{ij} = Y_{ji}^{-1} \qquad (3.14)$$

Since the rate of production of all substrates is negative a numerical sign is used in (3.14) to obtain yield coefficients, which are all positive. In this definition of the yield coefficient the first index j in Y_{ji} always refers to the reference rate.[1] Yield coefficients are measured in many consistent set of units, g g^{-1} or mole mole^{-1}, or for carbon containing compounds C-moles (C-mole)$^{-1}$, and as discussed in Section 2.2.3 they represent a very important set of design parameters for design and optimization of fermentation processes.

With the above definition of the yield coefficient a stoichiometric equation for growth of biomass X on glucose $C_6H_{12}O_6$ (or per C-mole: CH_2O) as the carbon and energy source can now be written:

$$-CH_2O - Y_{so}O_2 - Y_{sn}NH_3 - Y_{ss_1}S_1 - \ldots + Y_{sx}X + Y_{sc}CO_2 + Y_{sp_1}P_1 + \ldots + Y_{sw}H_2O = 0 \qquad (3.15)$$

Capital letters are used to denote substrates other than glucose, O_2 and the nitrogen source, which is taken to be NH_3. Likewise capital letters are used for metabolic products other than CO_2. In (3.15) the rates are all scaled by the rate of consumption of glucose in units of 1 C-mole = 30 g. Thus, if Y_{sn} = 0.05 then 0.05 moles of NH_3 is consumed every time 30 g = 1 C-mole glucose is consumed by the chemical reaction (3.15). From (3.15) other yield coefficients, for example the *respiratory quotient* RQ, which is frequently used in control of industrial scale aerobic reactions are easily found:

$$RQ = \left|\frac{q_c}{q_o}\right| = \frac{Y_{sc}}{Y_{so}} = Y_{oc} \qquad (3.16)$$

In all stoichiometric calculations in this textbook the chemical formula for the biomass X is written on the basis of one C-mole biomass:

$$X = CH_aO_bN_cS_dP_e\ldots \qquad (3.17)$$

Likewise all other carbon containing species $S_1, S_2 \ldots, P_1, P_2 \ldots$ are written on the basis of 1 C-mole. With this convention a carbon balance gives:

$$1 + Y_{ss_1} + Y_{ss_2} + \cdots = Y_{sx} + Y_{sc} + Y_{sp_1} + Y_{sp_2} + \cdots \qquad (3.18)$$

Water has been included in the general stoichiometry (3.15), but since the reaction is carried out in aqueous solution and since the concentration of substrates and products rarely exceeds 5 wt % the water produced adds so little to the total medium that it can be neglected when calculating the

---

[1] In the literature one often finds another notation for the yield coefficient, namely Y_{ij} for Y_{ji}.

Biochemical Reactions – A First Look

effluent concentrations. If substantial amounts of weak alkali or acid (1-2 M concentration) are added to keep pH constant this extra feed of water may of course have to be considered. The yield of water by the reaction is in itself of no importance and we shall in general not consider the term $Y_{sw}H_2O$.

The biomass formula (3.17) indicates that a long list of elements appear in the biomass. C, H, O and N are, however the dominant elements, while the content of S, P, and of all the trace elements such as Ca, Mg, Na, Fe, Co etc. is small compared to the N-content. These elements must, however, be present in the feed as e.g. Na_2SO_4 and K_2HPO_4 to synthesize an active biomass. The composition of X is determined by elemental analysis: A sample of dried biomass is ignited and the combustion products are analyzed. The residue is termed "ash", and it consists of inorganic compounds, mostly oxides. The ash content of biomass is usually in the range 4-8 wt %. It follows from the analytical procedure that the oxygen content of the sample cannot be measured, but must be calculated. In most examples of this book calculations are made on the basis of ash-free biomass, but in an experimental study the ash content of the biomass must of course be accounted for. Otherwise a significant error is introduced, and the calculation of rates is quite sensitive to the composition of the biomass. The composition of ash free biomass varies somewhat between different organisms, but the composition of a given organism varies much more with the growth conditions than between different organisms.

The yeast *S. cerevisiae* growing at glucose-limited conditions with an ample supply of nitrogen source has a macromolecular composition shown in Table 3.1. Based on the weight fractions of the 7 main groups of biomass components and their respective formula weight the average composition and formula weight of the biomass can be calculated

$$X = CH_{1.596}O_{0.396}N_{0.216}S_{0.0024}P_{0.017} \tag{3.19}$$

which corresponds to a formula weight of $M_x=23.57$ g (C-mole biomass)$^{-1}$. Table 3.2 shows the elemental composition of many organisms studied at different conditions and by different authors.

Table 3.1 Average composition of *S. cerevisiae*.

Macromolecule	Elemental composition	Percent by weight	g (C-mole)$^{-1}$
Protein	$CH_{1.58}O_{0.31}N_{0.27}S_{0.004}$	57	22.45
RNA	$CH_{1.25}O_{0.25}N_{0.38}P_{0.11}$	16	34.0
DNA	$CH_{1.15}O_{0.62}N_{0.39}P_{0.10}$	3	31.6
Carbohydrates	$CH_{1.67}O_{0.83}$	10	27.0
Phospholipids	$CH_{1.91}O_{0.23}N_{0.02}P_{0.02}$	10.8	18.5
Neutral fat	$CH_{1.84}O_{0.12}$	2.5	15.8
Pool of cellular metabolites	$CH_{1.8}O_{0.8}N_{0.2}S_{0.01}$	0.7	29.7

Table 3.2 Elemental composition of biomass for several microbial species.

Microorganism	Elemental composition	Ash content (w/w %)	Condition
Candida utilis	$CH_{1.83}O_{0.46}N_{0.19}$	7.0	Glucose limited, D=0.05 h^{-1}
	$CH_{1.87}O_{0.56}N_{0.20}$	7.0	Glucose limited, D=0.45 h^{-1}
	$CH_{1.83}O_{0.54}N_{0.10}$	7.0	Ammonia limited, D=0.05 h^{-1}
	$CH_{1.87}O_{0.56}N_{0.20}$	7.0	Ammonia limited, D=0.45 h^{-1}
Klebsiella aerogenes	$CH_{1.75}O_{0.43}N_{0.22}$	3.6	Glycerol limited, D=0.10 h^{-1}
	$CH_{1.73}O_{0.43}N_{0.24}$	3.6	Glycerol limited, D=0.85 h^{-1}
	$CH_{1.75}O_{0.47}N_{0.17}$	3.6	Ammonia limited, D=0.10 h^{-1}
	$CH_{1.73}O_{0.43}N_{0.24}$	3.6	Ammonia limited, D=0.85 h^{-1}
Saccharomyces cerevisiae	$CH_{1.82}O_{0.58}N_{0.16}$	7.3	Glucose limited, D=0.080 h^{-1}
	$CH_{1.78}O_{0.60}N_{0.19}$	9.7	Glucose limited, D=0.255 h^{-1}
Escherichia coli	$CH_{1.94}O_{0.52}N_{0.25}P_{0.025}$	5.5	Unlimited growth
	$CH_{1.77}O_{0.49}N_{0.24}P_{0.017}$	5.5	Unlimited growth
	$CH_{1.83}O_{0.50}N_{0.22}P_{0.021}$	5.5	Unlimited growth
	$CH_{1.96}O_{0.55}N_{0.25}P_{0.022}$	5.5	Unlimited growth
Pseudomonas fluorescens	$CH_{1.93}O_{0.55}N_{0.25}P_{0.021}$	5.5	Unlimited growth
Aerobacter aerogenes	$CH_{1.83}O_{0.55}N_{0.26}P_{0.024}$	5.5	Unlimited growth
Penicillium chrysogenum	$CH_{1.64}O_{0.52}N_{0.16}$	7.9	Unlimited growth
Aspergillus niger	$CH_{1.72}O_{0.55}N_{0.17}$	7.5	Unlimited growth
Average	$CH_{1.81}O_{0.52}N_{0.21}$	6.0	

Compared to the formula weight and composition of biomass calculated from the macromolecular composition (3.19) the compositions listed in Table 3.2 seem to contain an added 0.1 H$_2$O per C-atom. Tentatively this could be ascribed to a systematic error in the elementary analysis: many of the macromolecules contain strongly bound water, which is not completely released when the sample is dried before combustion. When the composition (3.19) is rescaled by adding 0.1 H$_2$O per C atom one obtains the average in Table 3.2 or

$$X = CH_{1.8}O_{0.5}N_{0.2} \qquad (3.20)$$

which corresponds to a formula weight of M_x=24.6 g (C-mole biomass)$^{-1}$. Unless otherwise stated we shall throughout the book use (3.20) as the "standard" formula for biomass. Eq. (3.20) is, however, only applicable in "normal" fermentations where the carbon source is the limiting growth factor.

As indicated in Table 3.2 the biomass composition is dependent on the growth conditions. This is further illustrated in Fig. 3.3, which shows results of an anaerobic yeast fermentation carried out in a batch reactor where nitrogen limitation sets in at 20 hours after the start of the fermentation. The cells cannot divide when the nitrogen source is used up, but they can still accumulate storage carbohydrates by polymerization of the substrate, glucose. Consequently the biomass concentration keeps increasing until all the glucose is used up. Assuming that no "active", i.e. N-containing

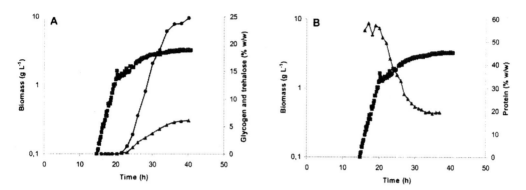

Figure 3.3 Anaerobic batch fermentation with *S. cerevisiae*. After 20 hours the nitrogen source is exhausted and nitrogen limitation sets in.
A. The biomass concentration and the cellular content of the storage carbohydrates trehalose and glycogen
B. The cellular content of protein and the biomass concentration. Anaerobic fermentation of *S. cerevisiae* with NH_3 as the nitrogen source. The data are taken from Schulze (1995) and Schulze et al. (1996).

biomass is synthesized after depletion of the N-source the original cells just become bloated with glycogen ($CH_{1.67}O_{0.83}$) and some trehalose ($CH_{1.78}O_{0.89}$). After 40 hours the biomass contains 25 wt % glycogen and 6 wt % trehalose. At the end of the batch fermentation the formula weight of X is 24.57 gL^{-1} and its composition has changed from (3.19) to

$$X = CH_{1.62}O_{0.52}N_{0.155}S_{0.0017}P_{0.012} \tag{3.21}$$

Accumulation of storage carbon compounds in procaryotes may amount to 70-80% of the total cell weight. A typical example is the accumulation of polyhydroxybutyrate (PHB) in *Ralstonia eutropha* (Lee and Choi, 2001) when nitrogen, phosphate or some other nutrient needed for growth of an active cell is denied to the culture. In these extreme cases the biomass composition approaches that of the storage polymer, e.g. $CH_{1.5}O_{0.5}$ for PHB.

3.3 Black Box Stoichiometries

Equation (3.15) is an attempt to represent all the chemical reactions by which the substrates are converted to biomass and metabolic products by a single chemical equation, a so-called "black box" model. This is clearly an immense simplification as all the biochemical reactions responsible for the overall conversion of substrates to metabolic products and biomass are lumped into a single overall reaction. For certain biological systems the yield coefficients may, however, be constant, even for growth at different growth conditions. Thus, Fig. 3.4 shows the rates of production of ethanol (e), CO_2 (c), biomass (x) and glycerol (g) in a continuous steady state anaerobic fermentation of *S. cerevisiae* as functions of the consumption of glucose (s) that was the limiting substrate in all the experiments. A vertical line through any point $(-q_sV)$ on the abscissa will give the amount of products synthesized per hour in the reactor of volume V at the dilution rate which corresponds to $(-q_sV)$. Increasing values of $(-q_sV)$ corresponds to increasing values of the dilution rate $D = v/V$. The four lines very nearly intersect at the origo of the diagram, and by regression the slopes are calculated to have the following values (all in C-mole per C-mole glucose):

$$\frac{q_eV}{-q_sV} = Y_{se} = 0.510 \;\; ; \;\; Y_{sc} = 0.275 \;\; ; \;\; Y_{sx} = 0.137 \;\; ; \;\; Y_{sg} = 0.077 \qquad (3.22)$$

Consequently the stoichiometry is

$$-CH_2O + 0.510CH_3O_{1/2} + 0.275CO_2 + 0.137X + 0.077CH_{8/3}O = 0 \qquad (3.23)$$

Since $Y_{se} + Y_{sc} + Y_{sg} + Y_{sx} = 0.999$ all significant carbon containing compounds are accounted for. The composition of ash free biomass was experimentally determined:

$$X = CH_{1.74}O_{0.60}N_{0.12} \qquad (3.24)$$

except for small amounts of S, P and minerals.

Only the carbon containing compounds are included in (3.23). In experiments where the nitrogen source appears in no products except the biomass the stoichiometric coefficient Y_{sn} is easily found. Thus, in the experiments of Duboc *et al.* (1998) the nitrogen source was NH_3 and $Y_{sn} = 0.12 \cdot Y_{sx}$ with the biomass composition in equation (3.24). The yield coefficient of H_2O, Y_{sw} is not included either. It can be found from either an O or an H balance (in mole H_2O per C-mole glucose):

$$Y_{sw} = 1 - \tfrac{1}{2} \cdot 0.510 - 2 \cdot 0.275 - 0.6 \cdot 0.137 - 0.077 = 0.0358$$
$$Y_{sw} = 0.5(2 + 3 \cdot 0.0164 - 3 \cdot 0.510 - 1.8 \cdot 0.137 - \tfrac{8}{3} \cdot 0.077) = 0.0377$$

The two values are identical to within the experimental error. As mentioned earlier the small amount of water produced by the overall reaction (3.23) has no perceptible influence on the molarity of the reactants. Thus, complete conversion of 1 C-mole glucose = 30 g glucose per liter medium leads to formation of 0.04 mole H_2O, a negligible amount compared to the 55 moles of water initially present per liter medium.

From (3.23) the yield coefficients on biomass produced are easily obtained using (3.14) (both on C-mole per C-mole biomass):

$$Y_{xs} = Y_{sx}^{-1} = 7.30 \;\; ; \;\; Y_{xe} = Y_{se}Y_{xs} = 3.72 \qquad (3.25)$$

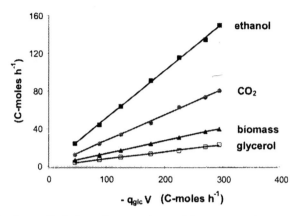

Figure 3.4 Anaerobic glucose-limited continuous culture of *S. cerevisiae*. Production rate as a function of glucose consumption rate at steady state: total ethanol (liquid and gaseous) (■); CO_2 (●); biomass (▲); glycerol (□). The lines are calculated by linear regression. The data are taken from Duboc *et al.* (1998).

Note 3.2 How to treat ions in the black box model.
Unless otherwise stated the stoichiometries discussed in this book are written with undissociated substrates and reactants, but the stoichiometry could easily be extended to consider charged compounds. Hereby there will be introduced an additional balance, namely the overall charge, but there will also be introduced an additional compound, namely protons. In reaction (3.15) and in the specific example of reaction (3.23) the nitrogen source (NH_3) is likely to be almost completely protonated at the usual medium pH of 5.5-6 in a yeast cultivation since pK_a is 9.25 for the acid NH_4^+. The ammonium ion is transported across the cell membrane by an active, ATP consuming mechanism (Section 2.1.2.3.) and delivered to the cytosol of the cell. The pH of the cytosol is close to neutral and again the dissociation of NH_4^+ is quite small. Nitrogen is, however, incorporated into the cell mass as NH_3 and with this drain of NH_3 away from the equilibrium all the protons of NH_4^+ are liberated and transported back to the medium.

This can be used to determine the specific growth rate μ. In a stoichiometry such as (3.23) with a negligible production of organic acids $q_x = \mu x$ is nearly proportional to the volumetric rate of proton production which can be determined by titration of the medium with (usually 1-5 M) NaOH or KOH to keep the pH of the medium constant at e.g. 5.5. Using a separate feed of NH_3 rather than the commonly applied mixed feed containing $(NH_4)_2SO_4$ or some other ammonium salt one may if desired avoid the high cation concentration of Na^+ or K^+ which is a side effect of titration with the strong bases.

At the end of this section we shall again emphasize that any given stoichiometry such as reaction (3.23) is true only for steady state operation and usually only for a given set of environmental conditions (pH, T, substrate concentrations). The substrate concentrations s_i in the bioreactor are time independent and equal to the effluent concentrations in the steady state, continuous bioreactor – and in no other situation. Since the value of s_i depends on the input variable D, in a fashion to be discussed in Chapter 7 it is very likely that a change in dilution rate leads to a different stoichiometry since the flux of carbon is going to be distributed differently in the many pathways of the metabolic network. The situation illustrated in Fig. 3.4 is, indeed, very rare. Here the stoichiometry remains unchanged over the wide range of dilution rates ($0.05 < D < 0.33$ h^{-1}), which corresponds to the substrate utilization rates 0.047 to 0.292 C-moles glucose h^{-1}. The change in stoichiometry around D_{crit} in the aerobic yeast fermentation schematically shown in Fig. 3.2 is a much more typical situation. When D increases above D_{crit} the yield coefficient Y_{se} jumps from virtually zero to a value much closer to the yield coefficient of the anaerobic yeast fermentation given in reaction (3.23) while Y_{sx} decreases rapidly. A complete change in stoichiometry when a "secondary" growth substrate such as NH_3 has been depleted has also been illustrated in Fig. 3.3 in connection with changes in the biomass composition.

3.4 Degree of Reduction Balances

All carbon containing substrates can be oxidized to CO_2 and relative to this end product of biochemical reactions the substrates and the carbon containing metabolic products are in a reduced state. A particular route through the metabolic network of a microorganism will convert the substrate, i.e. the input to the pathway, to a product that may be either reduced or oxidized relative to the substrate. Since all feasible pathways must be redox neutral in an organism working at steady state in a constant environment the net-reaction from substrate to product must be accompanied by consumption of or production of a separate chemical species in which redox power is stored. Looking one step further the totality of pathway reactions occurring in steady state in the cell must have a net production rate of zero for the redox carrying compound or compounds (see Section 5.2). As discussed in Section 2.1.3 there are several compounds, which can be used to transfer redox power from one pathway to another. These are:
- NADH (NAD$^+$)
- NADPH (NADP$^+$)
- FADH$_2$ (FAD)

In parenthesis are written the oxidized form of the compounds. The compounds NADH, NADPH and FADH$_2$ are related chemical species, but they occur as co-factors in different enzymatic reactions. As already mentioned in Section 2.1.3 NADH is mainly produced and again consumed in catabolic pathways whereas NADPH is largely synthesized in the pentose phosphate (PP) pathway and consumed in the anabolic pathway reactions. Thus, in Fig. 2.4 it is seen that NADPH is produced in the first two reactions of the PP pathway and that NADH is produced when glyceraldehyde-3 phosphate (G3P) is oxidized to 1,3-diphosphoglycerate in the EMP pathway. Sometimes a biochemical reaction is catalyzed by either one of several isoenzymes. Each isoenzyme may use a different cofactor, NADH or NADPH to absorb the redox power liberated by oxidation of the substrate. Thus, in *S. cervisiae* conversion of 2-oxoglutarate to glutamate can proceed by catalysis with two different glutamate dehydrogenases. In *S. cerevisiae* these two glutamate dehydrogenases are encoded by *GDH1* and *GDH2*, respectively, and the

Biochemical Reactions – A First Look

enzyme encoded by *GDH1* uses NADPH as cofactor, the other NADH. The stoichiometry of this reaction is:

$$-CH_{6/5}O - \tfrac{1}{3}NH_3 - \tfrac{1}{5}NAD(P)H + CH_{9/5}O_{4/5}N_{1/5} = 0 \quad (3.26)$$

where the parenthesis around P signifies that both co-factors can be used.

Many microorganisms such as *E. coli* (but not *S. cerevisiae*) are able to convert one of the redox co-factors to the other by a transhydrogenase catalyzed reaction.

$$-NADH - NADP^+ + NAD^+ + NADPH = 0 \quad (3.27)$$

For such microorganisms the two cofactors are in a sense equivalent, but usually they have to be treated as different entities in stoichiometric equations. In organisms with specialized organelles, such as the mitochondria in eucaryotes NADH in the cytosol can usually not be lumped with NADH in the mitochondria, and separate balances for each cofactor must be set up in each compartment of the cell. These facts of biochemistry must of course be considered in any serious quantitative study of cell metabolism as discussed in Section 5.4, but in the present context we shall lump different redox carrying co-factors (in Section 5.2.3 we are going to consider balances for the individual co-factors). Therefore, if a unit of redox power is defined as one H atom then all redox-carrying co-factors are equivalent to H_2 and carry 2 redox equivalents, i.e.

$$\text{Redox unit} \equiv H = 1 \;;\; NADH = NADPH = FADH_2 = \text{``}H_2\text{''} \quad (3.28)$$

We shall now introduce a systematic way of defining the redox level of different chemical compounds.

1. Define a redox neutral compound for each element of interest.
2. We choose H_2O, CO_2, NH_3, H_2SO_4, H_3PO_4 as the neutral compounds corresponding to the elements O, C, N, S and P. With this set of neutral compounds and with the unit of redox defined as H = 1 one obtains the following redox levels of the five listed elements:

$$O = -2 \;,\; C = 4 \;,\; N = -3 \;,\; S = 6 \;,\; P = 5 \quad (3.29)$$

3. Now the redox level of any reactant in a biochemical reaction can be calculated. With our convention to write all carbon containing compounds on a 1 C-atom basis one obtains the following *degrees of reduction* κ_i.
   CH_2O (glucose and other hexoses, acetic acid, lactic acid, formaldehyde), $\kappa = 4$.
   $CH_3O_{1/2}$ (ethanol), $\kappa = 6$.
   CH_2O_2 (formic acid), $\kappa = 2$.
   $CH_{1.8}O_{0.5}N_{0.2}$ (standard biomass), $\kappa = 4.20$

Table 3.3 compiles elemental composition of many different compounds and their corresponding degree of reduction

Table 3.3 Elemental composition and degree of reduction of typical compounds encountered in fermentation processes.

Compound	Formula	Degree of reduction per carbon, κ	Compound	Formula	Degree of reduction per carbon, κ
Formic acid	CH_2O_2	2	Alanine	$C_3H_7NO_2$	4
Acetic acid	$C_2H_4O_2$	4	Arginine	$C_6H_{14}N_4O_2$	3.67
Propionic acid	$C_3H_6O_2$	4.67	Aspargine	$C_4H_8N_2O_3$	3
Butyric acid	$C_4H_8O_2$	5.00	Aspartate	$C_4H_7NO_4$	3
Valeric acid	$C_5H_{10}O_2$	5.20	Cysteine	$C_3H_7O_2NS$	6
Palmitic acid	$C_{16}H_{32}O_2$	5.75	Glutamate	$C_5H_9NO_4$	3.60
Lactic acid	$C_3H_6O_3$	4	Glutamine	$C_5H_{10}N_2O_3$	3.60
Gluconic acid	$C_6H_{12}O_7$	3.67	Glycine	$C_2H_5NO_2$	3
Pyruvic acid	$C_3H_4O_3$	3.33	Leucine	$C_6H_{13}NO_2$	5
Oxalic acid	$C_2H_2O_4$	1	Isoleucine	$C_6H_{13}NO_2$	5
Succinic acid	$C_4H_6O_4$	3.50	Lysine	$C_6H_{14}N_2O_2$	4.67
Fumaric acid	$C_4H_4O_4$	3	Histidine	$C_6H_9N_3O_2$	3.33
Malic acid	$C_4H_6O_5$	3	Phenylalanine	$C_9H_{11}NO_2$	4.44
Citric acid	$C_6H_8O_7$	3	Proline	$C_5H_9NO_2$	4.40
Hexoses	$C_6H_{12}O_6$	4	Serine	$C_3H_7NO_3$	3.33
Disaccharides	$C_{12}H_{22}O_{11}$	4	Threonine	$C_4H_9NO_3$	4
Methane	CH_4	8	Tryptophane	$C_{11}H_{12}N_2O_2$	4.18
Ethane	C_2H_6	7	Tyrosine	$C_9H_{11}NO_3$	4.22
Propane	C_3H_8	6.67	Valine	$C_5H_{11}NO_2$	4.80
Methanol	CH_4O	6	Methionine	$C_5H_{11}O_2NS$	6
Ethanol	C_2H_6O	6	Urea	CH_4ON_2	6
iso-Propanol	C_3H_8O	6			
n-Butanol	$C_4H_{10}O$	6			
Glycerol	$C_3H_8O_3$	4.67			
Acetone	C_3H_6O	5.33			
Formaldehyde	CH_2O	4			
Acetaldehyde	C_2H_4O	5			

When the degree of reduction has been determined for all reactants and products in a stoichiometric equation it is possible to redox balance any of the metabolic pathways illustrated in Chapter 2. Table 3.4 shows some typical examples of pathway reactions. In some pathways CO_2 is split off by decarboxylation of a carboxylic acid. This has to be recognized by inspection of pathway diagrams in order to write the correct stoichiometry for the pathway reaction. Whereas the production or consumption of redox co-factors can be calculated from a redox balance the free energy yield associated with the reaction (in units of ATP) has to be looked up in pathway diagrams. Table 3.4 also specifies the ATP yield associated with the overall reactions. Except for a few overall reactions such as the conversion of glucose to lactate or to ethanol a redox carrying co-factor is either generated or consumed. Pathways that produce redox units must therefore balance pathways that consume redox units in order that the net production of redox units is zero in the steady state operation of the cell. This is going to be the foundation of metabolite balancing to be introduced in Chapter 5.

Biochemical Reactions – A First Look   63

Table 3.4 Stoichiometry of pathway reactions. The oxidized co-factor and H$_2$O are not included. Also shown is the ATP yield Y$_{sATP}$ (in ATP per C-mole glucose except for respiration where the yield is per NADH used). In some reactions this yield coefficient can have different, organism specific values.

Reaction	Stoichiometry	ATP gain
glucose → pyruvate	$-CH_2O + CH_{4/3}O + \frac{1}{3}NADH = 0$	1/3
glucose → lactate	$-CH_2O + CH_2O = 0$	1/3
glucose → acetaldehyde	$-CH_2O + \frac{1}{3}CO_2 + \frac{2}{3}CH_2O_{1/2} + \frac{1}{3}NADH = 0$	1/3
glucose → ethanol	$-CH_2O + \frac{1}{3}CO_2 + \frac{2}{3}CH_3O_{1/2} = 0$	1/3
glucose → glycerol	$-CH_2O + CH_{8/3}O - \frac{1}{3}NADH = 0$	-1/3
glucose → acetate	$-CH_2O + \frac{1}{3}CO_2 + \frac{2}{3}CH_2O + \frac{2}{3}NADH = 0$	1/3[#]
glucose → CO$_2$	$-CH_2O + CO_2 + 2NADH = 0$	1/3 or 2/3
Respiration	$-NADH - \frac{1}{2}O_2 = 0$	P/O
glucose → ribulose-5-P	$-CH_2O - \frac{1}{6}HPO_3^{-2} + \frac{5}{6}CH_2O(\frac{1}{5}HPO_3^{-2}) + \frac{1}{6}CO_2 + \frac{1}{3}NADPH = 0$	-1/6
glucose → glutamate	$-CH_2O - \frac{1}{6}NH_3 + \frac{5}{6}CH_{9/5}O_{6/5}N_{1/5} + \frac{1}{6}CO_2 + \frac{1}{2}NAD(P)H = 0$	1/3

[#] The value shown is for eukaryotes. For prokaryotes the ATP gain for acetate formation is 2/3 moles ATP per C-mole glucose (see also Fig. 2.6).

If oxygen is a reactant (i.e. in aerobic bioreactions) an arbitrary number of NADH generating pathways can be redox balanced by respiration for which the net stoichiometry is included in Table 3.4. Consequently a black box model for a net production of metabolites with lower degree of reduction than the carbon substrate can be written without involvement of redox carrying co-factor if O$_2$ is included as a reactant. The yield coefficients of the black box stoichiometry are then found by redox balancing the net reaction, using O$_2$ (or another oxidizing agent) to balance the difference in redox level between substrates and products. This will be illustrated in Example 3.2, as well as several examples will be used to illustrate the concept of redox balancing.

Example 3.1. Anaerobic yeast fermentation
We now examine the stoichiometry of reaction (3.23) a little closer. The pathway leading from glucose to ethanol is redox neutral, but some carbon is lost as CO$_2$. Production of biomass (κ = 4.18) is, however, not redox neutral since carbon is lost in connection with formation of precursor metabolites needed for biomass synthesis, e.g. ribulose-5-P that is a precursor metabolite for synthesis of nucleic acids and acetyl-CoA that is a precursor metabolite for lipid biosynthesis, and NADH (and FADH$_2$) is formed in the TCA cycle, which must be somewhat active even at anaerobic growth to produce precursors for amino acid biosynthesis. Conversion of some glucose to the more reduced glycerol counterbalances this biomass related production of redox equivalents. Using the stoichiometry of reaction (3.23) one obtains the following redox balance:

$$-(4 \cdot 1 + 0 \cdot 0.0164) + (6 \cdot 0.510 + 0 \cdot 0.275 + 4.67 \cdot 0.077 + 4.18 \cdot 0.137) = -0.0078 \approx 0 \quad (1)$$

The redox balance (1) is very nearly satisfied. Previously it was shown that the carbon balance is satisfied, and the combined tests confirm to a high degree of certainty that not only have the right compounds of the net reaction been found, but they are also determined in the right ratios.

Example 3.2. Aerobic growth with ammonia as nitrogen source
Let the stoichiometry of a general, aerobic process with one product P be

$$-S - Y_{sn}NH_3 - Y_{so}O_2 + Y_{sx}X + Y_{sc}CO_2 + Y_{sp}P = 0 \quad (1)$$

S is the carbon- and energy substrate $CH_{s_1}O_{s_2}$ with degree of reduction κ_s, X is biomass $CH_{x_1}O_{x_2}N_{x_3}$ with degree of reduction κ_x and P is a product $CH_{p_1}O_{p_2}$ with degree of reduction κ_p. With the above overall reaction stoichiometry the carbon, nitrogen and degree of reduction balances are:

$$Y_{sx} + Y_{sc} + Y_{sp} = 1 \quad (2)$$

$$Y_{sn} = x_3 Y_{sx} \quad (3)$$

$$\kappa_x Y_{sx} + \kappa_p Y_{sp} - \kappa_s - (-4)Y_{so} = 0 \quad (4)$$

Equations (2) and (4) are solved to obtain Y_{sx} and Y_{sp} in terms of Y_{sc} and Y_{so}

$$Y_{sx} = \frac{(4 - \kappa_p RQ)Y_{so} + \kappa_p - \kappa_s}{\kappa_p - \kappa_x} \quad (5)$$

$$Y_{sp} = \frac{(\kappa_x RQ - 4)Y_{so} + \kappa_s - \kappa_x}{\kappa_p - \kappa_x} \quad (6)$$

Where

$$RQ = \frac{Y_{sc}}{Y_{so}} \quad (7)$$

Let the substrate be glucose and ethanol the product. For a standard biomass with $\kappa_x = 4.20$:

$$Y_{sx} = \frac{(4 - 6RQ)Y_{so} + 2}{1.80} \quad ; \quad Y_{sp} = \frac{(4.20RQ - 4)Y_{so} - 0.20}{1.80} \quad (8)$$

In industrial yeast fermentation both Y_{so} and RQ are usually monitored more or less continuously, and Eq. (8) is used to calculate the biomass yield on glucose and the ethanol yield on glucose. Thus for RQ =

Biochemical Reactions – A First Look

3 and $Y_{so} = 0.1$ mole oxygen per C-mole glucose we find:
- $Y_{sx} = 0.333$ C-mole biomass per C-mole glucose
- $Y_{sp} = 0.367$ C-mole ethanol per C-mole glucose

For RQ = 1 and $Y_{so} = 1$ one obtains $Y_{sx} = Y_{sp} = 0$, which corresponds to complete combustion of glucose to carbon dioxide according to:

$$-CH_2O - O_2 + CO_2 + H_2O = 0 \qquad (9)$$

At these conditions neither biomass nor ethanol is produced.

Example 3.3 Anaerobic growth of yeast with ammonia as nitrogen source and ethanol as the only product

The stoichiometry of the overall reaction for anaerobic growth of *S. cerevisiae* with ammonia as the sole nitrogen source was given in Eq. (3.23). The degree of reduction balance yields

$$-4 + 4.20\, Y_{sx} + 6\, Y_{sp} = 0 \qquad (1)$$

for a standard biomass with $\kappa_x = 4.20$. Typically for yeast fermentation $Y_{sx} = 0.12$ g (g glucose)$^{-1}$ or 0.15 C mole biomass (C mole glucose)$^{-1}$. Consequently the maximum expectable yield of ethanol on glucose is 0.56 C mole (C mole glucose)$^{-1}$ even if no glycerol is produced as was the case in Example 3.1. The bacterium *Zymomonas mobilis* also produces ethanol, and the biomass yield is lower than for yeast fermentation. Typically Y_{sx} is 0.05 g (g glucose)$^{-1}$ or 0.06 C mole (C mole glucose)$^{-1}$. With the lower biomass yield a higher ethanol yield of $Y_{sp} = 0.624$ C mole ethanol (C mole glucose)$^{-1}$ or 94% of the theoretical yield is obtained. The bacterium is consequently a more efficient producer of ethanol than yeast – but yeast (*S. cerevisiae*) has other advantages, e.g. its extreme tolerance for ethanol.

Example 3.4. Biomass production from natural gas

The bacterium *Methylococcus capsulatus* grows aerobically with methane or methanol as the sole carbon and energy source. The catabolic pathways are different from those shown in Chapter 2 – in fact few organism are able to grow on C-1 carbon compounds, see Goldberg and Rokem (1991). The resulting biomass is an excellent protein source, which can be used directly as feed for domestic animals and as feed for fish, e.g. salmon. The protein content of the biomass can amount to 70% of the dry weight (DW). In hydrolyzed form the protein has considerable potential for use in human diets, as a daily supplement in school meals in poor countries or in aid packages sent to catastrophe stricken countries. The nucleic acids are broken down much faster than the proteins in a short, high temperature treatment of the spray-dried biomass, and consequently long term negative effects caused by the nucleic acids are avoided when the protein is used for human consumption.

Near Trondheim in Norway a 10,000 tons per year plant for production of biomass, so called single cell protein (SCP) from natural gas has been operating since 1999. The natural gas is a by-product from oil exploration in the North Sea. It is converted on land to methanol (nearly 900,000 tons per year), but a small fraction of the natural gas purified to 99% CH_4 is used as substrate in the SCP factory. Methanol could also be used as substrate for SCP production as was the case in an ICI plant (Billingham, GB), which operated for a short period in the 1970's, but was closed down partly as a result of sky rocketing prices of oil during the oil crisis of the 1970's. With the shortage of, and high price of fishmeal in todays marketplace there is no chance that the revived SCP production from methane will fail for economic reasons. The market is likely to expand to substitute a substantial fraction of the demand for high-grade animal protein, which in Europe alone is more than 10^6 tons per year.

Consider standard biomass (κ_x = 4.20) produced with a yield of 0.8 g (g CH$_4$)$^{-1}$. From the degree of reduction balance and a carbon balance the stoichiometry is calculated:

$$-CH_4 - 1.454 O_2 + 0.520 X + 0.480 CO_2 = 0 \qquad (1)$$

If the same yield of biomass could be obtained from methanol then

$$-CH_3OH - 0.954 O_2 + 0.520 X + 0.480 CO_2 = 0 \qquad (2)$$

As expected half a mole of O_2 is used to oxidize methane to methanol and consequently Y_{so} in equation (2) is equal to Y_{so} from equation (1) – 0.5.

The redox balance used in the examples is often referred to as the *generalized degree of reduction balance*, and it was first introduced by Roels (1983) as a generalization of the earlier work by Erickson *et al.* (1978). The degree of reduction balance can be specified as a linear combination of the elemental balances, where the multiplication factors in the linear combination are the redox levels (or the degree of reduction) of the elements. It is therefore also clear that the reduction balance does not introduce an additional balance to the system. When all the elemental balances are satisfied then the degree of reduction balance automatically closes. As discussed earlier the use of all the elemental balances is in practice hindered by lack of information on the yield coefficient for water, which cannot be experimentally determined, and it is therefore interesting to use the degree of reduction balance together with the carbon and nitrogen balances.

A major reason for applying these balances to a set of experimental rate data used to construct a stoichiometric equation (3.15) is to check the consistency of the experimental data. Eventually the stoichiometric equation is going to be used as a first basis for design of an industrial process and it may have catastrophic consequences if the steady state yield coefficients are wrong from the start.

Experiments can go wrong for many reasons of which some are listed below.

- Some of the products may go undetected because it was not suspected that they would be formed in the bioreaction.
- Products that were thought to remain in the liquid phase are in fact partly stripped to the gas phase.
- The instruments used to measure the rates are wrongly calibrated or they malfunction after some time.

Biochemical Reactions – A First Look

It may seem strange that a metabolic product is unexpectedly being produced, but the complexity of the metabolic network of microorganisms makes it quite possible to miss a product unless the biochemical potential of the particular organism used in the experiment has been thoroughly investigated. Thus, the unexpected conversion to gluconic acid of a large fraction of the glucose used as feed in penicillium production was for a long time puzzling (Nielsen et al., 1994). The appearance of oxalic acid at moderately high pH in citric acid fermentation (Example 3.6) can if not detected and thereafter avoided lead to a large economic loss.

Stripping of some ethanol (and the even lower boiling acetaldehyde) from both aerobic and anaerobic fermentations where N_2 is sparged through the medium to give a completely O_2 free environment is a common cause of error in the mass balances. Even at low liquid phase mole fraction ethanol has a high gas phase mole fraction and a substantial loss of ethanol results, especially at low D since the production rate is small, while the rate of stripping is more or less independent of D. Refluxing the medium, even with 1-2° C cooling water is of little help since the heat transfer coefficient is small on the mostly dry gas phase side of the heat exchanger. Feeding of substrate through the reflux condenser may be a practical way of avoiding this problem since in that case a wet surface is present on the vapor side of the heat exchanger (Duboc and von Stockar, 1998).

Finally instruments, especially flow meters may give rise to systematic errors. When the gas flow is low, e.g. 0.1-0.5 vvm (volume gas per volume medium per minute), in laboratory bioreactors calibration of flow meters can be difficult and the rate of gas flow v_g is in error. Here a high accuracy of the instrument that determines the concentration of the gas-phase reactant in the gas feed is of course of no help.

Examples 3.5 and 3.6 will illustrate how consistency tests of data may be of great help. These tests should clearly be made during the experimental program and not retrospectively when all the data have been collected and the experimental equipment is dismantled or is being used for other purposes.

Example 3.5. Consistency analysis of yeast fermentation
One of the first experimental investigations of continuous aerobic yeast fermentation where the data was of such generally high quality that they could be used - and still can be used - for quantitative physiological studies was by von Meyenburg (1969) who worked in professor Fiechters group at ETH, Zürich. Many later papers have used the data, e.g., Bijerk and Hall (1977) to set up a structured kinetic model for aerobic yeast growth. The data are shown in Fig. 3.5. The stirred tank bioreactor was operated as a chemostat, and the feed was 28 g L^{-1} sterile glucose solution with sufficient NH_3 (or rather NH_4^+) and other substrates to make the culture glucose limited. At low values of D no ethanol is produced and the biomass concentration is high and approximately constant at 14 g L^{-1}, except perhaps for a slight drop for the lowest D-values. At these D-values the metabolism is purely respiratory, and the yeast obtains sufficient ATP for growth by complete oxidation of glucose to CO_2. The critical dilution rate D_{crit} is in the vicinity of 0.25 h^{-1} and above D_{crit} the yeast starts to produce ethanol due to a bottleneck in the TCA pathway or in the respiration (see also Example 7.3). Above D_{crit} the mode of fermentation is called respiro-fermentative. RQ increases rapidly and Y_{sx} drops sharply.

The kinetics of the fermentation will be discussed in Example 7.3. Here we will analyze the stoichiometry for different D values, observe the sharp change of stoichiometry around D_{crit} and make a consistency test of the data.

The biomass composition is (for simplicity) assumed to be $X = CH_{1.83}O_{0.56}N_{0.17}$ and the ash content 8 wt % for all D-values. In a real situation this assumption must of course be checked. The overall stoichiometry is taken to be:

$$-CH_2O - Y_{sn}NH_3 - Y_{so}O_2 + Y_{sc}CO_2 + Y_{sp}CH_3O_{0.5} + Y_{sw}H_2O + Y_{sx}CH_{1.83}O_{0.56}N_{0.17} = 0 \quad (1)$$

Equation (1) combines one rate, $q_s = q_{glucose}$ and 6 yield coefficients. The elemental balances supplies four constraints, and hence 3 rates must be experimentally determined to identify the system. The data shown in Fig. 3.5 can be used to find the rates of production of glucose, biomass, ethanol, CO_2 and O_2. Hence the system is over-determined and we can make a consistency test of the data.

First consider Y_{sn}. The biomass is – according to reaction (1) – the only sink for the added nitrogen source. Y_{sn} can therefore be calculated based on Y_{sx}, but unless we measure q_{NH_3} there is no way to prove that $Y_{sn} = 0.17\ Y_{sx}$ or whether an unexpected N-containing product is produced. It does not help to include q_{NH_3} in the set of measurements together with q_x if we wish to explore the C-mass balances or the degree of reduction balance since these two measurements form a closed set, separate from the other measurements of rates which are possible with the stoichiometry in reaction (1).

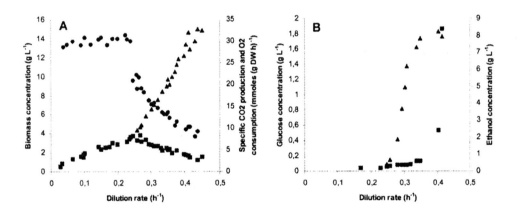

Figure 3.5. Chemostat cultures of *S. cerevisiae*.
A. Data for the biomass concentration (●), the specific oxygen uptake rate (■), and the specific carbon dioxide formation rate (▲).
B. Data for the glucose concentration (■) and the ethanol concentration (▲).

Biochemical Reactions – A First Look

Y_{sw} is – as mentioned earlier – uninteresting and it can probably not be measured accurately enough to be of any value. Water is therefore excluded from the analysis, and hereby all four elemental balances cannot be used. However, through using a carbon balance, a nitrogen balance and a degree of reduction balance the degrees of freedom is still 3, i.e. 3 rates must be experimentally determined to identify the system. Thus, when water is excluded from the stoichiometry one yield coefficient disappear, but also one of the constraints cannot be used.

By excluding water and assuming that Y_{sn} is directly determined from Y_{sx} the remaining variables in the system are Y_{so}, Y_{sc}, Y_{sp} and Y_{sx} together with q_s. Two constraints, the carbon – and the redox balances can be set up and three out of the five rates q_o, q_x, q_c, q_p and q_x must be used to identify the stoichiometry (except for Y_{sn} and Y_{sw} as explained above). All five rates are available from Fig. 3.5. Three values of D are chosen for further studies, $D = 0.15$ h^{-1} (or any D < 0.25 h^{-1}, at least within the accuracy of Fig. 3.5), $D = 0.30$ h^{-1} and $D = 0.40$ h^{-1}.

$D < 0.25$ h^{-1}
At these dilution rates we have: $x = 14$ g L^{-1}, $s = p \approx 0$, $s_f = 28$ g L^{-1}, $x_f = 0$.

$$Y_{sx} = \frac{14 \text{ g/L} \cdot 0.92}{25.17 \text{ (g/C-mole)}} \frac{30 \text{ (g/C-mole)}}{28 \text{ g/L}} = 0.548 \text{ C-moles DW (C-mole glucose)}^{-1} \quad (2)$$

From a carbon balance, we find

$$Y_{sc} = 1 - Y_{sx} = 0.452 \text{ moles CO}_2 \text{ (C-mole glucose)}^{-1} \quad (3)$$

and from a generalized degree of reduction balance

$$4 - 4Y_{so} = 4.20 Y_{sx} \Rightarrow Y_{so} = \tfrac{1}{4}(4 - 4.20 \cdot 0.548) = 0.425 \text{ moles O}_2 \text{ (C-mole glucose)}^{-1} \quad (4)$$

With these values for Y_{sc} and Y_{so}, RQ is calculated to be 1.06, which corresponds well with the data in Fig. 3.5. In the considered range of dilution rates it is observed that r_o and r_c both increase linearly with D and the slope is 30.0 mmoles of CO$_2$ per gram DW. This corresponds to $Y_{xc} = 0.82$ moles of CO$_2$ per C-mole of biomass, or

$$Y_{sc} = Y_{sx} Y_{xc} = 0.45 \text{ moles CO}_2 \text{ (C-mole glucose)}^{-1} \quad (5)$$

This value is nearly the same as that found from the carbon balance in (3), where the calculation is based on measurements of glucose and the biomass concentrations. Again it is concluded that there is a consistency in the experimental data. Observe that r_o and r_c are both larger than zero when $D = 0$. This is due to a consumption of glucose for maintenance purposes (see Section 5.2.1).

$D > 0.25$ h^{-1}
At the two dilution rates we find the following data from Fig. 3.5:
- $D = 0.3$ h^{-1} : $x = 7.1$ g L^{-1} and $p = 4.2$ g L^{-1}
- $D = 0.4$ h^{-1} : $x = 4.4$ g L^{-1} and $p = 8.3$ g L^{-1}

No attempt is made to smooth the biomass and the ethanol data, and the accuracy is not better than 0.1 – 0.2 g L^{-1}.

Both r_c and r_o appear to be well approximated by linear functions in D. By regression the following linear relationships are determined

$$r_c = 116\,D - 20.4 \quad \text{mmoles } CO_2 \text{ (g DW h)}^{-1} \tag{6}$$

$$r_o = -29.1\,D + 15.3 \quad \text{mmoles } O_2 \text{ (g DW h)}^{-1} \tag{7}$$

The decrease of the specific rate of oxygen consumption with increasing D is due to repression of the respiratory system at the increasing glucose concentrations (see Example 7.3 for further discussion).

For $D = 0.3\ h^{-1}$ one obtains

$$-q_s = (28 - 0.1) \cdot 0.3 / 30 = 0.279 \quad \text{C-mole } (L \cdot h)^{-1}$$

$$q_x = \frac{7.1 \cdot 0.92}{25.17} \cdot 0.3 = 0.07785 \quad \text{C-mole } (L \cdot h)^{-1}$$

$$q_p = \frac{4.2}{23} \cdot 0.3 = 0.0548 \quad \text{C-mole } (L \cdot h)^{-1}$$

$$q_c = 14.40 \cdot 7.1 \cdot 10^{-3} = 0.1022 \quad \text{mole } (L \cdot h)^{-1}$$

$$-q_o = 6.57 \cdot 7.1 \cdot 10^{-3} = 0.0466 \quad \text{mole } (L \cdot h)^{-1}$$

From these data the yield coefficients are calculated. The following table shows results for $D = 0.3\ h^{-1}$ and also for $D = 0.4\ h^{-1}$ (all in mole or C-mole per C-mole glucose)

D (h^{-1})	Y_{sx}	Y_{sp}	Y_{sc}	Y_{so}	Y_{sp_1}	Y_{se}
0.3	0.279	0.196	0.366	0.167	0.159	0.355
0.4	0.175	0.394	0.312	0.044	0.119	0.513

A carbon balance shows that some carbon is missing in the products for both $D = 0.3$ and $0.4\ h^{-1}$. The yield coefficient of the missing carbon is shown in the column Y_{sp_1} of the table. A degree of reduction balance yields:

$$D = 0.3\ h^{-1}\ :\ -4 + 4\ \ 0.167 + 0.279 \cdot 4.20 + 0.196 \cdot 6 + 0.159\, \kappa_{P_1} = 0$$
$$\Rightarrow\ \kappa_{P_1} = 6.19$$

and similarly at $D = 0.4\ h^{-1}$: $\kappa_{P_1} = 6.09$. Thus, for both cases the degree of reduction for P_1 is close to 6. There is good reason to believe that P_1 is ethanol. The final column of the table shows $Y_{se} = Y_{sp} + Y_{sp_1}$. For the low dilution rate almost half the produced ethanol has been stripped off; for the highest dilution rate only 23%. A loss of carbon of this magnitude is of course unacceptable.

Bijerk and Hall (1977) mention in their analysis of the von Meyenburg data that "12-13 wt % is missing from the carbon balance". In the analysis given here 16% is missing at $D = 0.3\ h^{-1}$ and 12 % at $0.4\ h^{-1}$ – both calculated on a C-mole basis. Taking the averages 12.5 wt % and 14% respectively a tentative

Biochemical Reactions – A First Look

formula weight of P_1 is calculated by

$$M_{P_1} = \frac{12.5}{14} \cdot 30 = 26.7 \text{ g (C-mole)}^{-1}$$

This does not contradict the strong indication from the redox balance that P_1 is ethanol but the loss in wt % ought to decrease with increasing D just as it has been argued that the loss in C-mole % should decrease with increasing D as was also confirmed in our analysis.

Example 3.6. Citric acid produced by *Aspergillus niger*

Citric acid $C_6H_8O_7$ is produced by fermentation of glucose with *Aspergillus niger* in an aerated tank at pH between 1.8 and 2. The biomass composition is $CH_{1.8}O_{0.5}N_{0.2}$. The nitrogen source is NH_4NO_3 and practically no CO_2 is formed when the production runs properly.

Typically 68 g citric acid is produced per 100 g glucose. The stoichiometry is

$$-CH_2O - Y_{so}O_2 - Y_{sn}NH_4NO_3 + Y_{sx}X + Y_{sp}CH_{4/3}O_{7/6} = 0 \tag{1}$$

$$Y_{sn} = \tfrac{1}{2} Y_{sx} \cdot 0.2 \tag{2}$$

The two remaining yield coefficients can be calculated in terms of the known value of Y_{sp}.

$$Y_{sp} = \frac{68/32}{100/30} = 0.6375 \text{ C-mole citric acid (C-mole glucose)}^{-1} \tag{3}$$

A carbon balance gives: $-1 + 0.6375 + Y_{sx} = 0$, and consequently $Y_{sx} = 0.3625$. From the nitrogen balance one immediately finds $Y_{sn} = 0.03625$. A degree of reduction balance gives:

$$-4 \cdot 1 - (-4) \cdot Y_{so} - 0.03625 \cdot \kappa_n + 0.3625 \cdot 4.20 + 0.6375 \cdot 3 = 0 \tag{4}$$

where $\kappa_n = 2 \cdot (-3) + 4 \cdot 1 + 3 \cdot (-2) = -8$. Thus

$$Y_{so} = \tfrac{1}{4}(4 + Y_{sn}\kappa_n - Y_{sx}\kappa_x - Y_{sp}\kappa_p) = 0.06875 \frac{\text{mole O}_2}{\text{C - mole glucose}} \tag{5}$$

If the experimental value of Y_{so} (which is quite easily found) does not match the calculated Y_{so} the set of experimental data is inconsistent. Thus, in an experiment conducted at a higher pH = 4-5 one obtains a lower citric acid yield 0.5375 C-mole (C-mole glucose)$^{-1}$ but the same biomass yield $Y_{sx} = 0.3625$. The oxygen demand is higher than expected: $Y_{so} = 0.11875$ mole O_2 (C-mole glucose)$^{-1}$. Clearly a new metabolic product is found besides citric acid, and from the carbon balance we find:

$$Y_{sp_1} = 1 - 0.3625 - 0.5375 = 0.1 \tag{6}$$

and we can use the degree of reduction balance in an attempt to identify the compound:

$$-4 - (-4) \cdot 0.11875 - (-8) \cdot 0.03625 + 0.3625 \cdot 4.20 + 0.5375 \cdot 3 + 0.1 \cdot \kappa_{pl} = 0 \quad (7)$$
$$\Rightarrow \kappa_{pl} = 1.0$$

which must be oxalic acid $(COOH)_2$ (or COOH on a C-mole basis).

The presence of even small amounts of the toxic oxalic acid in the citric acid will prohibit its use as a preservative in e.g. marmalades. Consequently, the producers make sure that pH never increases above 4 where the following side reaction occurs in parallel with the normal synthesis of citric acid in the TCA cycle (Fig. 2.5).

$$Pyruvate + CO_2 \xrightarrow{1} oxaloacetate \xrightarrow{2} acetate + oxalate \quad (8)$$

The enzymes in this pathway are: (1) pyruvate carboxylase and (2) oxaloacetate hydrolase. Pedersen *et al.* (2000) solved the problem in an elegant way: An oxalic acid non-producing strain of *A. niger* was constructed by deletion of the gene encoding for oxaloacetate hydrolase. The citric acid yield was the same as with the wild type strain, and the deletion strain had the same maximum specific growth rate $\mu = 0.20$ h^{-1} as the wild type strain.

Example 3.7. Growth with external electron acceptors other than oxygen
In the previous examples the electron acceptor was either oxygen or an intermediate in catabolism. There are, however, a number of external electron acceptors that can replace oxygen in some microorganisms (one group is the so-called obligate anaerobes). Examples of other electron acceptors are

1. Carbon dioxide, which is reduced to methane – see Example 4.5
2. Sulphate, which is reduced to sulfide
3. Nitrate, which is reduced to nitrite or molecular nitrogen

The concept of a degree of reduction balance may also be used to treat this type of growth. Consider the anaerobic growth of a microorganism with the composition $CH_{1.8}O_{0.5}N_{0.2}$ on a minimal medium containing glucose and ammonia. The electrons are donated to nitrate, which is reduced to molecular nitrogen. The overall stoichiometry is

$$Y_{sx}CH_{1.8}O_{0.5}N_{0.2} + \alpha N_2 + (1-Y_{sx})CO_2 + Y_{sw}H_2O - CH_2O - 0.2Y_{sx}NH_3 - 2\alpha HNO_3 = 0 \quad (1)$$

The degree of reduction balance gives

$$Y_{sx}\kappa_x - 6\alpha - 4 + 16\alpha = 0 \Leftrightarrow$$
$$\alpha = \frac{4 - Y_{sx}\kappa_x}{10} \quad (2)$$

If the yield of biomass from glucose is taken to be 0.5 C-moles of biomass per C-mole of glucose then $\alpha = 0.19$ and 0.38 moles of nitrate is used per C-mole of glucose metabolized. This process is important for removal of nitrate in biological waste-water treatment.

3.5 Systematic Analysis of Black Box Stoichiometries

In all the examples of Section 3.4 there was a single carbon and energy source, a single nitrogen source such as NH_3 (or NH_4NO_3) that was recovered in the biomass and usually a single metabolic product apart from CO_2. The yield coefficients could be calculated manually using a carbon – and a degree of reduction balance. In the general case with N substrates and M metabolic products, some of which might contain nitrogen, a more systematic procedure to calculate the stoichiometry is, however, needed.

Let the general stoichiometry be

$$-CH_{a_1}O_{b_1}N_{c_1} - \sum_{j=2}^{N-1} Y_{s_1 s_j} S_j - Y_{s_1 o} O_2 + Y_{s_1 x} X + \sum_{j=1}^{M-2} Y_{s_1 p_j} P_j + Y_{s_1 c} CO_2 + Y_{s_1 w} H_2 O = 0 \qquad (3.30)$$

For this stoichiometry the balance has to close for each of the four elements as illustrated in several examples in Section 3.4 where the carbon balance was set up using the yield coefficients. The elemental balances can, however, also be set up in terms of the volumetric production rates q, and generally this can be specified as:

$$\sum_{j=1}^{N} y_{s_j} q_{s_j} + y_x q_x + \sum_{j=1}^{M} y_{p_j} q_{p_j} = 0 \qquad (3.31)$$

where the coefficient y specifies the content of the element in the given compound, e.g. for glucose y is 1 for carbon, 2 for hydrogen and 1 for oxygen. For biomass given by eq. (3.20) y is 1 for carbon, 1.8 for hydrogen, 0.5 for oxygen and 0.2 for nitrogen. If an element does not appear in the compound y is zero, e.g. in water y is zero for carbon. Equation (3.31) is one relation between the $M + N + 1$ reaction rates, just as equation (3.30) is a chemical equation showing how chemically specified substrates are converted to different specified products. To generalize the four elemental balances we define a matrix E with 4 rows and $N + M + 1$ columns. In each column the elemental composition of one of the reaction species is written, i.e. the y's for the different reaction species are given. Now (3.31) can be written in compact notation

$$\mathbf{E\,q} = \mathbf{0} \qquad (3.32)$$

where $\mathbf{q}$ is the $N + M + 1$ column vector of volumetric reaction rates. Equation (3.32) provides four constraints between the $N + M + 1$ rates in $\mathbf{q}$, and 4 of these can consequently be calculated from the remaining $N + M - 3$ rates. If other elements – such as S - appear in some of the reactant compositions (cystein would be a case in point) the extension of (3.32) to contain more rows is obvious, but then there are typically an additional substrate included and the degrees of freedom is not affected.

Let the $N + M - 3$ measured rates be placed in the first $N + M - 3$ positions $\mathbf{q}_m$ of $\mathbf{q}$ and the remaining 4 elements in $\mathbf{q}_c$. Now equation (3.32) can be rewritten

74   Chapter 3

$$\mathbf{E}_m \mathbf{q}_m + \mathbf{E}_c \mathbf{q}_c = 0 \qquad (3.33)$$

Here $\mathbf{E}_c$ is a (4 x 4) matrix whereas $\mathbf{E}_m$ has 4 rows and $N + M - 3$ columns. Provided that $\det(\mathbf{E}_c) \neq 0$ the algebraic equation (3.33) is solved to give

$$\mathbf{q}_c = - (\mathbf{E}_c)^{-1} \mathbf{E}_m \mathbf{q}_m \qquad (3.34)$$

Where $(\mathbf{E}_c)^{-1}$ is the inverse of $\mathbf{E}_c$. The concept of the systematic procedure is illustrated in Example 3.8.

Example 3.8. Anaerobic yeast fermentation with CO_2, ethanol and glycerol as metabolic products.
Consider anaerobic fermentation of *S. cerevisiae* where the metabolic products are CO_2 (c) ethanol (e) and glycerol (g). Let the carbon source be glucose and the nitrogen source be NH_3. The biomass composition is $CH_{1.61}O_{0.52}N_{0.15}$. There are two substrates (CH_2O and NH_3), four products ($CH_3O_{1/2}$, $CH_{8/3}O$, CO_2 and H_2O) in addition to the biomass X. With 7 reacting species and four constraints one needs to measure 3 rates. The remaining rates can be calculated.

The measured rates are chosen as $\mathbf{q}_m = (q_s, q_x, q_g)$ and the calculated rates are then $\mathbf{q}_c = (q_c, q_n, q_e, q_w)$

$$\begin{pmatrix} 1 & 1 & 1 \\ 2 & 1.61 & \frac{8}{3} \\ 1 & 0.52 & 1 \\ 0 & 0.15 & 0 \end{pmatrix} \begin{pmatrix} q_s \\ q_x \\ q_g \end{pmatrix} + \begin{pmatrix} 1 & 0 & 1 & 0 \\ 0 & 3 & 3 & 2 \\ 2 & 0 & 0.5 & 1 \\ 0 & 1 & 0 & 0 \end{pmatrix} \begin{pmatrix} q_c \\ q_n \\ q_e \\ q_w \end{pmatrix} = \begin{pmatrix} 0 \\ 0 \\ 0 \\ 0 \end{pmatrix}$$

$\quad\quad\quad \mathbf{E}_m \quad\quad\quad\quad \mathbf{q}_m \quad\quad\quad\quad\quad \mathbf{E}_c \quad\quad\quad\quad\quad \mathbf{q}_c$

(1)

$$\begin{pmatrix} q_c \\ q_n \\ q_e \\ q_w \end{pmatrix} = -(\mathbf{E}_c)^{-1} \mathbf{E}_m \mathbf{q}_m = \begin{pmatrix} -\frac{1}{3} & -0.3133 & -\frac{2}{9} \\ 0 & -0.15 & 0 \\ -\frac{2}{3} & -0.6867 & -\frac{7}{9} \\ 0 & 0.45 & -\frac{1}{6} \end{pmatrix} \begin{pmatrix} q_s \\ q_x \\ q_g \end{pmatrix}$$

Based on the numbers in the resulting matrix above the yield coefficients in (3.30) are easily calculated:

Biochemical Reactions – A First Look

$$Y_{sc} = \frac{q_c}{-q_s} = \frac{-\frac{1}{3}q_s}{-q_s} + \frac{-0.3133q_x}{-q_s} - \frac{\frac{2}{9}q_g}{-q_s} = \frac{1}{3} - 0.3133Y_{sx} - \frac{2}{9}Y_{sg}$$

$$Y_{sn} = \frac{-0.15q_x}{-q_s} = -0.15Y_{sx}$$

$$Y_{se} = \frac{q_e}{-q_s} = \frac{2}{3} - 0.6867Y_{sx} - \frac{7}{9}Y_{sg}$$

$$Y_{sw} = \frac{q_w}{-q_s} = 0.45Y_{sx} - \frac{1}{6}Y_{sg}$$

(2)

Thus the black box model for this process becomes:

$$-CH_2O - 0.15Y_{sx}NH_3 + Y_{sx}CH_{1.61}O_{0.52}N_{0.15} + (\tfrac{1}{3} - 0.3133Y_{sx} - \tfrac{2}{9}Y_{sg})CO_2 + Y_{sg}CH_{8/3}O +$$
$$(\tfrac{2}{3} - 0.6867Y_{sx} - \tfrac{7}{9}Y_{sg})CH_3O_{1/2} + (0.45Y_{sx} - \tfrac{1}{6}Y_{sg})H_2O = 0$$

(3)

All stoichiometric coefficients are determined from the measured rates q_s, q_x and q_g – or from q_s, Y_{sx}, and Y_{sg}.

It is easily seen that the carbon balance is satisfied:

$$-1 + (\tfrac{1}{3} - 0.3133Y_{sx} - \tfrac{2}{9}Y_{sg}) + (\tfrac{2}{3} - 0.6867Y_{sx} - \tfrac{7}{9}Y_{sg}) + Y_{sx} + Y_{sg} = 0 \qquad (4)$$

The degree of reduction balance is automatically satisfied when H$_2$O is taken to be one of the reaction species – as is necessary when element balances are used:

$$-4 + 4.12Y_{sx} + 4\tfrac{2}{3}Y_{sg} + 6 \cdot (\tfrac{2}{3} - 0.6867Y_{sx} - \tfrac{7}{9}Y_{sg}) = 0 \qquad (5)$$

In examples as simple as this it is of course not necessary to set the whole mathematical machinery into action. The yield coefficient for NH$_3$ is immediately given to be 0.15 and if a degree of reduction balance plus a carbon balance is used one obtains

$$1 = Y_{sx} + Y_{sg} + Y_{se} + Y_{sc}$$
$$4 = 4.12Y_{sx} + 4\tfrac{2}{3}Y_{sg} + 6Y_{se}$$

(6)

$$\begin{pmatrix} Y_{sc} \\ Y_{se} \end{pmatrix} = \begin{pmatrix} 1 & 1 \\ 0 & 6 \end{pmatrix}^{-1} \begin{pmatrix} 1 - Y_{sx} - Y_{sg} \\ 4 - 4.12Y_{sx} - 4\tfrac{2}{3}Y_{sg} \end{pmatrix} = \tfrac{1}{6}\begin{pmatrix} 6 & -1 \\ 0 & 1 \end{pmatrix}\begin{pmatrix} 1 - Y_{sx} - Y_{sg} \\ 4 - 4.12Y_{sx} - 4\tfrac{2}{3}Y_{sg} \end{pmatrix} = \begin{pmatrix} \tfrac{1}{3} - 0.3133Y_{sx} - \tfrac{2}{9}Y_{sg} \\ \tfrac{2}{3} - 0.6867Y_{sx} - \tfrac{7}{9}Y_{sg} \end{pmatrix}$$

which is the result obtained previously. If desired Y_{sw} can be found from an H or an O balance.

Example 3.9. Production of lysine from glucose with acetic acid as byproduct

L-lysine ($CH_{7/3}O_{1/3}N_{1/3}$ – see Table 3.3.) can be produced from glucose in very high yields by aerobic fermentation using the bacterium *Corynebacterium glutamicum*. With NH_3 as nitrogen source and a standard biomass composition the stoichiometry of the total reaction is given in (1). Some carbon is lost to the undesired byproduct acetic acid.

$$-CH_2O - Y_{sn}NH_3 - Y_{so}O_2 + Y_{sx}CH_{1.8}O_{0.5}N_{0.2} + Y_{sp_1}CH_{7/3}O_{1/3}N_{1/3} + Y_{sp_2}CH_2O + Y_{sc}CO_2 + Y_{sw}H_2O = 0 \quad (1)$$

There are 8 rates and 4 constraints. 4 rates can be calculated, and the formal method (3.33) is needed to avoid algebraic errors.

Choose q_s, q_{p_1}, q_{p_2} and q_{o_2} as the four measured rates

$$\begin{pmatrix} q_n \\ q_x \\ q_c \\ q_w \end{pmatrix} = -\begin{pmatrix} 0 & 1 & 1 & 0 \\ 3 & 1.8 & 0 & 2 \\ 0 & 0.5 & 2 & 1 \\ 1 & 0.2 & 0 & 0 \end{pmatrix}^{-1} \begin{pmatrix} 1 & 1 & 1 & 0 \\ 2 & \tfrac{7}{3} & 2 & 0 \\ 1 & \tfrac{1}{3} & 1 & 2 \\ 0 & \tfrac{1}{3} & 0 & 0 \end{pmatrix} \begin{pmatrix} q_s \\ q_{p_1} \\ q_{p_2} \\ q_{o_2} \end{pmatrix} = \begin{pmatrix} 0.1905 & -\tfrac{1}{9} & 0.1905 & -0.1905 \\ -0.9524 & -\tfrac{10}{9} & -0.9524 & 0.9524 \\ -0.0476 & \tfrac{1}{9} & -0.0476 & -0.9524 \\ -0.4286 & 0 & -0.4286 & -0.5714 \end{pmatrix} \begin{pmatrix} q_s \\ q_{p_1} \\ q_{p_2} \\ q_{o_2} \end{pmatrix} \quad (2)$$

From equation (2) the yield coefficients in reaction (1) are obtained:

$$\begin{aligned} Y_{sn} &= 0.1905 + \tfrac{1}{9}Y_{sp_1} - 0.1905Y_{sp_2} - 0.1905Y_{so} \\ Y_{sx} &= 0.9524 - \tfrac{10}{9}Y_{sp_1} - 0.9524Y_{sp_2} - 0.9524Y_{so} \\ Y_{sc} &= 0.0476 + \tfrac{1}{9}Y_{sp_1} - 0.0476Y_{sp_2} + 0.9524Y_{so} \\ Y_{sw} &= 0.4286 \phantom{+ \tfrac{1}{9}Y_{sp_1}} - 0.4286Y_{sp_2} + 0.5714Y_{so} \end{aligned} \quad (3)$$

It is easily proved that the carbon, nitrogen and redox balances close:

$$\begin{aligned} 1 &= Y_{sx} + Y_{sp_1} + Y_{sp_2} + Y_{sc} \\ Y_{sn} &= 0.2Y_{sx} + \tfrac{1}{3}Y_{sp_1} \\ -4Y_{so} + 4 &= 4.20Y_{sx} + 4\tfrac{2}{3}Y_{sp_1} + 4Y_{sp_2} \end{aligned} \quad (4)$$

Note that the occurrence of a product (acetate) with the same formula per C-atom as a substrate (glucose) does not disturb the calculations. When stoichiometric calculations such as these are made as a preliminary to a process design it is of first importance to see whether a sufficiently high yield of the desired product can be obtained to make the process economically viable.

Based on our simple stoichiometric model and assuming that Y_{so} must be positive (oxygen is also used to provide energy through respiration of some carbon to CO_2) the maximum yield is obtained by setting $Y_{sx} = Y_{sp_2} = 0$ in the redox balance.

$$Y_{so} = \frac{4 - 4\frac{2}{3}Y_{sp_1}}{4} = 1 - \frac{7}{6}Y_{sp_1} \tag{5}$$

from which $(Y_{sp_1})_{max} = 6/7$ C-mole lysine (C-mole glucose)$^{-1}$ when $Y_{so} = 0$. This result is higher than the true theoretical maximum yield (0.75) which can only be found by analysis of individual pathways, but this is due to the fact that the black box model is too crude.

In actual lysine production the biomass yield is far from zero – it is in fact higher than for most aerobic processes. Consequently the theoretical limit for Y_{sp_1} is not at all approached in practice – but carbon yields of 0.30-0.35 are also economically acceptable. Furthermore, several essential amino acids must be supplied, but the main lysine production comes from uptake of NH_3 and conversion of the substrate to the amino acid. Several recent reviews treat this important fermentation process (e.g. de Graaf, 2000). The global production is several hundred thousand tons per year with a sales price in the range of 2.000 US $ per ton in bulk quantities (see also Table 2.1).

3.6 Identification of Gross Measurement Errors

From the previous sections it has been established that four out of the $N + M + 1$ rates can be calculated based on C, H, O and N balances – or for simple systems from a carbon and redox balance. It would, however, be catastrophic to use only the minimum number of rates when results of a fermentation (at a given set of environmental conditions) are to be interpreted in terms of a stoichiometric equation. First of all compounds may be missing from the stoichiometric equation as was the case in Example 3.6 or their rates of production or consumption may be misjudged due to gross experimental errors as was the case for ethanol which was stripped away in Example 3.5.

But even small – and quite unavoidable – experimental inaccuracies will lead to substantial errors in the calculated rates. The data of Duboc (1998), which were analyzed in Example 3.1, are very accurate. The sum of yield coefficients of products is 0.999 (Eq. 3.23) while the redox balance, Eq. (1) in Example 3.1 closes to within 0.008. Still, a calculation of two yield coefficients Y_{sx} and Y_{sg} based on Y_{sc}, Y_{se}, and using the carbon – and the redox balances leads to significant errors due to the ill-conditioned nature of the linear algebraic problem. Inserting Y_{sc} and Y_{se} from (3.22) one obtains

$$\text{Carbon balance}: 1 = 0.275 + 0.510 + Y_{sx} + Y_{sg}$$
$$\text{Redox balance}: 4 = 0.510 \cdot 6 + 4.18 Y_{sx} + 4.667 Y_{sg} \tag{3.35}$$
$$\begin{pmatrix} Y_{sx} \\ Y_{sg} \end{pmatrix} = \frac{1}{0.487} \begin{pmatrix} 4.667 & -1 \\ -4.18 & 1 \end{pmatrix} \begin{pmatrix} 0.215 \\ 0.940 \end{pmatrix} = \begin{pmatrix} 0.130 \\ 0.0848 \end{pmatrix}$$

whereas the calculated value of Y_{sx} is close to the measured value of 0.137 the calculated value of Y_{sg} is woefully different from the experimental value of 0.077. One can certainly determine the rate of glycerol formation with an experimental error, which is smaller than 10% relative.

Experimentally measured rates beyond the $N + M - 3$ rates necessary to determine the stoichiometric coefficients of (3.30) can be used either to validate the stoichiometry by a qualitative assessment of the difference between measured and calculated values of the four remaining stoichiometric coefficients – as was done in Examples 3.5 and 3.6 – or they may be used, partly or as a whole to obtain better values for all the stoichiometric coefficients in a so called reconciliation procedure. This is done by least squares fitting of the coefficients. Thus, if L measurements where $N + M - 3 < L \leq N + M + 1$ are used the remaining rates can be calculated as follows:

$$\mathbf{E}_c^T (\mathbf{E}_m \mathbf{q}_m + \mathbf{E}_c \mathbf{q}_c) = \mathbf{E}_c^T \mathbf{E}_c \mathbf{q}_c + \mathbf{E}_c^T \mathbf{E}_m \mathbf{q}_m = \mathbf{0} \qquad (3.36)$$

Here Equation (3.33) is multiplied from the left by $\mathbf{E}_c^T$, i.e. by the matrix $\mathbf{E}_c$ where rows and columns are transposed. When $L > N + M - 3$ the matrix $\mathbf{E}_c$ is not quadratic but contains 4 rows and less than 4 columns. But the product of $\mathbf{E}_c^T$ and $\mathbf{E}_c$ is a quadratic matrix of order less than 4 but larger than or equal to 0:

$$[(N + M + 1 - L) \times 4] \times [4 \times (N + M - 1 - L)] = [N + M + 1 - L] \times [N + M + 1 - L]$$

Consequently

$$\mathbf{q}_c = -\left[\mathbf{E}_c^T \mathbf{E}_c\right]^{-1} \left[\mathbf{E}_c^T \mathbf{E}_m\right] \mathbf{q}_m \qquad (3.37)$$

The requirement for application of Eq. (3.37) is that $\mathbf{E}_c^T \mathbf{E}_c$ has full rank, i.e., rank($\mathbf{E}_c^T \mathbf{E}_c$) = $N + M + 1 - L$, or det($\mathbf{E}_c^T \mathbf{E}_c$)≠0. By inserting Eq. (3.37) in Eq. (3.33), we get

$$\mathbf{R} \mathbf{q}_m = \mathbf{0} \qquad (3.38)$$

where

$$\mathbf{R} = \mathbf{E}_m - \mathbf{E}_c (\mathbf{E}_c^T \mathbf{E}_c)^{-1} \mathbf{E}_c^T \mathbf{E}_m \qquad (3.39)$$

The matrix $\mathbf{R}$ is called the *redundancy matrix* (Heijden et al., 1994a,b), and its rank specifies the number of independent equations in Eq. (3.38). Thus, if there are I elemental balances the redundancy matrix contains I - rank($\mathbf{R}$) dependent rows, and if these rows are deleted we obtain rank($\mathbf{R}$) independent equations relating the L measured rates, i.e.,

$$\mathbf{R}_r \mathbf{q}_m = \mathbf{0} \qquad (3.40)$$

Biochemical Reactions – A First Look

where $\mathbf{R}_r$ is the reduced redundancy matrix containing only the independent rows of $\mathbf{R}$. If rank($\mathbf{R}$) = 0, the system is obviously not over-determined.

The case $L = N + M + 1$ is obviously not of interest unless there are experimental errors ("inaccuracies") in the measured data. This case will be treated shortly for the general case of L measurements with L bounded as shown above. Usually some of the rates can, however, not be found with satisfactory precision. q_w is certainly one example, but q_x can be difficult to measure if the medium is strongly colored or contains solid particles as is often the case in industrial fermentations. Consequently equation (3.37) can be used to calculate a set of rates $\mathbf{q}_c$ where $\mathbf{q}_c$ contains less than 4 components when only C, H, O and N are considered.

We now assume that all the measured rates contain random errors. With $\mathbf{q}_m$ and $\overline{\mathbf{q}}_m$ being the vectors of, respectively, the true and the measured values, we have

$$\overline{\mathbf{q}}_m = \mathbf{q}_m + \boldsymbol{\delta} \qquad (3.41)$$

where $\boldsymbol{\delta}$ is the vector of measurement errors. It is now assumed that the error vector is normally distributed, with a mean value of zero and with a variance covariance matrix $\mathbf{F}$, i.e.,

$$\mathbf{E}(\boldsymbol{\delta}) = \mathbf{0} \qquad (3.42)$$

$$\mathbf{F} = \mathbf{E}\left[(\overline{\mathbf{q}}_m - \mathbf{q}_m)(\overline{\mathbf{q}}_m - \mathbf{q}_m)^T\right] = \mathbf{E}(\boldsymbol{\delta}\boldsymbol{\delta}^T) \qquad (3.43)$$

where $\mathbf{E}$ is a matrix with each element being represented by the expected value operator. If the model is correct and there are no measurement errors ($\boldsymbol{\delta} = \mathbf{0}$), all I - rank($\mathbf{R}$) equations in Eq. (3.40) will be satisfied, i.e., the elemental balances close exactly. In any real experimental investigation ($\boldsymbol{\delta} \neq \mathbf{0}$) there is a residual in each part of Eq. (3.41). The vector of residuals $\boldsymbol{\varepsilon}$ is given by

$$\boldsymbol{\varepsilon} = \mathbf{R}_r \boldsymbol{\delta} = \mathbf{R}_r (\overline{\mathbf{q}}_m - \mathbf{q}_m) = \mathbf{R}_r \overline{\mathbf{q}}_m \qquad (3.44)$$

since by Eq. (3.40) $\mathbf{R}_r \mathbf{q}_m = \mathbf{0}$. Since $\mathbf{E}(\boldsymbol{\delta}) = \mathbf{0}$, the residuals $\boldsymbol{\varepsilon}$ defined in Eq. (3.44) also have a mean of zero and the variance covariance matrix for $\boldsymbol{\varepsilon}$ is given by Eq. (3.45):

$$\mathbf{E}(\boldsymbol{\varepsilon}) = \mathbf{R}_r \mathbf{E}(\boldsymbol{\delta}) = \mathbf{0} \qquad (3.45)$$

$$\mathbf{P} = \mathbf{E}(\boldsymbol{\varepsilon}\boldsymbol{\varepsilon}^T) = \mathbf{R}_r \mathbf{E}(\boldsymbol{\delta}\boldsymbol{\delta}^T)\mathbf{R}_r^T = \mathbf{R}_r \mathbf{F} \mathbf{R}_r^T \qquad (3.46)$$

The minimum variance estimate of the error vector $\boldsymbol{\delta}$ is obtained by minimizing the sum of squared errors scaled according to the level of confidence placed on the individual measurements, i.e., to compute,

$$\underset{\boldsymbol{\delta}}{\text{Min}}(\boldsymbol{\delta}^T \mathbf{F}^{-1} \boldsymbol{\delta}) \qquad (3.47)$$

The solution to the minimization problem in Eq. (3.47) is given by

$$\hat{\delta} = FR_r^T P^{-1}\varepsilon = FR_r^T P^{-1} R_r \overline{q}_m \qquad (3.48)$$

where the "hat" specifies that it is an estimate. The estimate in Eq. (3.48) coincides with the maximum likelihood estimate, since δ is normally distributed.[2] Using Eq. (3.48) we find an estimate of the measured rates to be given by

$$\hat{q}_m = \overline{q}_m - \hat{\delta} = (I - FR_r^T P^{-1} R_r)\overline{q}_m \qquad (3.49)$$

where I is a unity matrix. In Note 3.3 it is shown that the estimate $\hat{q}_m$ given by Eq. (3.49) has a smaller standard deviation than the raw measurement $\overline{q}_m$, and the estimate is therefore likely to be more reliable than the measured data. Application of Eq. (3.49) to a set of experimental data is illustrated in Example 3.10.

Note 3.3 Variance covariance matrix of the rate estimates
The variance covariance matrix for the rate estimates is given by

$$\hat{F} = E\left[(\hat{q}_m - q_m)(\hat{q}_m - q_m)^T\right] = E\left[(\hat{\delta} - \delta)(\hat{\delta} - \delta^T)\right] \qquad (1)$$

and

$$E\left[(\hat{\delta} - \delta)(\hat{\delta} - \delta^T)\right] = E(\hat{\delta}\hat{\delta}^T) - E(\delta\hat{\delta}^T) - E(\hat{\delta}\delta^T) + E(\delta\delta^T) \qquad (2)$$

The first term represents the variance-covariance matrix for $\hat{\delta}$, which by using Eqs. (3.48) and (3.46) we find to be given by

$$E(\hat{\delta}\hat{\delta}^T) = FR_r^T P^{-1} E(\varepsilon\varepsilon^T)(FR_r^T P^{-1})^T = FR_r^T (P^{-1})^T R_r F^T \qquad (3)$$

For the second term in Eq. (2), we find

$$\begin{aligned} E(\delta\hat{\delta}^T) &= E\left[\delta(FR_r^T P^{-1}\varepsilon)^T\right] = E\left[\delta(FR_r^T P^{-1} R_r \delta)^T\right] \\ &= E(\delta\delta^T)R_r^T (P^{-1})^T R_r F^T = FR_r^T (P^{-1})^T R_r F^T \end{aligned} \qquad (4)$$

---

[2] When δ is normally distributed, the function to be minimized is the same for the least-square minimization problem and for the maximum-likelihood minimization problem. If the error vector is not normally distributed the estimate in Eq. (3.48) remains valid for the least-squares minimization problem, whereas it will no longer be the maximum-likelihood estimate (Wang and Stephanopoulos, 1983).

Biochemical Reactions – A First Look

and, for the third term

$$E(\hat{\delta}\delta^T) = E\left[(\mathbf{FR}_r^T\mathbf{P}^{-1}\varepsilon)\delta^T\right] = E\left[(\mathbf{FR}_r^T\mathbf{P}^{-1}\mathbf{R}_r\delta)\delta^T\right] = \mathbf{FR}_r^T\mathbf{P}^{-1}\mathbf{R}_r E(\delta\delta^T) = \mathbf{FR}_r^T\mathbf{P}^{-1}\mathbf{R}_r\mathbf{F} \quad (5)$$

while the fourth term is given by Eq. (3.43). By combination of these results we find that the contributions from the first and second term cancel and

$$\hat{\mathbf{F}} = \mathbf{F} - \mathbf{FR}_r^T\mathbf{P}^{-1}\mathbf{R}_r\mathbf{F} \quad (6)$$

Since the last term is positive, the variance-covariance matrix for the estimated rates is always smaller than that for the measured rates.

Example 3.10 Calculation of best estimates for measured rates
We return to the experimental data of von Meyenburg (1969), which were presented in Example 3.5 (see Fig. 3.5), and we want to find better estimates for the measured variables. We first consider the data for low dilution rates where no ethanol is formed and in Example 3.13 we will consider the data for higher dilution rates.

At low dilution rates the measured rates are q_s, q_o, q_c and q_x. There are two non-measured rates – ammonia utilization q_n and formation of water q_w. Thus with the biomass composition specified in Eq. (1) of Example 3.5 we have

$$\mathbf{E}_m = \begin{pmatrix} 1 & 0 & 1 & 1 \\ 2 & 0 & 0 & 1.83 \\ 1 & 2 & 2 & 0.56 \\ 0 & 0 & 0 & 0.17 \end{pmatrix} \quad ; \quad \mathbf{E}_c = \begin{pmatrix} 0 & 0 \\ 3 & 2 \\ 0 & 1 \\ 1 & 0 \end{pmatrix} \quad (1)$$

With a total of six components and four elemental balances, the number of degrees of freedom $F = 2$, and since four volumetric rates are measured ($K = 4$) the system is over-determined. From Eq. (3.39) the redundancy matrix is found to be

$$\mathbf{R} = \begin{pmatrix} 1 & 0 & 1 & 1 \\ 0 & -0.286 & -0.286 & 0.014 \\ 0 & 0.572 & 0.572 & -0.028 \\ 0 & 0.858 & 0.858 & -0.042 \end{pmatrix} \quad (2)$$

and rank($\mathbf{R}$) = 2. It is easily seen that the last two rows are proportional to the second row, and we therefore delete these two rows and find the reduced redundancy matrix

$$\mathbf{R}_r = \begin{pmatrix} 1 & 0 & 1 & 1 \\ 0 & -0.286 & -0.286 & 0.014 \end{pmatrix} \quad (3)$$

At $D = 0.15$ h^{-1} we find the measured volumetric rates to be (all in C-moles or moles; see Example 3.5)

$$\bar{\mathbf{q}}_m = \begin{pmatrix} -0.140 \\ -0.063 \\ 0.063 \\ 0.079 \end{pmatrix} \quad (4)$$

The vector or residuals is then

$$\varepsilon = \mathbf{R}_r \bar{\mathbf{q}}_m = \begin{pmatrix} 1 & 0 & 1 & 1 \\ 0 & -0.286 & -0.286 & 0.014 \end{pmatrix} \begin{pmatrix} -0.140 \\ -0.063 \\ 0.063 \\ 0.079 \end{pmatrix} = 10^{-3} \begin{pmatrix} 2.0 \\ 1.1 \end{pmatrix} \quad (5)$$

We see that there is fairly good consistency in the data, since the residuals are very small. We will, however, try to obtain even better estimates for the measured rates.

In order to find the variance-covariance matrix, we need to know something about the size of the measurement errors. We will here assume that these errors are uncorrelated and that the relative error in the gas measurements is 10%, whereas it is 5% for the glucose and the biomass measurements. The variance-covariance matrix is therefore:

$$\mathbf{F} = \begin{pmatrix} (0.05 \cdot 0.140)^2 & 0 & 0 & 0 \\ 0 & (0.10 \cdot 0.063)^2 & 0 & 0 \\ 0 & 0 & (0.10 \cdot 0.063)^2 & 0 \\ 0 & 0 & 0 & (0.05 \cdot 0.079)^2 \end{pmatrix}$$

$$\Downarrow \quad (6)$$

$$\mathbf{F} = 10^{-4} \begin{pmatrix} 0.490 & 0 & 0 & 0 \\ 0 & 0.397 & 0 & 0 \\ 0 & 0 & 0.397 & 0 \\ 0 & 0 & 0 & 0.157 \end{pmatrix}$$

Thus the **P** matrix is found from Eq. (3.46):

$$\mathbf{P} = \mathbf{R}_r \mathbf{F} \mathbf{R}_r^T = 10^{-4} \begin{pmatrix} 1.044 & -0.111 \\ -0.111 & 0.065 \end{pmatrix} \; ; \; \mathbf{P}^{-1} = 10^5 \begin{pmatrix} 0.117 & 0.201 \\ 0.201 & 1.887 \end{pmatrix} \quad (7)$$

The estimate for the vector of measurement errors can now be calculated by using Eq. (3.48):

$$\hat{\delta} = \mathbf{FR}_r^T \mathbf{P}^{-1} \mathbf{R}_r \overline{\mathbf{q}}_m = 10^{-3} \begin{pmatrix} 2.2 \\ -2.8 \\ -1.0 \\ 0.8 \end{pmatrix} \qquad (8)$$

and the estimate for the measured rates is therefore

$$\hat{\mathbf{q}}_m = \overline{\mathbf{q}}_m - \hat{\delta} = \begin{pmatrix} -0.142 \\ -0.060 \\ 0.064 \\ 0.078 \end{pmatrix} \qquad (9)$$

It is concluded that the measurements are very good, since the estimated rates differ only slightly from the raw data. It is, however, justified to make the small corrections, and we can next calculate the variance covariance matrix for the estimated rates by using Eq. (6) of Note 3.3.

$$\hat{\mathbf{F}} = 10^{-4} \begin{pmatrix} 0.209 & 0.112 & 0.116 & -0.092 \\ 0.112 & 0.154 & -0.152 & -0.041 \\ -0.116 & -0.152 & 0.151 & -0.034 \\ -0.092 & 0.041 & -0.034 & 0.127 \end{pmatrix} \qquad (10)$$

It is observed that the diagonal elements in $\hat{\mathbf{F}}$ are smaller than the diagonal elements in $\mathbf{F}$, and the variance of the estimates is therefore smaller than for the raw measurements. The estimates $\hat{\mathbf{q}}_m$ are therefore likely to be more accurate and reliable. Whereas the errors of the raw measurements are uncorrelated, it is observed from Eq. (10) that the errors of the estimated rates are correlated, since the rates are correlated through the constraints in Eq. (3.38).

In the analysis we have used the volumetric rates, whereas yield coefficients were used in Example 3.5. The present error analysis could, however, just as well be carried out using the yield coefficients (simply replace the volumetric rate vector with a vector containing the yields).

Normally the variance-covariance matrix is assumed to be diagonal, i.e., the measurements are uncorrelated. However, the volumetric rates are seldom measured directly, but they are based on measurements of so-called primary variables, which may influence more than one of the measured volumetric rates. An example is measurement of the oxygen uptake rate and the carbon dioxide production rate, which are based on measurement of the gas flow rate through the bioreactor together with measurement of the partial pressure of the two gases in the exhaust. If there is an error in the measured gas-flow rate, this influences both of the above-mentioned rates, and errors in the measured rates are therefore indirectly correlated. The same objection holds for measurements of many other volumetric rates, which are in reality obtained by combination of a concentration and a flow-rate measurement. In all these cases of indirect error correction it is difficult to specify the true variance-covariance matrix **F**. Madron *et al.* (1977) describe a method by which the variance-covariance matrix can be found from knowledge of the errors in the primary variables (see Note 3.4), but even if we know that there is often a certain coupling between measurement errors we will use a diagonal **F** matrix since we can rarely do anything better. The diagonal **F** matrix is preferable, partly because the calculations are simplified but also because we usually know little beyond the order of magnitude of the measurement errors. Thus in some cases **F** is expressed as

$$\mathbf{F} = c\mathbf{I} \tag{3.50}$$

where **I** is the unity matrix and, inserting Eq. (3.50) in Eq. (3.49), we obtain

$$\hat{\mathbf{q}}_m = (\mathbf{I} - \mathbf{R}_r^T (\mathbf{R}_r \mathbf{R}_r^T)^{-1} \mathbf{R}_r) \bar{\mathbf{q}}_m \tag{3.51}$$

which is the classical least-squares estimate. Equation (3.51) may be useful in many situations, and its application to the data analyzed in Example 3.10 is illustrated in Example 3.11.

Note 3.4 Calculation of the variance-covariance matrix from the errors in the primary variables

Normally the measured rates are determined from several measurements of so-called primary variables, e.g., the volumetric glucose utilization rate in a chemostat is measured from the glucose concentration in the exit stream and the feed flow rate. Specification of the variance-covariance matrix is therefore not straightforward, but Madron *et al.* (1977) describe a simple approach to find **F**. The measured rates are specified as functions of the primary variables, which are collected in the vector **v** (dimension L^*):

$$q_{m,i} = f_i(\mathbf{v}) ; \quad i = 1,...,L \tag{1}$$

Generally the functions f_i are nonlinear, but in order to obtain an approximate estimate of the variance and the co-variances the functions are linearized. Thus the errors δ_i in the L measured rates are expressed as linear combinations of the errors δ_j^* of the L^* primary variables:

$$\delta_i = \sum_{j=1}^{L^*} \left(\frac{\partial f_i}{\partial v_j} \right) \delta_j^* = \sum_{j=1}^{L^*} g_{ij} \delta_j^* ; \quad i = 1,...,L \tag{2}$$

The sensitivities g_{ji} are collected in the matrix **G** (of dimension $L \times L^*$), and the variance-covariance matrix **F** is calculated from

$$\mathbf{F} = \mathbf{G}\mathbf{F}^*\mathbf{G}^T \tag{3}$$

where **F*** is the diagonal matrix containing the L^* variances of the primary variables. The accuracy of the computed variances is limited by the accuracy of the linear approximation in Eq. (2) involved in the computation of the sensitivities.

Biochemical Reactions – A First Look

Example 3.11 Application of the least-squares estimate
Using Eq. (3.51) on the data analyzed in Example 3.10, i.e., $\mathbf{R}_r$ and $\mathbf{q}_m$ are given respectively by Eqs. (3) and (4) of Example 3.10, we find

$$\hat{\mathbf{q}}_m = \begin{pmatrix} -0.142 \\ -0.060 \\ 0.064 \\ 0.077 \end{pmatrix} \qquad (1)$$

which is almost identical with the estimates for the rates found in Example 3.10, where information concerning the measurement errors of the rates was included.

If any components in the residual vector are significantly different from zero, either there must be a significant error in at least one of the measurements or the applied model is not correct. To quantify what is meant by residuals significantly different from zero, we introduce the test function h given by the sum of weighted squares of the residuals, i.e., the residuals are weighted according to their accuracy:

$$h = \varepsilon^T \mathbf{P}^{-1} \varepsilon \qquad (3.52)$$

When the raw measurements are uncorrelated, the test function h is chi-square distributed [see, e.g., Wang and Stephanopoulos (1983)], and Heijden *et al.* (1994b) proved that this is also the case of correlated raw data. The degrees of freedom of the distribution is equal to rank($\mathbf{P}$) = rank($\mathbf{R}$).

The calculated value of h is compared with values of the χ^2 distribution at the given value of rank($\mathbf{R}$). If at a high enough confidence level $1-\theta$ one obtains that h is larger than χ^2, then there is something wrong with the data or the model. Table 3.5 is an extract from a table of χ^2 values. Normally a confidence level of at least 95% should be used.

Table 3.5 Values of the χ^2 distribution.

Degrees of freedom	Confidence level (1 - θ)					
	0.500	0.750	0.900	0.950	0.975	0.990
1	0.46	1.32	2.71	3.84	5.02	6.63
2	1.39	2.77	4.61	5.99	7.38	9.21
3	2.37	4.11	6.25	7.81	9.35	11.30
4	3.36	5.39	7.78	9.49	11.10	13.30
5	4.35	6.63	9.24	11.10	12.80	15.10

Example 3.12 Calculation of the test function h
With the data in Example 3.10, the vector of residuals is given by Eq. (5) of Example 3.10 and the test function is found to be

$$h = \varepsilon^T \mathbf{P}^{-1} \varepsilon = 10^{-1} \begin{pmatrix} 2.0 & 1.1 \end{pmatrix} \begin{pmatrix} 0.117 & 0.201 \\ 0.201 & 1.887 \end{pmatrix} \begin{pmatrix} 2.0 \\ 1.1 \end{pmatrix} = 0.364 \quad (1)$$

Since rank($\mathbf{R}$) = 2, the number of degrees of freedom for the χ^2 distribution is 2, and from Table 3.5 it is seen that h is far too small to raise any suspicion about the quality of the measurements.

The test of the data quality is easily carried out, but one cannot from the result that $h > \chi^2$ at, say, the 97.5% level conclude whether the unsatisfactorily large errors are due to a systematic defect of the data or to large random errors. However, in many cases one may suspect that one of the measurements has a systematic error, and if by leaving this measurement out of the analysis the errors become non-significant it is reasonable to assume that this particular measurement contains a systematic error. Thus the approach illustrated above can be used for error diagnosis, but only when the system is over-determined by at least two measurements, i.e., rank($\mathbf{R}$) $\geq$ 2. Otherwise, when one measurement is left out the system is no longer over-determined, and the analysis cannot be carried out. The concept of eliminating one measurement at a time is very simple, as illustrated in Example 3.13, but it may be time-consuming if no information about the possible source of error is available; i.e., one has to repeat the calculations with each of the measured rates left out of the analysis. A more systematic approach for error diagnosis is found in Heiiden *et al.* (1994a,b).

Example 3.13 Error diagnosis of yeast fermentation
We now consider the data of von Meyenberg (1969) for high dilution rates. In Example 3.5 it was concluded that some ethanol was probably missing. As in Example 3.10 we have two non-measured rates, ammonia utilization and formation of water, whereas there are five measured rates: consumption of glucose and oxygen and production of carbon dioxide, ethanol and biomass. Thus

$$\mathbf{E}_m = \begin{pmatrix} 1 & 0 & 1 & 1 & 1 \\ 2 & 0 & 0 & 3 & 1.83 \\ 1 & 2 & 2 & 0.5 & 0.56 \\ 0 & 0 & 0 & 0 & 0.17 \end{pmatrix}; \quad \mathbf{E}_c = \begin{pmatrix} 0 & 0 \\ 3 & 2 \\ 0 & 1 \\ 1 & 0 \end{pmatrix} \quad (1)$$

and the redundancy matrix is found to be

$$\mathbf{R} = \begin{pmatrix} 1 & 0 & 1 & 1 & 1 \\ 0 & -0.286 & -0.286 & 0.143 & 0.014 \\ 0 & 0.571 & 0.571 & -0.286 & -0.029 \\ 0 & 0.857 & 0.857 & -0.429 & -0.043 \end{pmatrix} \quad (2)$$

Biochemical Reactions – A First Look

R has the rank 2; the last two rows are multiples of the second row. The reduced redundancy matrix is thus found to be

$$\mathbf{R}_r = \begin{pmatrix} 1 & 0 & 1 & 1 & 1 \\ 0 & -0.286 & -0.286 & 0.143 & 0.014 \end{pmatrix} \quad (3)$$

At $D = 0.3$ h^{-1} we have the following measured rates from Example 3.5

$$\bar{\mathbf{q}}_m = \begin{pmatrix} -0.279 \\ -0.047 \\ 0.102 \\ 0.055 \\ 0.078 \end{pmatrix} \quad (4)$$

The vector of residuals is then

$$\varepsilon = \mathbf{R}_r \bar{\mathbf{q}}_m = \begin{pmatrix} -0.0440 \\ -0.0067 \end{pmatrix} \quad (5)$$

For the variance-covariance matrix we use the same measurement errors as in Example 3.10, i.e. 10% in the gas measurements and 5% in the medium-based measurements. Consequently

$$\mathbf{F} = \begin{pmatrix} (0.05 \cdot 0.279)^2 & 0 & 0 & 0 & 0 \\ 0 & (0.1 \cdot 0.047)^2 & 0 & 0 & 0 \\ 0 & 0 & (0.1 \cdot 0.102)^2 & 0 & 0 \\ 0 & 0 & 0 & (0.05 \cdot 0.055)^2 & 0 \\ 0 & 0 & 0 & 0 & (0.05 \cdot 0.078)^2 \end{pmatrix} \quad (6)$$

and

$$\mathbf{P} = \mathbf{R}_r \mathbf{F} \mathbf{R}_t^T = 10^{-3} \begin{pmatrix} 0.321 & -0.028 \\ -0.028 & 0.010 \end{pmatrix}; \quad \mathbf{P}^{-1} = 10^5 \begin{pmatrix} 0.041 & 0.111 \\ 0.111 & 1.260 \end{pmatrix} \quad (7)$$

Now we calculate the test function

$$h = \varepsilon^T \mathbf{P}^{-1} \varepsilon = 20.27 \quad (8)$$

Comparison of the calculated h with the values of the χ^2 distribution for rank(**R**)= 2 shows that there is an overwhelming probability that the data set contains significant deficiencies.

Since rank(**R**)= 2, it is possible to drop one of the measurements and still have an over-determined system. We can therefore carry out the error diagnosis described above, i.e. delete one of the measured rates at a time and calculate the test function. The results of this analysis give the test function for each case, i.e., with each of the measured rates deleted from the analysis:

- Ethanol $h = 1.87 \cdot 10^{-6}$
- Carbon dioxide $h = 19.89$
- Oxygen $h = 9.26$
- Biomass $h = 3.36$
- Glucose $h = 4.41$

When one of the measurements is left out of the analysis, rank($\mathbf{R}$) = 1, and the test function should therefore be compared with the χ^2 distribution with one degree of freedom. It is evident that when the ethanol measurement is left out, all signs of trouble disappear from the data set. It seems beyond doubt that there is a systematic error in the ethanol measurement, and we can obtain a nice statistical confirmation of the somewhat more qualitative argument of Example 3.5, which strongly indicated that some ethanol was missing from the mass balance. Using equation (3.48) we find good estimates for the measurement errors

$$\hat{\delta} = 10^{-3} \begin{pmatrix} 0.506 \\ 0.115 \\ 0.811 \\ 0.035 \end{pmatrix} \qquad (9)$$

These are small, and correction of the measured rates (not including that of ethanol) is therefore not necessary. Using Eq. (3.37), we calculate the three non-measured rates (including that of ethanol)

$$\mathbf{q}_c = \begin{pmatrix} q_n \\ q_p \\ q_w \end{pmatrix} = \begin{pmatrix} -0.012 \\ 0.100 \\ 0.076 \end{pmatrix} \qquad (10)$$

The calculated volumetric rate of ethanol production is thus 0.100 C-moles of ethanol per liter per hour, which corresponds to 0.357 C-moles of ethanol per C-mole of glucose. This is identical to the sum of Y_{sp} and the calculated Y_{sp1} found in Example 3.5. The method for error diagnosis illustrated here is very simple and quite powerful. It is, however, advisable not to embark on a mechanical error analysis without first using the intuitively simple engineering approach of Example 3.5.

PROBLEMS

Problem 3.1. Rate of uptake of gas phase substrates. Experimental errors in determination of rates.
Consider a well stirred laboratory reactor of volume 2 L which is used for continuous cultivation of *S. cerevisiae*. The reactor is sparged with v_{gf} = 1.3 L air min^{-1}. The volume fraction of O_2 in v_{gf} is 20.96 %. T = 30° C, and p = 1 atm. Assume, that $v_g = v_{gf}$.

a. Determine the rate of oxygen transfer q_o^t when the volume fraction of oxygen in the exhaust gas is 8.15 %.

b. The oxygen tension in the reactor is measured to 25µ M. s_o^* is 1.20 mM when π_o = 1 atm. Determine the mass transfer coefficient $k_l a$.

Biochemical Reactions – A First Look

c. Determine the difference between q_o' and $-q_o$ when s_{of} is $\cong 0$ and $v = 600$ mL h^{-1}

d. The ethanol concentration in the liquid effluent from the reactor is measured to 4.2 g L^{-1}. The inlet concentration is zero. Determine the rate of ethanol production (C-moles L^{-1} h^{-1}). Also determine the mole fraction of ethanol in the liquid.

e. From physical chemistry the following relation is obtained between the liquid mole fraction x and the gas phase mole fraction y.

$$y_i = \gamma_i x_i \frac{\pi_i^{sat}}{p} = \frac{\pi_i}{p} \qquad (1)$$

γ_i is the activity coefficient, p is the total pressure, and π_i is the partial pressure of component i in the gas phase. π_i^{sat} is the vapor pressure of ethanol at the system at the system temperature = 78.4 mm Hg at 30° C The activity coefficient for ethanol in water is quite high. Rarey and Gmehling (1993) give the value 6.61 for γ_i when $x_i < 0.01$ mol mol^{-1}. Assuming that thermodynamic equilibrium is established between ethanol in the liquid phase and ethanol in the gas phase determine how much ethanol is being stripped to the gas phase (C-moles L^{-1}h^{-1}) when $v_g = v_{gf} = 1.3$ L min^{-1}. Compare with the ethanol produced by the bioreaction (question 4). Will the evaporated ethanol have a significant effect on v_g – i.e. will it invalidate the assumption that $v_g = v_{gf}$?

f. π_{H2O}^{sat} is 31.7 mm Hg at 30° C. Assuming that dry air, 20.96 % oxygen is used as feed to the reactor. What is the difference between v_{gf} and v_g due to water evaporation? Again assume thermodynamic equilibrium between liquid and gas phase and $\gamma_{H2O} = 1$. Will the difference between v_g and v_{gf} influence the answer to question a?

g. In question 1 the same oxygen transfer rate would be obtained if v_{gf} was 6.5 L min^{-1} and the exit partial pressure of O$_2$ was 0.1840 atm.
Reconsider the answer to question 6 if we also here use the assumption $v_g = v_{gf}$. The answer to this question explains why
* One should try to have a significant difference in oxygen pressure between inlet and outlet .
* The off-gas should be dried before measuring π_{O2}. (Christensen et al., 1995)

The present problem is inspired by the experimental investigation of errors in measurement of oxygen uptake rates and ethanol production rates in bioreactors performed by Duboc and von Stockar (1998). This reference deserves a careful study as a preliminary to an experimental study of bioreaction rates.

Problem 3.2 Single-cell protein from ethane
The microorganism *Mycobacterium vaccae* is able to grow with ethane as the sole source of carbon and energy and with NH$_3$ as the nitrogen source. The limiting substrate is ethane, and the yield $Y_{sx} = 23.7$ gram dry weight per mole ethane.

a. Except for small amounts of S and P an elemental analysis of dry cell mass is C, 47.4 wt%; N, 8.30 wt%; H, 7.43wt%. The ash content is 4.00 wt%, while the oxygen content (which cannot be obtained by the analysis) must be found from a total mass balance.

Determine the elemental composition of the ash-free biomass and its formula weight per C-atom. Calculate Y_{sx} in C-moles (C-mole)$^{-1}$.

b. Calculate the oxygen consumption Y_{so} (moles O_2 (C-mole ethane)$^{-1}$ when it is assumed that CO_2 is the only metabolic product (apart from H_2O). Write the stoichiometry for conversion of ethane to SCP.

c. Measurements show that 10.44 g DW is formed per mole O_2 consumed. The yield Y_{sx} is that calculated above, but the value of RQ is 0.3805. Show that an extra metabolic product must be formed besides CO_2 (and H_2O). Determine the degree of reduction of this extra metabolic product. Using your general chemical knowledge you should be able to identify the extra metabolic product.

Problem 3.3 Production of Penicillin V
Fig. 2.8 recaptures the biosynthesis of penicillin. In the last step of the pathway phenoxyacetic acid forms an amide with the amino group of 6-APA. Penicillin V can also be formed directly from isopenicillin N by an exchange of the side chain L-α aminoadipic acid with phenoxyacetic acid. Glucose (s) is the limiting substrate. In a long series of metabolic reactions the three precursor metabolites shown on Fig. 2.8 at the entry to the penicillin synthesis pathway are finally produced from glucose. Another substrate is phenoxyacetic acid (s_1) which is fed continuously to the reactor with the other substrates (glucose, ammonia, SO_4^{2-}, PO_4^{3-}, minerals etc. in an ideal case. In industry a much cheaper carbon and nitrogen source is often in part of the fermentation as described in Section 2.2.2). The molecular formula for penicillin is $C_{16}H_{18}O_4N_2S$. What is the molecular formula for phenoxyacetic acid?

a. On a C-mole basis what is the yield coefficient Y_{ps2}? Rewrite Y_{ps2} in g g^{-1}

b. The composition of *Penicillium chrysogenum* is $X = CH_{1.64}O_{0.52}N_{0.16}S_{0.0046}P_{0.0054}$. Determine the oxygen requirement Y_{po} if $Y_{sx} = 0.127$ and $Y_{sp} = 0.194$. Also find RQ.

c. Make the assumption that during a fed-batch cultivation of penicillin r_p will always be proportional to $r_x = \mu$. Discuss whether this assumption appears to be reasonable. If the assumption holds then what would be the effect on the specific penicillin production rate if during the cultivation Y_{so} starts to increase.

d. On line RQ measurement show that during the first 120 hours of the production phase (after biomass has been formed during an initial batch phase) RQ = 0.95 to 1.15 with considerable scatter of the data. After 150 hours when growth of the fungus has virtually ceased RQ increases rapidly above 1.20. Discuss probable reasons for this observation. Is it due to break-down of the assumption in question c? Are other products formed? Could there be some straightforward experimental error?

Problem 3.4 Checking an anaerobic mass balance for cultivation of *S. cervisiae*.
In a steady state, strictly anaerobic chemostat with $V = 1$ L the yeast *S. cerevisiae* grow on a sterile medium with 23 g L^{-1} glucose and 1 g L^{-1} NH_3. The feed rate is $v = 0.1$ L h^{-1}. The biomass composition is $X = CH_{1.76}O_{0.56}N_{0.17}$.

The effluent from the bioreactor has the following composition in g L^{-1}: Glucose (*s*), 0.048; ammonia

Biochemical Reactions – A First Look

(s_n), 0.728; biomass (x), 2.36; ethanol (p_e), 8.85; glycerol (p_g), 2.39. The CO_2 production is measured by a photoacustic method. The CO_2 concentration in the effluent gas (in the head space) is found to be 2.88 vol %. On a flow-meter the total gas flow (30°C, 1atm) is determined to 0.25 L min^{-1}. v_{gf} is assumed to be equal to v_g.

a. Use the data to set up a carbon balance. Show that the data are not consistent, and calculate the yield coefficient of a missing metabolite.

b. Calculate the degree of reduction of the missing metabolite. In your opinion what is the source of the error?

Problem 3.5 Production of poly β-hydroxy butyrate (PHB)
van Aalst et al. (1997) studied the formation of PHB by the bacterium *Paracoccus pantotrophus* in a continuous stirred tank reactor operated at steady state. When growing under N-limitation but with an excess of the carbon source the organism can build up a large carbon storage that can be used to support active growth when, or if the N-limitation is lifted. This is part of the clean-up cycle in waste water treatment plants, but PHB has an independent and increasing interest as a biodegradable polymer – in competition with particularly poly-lactide. Although PHB is a part of the biomass it can just as well be regarded as a metabolic product. Its formula is $P = CH_{1.5}O_{0.5}$ per carbon atom. The remainder of the biomass (the "active" biomass) has the formula $X = CH_{1.73}O_{0.44}N_{0.24}$ per carbon atom. The organism uses acetic acid, CH_3COOH as carbon and energy source. It grows aerobically, and some CO_2 is produced together with X and P.

a. The result of two experiments, one with N-limitation and one with C-limitation are given below.

Dilution rate and effluent concentration of substrates (D h^{-1}, s_i mg L^{-1})	Production rates in mg L^{-1}h^{-1}		
	-q_s	q_x	q_p
$D = 0.5$			
$s_{HAc} = 70$	3065	1050	75
$s_{NH3} = 95$			
$D = 0.1$			
$s_{HAc} = 750$	1625	260	420
$s_{NH3} = 10.5$			

For each of the two experiments determine the yield coefficients in the black box model

$$CH_2O \text{ (HAc)} - Y_{sn} NH_3 - Y_{so} O_2 + Y_{sx} X + Y_{sp} P + Y_{sc} CO_2 + Y_{sw} H_2O = 0$$

b. For each of the two experiments calculate thye exit concentrations of X and P when there is no X or P in the feed. Also calculate the inlet concentrations of acetic acid and ammonia.

c. Could any single kinetic expression describe the results of both experiments?

Problem 3.6 Comparison between two strains of *S. cerevisiae*.
Nissen *et al.* (2001) studied anaerobic cultivation of three different strains of *S. cerevisiae*, TN1, TN26 and TN21 on a defined medium with glucose as carbon and energy source. TN1 is the "standard" haploid strain used in a number of studies by the authors. TN21 is a transformant of TN1 which expresses the transhydrogenase gene *tdh* from *Azotobacter vinlandii* to a high level, 4.53 units mg^{-1}. TN26 is also a transformant of TN1 with the same multicopy expression vector as in TN1, but without ligation of the *tdh* gene. TN26 is consequently genetically identical to TN21, except that it cannot express transhydrogenase. It is therefore unable to equilibrate the reversible reaction between $NADPH/NAD^+$ and $NADH/NADP^+$.

The table below shows product yields from steady state, glucose limited continuous cultivations of the three strains. Note the high accuracy of the data, and also that TN26 behaves exactly as TN1. There are, however, small but significant differences between TN21 and TN26. All data are given as C-mole (C-mole glucose)$^{-1}$.

Product	TN1	TN26	TN21
Ethanol	0.494 ± 0.012	0.493 ± 0.011	0.451 ± 0.011
Glycerol	0.091 ± 0.001	0.093 ± 0.001	0.118 ± 0.001
CO_2	0.275 ± 0.007	0.271 ± 0.007	0.262 ± 0.007
Biomass	0.111 ± 0.002	0.112 ± 0.002	0.102 ± 0.002
Succinate	0.005 ± 0.0003	0.005 ± 0.0003	0.006 ± 0.0003
Pyruvate	0.005 ± 0.0003	0.005 ± 0.0003	0.001 ± 0.0003
Acetate	0.005 ± 0.0003	0.005 ± 0.0003	0.007 ± 0.0003
2-Oxaloacetate	0.002 ± 0.0003	0.007 ± 0.0003	0.045 ± 0.0003
Total	0.988 ± 0.022	0.991 ± 0.021	0.992 ± 0.021

a. Check the redox balance for TN26 and TN21.

b. Assume that the missing carbon is ethanol. Repeat the calculation of a with this assumption.

c. Discuss qualitatively the difference between the product profiles for TN26 and TN21. What could be the reason for the higher yield of 2-oxaloacetate (=α ketoglutarate)? The dominant cofactor for conversion of 2-oxoglutarate to glutamic acid is NAPDH in all three strains.

REFERENCES

van Aalst, M.A., van Leeuwen, M.A., Pot, M.C, van Loosdrecht, M. and Heijnen, J.J (1997). Kinetic modelling of PHB production and consumption by *Paracoccus pantotrophus* under dynamic substrate supply. *Biotechnol.Bioeng*, 55, 773-782.

Andersen, M.Y., Pedersen, N., Brabrand, H., Hallager, L., and Jørgensen, S.B. (1997). Regulation of a continuous yeast bioreactor near the critical dilution rate using a productostat. *J. Biotechnol.*, 54, 1-14.

Bijkerk, A.H.E. and Hall, R.J. (1977). A mechanistic model of the aerobic growth of *Saccharomyces cerevisiae*. *Biotechnol. Bioeng.* 19, 267-296.

Christensen ,L.H., Schulze, U., Nielsen, J., and Villadsen, J. (1995). Acoustic gas analysis for fast and precise monitoring of bioreactors. *Chem. Eng. Sci.* 50, 2601-2610.

Cook, A. H. (1958). *The Chemistry and Biology of Yeasts*, Academic Press, New York.

Duboc, P., von Stockar, U. (1998). Systematic errors in data evaluation due to ethanol stripping and water vaporization. *Biotechnol. Bioeng.* 58, 428-439

Duboc, P., von Stockar, U. and Villadsen, J. (1998). Simple generic model for dynamic experiments with *S. cerevisiae* in continuous culture: decoupling between anabolism and catabolism. *Biotechnol. Bioeng.* **60**, 180-189.

Erickson, L. E., Minkevich, I. G., and Eroshin, V. K. (1978). Application of mass and energy balance regularities in fermentation. *Biotechnol. Bioeng.* **20**, 1595-1621

de Graaf, A. (2000). Metabolic Flux Analysis of *Corynebacterium glutamicum*. In *Bioreaction Engineering*, ed. K. Schügerl and K.H. Bellgradt, 506-555. Springer Verlag, Berlin.

Goldberg, I. and Rokem, J.S. (1991). *Biology of Methylotrophs*, Butterworth-Heinemann, Boston.

Heijden, R.T. J. M. van der, Heijnen, J. J., Hellinga, C., Romein, B., and Luyben, K., Ch. A. M. (1994a). Linear constraint relations in biochemical reaction systems: I, Classification of the calculability and the balanceability of conversion rates. *Biotechnol. Bioeng.*, **43**, 3-10.

Heijden, R.T. J. M. van der, Romein, B., Heijnen, J. J., Hellinga, C., and Luyben, K. Ch. A. M. (1994b). Linear constraint relations in biochemical reaction systems: II, Diagnosis and estimation of gross errors, *Biotechnol. Bioeng.*, **43**, 11-20.

Lee, S.Y., Choi, J.I. (2001). Production of microbial polyesters by fermentation of recombinant microorganisms. *Advances in Biochem. Eng.* **71**, 183-207.

Lei, F. (2001). Dynamics and nonlinear phenomena in continuous cultivations of *Saccharomyces cerevisiae*. Ph.D. thesis, Department of Chemical Engineering, DTU.

Madron, F., Veverka, V., and Vanecek, V. (1977). Statistical analysis of material balance of a chemical reactor. *A.I.Ch.E.Journal*, **23**, 482-486.

Meyenburg, K. von (1969). Katabolit-Repression und der Sprossungszyklus von *Saccharomyces cerevisiae*, Dissertation, ETH, Zürich.

Nielsen, J., Johansen, C.L., Villadsen, J. (1994). Culture fluorescense measurements during batch and fed-batch cultivations with *Penicillium crysogenum*. *J. Biotechnol.*, **38**, 51-62.

Nissen, T.L., Anderlund, M., Nielsen, J., Villadsen, J., and Kjelland-Brandt,.M.C. (2001). Expression of a cytoplasmic transhydrogenase in *Saccharomyces cerevisiae* results in formation of 2-oxoglutarate due to depletion of the NADPD-pool. *Yeast* **18**, 19-32.

Oura, E. (1983). Biomass from carbohydrates. In *Biotechnology*, H. Dellweg, ed., Vol. 3, 3-42. VCH Verlag, Weinheim, Germany.

Pedersen, H., Christensen, B., Hjort, C. and Nielsen, J. (2000) Construction and characterization of an oxalic acid nonproducing strain of *Aspergillus niger*. *Metabol. Eng.*, **2**, 34-41.

Postma, E., Verduyn, C., Scheffers, A. and van Dijken, J.(1989). Enzymatic analysis of the Crabtree-effect in glucose limited chemostat cultures of *Saccharomyces cerevisiae*. *Appl. Environ. Microbiology*, **55**, 468-477.

Rarey, J.R., and Gmehling, J. (1993) Computer – operated differential static apparatus for the measurement of vapor-liquid equilibrium data. *Fluid Phase Equilib.* **83**, 279-287.

Roels, J. A. (1983). *Energetics and Kinetics in Biotechnology*. Elsevier Biomedical Press, Amsterdam.

Schulze, U., Lidén, G., Nielsen, J. and Villadsen, J. (1996). Physiological effect of N-starvation in an anaerobic batch culture of *S. cerevisiae*. *Microbiology* **142**, 2299-2310.

Schulze, U., Anaerobic physiology of *S. cerevisiae*. (1995). Ph.D. thesis, Department of Biotechnology, DTU.

Wang, N. S. and Stephanopoulos, G. (1983). Application of macroscopic balances to the identification of gross measurement errors. *Biotechnol. Bioeng.*, **25**, 2177-2208.

4

Thermodynamics of Bioreactions

Chapter 2 gives an overview of the many biochemical reactions occurring in a living cell, and as discussed these biochemical reactions are exploited in the biotech industry to produce many interesting compounds. The pathway activity is to a large extent determined by the kinetics of the enzymes catalyzing the individual reactions as will be discussed in Chapter 6. However, of equal importance is thermodynamic constraints, since all reactions, independent of the properties of the enzymes, have to operate according to the laws of thermodynamics. Any treatment of bioreactions is therefore not complete unless one considers the thermodynamic constraints imposed on individual bioreactions. Since usually a bioreaction is accompanied by generation of heat the thermodynamic principles are also of importance in connection with design of the heat exchangers to be installed in bioreactors. In this chapter we will take a brief look at the thermodynamics involved in bioreactions. Classical thermodynamics is a discipline, which is extensively described in many textbooks, and it is not the purpose of this chapter to give an in-depth explanation of all the fundamental thermodynamic aspects involved in cellular processes. Still, to give some understanding of the relation between growth of a microbial culture and the energy generation and consumption in the process, certain topics of thermodynamics will be briefly reviewed here. As mentioned above this will also provide a tool for an important part of bioreactor design.

4.1 Chemical Equilibrium and Thermodynamic State Functions

In thermodynamics, a *system* is defined as that part of the universe that is being studied, such as a bioreactor or a cell, whereas the rest of the universe is referred to as its surroundings. A system is said to be *open* or *closed* according to whether it can exchange matter and energy with its surroundings. Because living cells take up nutrients, release metabolites, and generate work and heat, they are open systems. The state of a system is defined by a set of state functions, which include the enthalpy (H, equal to the heat absorbed at constant pressure when the only type of work is due to volume change) and entropy (S, a measure of the degree of order in the system).

Biochemical reactions occurring within a cell have to satisfy the laws of thermodynamics, and according to the second law of thermodynamics "spontaneous processes occur in a direction that increases the overall disorder (or entropy) of the *universe* or, mathematically, $\Delta S > 0$". Thus, spontaneity of a process is determined from the *overall* change in entropy. In the study of cellular processes spontaneity is indeed an important issue, but disordering of the universe by spontaneous processes is an impractical criterion to use for assessing spontaneity, as it is impossible to determine changes in the entropy of the entire universe. Furthermore, the spontaneity of a process cannot be decided from the entropy change of the system in question alone, because an exothermic process ($\Delta H_{system} < 0$, *i.e.,* heat is evolved from the system) may be spontaneous even if it is accompanied by a decrease in the entropy of the system, $\Delta S_{system} < 0$ (of course, the *total* entropy change is positive in this case too, due to an increase in the entropy of the environment that more than counterbalances the decrease in the entropy of the system). An example is the spontaneous folding of denatured proteins to their highly ordered (*i.e.*, $\Delta S_{system} < 0$) native conformation.

Due to these difficulties with the use of entropy, spontaneity is determined by using another state function, the *Gibbs free energy*:

$$\Delta G = \Delta H - T\Delta S \qquad (4.1)$$

The meaning of the free energy is that, for a process at constant temperature and pressure, the maximum work that can be done by the system (but not including the work of displacement) is equal to the decrease in the free energy of the system. For constant temperature and pressure processes, which describe the vast majority of biological systems, the criterion for spontaneity is $\Delta G \leq 0$.

Spontaneous processes, *i.e.,* those with $\Delta G \leq 0$, are said to be *exogenic*, and they can be utilized to do work. Processes that are not spontaneous, *i.e.,* those that have positive ΔG values, are termed *endergonic*, and they must be driven by the input of free energy. Processes at equilibrium, *i.e.* the forward and backward processes are exactly balanced, are characterized by $\Delta G = 0$. Notice that the Gibbs free energy varies with temperature, which must therefore always be specified. This dependence on temperature explains the spontaneous denaturation of proteins above a certain temperature. Formation of a native protein from its denatured form has both a negative ΔH and a negative ΔS. Above the temperature where ΔH equals $T\Delta S$, the Gibbs free energy of the folding process is positive and the reverse reaction (*i.e.,* denaturation) is a spontaneous process, *i.e.* the native protein will tend to denature.

It is important to note that a large negative value of ΔG does not necessarily imply that a chemical reaction will proceed at a measurable rate. Thus, the free energy change of the phosphorylation of glucose to glucose-6-phosphate by ATP is large and negative, but this reaction does not occur just by mixing glucose and ATP. Only when the enzyme hexokinase is added does the reaction proceed. Similarly, most biological molecules, including proteins, nucleic acids, carbohydrates, and lipids, are thermodynamically unstable to hydrolysis, but their spontaneous hydrolysis is insignificant. Only when hydrolytic enzymes are added do the hydrolysis reactions proceed at a measurable rate. Despite their importance in accelerating a reaction, enzymes do not change the ΔG for the reaction. As catalysts they can only speed up the attainment of thermodynamic equilibrium, but they do not allow a reaction with a positive ΔG to proceed.

4.1.1 Changes in Free Energy and Enthalpy

The change of Gibbs free energy by a chemical reaction is calculated from

$$(\Delta G)_{reaction} = (\sum_i Y_{p_i} \Delta G_{f,i})_{products} - (\sum_j Y_{s_j} \Delta G_{f,j})_{substrates} \qquad (4.2)$$

where Y_{p_i} and Y_{s_j} are the absolute values of stoichiometric coefficients for products and substrates respectively. In eq. (4.2) ΔG_f is the free energy of formation of the reactants at the state at which they are present in the reaction. Conventionally elements in their stable form at 25°C and 1 atm are defined to have zero free energy. ΔG_f^o is then the standard free energy of formation of the compound from its constituent elements at the reference temperature; here chosen as 25°C, and 1 atm.

In analogy with eq. (4.2) the enthalpy of reaction $(\Delta H)_{reaction}$ can be found by using ΔH_f instead of ΔG_f, where ΔH_f is the enthalpy of formation of the reactant. Similarly the entropy change ΔS that accompanies the reaction can be calculated. All chemical compounds are defined to have zero entropy at $T = 0$ K = - 273°C. At standard conditions 298 K, 1 atm the entropy S^0 is consequently positive for all compounds, reflecting the increase of "disorder" which accompanies the change in reference temperature for S from 0 to 298K.

Extensive tables of ΔG_f^o, ΔH_f^o and S^0 are available, and using data from these tables ΔG^0 and ΔH^0 for the reaction can be calculated using (4.1). When calculating ΔG^0 and ΔH^0 at temperatures in the vicinity of 25°C the change in ΔH_f^o and in S^o from $(\Delta H_{f,298}^o, S_{298}^o)$ can often be neglected while ΔG_f^o is a stronger function of T.

Example 4.1 Thermodynamic data for H_2O
In Handbook of Chemistry and Physics one obtains the following standard values for O_2, H_2 and H_2O – all in the gas phase

	ΔG_f^o	ΔH_f^o	S^o
O_2	0	0	49
H_2	0	0	31.21
H_2O	-54640	-57800	45.11

ΔG_f^o and ΔH_f^o are in cal mole^{-1}, S^o is in cal (mole K)$^{-1}$. Since eq. (4.1) is a relation between ΔG, ΔH and ΔS we can use two of these quantities to calculate the third. Thus, for water

$$\Delta H_f^o = -54640 + (45.11 - 31.21 - 0.5 \cdot 49) \cdot 298 = -57800 \text{ cal mole}^{-1}$$

Liquid water has much lower entropy 16.72 cal (mol K)$^{-1}$ than gaseous H$_2$O at the same conditions. Consequently for formation of liquid water from O$_2$ and H$_2$ at 25°C:

$$\Delta H_f^o = -57800 - 582 \cdot 18 = -68280 \text{ cal mole}^{-1}$$

$$\Delta G_f^o = -68280 - 298 \,(16.72 - 31.21 - 24.5) = -56661 \text{ cal mole}^{-1}$$

The heat of vaporization of H$_2$O at 25°C is 582 cal g^{-1}.

For any chemical reaction the free energy of reaction is related to the free energy of reaction at standard conditions by

$$\Delta G = \Delta G^0 + RT \ln K \tag{4.3}$$

where the mass action fraction K for the reaction is defined by:

$$K = \frac{\Pi\, p_i^{Y_{p_i}}}{\Pi\, s_i^{Y_{s_i}}} \tag{4.4}$$

$\Pi\, s_i^{Y_{s_i}}$ is the product of substrate activities where each activity is raised to the power of the corresponding stoichiometric coefficient in the reaction to be studied. $\Pi\, p_i^{Y_{p_i}}$ is the corresponding quantity for the products. For reactants in dilute aqueous solutions the activities can be set equal to their molar concentrations. For gas-phase reactants at low total pressure the activities are equal to the partial pressures in atm.

At equilibrium $\Delta G = 0$ and therefore

$$\Delta G^0 = -RT \ln K_{eq} \tag{4.5}$$

where K_{eq} is the *equilibrium constant* for the reaction.

Example 4.2 Equilibrium constant for formation of H$_2$O
For the reaction between O$_2$ and ½H$_2$ to form gaseous water at 25°C, 1 atm:

$$\ln K_{eq} = \frac{-\Delta G^0}{RT} = \frac{54640}{298 \cdot 1.987} = \ln \left[\frac{p_{H_2O}}{p_{O_2}^{1/2}\, p_{H_2}} \right]_{eq} = 92.28 \tag{1}$$

Thermodynamics of Bioreactions

For the corresponding reaction where the produced water is in equilibrium with liquid water at 25°C

$$\tfrac{1}{2}O_2 + H_2 \rightarrow (H_2O)_{gas} \quad ; \quad \Delta G_1$$
$$(H_2O)_{gas} \leftrightarrow (H_2O)_l \quad ; \quad \Delta G_2 = 0 \tag{2}$$

The total reaction to form water in the liquid phase has $\Delta G = \Delta G_1 + \Delta G_2 = \Delta G_1$

$$\Delta G = \Delta G^0 + RT\ln\frac{23.7}{760} = -54640 - 2053 = -56693 \text{ cal mole}^{-1} \tag{3}$$

since the partial pressure of H_2O in equilibrium with liquid water at 25°C is 23.7 mm Hg. Apart from round-off errors this is the same result as was obtained in Example 4.1 where a standard free energy ΔG^0 was defined in a reference frame of liquid water at 25°C rather than gaseous water. Similarly, if in a reaction with no liquid water present we can remove the produced water vapor to a partial pressure of 23.7 mm Hg while the partial pressures of O_2 and H_2 are kept at 1 atm we can shift the free energy of the gas phase reaction from –54.64 to –56.69 kcal mole^{-1} H_2O formed.

The two simple examples 4.1 and 4.2 are given to show how thermodynamic properties can be calculated using eqs. (4.1)-(4.4). It is demonstrated how the value of ΔG can be pushed in a desired direction by fixing the concentrations (or for gas phase reactions the partial pressures) of reactants at levels different from those at standard conditions. Since the reaction only proceeds spontaneously when ΔG is negative we may force it thermodynamically by decreasing the product concentrations and increase the substrate concentrations. For reactions to run inside a living organism it is therefore important that the levels of substrates and products are allowed to vary without affecting the overall thermodynamic feasibility of the reaction. For this reason the first reaction in a pathway is typically designed to have a large and negative ΔG as it may hereby be thermodynamically feasible even when the substrate concentration gets very low. Similarly the last reaction in a pathway also typically has a large and negative ΔG to make it thermodynamic feasible even when the product concentration gets very high. This is illustrated for the EMP pathway in Example 4.3.

The typical way in which a pathway reaction is forced to proceed in a thermodynamically unfavored direction is by coupling it to a thermodynamically favoured reaction. This is where ATP plays a crucial role as a co-factor in pathway reactions. Thus, phosphorylation of glucose to glucose-6 phosphate (G6P) is coupled to the hydrolysis of one ATP molecule to ADP

$$-ATP + ADP + P_i = 0 \quad ; \quad \Delta G^0 = -30.5 \text{ kJ mole}^{-1} \tag{4.6}$$

$$-\text{glucose} - P_i + G6P = 0 \quad ; \quad \Delta G^0 = 13.8 \text{ kJ mole}^{-1} \tag{4.7}$$

Addition of (4.6) and (4.7) yields

$$-\text{glucose} - ATP + ADP + G6P = 0 \quad ; \quad \Delta G^0 = -30.5 + 13.8 = -16.7 \text{ kJ mole}^{-1} \tag{4.8}$$

Further down in the EMP pathway ATP is regenerated by conversion of phosphoenolpyruvate (PEP) to pyruvate (PYR):

$$-\text{PEP} + \text{PYR} - \text{ADP} - P_i + \text{ATP} = 0 \ ; \ \Delta G^0 = -61.9 - (-30.5) = -31.4 \text{ kJ mole}^{-1} \quad (4.9)$$

The many individual reactions, which together constitute the EMP pathway net-reaction

$$-(\text{glucose} + 2\text{NAD}^+ + 2\text{ADP} + 2P_i) + (2\text{PYR} + 2\text{NADH} + 2\text{ATP}) = 0 \quad (4.10)$$

will, by careful harnessing of the large release of free energy in some of the reactions, proceed voluntarily although some of the reaction steps have a positive ΔG^0 (see Example 4.3).

The ATP that participates in the reactions is not to be construed as a stationary pool. ATP is formed and consumed all the time, and the turnover frequency is high. Left alone ATP has a half life time of seconds or at most a few minutes, depending on the cell type and on the environment. It is tacitly assumed that in steady state fermentations the ATP generation in catabolic pathways is always balanced by its consumption in anabolic pathways.

It should finally be remarked that the $\Delta G^0 = -30.5$ kJ mole^{-1} hydrolyzed ATP is a more or less empirical value. In living cells concentrations of ions, coenzymes and metabolites might vary by

---

Example 4.3. Free energy changes of reactions in the EMP pathway
To determine the free energy changes of cellular reactions, it is necessary to know the concentration of all metabolites and co-factors participating in these reactions. Such data are available only for a few pathways, and thermodynamic considerations are therefore often based on evaluation of standard free energy changes. This may, however, lead to erroneous conclusions since the use of standard free energy changes assumes certain fixed concentrations for reactants and products (those of the standard state) that may be different from the actual intracellular metabolite concentrations. To illustrate this point, we calculate the free energy change for some of the reactions in the EMP pathway (see Fig. 2.4). Table 4.1 lists the measured intracellular concentrations of some of the intermediates, ATP, ADP, and orthophosphate in the human erythrocyte, and Table 4.2 lists the calculated free energy changes.

Table 4.1 Concentrations of intermediates and co-factors of the EMP pathway in the human erythrocyte[a]

Metabolite/co-factor	Concentration (μM)	Metabolite/co-factor	Concentration (μM)
Glucose (GLC)	5000	2-Phosphoglycerate (2PG)	29.5
Glucose-6-P (G6P)	83	Phosphoenolpyruvate (PEP)	23
Fructose-6-P (F6P)	14	Pyruvate (PYR)	51
Fructose-1,6-bisP (FDP)	31	ATP	1850
Dihydroxyacetone P (DHAP)	138	ADP	138
Glyceraldehyde-3-P (GAP)	18.5	P_i	1000
3-Phosphoglycerate (3PG)	118		

[a]The data are taken from Lehninger (1975).

Thermodynamics of Bioreactions

several orders of magnitude across the membranes that separate organelles, and the actual $\triangle G$ may therefore be much different from $\triangle G^0$. This also holds for most other biochemical reactions, and one must be extremely careful when evaluating thermodynamic feasibility based exclusively on the $\triangle G^0$ for the reaction. This is further illustrated in Example 4.3.

From the calculated free energy changes (see Table 4.2) it is observed that, except for the hexokinase-, phosphofructokinase-, pyruvate kinase-, and, possibly, the triose-P-isomerase-catalyzed reactions, all the reactions of the EMP pathway are close to equilibrium (the free energy change for the reaction series 3-P-glyceraldehyde dehydrogenase and 3-phosphoglycerate kinase is also close to zero). Thus, the *in vivo* activity of several of the enzymes is sufficiently high to equilibrate the conversions, or, in other words, the forward and backward reactions of these conversions are much faster than the net flux through the pathway. Obviously, these equilibrium reactions are very sensitive to changes in the concentration of pathway intermediates, and therefore they rapidly communicate changes in flux generated by one of the reactions with a high negative free energy change throughout the rest of the pathway.

The three reactions with large negative free energies are *thermodynamically irreversible* and are often considered as key control points in the pathway. Obviously, the *in vivo* activity of the three enzymes hexokinase, phosphofructokinase, and pyruvate kinase is too low to equilibrate the reactions they catalyze. This may be the result of either too low gene expression, *i.e.*, the *in vivo* r_{max} of the enzymes is too low, or regulation at the enzyme level, *e.g.*, allosteric regulation or covalent enzyme modifications. With the large and negative free energies of hexokinase and pyruvate kinase the first and last reaction of the pathway may be thermodynamically favored even when the glucose concentration becomes very low and even when the pyruvate concentration increases. Thus, the pathway is designed to operate for a wide

Table 4.2 Free energy changes over reactions of the EMP pathway in the human erythrocyte

Reaction	ΔG^0 (kJ mol^{-1})	ΔG (expression)	ΔG (kJ mol^{-1})
Hexokinase	-16.74	$\Delta G^0 + RT \ln \frac{[G6P][ADP]}{[GLC][ATP]}$	-33.3
Glucose-6-P isomerase	1.67	$\Delta G^0 + RT \ln \frac{[F6P]}{[G6P]}$	-2.7
Phosphofructokinase	-14.22	$\Delta G^0 + RT \ln \frac{[FDP][ADP]}{[F6P][ATP]}$	-18.7
Aldolase	23.97	$\Delta G^0 + RT \ln \frac{[DHAP][GAP]}{[FDP]}$	0.7
Triose-P-isomerase	7.66	$\Delta G^0 + RT \ln \frac{[GAP]}{[DHAP]}$	2.7
Phosphoglycerate mutase	4.44	$\Delta G^0 + RT \ln \frac{[2PG]}{[3PG]}$	1.0
Enolase	1.84	$\Delta G^0 + RT \ln \frac{[PEP]}{[2PG]}$	1.2
Pyruvate kinase	-31.38	$\Delta G^0 + RT \ln \frac{[PYR][ATP]}{[PEP][ADP]}$	-23.0

range of concentrations of the substrate and the end product of the pathway, and this allows the pathway to function at most conditions the cells may experience in practice.

4.1.2 Combustion – A Change in Reference State

When an organic compound is burnt in oxygen the reaction products are always CO_2, H_2O and sometimes N_2 for nitrogen containing compounds. The thermodynamic data for combustion are found in the same way as data for free energy of formation, enthalpy of formation and entropy of formation, except that in chemical thermodynamics CO_2 and H_2O are defined to have zero free energy and enthalpy of combustion at standard conditions 25°C (or 20°C in some tables) and 1 atm, with liquid water and with CO_2 in gas form. The physical state of the compound to be combusted is taken as its natural state at the reference temperature unless otherwise specified in the large tables of heats of combustion that are available.

Thus for combustion of H_2 with ½O_2 the heat of combustion $-\triangle H_c$ is 68,280 cal mole^{-1} = 4.186 · 68.28 = 286 kJ mole^{-1} as derived from Example 4.1 with liquid water as the product. Similarly we have: $-\triangle G_c$ = 56661 · 4.186 = 237 kJ mole^{-1}. $(-\triangle H_c)$ is larger than $(-\triangle G_c)$ due to the decrease in entropy when the "ordered" H_2O molecule is formed from the "less ordered" O_2 and H_2. Normally the relative difference between $\triangle G_c$ and $\triangle H_c$ is much smaller than is the case for formation of H_2O, and the two quantities have approximately the same values. As will be seen in Example 4.6 there can, however, be very large differences between $\triangle G_c$ and $\triangle H_c$.

Example 4.4 Calculation of $\triangle G_c$ for ethanol combustion at 25°C, 1 atm.
Ethanol combustion to the final combustion products of liquid H_2O and gaseous CO_2 occurs by the stoichiometry (1)

$$-C_2H_5OH - 3O_2 + 2(CO_2)_g + 3(H_2O)_l = 0 \quad (1)$$

The following data are given for formation of CO_2 from $C_{graphite}$ and O_2 and formation of ethanol according to

$$-(2C + 3H_2 + \tfrac{1}{2}O_2) + C_2H_5OH = 0 \quad (2)$$

Compound	$\triangle H_f^o$ (kcal mole^{-1})	$\triangle G_f^o$ (kcal mole^{-1})	S^o (cal mole^{-1} K^{-1})
CO_2	-94.05	-94.26	51.06
C_2H_5OH	-66.36	-41.79	38.4

Thermodynamics of Bioreactions

Using $S^o = 1.36$ cal (mole K)$^{-1}$ for $C_{graphite}$ ΔG_f^o can be calculated from ΔH_f and S^o just as in Example 4.1

$$CO_2: \quad \Delta G_f^0 = -94.05 - 0.298 \cdot (51.06 - 1.36 - 49) = -94.26 \text{ kcal mole}^{-1} \quad (3)$$

$$C_2H_5OH: \quad \Delta G_f^0 = -66.36 - 0.298 \cdot (38.4 - 2 \cdot 1.36 - 3 \cdot 31.21 - 0.5 \cdot 49) = -41.79 \text{ kcal mole}^{-1} \quad (4)$$

The heat of combustion of ethanol by (1) is now calculated as the sum of the heat of formation of the $(CO_2)_g$ and the $(H_2O)_l$ that make up the ethanol molecule less the heat of formation of ethanol.

$$Q_c = -\Delta H_c = -66.36 + 2 \cdot 94.05 + 3 \cdot 68.28 = 326.6 \text{ kcal mole}^{-1} \quad (5)$$

or 1367 kJ (mole ethanol)$^{-1}$

$$\Delta G_c = -327 - 0.298 \cdot (2 \cdot 51.06 + 3 \cdot 16.72 - 38.4 - 3 \cdot 49.0) = 317 \text{ kcal mole}^{-1} \quad (6)$$

or 1325 kJ (mole ethanol)$^{-1}$

4.2 Heat of Reaction

For design of heat exchangers it is important to calculate the heat of reaction, not so much of the individual biochemical reactions occurring within the cell but rather for the overall conversion of substrates to biomass and products. Table 4.3 collects heat of combustion data for many of the compounds involved in bioprocesses. The reference state is changed slightly compared to that used in the previous examples: CO_2 is not in the gas phase, but dissolved in water at pH = 7, and for CO_2 as well as other compounds which may take part in acid-base reactions the sum of the ions which are formed at pH = 7 is used as the reference concentration unit. The difference between the "biological" thermodynamic data such as $-\Delta H_c$ in Table 4.3 and those used in chemical thermodynamics is small. Since the values of Table 4.3 are going to be used to find heats of reaction when e.g. glucose is converted to ethanol and CO_2 the difference in reference state means nothing, or hardly anything. The calculations of Examples 4.2 and 4.4 illustrate how the values in the table may be obtained.

Also included in the table is a column showing the degree of reduction κ^* of the compounds. κ of Chapter 3 is identical to κ^* for all compounds which do not contain N, but in relation to combustion calculations it seems more natural to use N_2 rather than NH_3 as the "redox neutral" N containing compound. If the elements S and P appear the degree of reduction of S and P is set to 6 and 5 respectively just as in κ. Note in particular that κ^* for "standard" biomass $CH_{1.8}O_{0.5}N_{0.2}$ is 4.80 whereas $\kappa = 4.20$.

Table 4.3 Heat of combustion for various compounds at standard conditions[a] and pH 7.

Compound	Formula	Degree of reduction (κ^*) per carbon	$-\Delta H_c^0$ (kJ/mole)	$-\Delta H_c^0$ (kJ/C-mole)	$-\Delta H_c^0/\kappa^*$ (kJ/C-mole)
Formic acid	CH_2O_2	2	255	255	127.5
Acetic acid	$C_2H_4O_2$	4	875	437	109.5
Propionic acid	$C_3H_6O_2$	4.67	1527	509	109.0
Butyric acid	$C_4H_8O_2$	5.00	2184	546	109.2
Valeric acid	$C_5H_{10}O_2$	5.20	2841	568	109.2
Palmitic acid	$C_{16}H_{32}O_2$	5.75	9978[b]	624	108.5
Lactic acid	$C_3H_6O_3$	4	1367	456	114.0
Gluconic acid	$C_6H_{12}O_7$	3.67			
Pyruvic acid	$C_3H_4O_3$	3.33			
Oxalic acid	$C_2H_2O_4$	1	246	123	123.0
Succinic acid	$C_4H_6O_4$	3.50	1491	373	106.6
Fumaric acid	$C_4H_4O_4$	3	1335	334	111.3
Malic acid	$C_4H_6O_5$	3	1328	332	110.7
Citric acid	$C_6H_8O_7$	3	1961	327	109.0
Glucose	$C_6H_{12}O_6$	4	2803	467	116.8
Fructose	$C_6H_{12}O_6$	4	2813	469	117.2
Galactose	$C_6H_{12}O_6$	4	2805	468	117.0
Sucrose	$C_{12}H_{22}O_{11}$	4	5644	470	117.5
Lactose	$C_{12}H_{22}O_{11}$	4	5651	471	117.8
Hydrogen	H_2	2	286		
Methane	CH_4	8	890[c]	890	111.3
Ethane	C_2H_6	7	1560[c]	780	111.4
Propane	C_3H_8	6.67	2220[c]	740	110.9
Methanol	CH_4O	6	727	727	121.2
Ethanol	C_2H_6O	6	1367	683	113.8
iso-Propanol	C_3H_8O	6	2020	673	112.2
n-Butanol	$C_4H_{10}O$	6	2676	669	111.5
Ethylene glycol	$C_2H_6O_2$	5	1179	590	118.0
Glycerol	$C_3H_8O_3$	4.67	1661	554	118.6
Acetone	C_3H_6O	5.33	1790	597	112.0
Formaldehyde	CH_2O	4	571[c]	571	142.8
Acetaldehyde	C_2H_4O	5	1166	583	116.6
Urea	CH_4ON_2	6	632	632	105.3
Ammonia	NH_3	3	383[c]		(127.7)
Biomass	$CH_{1.8}O_{0.5}N_{0.2}$	4.80	560	560	116.7

[a] 298 K and 1 atm, liquid water, CO_2 dissolved and dissociated in H_2O.
[b] Solid form.
[c] Gaseous form.

The average value of $(-\Delta H_c)/\kappa^*$ for the compounds in Table 4.3 is 115 kJ mole^{-1} with a standard deviation of 6-7 kJ mole^{-1}.

From the additive property of ΔH (and ΔG) one obtains equation (4.10) for the enthalpy change in

Thermodynamics of Bioreactions

the system due to the bioreaction

$$\Delta H = \sum_i Y_{p_i}(-\Delta H_{ci}) - \sum_j Y_{s_j}(-\Delta H_{cj}) \quad (4.10)$$

The heat delivered to the surroundings, the heat of reaction Q, is given by:

$$Q = \sum_j Y_{s_j}(-\Delta H_{cj}) - \sum_i Y_{p_i}(-\Delta H_{ci}) = \sum_j Y_{s_j} Q_{cj} - \sum_i Y_{p_i} Q_{ci} \quad (4.11)$$

where Q_{ci} is the heat of combustion of reactant i. When the approximation $-\Delta H_{ci} = Q_{ci} = 115\, \kappa_i^*$ is introduced we find:

$$Q = \left[\sum_j Y_{s_j} \kappa_j^* - \sum_i Y_{p_i} \kappa_i^* \right] \cdot 115 \text{ kJ (C-mole)}^{-1} \quad (4.12)$$

If the relation between Q_{ci} and κ^* had been exactly true (4.12) would be identical to (4.11). The small errors introduced with (4.12) are of no importance in engineering calculations leading for example to the design of a heat exchanger in a bioreactor – and (4.11) is not either exact: active biomass may have a degree of reduction κ from just over 4 to perhaps 4.40 and still the heat of combustion of all biomass is set to 560 kJ (C-mole)$^{-1}$ in Table 4.3.

Since the reference level for Q_c as well as for the degree of reduction of an element is arbitrary one may just as well apply κ_i instead of κ_i^* in (4.12). Thus (4.12) and (4.13) gives identical values for Q

$$Q = \left[\sum_j Y_{s_j} \kappa_j - \sum_i Y_{p_i} \kappa_i \right] \cdot 115 \text{ kJ (C-mole)}^{-1} \quad (4.13)$$

It is perhaps convenient to work with only one set of degrees of reduction and (4.13) may be mnemotechnically easier to apply than (4.12).

If Y_{so} is determined by a degree of reduction balance

$$(-4)Y_{so} = \sum_i Y_{p_i} \kappa_i - \sum_j Y_{s_j} \kappa_j \quad (4.14)$$

then (4.13) is identical to (4.15), which is very easy to use for aerobic processes when Y_{so} is a calculated quantity:

$$Q = 4 \cdot 115 Y_{so} = 460 Y_{so} \text{ kJ (C-mole)}^{-1} \quad (4.15)$$

It should finally be remarked that similar to the relation $Q_{ci} \cong 115 \kappa_i^*$ there exists an approximate relation between $-\Delta G_{ci}$ and κ_i^*, that holds for any compound with a positive κ_i^* (i.e. not CO_2 and H_2O).

$$-\Delta G_{ci} = 94.4 \kappa_i^* + 86.6 \quad \text{kJ (C-mole)}^{-1} \tag{4.16}$$

This relation may be used in approximate calculations of the free energy change in a given (bio-) chemical reaction. For the compounds listed in Table 4.3 the standard deviation of the average (4.16) is also about 6 kJ (C-mole)$^{-1}$.

Example 4.5 Heat of reaction for aerobic growth of yeast
Consider the stoichiometry for aerobic yeast fermentation at $D = 0.3$ h^{-1} derived in Example 3.5

$$\begin{aligned}&-(CH_2O + 0.279 \cdot 0.17 NH_3 + 0.167 O_2) + \\ &(0.279 CH_{1.83}O_{0.50}N_{0.17} + 0.366 CO_2 + 0.355 CH_3O_{1/2} + 0.275 H_2O) = 0\end{aligned} \tag{1}$$

We now calculate the heat of reaction using eqs. (4.11), (4.12), (4.13) and (4.15) respectively:

Eq. (4.11): $\quad Q = 467 + 0.0474 \cdot 383 - 0.279 \cdot 560 - 0.355 \cdot 683 = 86.4$ kJ (C-mole glucose)$^{-1}$

Eq. (4.12): $\quad Q = (4 + 3 \cdot 0.0474 - 4.71 \cdot 0.279 - 6 \cdot 0.355) \cdot 115 = 80.3$ kJ (C-mole glucose)$^{-1}$

Eq. (4.13): $\quad Q = (4 - 4.20 \cdot 0.279 - 6 \cdot 0.355) \cdot 115 = 80.3$ kJ (C-mole glucose)$^{-1}$

Eq. (4.15): $\quad Q = 460 \cdot 0.167 = 76.8$ kJ (C-mole glucose)$^{-1}$

The last result is not equal to the two previous results. The reason is that Y_{so} was determined by an independent measurement of the rate of oxygen uptake, and not from a degree of reduction balance. As shown in Example 3.5 the redox balance based on the measured yield coefficients closes only to within 98%. More than this cannot be expected except in very high quality experiments, and the difference between the calculated Q values reflects the uncertainty which must be expected when κ_i are used rather than the tabulated values of Q_{ci}.

Using eq. (4.16) the free energy change accompanying reaction (1) can also be calculated, at least approximately:

$$-\Delta G = 94.4 \, (4 \cdot 1 + 3 \cdot 0.0474 - 4.71 \cdot 0.279 - 6 \cdot 0.355)$$
$$+ 86.6 \, (1 + 0.0474 - 0.279 - 0.355) = 101.5 \text{ kJ (C-mole)}^{-1}$$

Using thermodynamic tables one obtains the result $-\Delta G = 107.8$ kJ (C-mole)$^{-1}$. The approximate result is certainly good enough for engineering calculations – first of all it shows that the reaction is strongly thermodynamically favored.

Thermodynamics of Bioreactions

When (4.11) and (4.16) are used to calculate Q and $\triangle G$ for all three D-values of example 3.5 the following results are obtained

D	0.15 h^{-1}	0.30 h^{-1}	0.40 h^{-1}
Q	195.8	86.4	30.0
	357	310	172
-$\triangle$G	208	102	47
	380	366	269

the upper numbers are in kJ (C-mole glucose)$^{-1}$, the lower numbers in kJ (C-mole biomass)$^{-1}$. Both Q and -$\triangle G$ decrease drastically when calculated as kJ (C-mole glucose)$^{-1}$, but especially -$\triangle G$ is much less dependent on D when calculated on a C-mole biomass basis.

It is seen in Example 4.5 that the heat of reaction drastically declines when the process becomes less aerobic (Y_{so} decreases). According to (4.12), (4.13) or (4.15) the heat of reaction should decrease to zero in a fully anaerobic process. This is of course not quite true even for an anaerobic process using a sugar as substrate, and the result is totally wrong in other systems e.g. when H_2 is used as energy source and CO_2 as carbon source - see Example 4.6.

Using a microcalorimeter heat production rates $q_Q = dQ/dt$ of less than 1W per L medium can be measured quite accurately, and q_Q can be included in the database as an independent rate measurement. q_Q is, however, strongly correlated to q_{O_2}, and the two measurements should not be included together in the set of independent rate measurements. Several research groups have been using microcalorimetry to measure the heat production rate of fermentations, e.g. at Gothenburg University (Sweden) and at EFPL in Lausanne (Switzerland). A joint publication by the two groups (von Stockar et al., 1993) gives several examples of physiological phenomena studied by means of microcalorimetry.

Example 4.6 Anaerobic growth on H_2 and CO_2 to produce CH_4
The archae *Methanobacterium thermoautotrophicum* uses methanogenesis as its catabolic pathway, i.e. it converts H_2 and CO_2 to CH_4, creating energy in the process. It has an extremely low biomass yield coefficient of about 0.02 C-mol (mole H_2)$^{-1}$, but a large rate of heat production r_Q, up to 13 W (g biomass)$^{-1}$. The biomass composition of the organism has been experimentally determined to:

$$X = CH_{1.68}O_{0.39}N_{0.24} \ (\kappa = 4.18),$$

and with $Y_{H_2 X} = 0.02$ C-mole (mole H_2)$^{-1}$ the stoichiometry for a black box model is determined from a carbon and a redox balance:

$$-H_2 - 0.260\,CO_2 - 0.0048\,NH_3 + 0.02\,X + 0.240\,CH_4 + 0.511\,H_2O = 0 \qquad (1)$$

Schill et al. (1999) made an experimental and theoretical study of this rather strange microorganism for which the growth associated heat production is inordinately high – as implied by the name of the organism. The reaction temperature was 60°C and to calculate the heat of reaction and the free energy change associated with eq. (1) the thermodynamic data of Table 4.3 must be corrected.

The change in heats of combustion with modest changes of temperature is negligible – for example for the heat of combustion of H_2 to form liquid water:

$$(-\Delta H_c)_{333K} - (-\Delta H_c)_{298K} = (7 + \tfrac{1}{2} \cdot 7 - 9)(333 - 298) + (563 - 582) \cdot 18 = -289 \text{ cal mole}^{-1} \quad (2)$$

The first term in (2) is $\Delta\alpha\,\Delta T$, where $\Delta\alpha$ is the change in molar specific heat by the reaction. The second term derives from the change in heat of condensation of H_2O between 298 and 333 K

Using the result of Example 4.1

$$(-\Delta H_c)_{333K} = (68280 - 289) \text{ cal (mole } H_2)^{-1} = 285 \text{ kJ (mole } H_2)^{-1}$$

The change in free energy is much larger. Thus for the combustion of H_2 which at the reaction temperature 333K is present in a measured concentration $c_{H_2} = 8 \cdot 10^{-6}\,M$ in the cytosol of the organism

$$(-\Delta G_c)_{333K} - (-\Delta G_c)_{298K} = -\Delta T \Delta S + RT \ln(H \cdot c_{H_2}) \quad (3)$$

H is the Henry's law coefficient 1250 atm L mole^{-1} for solution of H_2 in water at 60°C, $\Delta T = 35K$, and ΔS is 39 cal (mol K)$^{-1}$ from Example 4.1 (liquid water).

$$(-\Delta G_c)_{333} = 56661 - 35 \cdot 39 - 1.987 \cdot 333 \ln 100 = 52.2 \text{ kcal mole}^{-1} = 219 \text{ kJ mole}^{-1} \quad (4)$$

The table below summarizes the thermodynamic data (in kJ (C-mole)$^{-1}$) used by Schill et al. (1999) to calculate the heat of reaction and free energy change associated with (1). The measured concentration of the dissolved gases at 333K is also shown.

Compound	$-\Delta H_c$	$-\Delta G_c$
H_2 (8·10^{-6}M)	285	219
CH_4 (5.7·10^{-4}M)	892	808
CO_2 (1.6·10^{-4}M)	0	-12.8
NH_3 (50 mM)	383	401
Biomass	458	541

Except for the rather low value of ($-\Delta H_c$) for biomass – the value is calculated from a calorimetric experiment by the authors – the heats of combustion are practically the same as those shown in Table 4.3. We can now calculate ΔH and ΔG for the stoichiometry (1)

$$Q = -\Delta H = 285 + 0.0048 \cdot 383 - 0.02 \cdot 458 - 0.240 \cdot 892 = 63.6 \text{ kJ (mole } H_2)^{-1}$$

$$-\Delta G = 219 - 12.8 \cdot 0.260 + 0.0048 \cdot 401 - 0.02 \cdot 541 - 0.240 \cdot 808 = 12.9 \quad \text{kJ (mole H}_2)^{-1}$$

It is seen that the heat of reaction is 5 times higher than the free energy change, and this is quite unusual. The reason for the small ΔG is that much of the free energy contained in the substrate (H_2) is recovered in the product (CH_4). The entropy change by (4.1) is consequently large and negative.

When ΔH and ΔG are calculated on the basis of biomass formed

$$-\Delta H = 3160 \text{ kJ (C-mole biomass)}^{-1}, \qquad (-\Delta G) = 645 \text{ kJ (C-mole biomass)}^{-1} \qquad (4)$$

for a yield coefficient of $Y_{sx} = 0.02$ the heat of reaction is about 9 times higher than for respiratory growth of yeast on glucose when compared on a C-mole biomass basis (see Example 4.5) while $-\Delta G$ is only 70% higher than for respiratory growth of yeast on glucose. One must conclude that *Methanobacterium thermoautotrophicum* grows on CO_2 and H_2 with a very poor thermodynamic efficiency – and also that anaerobic processes are certainly not always associated with a small heat of reaction.

4.3 Non-equilibrium Thermodynamics

As mentioned earlier cellular systems are open systems where many processes operate far from equilibrium. If they were at equilibrium there would be no flow through the many different cellular pathways and the cells would stop functioning. For many cellular processes thermodynamic driving forces are used directly, e.g. in passive diffusion of substrates across the cytoplasmic membrane, and even if the transport is mediated by a carrier as in facilitated diffusion the driving force is still based on thermodynamics. When the flow is in the direction of the thermodynamic force gradient it is referred to as *conjugate flow*. There are, however, also some important cellular processes where the flow is *non-conjugate*, i.e. the flow is against a thermodynamic driving force. Clearly non-conjugate flow does not occur on its own, but through tight coupling with a conjugate flow it is possible to drive processes against a thermodynamic driving force. Non-equilibrium thermodynamics is an extension of classical thermodynamics to non-equilibrium states. It supplies relationships between flows and thermodynamic driving forces, for both conjugate and non-conjugate flows. Additionally, non-equilibrium thermodynamics allows for a description of processes where conjugate flows are tightly coupled to non-conjugate flows.

An example of a cellular process where conjugate and non-conjugate flows are coupled is the oxidative phosphorylation (see Fig. 4.1), which is an essential life process for all animals and an option for many microorganisms. In this process protons are transferred across the mitochondrial membrane (the cytosolic membrane in bacteria) against a proton gradient. This transfer of protons is driven by the oxidation of NADH by oxygen, and there is consequently a tight coupling of the non-conjugate flow of protons across the membrane and the oxidation of the co-factor NADH. Furthermore, the process is closely coupled to the phosphorylation of ATP, which takes place by an ATPase that converts ADP and free phosphate into ATP. The process of ATP generation is thermodynamically driven by the translocation of protons down the concentration gradient.

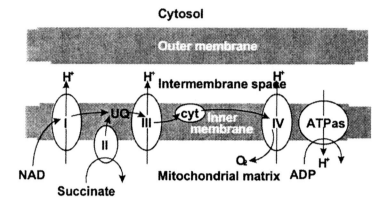

Figure 4.1 The electron transport chain and oxidative phosphorylation in eukaryotes. Electrons are transported from NADH or succinate, through the electron transport chain, to oxygen. The elements of the electron transport chain are organized in large complexes located in the inner mitochondrial membrane. Electrons from NADH are first donated to complex I, the NADH dehydrogenase. Electrons from succinate are first donated to FAD, which is integrated in complex II, the succinate dehydrogenase (one of the TCA cycle enzymes). Electrons from complexes I and II (or other flavoproteins located in the inner mitochondrial membrane) are transferred to ubiquinone (UQ), which diffuses freely in the lipid membrane. From UQ, electrons are passed on to the cytochrome system. First, the electrons are transferred to complex III, which consists of two b-type cytochromes *(b566* and *b562)* and cytochrome *c1*. Electrons are then transferred to complex IV via cytochrome *c*, which is only loosely attached to the outside face of the membrane. Complex IV (or cytochrome oxidase), which contains cytochromes *a* and *a3*, finally delivers the electrons to oxygen. Complexes I, III, and IV span the inner mitochondrial membrane, and when two electrons are transported through these complexes, protons (four at each complex) are released into the intermembrane space. These electrons may be transported back into the mitochondrial matrix by a proton-conducting ATP-synthase (or the F_1F_0-ATPase complex). In this complex, one ATP is generated when three protons pass through the ATPase. One additional proton is transported into the mitochondrial matrix in connection with the uptake of ADP and ~P and the export of ATP, so that a total of four protons are required per ATP generated by this mechanism. Thus, for each electron pair transferred from complex I all the way to complex IV, 12 protons (four protons for each of the three complexes) are pumped from the mitochondrial matrix into the intermembrane space. Upon re-entry of the protons into the mitochondrial matrix through the ATP synthase, 3 moles of ATP are generated (12 protons/4 protons per ATP). The theoretical stoichiometry of the oxidative phosphorylation is, therefore, 3 mol of ATP synthesized per mole NADH oxidized and 2 mol of ATP synthesized per mole of succinate oxidized.

Some organisms have a more or less defective respiration system. Lactic bacteria lack the transfer agent cytochrome c between complexes III and IV, and cannot gain any energy by respiration, although other advantages, especially the possibility to oxidize NADH by means of NADH oxidases, are obtained in aerobic lactic fermentations. *Saccharomyces cerevisiae* has a defective stage 1 of the chain, which means that NADH and $FADH_2$ are energetically equivalent in this organism. The theoretical P/O ratio for this yeast is therefore 2. Another yeast *Candida utilis*, can use all three stages to produce ATP from oxidation of NADH, and the theoretical P/O ratio for this yeast is therefore 3.

Fermentation physiologists have studied – as ardently as the theoretical bioscientists – the ATP yield on redox equivalents oxidized in respiration. Their goal has been to get a true (rather than a theoretical) value of ATP gained in catabolism in order to find biomass yield on different substrates, i.e. to find the ATP required for balanced growth, typically in steady state continuous cultures. Using aerobic cultivations of the yeast *S. cerevisiae* growing on different substrates (acetate, lactate, ethanol, glucose etc.) an average value of 1.2-1.3 for the effective P/O ratio was derived by van Gulik and Heijnen (1995). In their study they used a rather complete metabolic network model for *S. cerevisiae* to obtain the distribution of metabolites (and hence the ATP gained in catabolism). Vanrolleghem and Heijnen (1998) used mixtures of glucose and ethanol and measured the variation in Y_{sx} and Y_{ox} with changing ratio of ethanol and glucose in the feed and compared this with simulations, again based on a large metabolic network. A rather low value for P/O of 1.05-1.15 was obtained by least squares fitting of the model parameters to the data. In Section 5.2.3 we will look further into the energetics of aerobic processes and illustrate how the P/O ratio can be derived from fermentation data.

One may ask why the experimentally obtained P/O ratios are so much smaller than the theoretical values. A sound theoretical explanation of the ATP production process coupled with oxidation of redox equivalents might presumably give some hints for construction of strains which produce ATP more effectively than those used today (whether this is beneficial for the organism or for the user of the organism is of course an open question). The chemoosmotic hypothesis is one of at least two competing explanations of oxidative phosphorylation. In Note 4.1 we review some of the main steps in the mechanism, which in a remarkably efficient way is able to generate free energy rather than heat out of the oxidation of H_2 – the real chemical process that occurs when the redox cofactors are converted to their oxidized form.

Note 4.1 What is the operational P/O ratio?
The mechanism of the oxidative phosphorylation is outlined in Fig. 4.1. It has been studied by some of the most innovative physiologists over the last 40-50 years and is reviewed by Senior (1988). This reference and semiquantitative treatments by Rottenberg (1979) and by Stucki (1980) – both referred to by Roels (1983) – of the thermodynamic efficiency of the overall process were used in the preparation of the present note. Basically, the huge amount of free energy made available by the redox process Eq. (1) is used to drive the phosphorylation process Eq. (2).

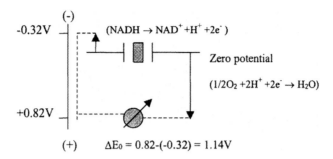

Figure 4.2 Galvanic element with the two single-electrode processes involved in the oxidation of NADH.

The synthesized ATP is used almost immediately by free energy-requiring cell processes (e.g., polymerization processes), and the pool of ATP is in pseudo steady state on the time scale of the growth process

$$NAD^+ + H_2O - NADH - \tfrac{1}{2}O_2 = 0 \quad ; \quad \Delta G_o^0 = -220 \text{ kJ mole}^{-1} \quad (1)$$

$$ATP + H_2O - ADP - \sim P = 0 \quad ; \quad \Delta G_p^0 = 30.5 \text{ kJ mole}^{-1} \quad (2)$$

The standard free energy associated with the oxidation of 1 mole of NADH is derived by consideration of the two single-electrode processes that are combined to a galvanic element in Fig.4.2. The standard free energy released by reduction of ½O$_2$ by NADH is calculated from the electromotoric force ΔE_o of the galvanic element:

$$-\Delta G_o^0 = nF\Delta E_o = 2 \cdot 96.5 \cdot 1.14 = 220 \text{ kJ mole}^{-1} \quad (3)$$

Synthesis of ATP is mediated by a membrane-bound ATPase (see Fig. 4.1). Its correct name is "F$_0$F$_1$-type proton ATPase" or "ATP synthase". Reaction between ADP and ~P on the knob-like F$_1$ part of the protein, which sticks into the cytoplasm (for procaryotes) or into the mitochondria (for eucaryotes), proceeds readily enough, but unless a steady stream of protons flows through the stem-like F$_0$ part of the protein, which transcends the membrane (cytoplasmic or inner mitochondrial), the synthesized ATP does not detach from F$_1$ and ATP production stops. The flow of protons to F$_1$ is necessary to accomplish the release of ATP from F$_1$, and the overall effect of the proton flow is a considerable release of free energy, enough to drive the chemical reaction in Eq. (2) against a negative affinity A_p (which is equal to $-\Delta G_p^0$). The driving force - also called the *protonmotive force* Δp by Peter Mitchell (1961), who introduced the so-called chemoosmotic hypothesis to explain the coupling between Eqs. (1) and (2) – is given by

$$\frac{\Delta \mu_{H^+}}{F} = \Delta p = \Delta \Psi - 2.303 \frac{RT}{n} \Delta pH = \Delta \Psi - 0.060 \Delta pH \tag{4}$$

Δp has the unit volts per mole. $\Delta \psi$ is the membrane potential (more negative on the inside, hence $\Delta \psi$ is positive and has been found to be approximately 0.14 V). ΔpH is the difference in pH between the outside and the inside of the membrane (about 0.05 pH units and negative). The total value of Δp is about 0.143 V per mole of H^+, equivalent to a release of free energy of 96.5 · 0.143 = 13.8 kJ per mole of protons transported. The required inward flow of protons is provided for by the redox reaction in Eq (1), which expends some of its free energy in pumping protons in the opposite direction, against the steep electrochemical gradient. The two flows are vectorially arranged, and they are mediated by different and mutually isolated transmembrane proteins. Thus the total oxidative phosphorylation process is conceived as a cyclic process with four steps, two scalar chemical reactions and two vectorial transport processes. The affinity of one transport process and of one chemical process is negative, whereas the other two processes are energetically coupled by the requirement that the total rate of free energy dissipation D must be non-negative. D is given as the sum of four terms $v_i A_i$, where v_i is the "flow" (moles per second converted or transported) and A_i is the corresponding drive force, i.e. the affinity of the ith process.

$$D = \sum_i^4 v_i A_i \geq 0 \tag{5}$$

It is convenient to combine the four reactions into two pairs. One is the redox process and the associated outward proton transport; the other is the phosphorylation and its associated downhill flow of protons. Let A_o and A_p be the total affinity of each of these two process pairs:

$$A_o = A_o^0 - n_o \Delta_{H^+} \tag{6}$$

$$A_p = A_p^0 + n_p \Delta_{H^+} \tag{7}$$

A_o^0 is large and positive, and A_p^0 is negative. n_o and n_p denote the moles of protons excluded or taken up by the cell (or mitochondria). From Eq. (5) it follows that $A_o v_o + A_p v_p$ is nonnegative, v_o and v_p being the overall rates of the oxidation and phosphorylation reactions, respectively.

Now, by the theory of non-equilibrium thermodynamics (see *e.g.* Katchalsky and Curran (1965)) – the driving forces and the flows of two energetically coupled processes are related by

$$v_o = c_{oo} A_o + c_{op} A_p \tag{8}$$

$$v_p = c_{po} A_o + c_{pp} A_p \tag{9}$$

The linear equations hold very close to equilibrium where microscopic reversibility leads to Onsager's reciprocal relations, which state that $c_{op} = c_{po}$ (or more generally that the c_{ij} matrix is symmetric). Whether enzymatic processes that may work far from equilibrium really conform to the assumptions that permit linearization of much more complicated relations between v_i and A_j is a moot question, which Rottenberg (1979) purports to answer by reference to the near constancy of overall enzymatic conversion rates when the substrate level is much higher than the K_m of the enzyme.

If $c_{op}(=c_{po}) = 0$, the processes are energetically uncoupled and oxidation will not drive phosphorylation.

The ratio between v_p and v_o is the operational P/O ratio:

$$\frac{v_p}{v_o} = \frac{c_{po}A_o + c_{pp}A_p}{c_{oo}A_o + c_{op}A_p} = \left(\frac{c_{pp}}{c_{oo}}\right)^{1/2}\left(\frac{c_{pp}}{c_{oo}}\right)^{1/2}\frac{c_{po}/c_{pp}A_o + A_p}{A_o + c_{op}/c_{oo}A_p}$$

$$= \left(\frac{c_{pp}}{c_{oo}}\right)^{1/2}\frac{(c_{pp}/c_{oo})^{1/2}(A_p/A_o) + (c_{op}^2/c_{oo}c_{pp})^{1/2}}{(c_{op}/c_{oo})(A_p/A_o) + 1} = z\frac{z(A_p/A_o) + q}{qz(A_p/A_o) + 1} \quad (10)$$

where

$$z = \left(\frac{c_{pp}}{c_{oo}}\right)^{1/2} \quad \text{and} \quad q = \frac{c_{op}}{(c_{oo}c_{pp})^{1/2}} \quad (11)$$

q describes the degree of coupling between oxidation and phosphorylation. Since it can be proved that the energy dissipation D is nonnegative if and only if $c_{oo}c_{pp} - c_{op}^2 > 0$, it follows that $0 < q < 1$ when $c_{op} < 0$, i.e. when oxidation drives phosphorylation. The case $q \to 1$ is interesting. Here $v_p/v_o \to z$. If no other processes scavenge the proton gradient set up by the oxidation process, the ratio between the overall rates can be shown to be $v_p/v_o = z = n_o/n_p$. The numbers n_o and n_p are not theoretically given, but most researchers agree that the ratio P/O is between 2 and 3 unless, e.g. the inward flow of protons is used to support transport of substrates or other cations by active transport processes.

The ratio between output and input energy is the thermodynamic efficiency η of the energy coupling system

$$\eta = \left(\frac{r_p}{r_o}\right)\left(\frac{-A_p}{A_o}\right) = -z\left(\frac{A_p}{A_o}\right)\frac{z(A_p/A_o) + q}{qz(A_p/A_o) + 1} = -f\frac{f + q}{qf + 1} \quad (12)$$

where $f = z(A_p/A_o)(<0$, since A_p is negative). The maximum efficiency is obtained for $qf^2 + 2f + q = 0$, or

$$f = \left(-\frac{1}{q}\right)(1 - \sqrt{1 - q^2}) \quad (13)$$

$$\eta_{max} = \frac{q^2}{(1 + \sqrt{1 - q^2})^2} \quad (14)$$

Experimental values for the force ratio f obtained for living microorganisms are fairly constant at -0.7 to -0.8 (Roels, 1983). Thus with a coupling coefficient of $q = 0.95$ it follows from Eq. (10) that

Thermodynamics of Bioreactions

$$\frac{v_p}{v_o} \approx z \frac{-0.75 + 0.95}{0.95(-0.75) + 1} = 0.696z \tag{15}$$

The maximum value for z is equal to the theoretical P/O ratio, i.e. 3 for oxidation of NADH if all constituents of the respiratory system are present, and it is 2 for oxidation of some $FADH_2$. When $q<1$, z is a complicated function of the rates of the proton flows that exist in parallel with the main proton flow through the protein F_0. Hence the experimentally derived effective P/O ratios in the range 1.2-1.3 are not at all improbable.

The thermodynamic efficiency of the process is (-0.696) (-0.75), or only slightly above 50%. It is close to the optimal value (f = -0.72 and η_{max} = 0.529) obtained by Eq. (13). The decrease of thermodynamic efficiency by a factor 2 when q decreases from 1 to 0.95 is quite dramatic. It may be interesting to speculate why q is less than 1. Roels (1983) proposed an answer: The less-than-perfect coupling between oxidation and phosphorylation is due to the extraneous processes driven by the proton gradient across the membrane. Looking at the total process of cell growth, some of the thermodynamic efficiency lost in the oxidative phosphorylation is regained since, e.g., membrane energetization is maintained (no ATP is needed for this), as is sugar transport by secondary active transport. This yields an intracellular phosphorylated sugar which unloads its internal free energy in further cell processes.

PROBLEMS

Problem 4.1 Thermodynamic properties of NH_3 and of NH_4OH.

From a table of thermodynamic data one obtains
 S^0 = 45.77 cal mol^{-1} K^{-1} for N_2
 S^0 = 46.01 for NH_3
 ΔH_f^0 = - 11.04 kcal mol^{-1} for NH_3
all at standard conditions (298 K) and gaseous NH_3. Other data are to be taken from the examples in this chapter.

a. Calculate (ΔG_f^0) for $(NH_3)_g$ at standard conditions.

b. Calculate the heat of combustion and the free energy of combustion for $(NH_3)_g$ at the standard state. Compare with table 4.3 for ΔH_c^0.

  The following relation holds for $x \leq 4$ wt % aqueous solutions of NH_3:

$$\pi_{NH3} \text{ (mm Hg)} = 8.24 \, x \tag{1}$$

  The heat of absorption of NH_3 in water is 446 cal g^{-1}.

c. Calculate ΔH_f, ΔH_c, ΔG_f and ΔG_c for a 1 M solution of NH_3 in water.

d. Assume that NH_4OH is fully dissociated to NH_4^+ and OH^-.

(ΔH_f, ΔG_f) for OH^- is -54.97 and -39.59 kcal mol^{-1} respectively.

Calculate ΔH_c and ΔG_c for NH_4^+

Problem 4.2 Is a lysine yield of 6/7 on glucose possible from a thermodynamic point of view?

In Example 3.9 a maximum yield of 6/7 C-mole lysine (C-mole glucose)$^{-1}$ was predicted by imposing the constraint that all yield coefficients in Eq (3.9.5) should be non-negative. Stephanopolous and Vallino (1991) and (1993), and Marx *et al.* (1996) report that the maximum theoretical yield is only 3/4. The reason could be that ΔG is positive for the reaction:

$$- CH_2O - (2/7) NH_3 + (6/7) CH_{7/3}O_{1/3}N_{1/3} + (1/7) CO_2 + (3/7) H_2O = 0 \qquad (1)$$

a. Use eq (4.16) to calculate an approximate value for $(-\Delta G_c^0)_{lysine}$.

b. Use this value to determine ΔG^0 for reaction (1).

c. Experimental values for ΔG_c and for ΔH_c are difficult to find for lysine, but Morrero and Gani (2001) have calculated ΔG_f^0 and ΔH_f^0 for L-lysine by a group contribution method. Their calculated values are :

$$-\Delta G_f^0 = 214 \text{ kJ mole}^{-1} \text{ and } -\Delta H_f^0 = 450 \text{ kJ mole}^{-1}. \qquad (2)$$

Use these data to calculate ΔG_c^0 and ΔH_c^0 for L-lysine. Data for CO_2 and H_2O are taken from the examples of chapter 4. Note that the difference between ΔG_c calculated in question 1 and question 3 is quite small.

d. Will reaction (1) proceed voluntarily in the direction indicated in (1) ?

Problem 4.3 The effect of a transhydrogenase reaction in *Saccharomyces cerevisiae*.

In Problem 3.6 three different strains of *S. cerevisiae* were compared. The purpose of constructing strain TN21 was to provide the anaerobically growing yeast culture with a sink for NADH. If NADH was converted to NADPH it was thought that less glycerol would be formed – in TN1 and TN26 glycerol production is the only means of balancing the excess NADH formed in connection with the formation of precursor metabolites like pyruvate and acetyl-CoA. Furthermore the produced NADPH might substitute NADPH produced in the PP pathway and consequently diminish the glucose loss in this pathway.

a. Consider the reaction

$$- NADH - NADP^+ + NADPH + NAD^+ = 0 \qquad (1)$$

which is catalysed by the transhydrogenase in TN21. Is it reasonable to assume that ΔG^0 is almost zero for reaction (1)?

b. Nissen et al. (2001) measured the intracellular concentration of each of the four reactants in (1). Actually only the cytosolic concentrations should have been measured since the inserted transhydrogenase operates only in the cytosol, but the measurements in the table below are thought to give an adequate representation of the relative values of the four nucleotides in the cytosol.

Intracellular concentrations of NAD(H) and NADP(H) in μ mol (g DW)$^{-1}$ for the two control strains TN1 and TN26, and for the transhydrogenase-containing strain TN21, sampled during exponential growth in anaerobic batch cultivations. Each measurement was carried out twice.

Strain	NAD^+	$NADP^+$	NADH	NADPH	$NADH/NAD^+$	$NADPH/NADP^+$
TN1	2.87 ± 0.09	0.23 ± 0.01	0.44 ± 0.01	1.21 ± 0.07	0.15 ± 0.01	5.26 ± 0.55
TN26	2.85 ± 0.11	0.24 ± 0.01	0.43 ± 0.01	1.19 ± 0.07	0.15 ± 0.01	4.96 ± 0.52
TN21	2.17 ± 0.07	0.27 ± 0.02	0.54 ± 0.02	0.80 ± 0.10	0.17 ± 0.01	2.96 ± 0.60

Calculate ΔG for reaction (1) using the very consistent data of the table for TN 1 and TN 26 (which should be almost identical).

c. Based on the results of question 2: what is the expected value of the ratio between the numbers in the last two columns of the table? Does the result for TN 21 correspond to the expected value of the ratio? (A tentative explanation for the answer may be that the ratio $NADH/NAD^+$ is highly controlled by other mechanisms, and that it is not permitted to increase to its equilibrium value) In any event the result of the genetic manipulation of the yeast strain is a total fiasco. The secretion of 2-oxoglutarate observed in problem 3.6 for TN 21 could be a result of the lower NADPH concentration in this strain, since NADPH is the preferred cofactor for conversion of 2-oxaloacetate to glutamic acid.

Problem 4.4 Nitrification by *Nitrosomonas sp.*

Fixation of CO_2 to biomass can be driven by a simultaneous oxidation of NH_3 to nitrite. The overall reaction is

$$-CO_2 - (a + 0.2) NH_3 - (1.5 a - 1.05) O_2 + CH_{1.8}O_{0.5}N_{0.2} + a\, HNO_2 + (a - 0.6) H_2O = 0 \quad (1)$$

Reaction (1) can be divided into two separate chemical reactions which run in parallel::

$$- CO_2 - 0.2\, NH_3 - 0.6\, H_2O + CH_{1.8}O_{0.5}N_{0.2} + 1.05\, O_2 = 0 \quad (2)$$

$$- NH_3 - 1.5\, O_2 + HNO_2 + H_2O = 0 \quad (3)$$

Determine ΔG for each of the two reactions (1) and (2). Determine the minimum value of parameter a in (1) for the whole process to be thermodynamically feasible. Chemolitotrophic bacteria such as *Nitrosomonas* are thermodynamically quite inefficient. Assume a thermodynamic efficiency of 30 % for fixation of CO_2 and calculate the maximum, yield of biomass on NH_3.

This problem is inspired by work on Chemolitotrophs by dr. Lars K. Nielsen at University of Queensland, Australia.

REFERENCES

Katchalsky, A. and Curran, F. P. (1965). Non-equilibrium Thermodynamics in Biophysics. Harvard U. Press, Cambridge, MA

Lehninger, A. E. (1975). *Biochemistry, 2nd ed.*, Worth, New York.

Morrero, J., Gani, R. (2001). "Group contribution based estimation of pure component properties." *Fluid Phase Equil.* **183/184** 183-208.

Marx, A., de Graaf, A.A., Wiechert, W., Eggeling, L., Sahm, H. (1996). "Determination of the fluxes in the central metabolism of *Corynebacterium glutamicum* by NMR combined with metabolite balancing." *Biotechnol. Bioeng.* **49**, 111-129.

Mitchell, P. (1961). Coupling of phosphorylation to electron and hydrogen transfer by a chemiosmotic type of mechanism. *Nature* **191**, 144-148

Nissen, T.L., Anderlund, M., Nielsen, J., Villadsen, J., and Kjelland-Brandt,.M.C. (2001). Expression of a cytoplasmic transhydrogenase in *Saccharomyces cerevisiae* results in formation of 2-oxoglutarate due to depletion of the NADPD-pool. *Yeast* **18**, 19-32.

Roels, J. A. (1983). Energetics and Kinetics in Biotecnology. Elsevier Biomedical Press, Amsterdam.

Rottenberg, H. (1979). "Non-equilibrium thermodynamics of energy conversion in bioenergetics". *Biochem. Biophys. Acta* **549**, 225-253.

Schill, N. A., Liu, J.-S., von Stockar, U. (1999). "Thermodynamic analysis of growth of *Methanobacterium thermoautotrophicum*". *Biotechnol. Bioeng.* **64**, 74-81.

Senior, A. E. (1988). "ATP synthesis by oxidative phosphorylation". *Physiol, Rev.* **68**, 177-231.

Stephanopolous, G., Vallino, J.J. (1991). "Network rigidity and metabolic engineering in metabolic overproduction." *Science* **252**, 1675-1680.

Stucki, J. W. (1980). "The optimal efficiency and the economic degrees of coupling of oxidative phosphorylation". Eur. J. Biochem. **109**, 269-283.

Vallino, J.J., Stephanopolous, G. (1993). Metabolic flux distributions in *Corynebacterium glutamicum* during growth and lysine overproduction. *Biotechnol.Bioeng.* ,**41** 633-646.

van Gulik, W. M., Heijnen, J. J. (1995). " A metabolic network stoichiometry analysis of microbial growth and product formation". *Biotechnol. Bioeng.* **48**, 681-698.

Vanrolleghem, P. A., Heijnen, J. J. (1998) "A structured approach for selection among candidate metabolic network models and estimation of unknown stoichiometric coefficients". *Biotechnol. Bioeng.* **58**, 133-138.

von Stockar, U., Gnaiger, E., Gustafsson, L., Larsson, C., Marison, I., Tissot, P. (1993). "Thermodynamic considerations in constructing energy balances for cellular growth". *Biochim. Biophys. Acta* **1833**, 221-240.

5

Biochemical Reaction Networks

Evolution has resulted in a complex metabolic network to ensure the proper function of living cells. In many biotech processes the metabolic networks of living cells are exploited for production of desired compounds. As discussed in Chapter 2 the metabolism can roughly be divided into catabolism and anabolism. The catabolism ensures supply of Gibbs free energy in the form of high-energy phosphate bonds in ATP, co-factors like NADH and NADPH, and a set of 12 precursor metabolites – everything that is needed for biosynthesis of cell mass. In Chapter 3 we lumped the myriad of reactions in the metabolic network into a single overall reaction – the black box model. This is very useful when the aim is to check the overall balances of carbon flowing in and out of the cell, but it does not supply any information about the processes occurring inside the cells. Clearly the complexity increases when intracellular reactions are considered in the analysis, but as will be discussed in this chapter there is a constant balancing of the formation and consumption of intracellular metabolites, and this imposes a large number of constraints on the fluxes through the different branches of the metabolic network. Thus, even though the complexity increases, the degrees of freedom do not necessarily increase, and expansion of the analysis to include intracellular reactions – in some cases lumped together into a few overall reactions – may in many cases be very useful. In this chapter we are going to look into the basic concepts of metabolic balancing (Section 5.1). In the rest of the chapter we will consider biochemical reaction models of increasing complexity – first simple models balanced only with respect to ATP, NADH and possibly NADPH (Section 5.2); thereafter simple metabolic networks where the metabolic network is lumped into a few overall pathway routes (Section 5.3), and finally more complete metabolic networks that permit a deeper analysis of the activities in the different branches of the metabolic network at varying growth conditions (Section 5.4).

5.1 Basic Concepts

When calculating the fluxes through the different branches of a metabolic network we shall use information about the net flows in and out of the cells, i.e. the net rate of production of substrates and metabolic products that can be measured as described in section 3.1. These are the only fluxes that can be directly measured, since it is not yet possible to measure fluxes within the cell. If a cell is pulled apart in order to measure the enzyme activities the controls imposed on the individual enzymes is lost, and the enzyme activity measured *in vitro* will therefore not say much about the *in vivo* flux. Furthermore, in enzyme assays the maximum enzyme activity is typically measured and in order to identify the actual rate of an enzyme catalyzed reaction it is necessary to know the substrate concentration (as well as the concentrations of all other metabolites that affect the enzyme reaction rate). Through material balances around intracellular metabolites it is, however, possible to calculate the *in vivo* fluxes through different branches of the metabolic network as will be illustrated in the following. This concept is referred to as *metabolite balancing*.

Consider a network in which we have identified J fluxes (or rates of pathways reactions) $v_1,...,v_J$ which we desire to calculate. The calculation relies on measurement of specific rates of substrate uptake and product secretion, and in the network we consider N substrates and M metabolic products (see Fig. 5.1). Since we can also measure the specific growth rate of the biomass there are therefore $r_1,...,r_{N+M+1}$ measurable rates. In addition to the measurable rates there are a set of constraints imposed by mass balances around the individual metabolites, each of which express that the concentration of a certain metabolite is constant, i.e., that the net rate of production of the metabolite is zero. Formulation of the constraint is done rigorously in the following way (see also Fig. 5.2):

For all intracellular metabolites the fluxes leading to a given metabolite are balanced with the fluxes leading away from the metabolite. Hereby there is no net accumulation of the metabolite.

As a corollary to this definition we again emphasize that metabolites in un-branched pathways do not provide a constraint of much use, as illustrated in Fig. 5.2. Clearly the two first reactions could have been lumped into a single overall conversion of the substrate to the metabolite B. To illustrate with a real example consider the EMP pathway in Fig. 2.4. In this pathway 3-phosphoglycerate (3PG) is produced and consumed at the same rate, but it is not involved in any other reaction, and the constraint $r_{3PG}=0$ is of no use. Conclusions based on only part of the metabolic network of an organism can, however, prove to be wrong. Actually 3PG is also used as precursor for synthesis of the amino acid serine, and if the flux towards this amino acid is significant compared with the flux through the EMP pathway the balance around 3PG may suddenly be important in the analysis. As long as only parts of the metabolic network are analyzed simplified pathway diagrams can, however, be used with confidence.

Similar to balancing of pathway intermediates redox equivalents formed in one pathway (see Table 3.4) as a result of a net oxidation of a substrate to a product will have to be consumed in other pathways. As long as only one type of redox carrying co-factor, e.g., NADH, is involved the constraint $r_{NADH}=0$ is immediately applicable, but some reactions in the cell may apply different co-factors, e.g., both NADH and NADPH may be used as co-factor in the conversion of acetaldehyde to acetate in *S. cerevisiae* (see Fig. 2.6). Compartmentation of the cell also leads to difficulties, since e.g. NADH in the cytosol of yeast is not interchangeable with NADH in the mitochondria. Thus, the finer details of redox balancing require a sophisticated treatment, but in simple approximate calculations, especially in calculations of the distribution of metabolic products in fermentative pathways it is sufficient to work with only one redox carrier as illustrated in Section 5.3.

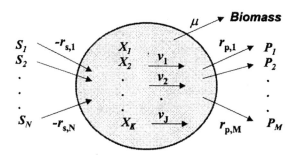

Figure 5.1 A general representation of reactions considered in a metabolic network. N substrates enter the cell and are converted into M metabolic products via a total of K intracellular metabolites. The conversions occur via J intracellular reactions for which the rates are given by $v_1...v_J$. Rates of substrate formation $(r_{s,1},...,r_{s,N})$ and product formation $(r_{p,1},...,r_{p,M})$ are also shown.

Whereas the NADH requirement of a given pathway reaction can be calculated as discussed in Section 3.4 this is not the case for the "energy equivalents" which are often represented solely by ATP. There is really no way of telling if the conversion of one metabolite to another with less free energy is accompanied by release of ATP or the difference in Gibbs free energy is dissipated as heat. Thus, the hydrolysis of glycerol-3-phosphate (G3P) to glycerol is not accompanied with formation of ATP whereas the corresponding reaction from acetyl-phosphate to acetate does lead to formation of ATP (see Fig. 2.6). In order to identify the involvement of ATP in specific reactions it is therefore necessary to consult the biochemistry literature (or reaction databases on the internet, e.g., www.genome.ad.jp). For the main pathways such as the EMP pathway leading from glucose to pyruvate the release of ATP is the same (2 moles of ATP per mole glucose) for all organisms using this pathway.

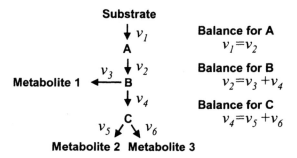

Figure 5.2 A schematic illustration of the concept of metabolite balancing. For the simple metabolic pathways the fluxes through the different enzymatic reactions are related through the three algebraic equations that represent balances for the three intracellular metabolites A, B and C.

When it comes to the consumption of ATP in anabolic pathways the picture is much less clear. First of all experimental studies show that copious amounts of ATP are consumed in the biosynthetic reactions, far more than what can be calculated based on the free energy harbored in the macromolecules which make up the cell. This is a consequence of the process where Gibbs free energy is transferred from the catabolism to the anabolism as "discrete" packages in the form of high-energy phosphate bonds in ATP. The ATP costs for synthesizing cell mass also varies significantly with the environmental conditions – both due to variations in the macromolecular composition as discussed in Section 2.1.4 and due to variations in ATP costs for maintenance of cellular activity as will be discussed in Section 5.2.1.

In order to formalize the concept of metabolite balancing we first specify the stoichiometry for the individual pathway reactions to be considered in the analysis. The stoichiometry for the individual reactions is written directly into the rows of the total stoichiometric matrix **T**. The stoichiometric coefficients for the reacting species are collected in the columns of **T**. The number of rows in **T** is equal to J, the number of fluxes in the model. In the stoichiometric matrix the compounds are organized such that the first N columns contain the stoichiometric coefficients for the substrates, in the columns $N+1$ to $N+M$ are given the stoichiometric coefficients for the metabolic products. Columns $N+M+1$ to $N+M+K$ contain the stoichiometric coefficients for the K intracellular metabolites. With this organization of the stoichiometric coefficients it is possible to relate the intracellular reaction rates represented by the flux vector **v** to the vector **r** of measurable rates given in Eq. (3.12) through:

$$\begin{pmatrix} \mathbf{r} \\ \mathbf{0} \end{pmatrix} = \mathbf{T}^T \mathbf{v} \tag{5.1}$$

In this matrix equation the last K rows represent the balances for the intracellular metabolites.

Example 5.1 Simple metabolic model
To illustrate the matrix equation (5.1) we consider the simple metabolic network in Fig. 5.2. From this network the stoichiometric matrix is directly set up:

$$\mathbf{T} = \begin{pmatrix} S & P_1 & P_2 & P_3 & A & B & C \\ -1 & 0 & 0 & 0 & 1 & 0 & 0 \\ 0 & 0 & 0 & 0 & -1 & 1 & 0 \\ 0 & 1 & 0 & 0 & 0 & -1 & 0 \\ 0 & 0 & 0 & 0 & 0 & -1 & 1 \\ 0 & 0 & 1 & 0 & 0 & 0 & -1 \\ 0 & 0 & 0 & 1 & 0 & 0 & -1 \end{pmatrix} \tag{1}$$

Here the stoichiometric coefficients for the single substrate considered in the network are given in the first column of **T** and the stoichiometric coefficients for the three metabolic products are given in columns 2 to 4 of **T**. Finally the stoichiometric coefficients for the three intracellular metabolites A, B and C are listed in the columns 5 to 7 of **T**. The rows of the stoichiometric matrix clearly specify the stoichiometry for the six reactions considered in the network, with the first reaction being the conversion of the substrate to the intracellular metabolite A and the second reaction being the conversion of A to B *etc.* With the stoichiometric matrix specified the balances are easily derived by transposing the matrix (changing columns and rows) and multiplying with the flux vector **v**:

$$\begin{pmatrix} r_s \\ r_{p,1} \\ r_{p,2} \\ r_{p,3} \\ 0 \\ 0 \\ 0 \end{pmatrix} = \begin{pmatrix} -1 & 0 & 0 & 0 & 0 & 0 \\ 0 & 0 & 1 & 0 & 0 & 0 \\ 0 & 0 & 0 & 0 & 1 & 0 \\ 0 & 0 & 0 & 0 & 0 & 1 \\ 1 & -1 & 0 & 0 & 0 & 0 \\ 0 & 1 & -1 & -1 & 0 & 0 \\ 0 & 0 & 0 & 1 & -1 & -1 \end{pmatrix} \begin{pmatrix} v_1 \\ v_2 \\ v_3 \\ v_4 \\ v_5 \\ v_6 \end{pmatrix} = \begin{pmatrix} -v_1 \\ v_3 \\ v_5 \\ v_6 \\ v_1 - v_2 \\ v_2 - v_3 - v_4 \\ v_4 - v_5 - v_6 \end{pmatrix} \quad (2)$$

The top four rows of this matrix equation relate the net flows in and out of the cells to the intracellular flux vector and the bottom three rows specify balances for the intracellular metabolites. It is seen that the balances for the intracellular metabolites impose a set of constraints on the six fluxes, and the degrees of freedom is three, i.e., only three of the measurable rates (r_s, $r_{p,1}$, $r_{p,2}$ and $r_{p,3}$) needs to be measured in order to calculate all the fluxes.

In the matrix equation (5.1) there are *J* unknowns, namely the elements of the flux vector **v**, and there are *K* constraints imposed by the mass balances for the intracellular metabolites. Thus, the degrees of freedom in the equation system is $F=J-K$, and if *F* of the measurable rates are available from experiments the remaining $N+M-F$ rates can be calculated. A systematic procedure to do this will be presented in Section 5.3.

The matrix equation (5.1) implies that the formation of intracellular metabolites exactly balances the consumption, i.e., the net formation rate of all intracellular metabolites is zero. During cell growth the biomass will, however, expand and if the net formation rate of an intracellular metabolite is zero this will result in a decreasing concentration as a consequence of dilution. For most metabolites this dilution effect is negligible, but in a few cases one must consider this dilution effect as discussed in Note 5.1.

Note 5.1 Dilution effect on intracellular metabolites
The correct material balance for an intracellular metabolite is given by:

$$\frac{dc_{met}}{dt} = r_{met} - \mu c_{met} \quad (1)$$

where r_{met} is the net rate of formation of the metabolite in all the reactions within the cell. In a steady state metabolite concentration c_{met} (unit: moles (g DW)$^{-1}$) must stay constant, and the balance reduces to:

state metabolite concentration c_{met} (unit: moles (g DW)$^{-1}$) must stay constant, and the balance reduces to:

$$0 = r_{met} - \mu c_{met} \tag{2}$$

r_{met} combines all the rates of formation and all the rates of consumption of the intracellular metabolite:

$$0 = \sum r_{met,for} - \sum r_{met,con} - \mu c_{met} \tag{3}$$

To keep the metabolite concentration constant when the cell mass increases by growth the total rate of formation must be larger than the total rate of consumption by other pathway reactions. The dilution term is, however, normally negligible, as illustrated below for two different types of intracellular compounds: intermediates in the glycolytic pathway and ATP.

In aerobically grown *S. cerevisiae*, the intracellular level of metabolites in the glycolytic pathway has been measured in continuous cultures (dilution rate = 0.1 h^{-1}) to be in the range of 0.05-1.0 μmol (g DW)$^{-1}$ (Theobald *et al.*, 1997). At these growth conditions, the flux through the EMP pathway is about 1.1 mmol (g DW)$^{-1}$ h^{-1}, and it is therefore seen that the flux through a metabolite pool, i.e., each of the two first terms in Eq. (3), is much higher than the dilution term of Eq. (2), which is about 0.005-0.1 μmol (g DW)$^{-1}$ h^{-1}. Thus, even for much lower glycolytic fluxes the dilution term is negligible.

The ATP pool in *S. cerevisiae* is about 8.0 μmol (g DW)$^{-1}$. Because ATP is involved in many reactions, it is difficult to assess the flux through the pool. However, the total ATP requirements for cell growth is about 7.0 mmol (g DW)$^{-1}$ h^{-1} at a specific growth rate of 0.1 h^{-1} (see Section 5.2). Thus, the flux through the pool is a factor of 10,000 higher than the dilution term, and this term obviously can be neglected in a balance for ATP. Even for cells that have a higher pool of ATP, *e.g.*, lactic acid bacteria, this conclusion still holds. A similar conclusion can be drawn for other cofactors like NADH and NADPH.

For intracellular metabolites present at relatively high concentrations (or rather with relatively low turnover frequency), e.g., some amino acids, the dilution term may account for up to 10% of the total fluxes leaving the pool (the sum of the consumption terms in Eq. (3) and the dilution term). Here the dilution term may be included in the model simply by including an additional reaction consuming the metabolite. The product of the specific growth rate and the (experimentally determined or estimated) concentration of the metabolite determines the flux of this reaction.

5.2 Growth Energetics

In order to harness the free energy produced by catabolic processes in terms of high-energy phosphate bonds – in particular in the form of ATP – for subsequent use in the biosynthesis of biomass constituents in the anabolism the cellular content of ATP (and ADP) must be controlled quite rigorously. As discussed in Section 2.1.4 the turnover time of ATP is low (see Fig. 2.7), and there must be tight balancing of the energy-forming reactions (catabolism) and the energy-utilizing reactions (the anabolism) inside the cell. This can be formulated as:

$$\mathbf{T}_{ATP}^{T} \mathbf{v} = 0 \tag{5.2}$$

column in the total stoichiometric matrix **T** that contains all the stoichiometric coefficients for ATP.

In analogy to the tight balancing of synthesis and consumption of ATP the cell needs to balance the synthesis and consumption of the co-factors NADH or NADPH, which as mentioned in Section 2.1.4 also have a small turnover time. Consequently the cell must exercise a strict control of the level of these compounds as well, and balances similar to Eq. (5.2) can be set up for these co-factors:

$$\mathbf{T}_{NADH}^T \mathbf{v} = 0 \tag{5.3}$$

$$\mathbf{T}_{NADPH}^T \mathbf{v} = 0 \tag{5.4}$$

The balances for ATP, NADH and NADPH in Eqs. (5.2)-(5.4) may be used to relate the fluxes through different parts of the metabolic network, and as we will see in Sections 5.2.2 and 5.2.3 this can be used to derive simple linear rate equations. However, in order to apply the ATP balance it is important that all ATP forming and consuming reactions are considered, and we therefore first consider an important group of energy-consuming processes inside the cell – namely *maintenance processes*.

5.2.1 Consumption of ATP for Cellular Maintenance

In 1959 it was shown by Herbert that it is necessary to consider what he called "endogenous metabolism" when the substrate utilization for biomass growth is to be calculated. He assumed that this endogenous metabolism results in a decrease of the amount of biomass, and he described the degradation as a first-order process with a specific rate of biomass degradation μ_e. Restitution of the degraded biomass requires substrate, and the total substrate consumption is therefore:

$$(-r_s) = Y_{xs}^{true} \mu + Y_{xs}^{true} \mu_e = Y_{xs}^{true} \mu + m_s \tag{5.5}$$

Equation (5.5) shows that there are two contributions to the substrate utilization: one term which is proportional to the observed, net specific growth rate (i.e., a growth-associated part) and another term which counteracts the continuous degradation of the cell mass due to endogenous metabolism. The so-called true yield coefficient specifies the yield in the conversion of substrate into biomass. In 1965 Pirt introduced an empirical correlation identical in form to Eq. (5.5), but he collected the product of Y_{xs}^{true} and μ_e in the empirical constant m_s, as shown in the last expression in Eq. (5.5). The empirical constant was called the *maintenance coefficient*.

In Section 7.3.2 we are going to discuss the application of Eq. (5.5) for description of cellular processes and show that despite its empirical nature it gives a good description of the specific substrate uptake rate for many cellular systems. However, the simple linear rate equation does not in a biologically satisfactory way explain what the extra substrate consumed for maintenance is in fact used for, i.e., which energy-requiring processes inside the cell do not lead to net formation of biomass. It is not at all clear which cellular processes should be categorized as maintenance processes, but below we list some of the most important processes that are customarily regarded as maintenance processes:

- **Maintenance of gradients and electrical potential.** In order to ensure proper function of the cell it is necessary to maintain concentration gradients, e.g., a proton gradient across the cellular membrane. Furthermore, it is necessary to maintain an electrical potential across the cellular membrane. These processes require energy (and consequently substrate), but they do not lead to formation of any new biomass, and they are therefore typical examples of maintenance processes. We shall see (Note 5.2) that the major part of the maintenance requirement originates in these processes.
- **Futile cycles.** Inside the cells there are pairs of reactions that results in the net hydrolysis of ATP. An example is the conversion of fructose-6-phosphate (F6P) to fructose 1,6-bisphosphate (a reaction that requires ATP) followed by its hydrolysis back to F6P by a phosphatase (a reaction that does not result in ATP formation). This two step futile cycle represents a very simple situation and in practice the regulatory system of the cell ensures that this futile cycle does not operate, e.g., the phosphatase is repressed in the presence of glucose. However, there are more complex futile cycles that involve a large number of reactions. The result is always a net hydrolysis of ATP. The exact function of such futile cycles is not known, but they may serve to generate heat by the hydrolysis of ATP and hence establish a higher temperature than that of the environment. Since futile cycles result in utilization of energy without net formation of biomass, they may also be considered as maintenance processes.
- **Turnover of macromolecules.** Many macromolecules (e.g., mRNA) are degraded and synthesized continuously inside the cell. This does not result in net formation of biomass either, but substantial amounts of Gibbs free energy are used, and turnover of macromolecules is therefore another typical example of a maintenance process.

Utilization of energy, and consequently substrate, in each of the three processes listed above is likely to be a function of the specific growth rate. When the specific growth rate is high there is a high turnover of macromolecules, and with increasing activity level in the cell it is for example necessary to pump more protons out of the cell. Furthermore, with a higher flux through the cellular pathways there is a higher loss of energy in the futile cycles. This is biologically reasonable since when the cells grow under limited conditions (a low specific growth rate) they will try to use the substrate as efficiently as possible and the maintenance processes are therefore curtailed. The energy expenditure in maintenance processes is therefore likely to be an increasing function of the specific growth rate. Thus, part of the Gibbs free energy spent in these maintenance processes may be included in the overall yield coefficient Y_{xs}^{true}, and only the part of free energy that is spent at zero-growth rate are included in the maintenance coefficient.

In 1960 Bauchop and Elsden introduced the concept of ATP requirements for biomass synthesis via the yield coefficient Y_{xATP} (unit: mmoles ATP (g DW)$^{-1}$), and proposed a balance equation that is analogous to Eq. (5.5):

Table 5.1 Experimentally determined values of Y_{xATP} and m_s for various microorganisms grown under anaerobic conditions with glucose as the energy source.

Microorganism	Y_{xATP} (mmoles ATP (g DW)$^{-1}$)	m_{ATP} (mmoles ATP (g DW h)$^{-1}$)	Reference
Aerobacter aerogenes	71	6.8	Stouthamer and Bettenhaussen (1976)
	57	2.3	Stouthamer and Bettenhaussen (1976)
Escherichia coli	97	18.9	Hempfling and Mainzer (1975)
Lactobacillus casei	41	1.5	de Vries et al. (1970)
Lactobacillus delbruckii	72	0	Major and Bull (1985)
Lactococcus cremoris	73	1.4	Otto et al. (1980)
	53	-	Brown and Collins (1977)
	15-50	7-18	Benthin et al. (1994)[#]
Lactococcus diacetilactis	47	-	Brown and Collins (1977)
Saccharomyces cerevisiae	71-91	<1	Verduyn et al. (1990)

[#] In their analysis, Benthin et al. (1994) found a large variation in the energetic parameters depending on the medium composition, see Example 5.2.

$$r_{ATP} = Y_{xATP}\mu + m_{ATP} \qquad (5.6)$$

Here r_{ATP} specifies the total formation rate of ATP in catabolic pathways (different from the net formation rate of ATP, which is implicitly assumed to be zero in the equation). From precise measurements of the metabolic products of the anaerobic metabolism it is possible to calculate the specific formation rate of ATP, i.e., r_{ATP}. This may be used to find experimental values for Y_{xATP} and m_{ATP}, as shown in many studies, e.g., Benthin et al. (1994) for lactic acid bacteria. In aerobic processes a major part of the ATP formation originates in the respiration, and the yield of ATP in this process is given by the P/O ratio. As discussed in Section 4.3 it is difficult to estimate the operational value of the P/O ratio. Detailed empirical studies, e.g., van Gulik and Heijnen (1995), have indicated an operational P/O ratio of 1.2-1.3, but this will depend on the microorganism and perhaps also on the environmental conditions. The ATP production can therefore not be calculated with the same accuracy in organisms that gain ATP by respiration as for organisms that are only able to produce ATP in fermentative pathways (substrate level phosphorylation).

In Table 5.1 experimentally determined values for Y_{xATP} and m_{ATP} are collected for a number of microorganisms growing at anaerobic conditions where r_{ATP} could be precisely determined. There is a large variation in the experimentally found values. This is explained by the fact that Y_{xATP} depends both on the applied medium and on the macromolecular composition of the biomass, as illustrated by the lower values obtained with growth on a complex medium compared with growth on a minimal medium.

One may also calculate theoretical values for Y_{xATP} from the metabolic network. In Table 2.5 a value of 41 mmoles of ATP per gram dry weight was given for cell synthesis on a minimal medium. If we add 6 mmoles of ATP per gram dry weight used in the transport processes (Stouthamer, 1979), we obtain a theoretical value for Y_{xATP} of 47 mmol of ATP per gram dry weight for *E. coli*. By comparison it is seen that the experimental value of Y_{xATP} for *E. coli* in Table 5.1 is 2.0-2.5 times larger than the theoretical value. This is a general observation made also for other microorganisms. The reason is that energy used in the maintenance processes is included in Y_{xATP}, as discussed in Note 5.2.

Note 5.2 Calculation of the total ATP consumption for maintenance
We now want to evaluate the total ATP consumption for maintenance reactions in *E. coli* with growth on a minimal medium. As discussed in the text there is a substantial deviation between the value of Y_{xATP} for *E. coli* in Table 5.1 and the value of 41 mmoles (g DW)$^{-1}$ for synthesis of an *E. coli* cell found in Table 2.5. Even if transport of substrates is considered there is a large deviation as mentioned above. The difference between the "theoretically" calculated value and the experimentally determined value must be due to the three types of maintenance processes.

Many of the cellular macromolecules are very stable, and it is mainly enzymes and mRNA that are degraded and resynthesized inside the cell. The half-life of mRNA is on the order of a few minutes, and there are good reasons for this low value. In order to control the synthesis of proteins at the genetic level it is important that mRNA be quite unstable, since otherwise translation of the mRNA could continue even when the enzyme is not needed. Since it is much cheaper from an energetic point of view to synthesize mRNA than protein, it is better for the cell to have a high turnover rate of mRNA rather than synthesize unnecessary protein. The turnover rate of enzymes is not known exactly, and probably it is heavily dependent on the cellular function of the enzyme. If the degradation of mRNA and protein are first-order processes, the rate of turnover depends on the content of these two components inside the cell. Since the content of enzymes and mRNA increases with the specific growth rate, the ATP requirement for turnover of macromolecules is therefore increasing with the specific growth rate. Using a half-life for mRNA of 1 min and a protein half-life of 10 hours, the ATP requirement for macromolecular turnover can be calculated to be in the order of 6 mmoles ATP (g DW h)$^{-1}$. Consequently, only a minor part of the ATP requirement for maintenance processes can be accounted for by turnover of macromolecules.

Stouthamer (1979) states that up to 50% of the total energy production during anaerobic growth of *E. coli* is used for maintaining membrane potentials, i.e., maintenance of the proton and electrochemical gradient across the cytoplasmic membrane. This corresponds to 49-59 mmoles of ATP per gram dry weight, and only a minor fraction of the ATP is therefore "lost" in the futile cycles. During aerobic growth, membrane energetization is ensured by the respiration (i.e., the oxidation of NADH), and this at least partly explains why the operational P/O ratio is below the theoretical value, as discussed in Section 4.3.

5.2.2 Energetics of Anaerobic Processes

In anaerobic processes there is no consumption of oxygen, and microorganisms grown under anaerobic conditions therefore have to rely on so-called substrate-level phosphorylation to obtain ATP for growth (or they use external electron acceptors other than oxygen). *Obligate anaerobes*

Biochemical Reaction Networks

such as methanogenic bacteria can not grow in the presence of oxygen, but many organisms, including *S. cerevisiae* and *E. coli* are *facultative anaerobes*, i.e., they grow both anaerobically and aerobically. Both *S. cerevisiae* and *E. coli* can gain ATP by respiration, but many facultative anaerobes for one reason or another lack the oxidative phosphorylation pathway. Thus, in lactic acid bacteria the lack of Cytochrome C (a constituent of the electron transport chain) impairs the functioning of the electron transport chain in oxidative phosphorylation. Although *Lactococcus lactis* is unable to gain free energy by respiration growth at aerobic conditions can give the bacteria advantages. For example the presence of oxygen can help to keep a low NADH level through the action of NADH-oxidases that use oxygen as electron acceptor.

During anaerobic growth the products of the catabolism are typically ethanol, lactic acid, and acetate, as illustrated in Fig. 2.6. Many other products may, however, be formed by other microorganisms grown at anaerobic conditions, e.g., acetoin, acetone and butanol by *Clostridium acetobutylicum* (see Problem 5.4). Microorganisms that are used industrially to produce metabolites as products of the catabolism may be divided into two groups, the *homofermentative* and the *heterofermentative*. A homofermentative microorganism will under most operating conditions produce a single product, whereas a heterofermentative microorganism produces many different products. In most cases it is necessary to form several metabolic products in order to balance the co-factors NADH and NADPH, except if growth is on a rich medium where there is very little net requirement of NADPH and formation of NADH in connection with biomass synthesis. At these conditions the ATP balance can be used to derive the linear rate equation (5.5) as shown in Example 5.2. When several different products are formed in order to balance the co-factors NADH and NADPH the metabolic network becomes somewhat more complex as illustrated in Example 5.4, but linear rate equations can be derived also in this case.

Example 5.2 ATP requirements for growth of *Lactococcus cremoris*

Lactococcus cremoris is used as a starter culture in the dairy industry for the production of butter, yogurt, and cheese. It is also used in some types of fermented sausage and sour bread. *L. cremoris* is characterized as a homofermentative Gram-positive bacteria mainly producing lactic acid, but it may produce many other products at conditions of very low sugar concentrations. *L. cremoris* is a multiple-amino acid auxotroph; i.e., it requires a supply of several amino acids for growth, and it is therefore normally grown on a rich medium. This is a drawback when a detailed analysis of the growth process is to be carried out, since it is difficult to identify the growth-limiting substrate. However, it also means that there is no net production/consumption of NADH and NADPH in connection with biomass synthesis and the overall biomass synthesis reaction can therefore be specified as:

$$\text{biomass} - a \text{ glucose} - b \text{ nitrogen source} - Y_{xATP} ATP = 0 \tag{1}$$

where the nitrogen source is a complex mixture of amino acids *etc.* a and b are yield coefficients in the overall biomass synthesis process. Since the stoichiometric coefficient for biomass is 1 the forward rate of this reaction is given by the specific growth rate μ.

The ATP required for biomass synthesis is supplied by the catabolic reactions, which at homofermenative conditions are limited to conversion of glucose to lactic acid. The overall stoichiometry for this process is (on a C-mole basis):

$$\text{lactic acid} + 0.333\,\text{ATP} - \text{glucose} = 0 \qquad (2)$$

Since the stoichiometric coefficient for lactic acid is 1 the forward rate of this reaction is equal to the specific rate of lactic acid production r_p (C-moles of lactic acid (C-mole biomass h)$^{-1}$). In addition to these two reactions ATP is consumed for maintenance, and this may be considered simply as a hydrolysis reaction of ATP with forward reaction rate m_{ATP}.

With the above reactions a balance for ATP directly gives:

$$0 = -Y_{xATP}\mu + r_p - m_{ATP} \qquad (3)$$

which can easily be re-written to Eq. (5.6) since r_p equals 3 r_{ATP}. The balance equation (3) can also be rewritten as a linear rate equation for the specific product formation rate, and in fact the first presentation of a linear rate equation like Eq. (5.5) was empirically derived from analysis of the production of lactic acid by *Lactococcus delbrueckii* at different specific growth rates (Luedeking and Piret, 1959).

With the simple model given above the glucose uptake rate can also be derived:

$$-r_s = a\mu + r_p \qquad (4)$$

which upon elimination of r_p by using Eq. (3) gives:

$$r_s = (a + Y_{xATP})\mu + m_{ATP} = Y_{xs}^{true}\mu + m_s \qquad (5)$$

Thus, the linear rate equation (5.5) is seen to be a consequence of the balancing of ATP in the cell, and this explains the general applicability of this linear relationship between the specific substrate uptake rate and the specific growth rate.

To illustrate the validity of the simple model used here we will consider experimental data by Benthin *et al.* (1994), who made a meticulous analysis of the growth and product formation of *L. cremoris*. Fig. 5.3A shows results from a batch fermentation of *L. cremoris*. The medium contained 20 gram per liter of glucose and 7 gram per liter of a complex N source (a mixture of 50% yeast extract and 50% caseine peptone). The biomass concentration was monitored by flow injection analysis (Benthin *et al.*, 1991), and the specific total acid production was measured by monitoring the amount of alkali added to keep the pH constant in the bioreactor. Throughout the batch fermentation the glucose concentration is high, and consequently only lactic acid is produced. It is observed that there are two distinct growth phases. First a phase with rapid growth, about µ=0.3 - 0.7 h^{-1}. This is followed by a phase where the specific growth rate first decreases to about 0.3 h^{-1} and then recovers to about 0.45 h^{-1}. After 12 hours the specific growth rate decreases monotonously to a very low value.

By plotting the specific production rate r_p versus the specific growth rate µ (Fig. 5.3B) it is observed that two straight lines appear, one for µ>0.3 h^{-1} and one for µ<0.3 h^{-1}. The lactic acid production is closely correlated to the ATP production (1 mole of ATP per mole of lactic acid produced), and the plot in Fig. 5.3B can therefore be used to estimate Y_{xATP} and m_{ATP} from, respectively, the slope and the intercept. The results are:

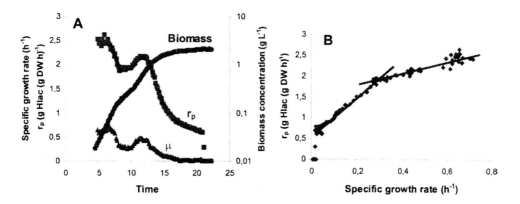

Figure 5.3 Batch fermentation of L. cremoris on a medium containing glucose and a complex nitrogen source.
A. Measurements of the biomass concentration (●), the specific acid production rate (■), and the specific growth rate (▲).
B. Plot of the specific acid production rate versus the specific growth rate.

- $\mu > 0.3$ h^{-1} Y_{xATP} = 15 mmoles ATP (g DW)$^{-1}$ m_{ATP} = 18 mmoles ATP (g DW h)$^{-1}$
- $\mu < 0.3$ h^{-1} Y_{xATP} = 50 mmoles ATP (g DW)$^{-1}$ m_{ATP} = 7 mmoles ATP (g DW h)$^{-1}$

In the first growth phase, Y_{xATP} is very low and m_{ATP} is high. Here the medium is very rich; i.e., all building blocks are supplied from the medium, and the cell therefore uses ATP mainly for polymerization reactions. In the second growth phase, Y_{xATP} is higher and m_{ATP} is low. Here one (or more) compound(s) in the complex nitrogen source has been exhausted, and the cells have to synthesize this (or these) compound(s). The ATP requirement for growth is therefore higher. With the higher ATP requirement for cellular growth, the cell is apparently able to get some savings by reducing the ATP consumption in maintenance processes; i.e., it is able to adjust its behavior to the more hostile environment.

It has not been possible to identify the (probably N-containing) compound(s) which cause(s) the change in metabolism at about 7 hours, but analysis of the medium throughout the fermentation showed that mainly peptides are metabolized during the first growth phase, whereas both peptides and free amino acids are metabolized in the second growth phase. L. cremoris has the ability to take up peptides, and it normally grows better on a medium containing peptides than on a medium that contains only free amino acids. The first shift in metabolism may be explained by exhaustion of small peptides, which may support a very rapid growth.

Based on average macromolecular composition of L. cremoris, Benthin (1992) calculated that the theoretical ATP requirement for cellular synthesis with growth on a complex medium is 26 mmole of ATP per gram dry weight. It is observed that during rapid growth this value is higher than the experimentally found Y_{xATP}, and this can be explained only by the ability of the cell to extract free energy from the complex nitrogen source. At the end of the fermentation the experimentally found Y_{xATP} is almost twice as high. Here the maintenance processes become relatively expensive compared to the growth process, as was discussed in Note 5.2.

5.2.3 Energetics of Aerobic Processes

During aerobic growth it is possible to oxidize part of the energy source completely to carbon dioxide in the TCA cycle. Since much more Gibbs free energy (and consequently ATP) is gained by complete oxidation of the energy source, it is possible to obtain much higher biomass yields from the energy source in an aerobic process than in an anaerobic process. In aerobic processes most of the ATP is formed in the oxidative phosphorylation, and therefore this process has a central position when energetic balances are to be set up. Unfortunately, the value of the P/O ratio is not exactly known as discussed in Section 4.3, and it may also vary with the operating conditions.

From an energetic analysis of a growth process with carbon dioxide as the only metabolic product we may, however, calculate the operational stoichiometry of the oxidative phosphorylation as illustrated in the following. Consider a simple, but still quite general model for aerobic growth without metabolite formation. As described in Section 2.1, the growth process starts with formation of precursor metabolites. Next, building blocks are synthesized from the precursor metabolites, and finally the building blocks are polymerized into macromolecules. In the synthesis of the precursor metabolites, both carbon dioxide and NADH are produced as by-products. In the formation of building blocks from the precursor metabolites, NADPH and ATP are required, and these compounds are produced in the catabolism. Finally, the polymerization requires ATP. The overall synthesis of biomass can therefore be described by:

$$\text{biomass} + Y_{xc}\,CO_2 + Y_{xNADH}\,NADH - (1+Y_{xc})\,CH_2O$$
$$- Y_{xATP}\,ATP - Y_{xNADPH}\,NADPH = 0 \quad (5.7)$$

As argued in Section 3.3 we need not consider the nitrogen source (and other nutrients) in the analysis. In Eq. (5.7) the stoichiometric coefficients are given relative to the formation of biomass and are therefore the yield coefficients. Thus, Y_{xc} specifies the moles of carbon dioxide formed per C-mole biomass produced. If a C-mole basis is applied the carbon balance directly gives the stoichiometric coefficient for glucose as $1+Y_{xc}$. Notice that in the overall reaction NADH is formed and NADPH is consumed in connection with biomass formation, which is the typical situation. The stoichiometric coefficients for NADH and NADPH can be calculated from detailed information about the biosynthesis of biomass, i.e., if the exact requirement for the individual building blocks is known and the biosynthetic routes to the individual building blocks have been unraveled. This has been illustrated for different microbial cells, and Table 5.2 collects some of these results. The values in Table 5.2 are given on a gram dry weight basis for the biomass, but using a molecular mass of 25 g (C-mole)$^{-1}$ the yield coefficients Y_{xNADH} and Y_{xNADPH} are found for *P. chrysogenum* to be 0.458 mmoles (C-mole)$^{-1}$ and 0.243 mmoles (C-mole)$^{-1}$, respectively[1].

---

[1] Note that the NADPH requirement for growth is considerably higher for the bacterium *E. coli* than for the two fungi. Prokaryotes have a higher lipid and protein content than eukaryotes and a substantial portion of the NADPH is spent in the sythesis of these cell constituents. In a more recent textbook Lengeler *et al.* (1999) gives a figure of about 19 mmoles NADPH per g DW, which is slightly higher than the value specified in Table 5.2. The calculations are complicated by the presence of isoenzymes using different co-factors for some of the steps. Therefore the calculated values should in reality be presented as ranges [see e.g. Albers *et al.* (1996)].

Biochemical Reaction Networks

The catabolic pathways supply the required ATP and NADPH for biomass synthesis. Excess NADH formed in the biosynthetic reactions and NADH formed in the catabolic pathways are both reoxidized by transfer of electrons to oxygen via the electron transport chain. Reactions (5.8)-(5.10) summarize the overall stoichiometry for the catabolic pathways. Reaction (5.8) specifies NADPH formation in the pentose phosphate pathway, when glucose is completely oxidized to CO_2. A complete oxidation of glucose to CO_2 is in fact possible in the pentose phosphate pathway (Fig. 2.4) since the pentoses can be fed back to the fructose-6-P/glucose-6-P pool and hereby be returned to the pentose phosphate pathway. Six cycles in the oxidative pentose phosphate pathway will result in complete oxidation of a glucose molecule. Reaction (5.9) is the overall stoichiometry for the combined EMP pathway and the TCA cycle. Finally, reaction (5.10) is the overall stoichiometry for oxidative phosphorylation. Compartmentation is not considered in the stoichiometry *i.e.*, no differentiation is made between cytosolic and mitochondrial NADH, and $FADH_2$ formed in the TCA cycle is pooled together with NADH. The P/O ratio in reaction (5.10) is therefore the overall (or operational) P/O ratio for oxidative phosphorylation.

$$CO_2 + 2\ NADPH - CH_2O = 0 \qquad (5.8)$$

$$CO_2 + 2\ NADH + 0.667\ ATP - CH_2O = 0 \qquad (5.9)$$

$$P/O\ ATP - 0.5\ O_2 - NADH = 0 \qquad (5.10)$$

As argued in Section 3.4 we have omitted ADP, NAD^+ and $NADP^+$ in the stoichiometry. Finally, consumption of ATP for maintenance is included as a separate reaction where ATP is hydrolyzed to less energy containing adenylates:

$$- ATP = 0 \qquad (5.11)$$

Note that, with the stoichiometry defined on a C-mole basis, the stoichiometric coefficients extracted from the biochemistry, *e.g.*, formation of 2 mol ATP per mole of glucose in the EMP pathway, are divided by 6, because 1 mole of glucose contains 6 moles of carbon.

Table 5.2 Calculated values of the requirements of NADPH for biomass synthesis and the amount of NADH formed in connection with biomass synthesis

Organism	Y_{xNADPH} mmoles NADPH (g DW)$^{-1}$	Y_{xNADH} mmoles NADH (g DW)$^{-1}$	Reference
E. coli	13.91	16.97	Ingraham et al. (1983)
S. cerevisiae	8.24	15.43	Oura (1983)[a]
P. chrysgogenum	8.49	16.00	Nielsen (1997)

[a] See also Albers et al. (1998) for a detailed calculation of the NADH formation in connection with biomass synthesis. The calculations are based on growth on one specific carbon source (glucose) and one specific nitrogen source (ammonia). Furthermore, a certain composition of the biomass in terms of protein, lipid content etc. has been used. The original references should be consulted for detailed information.

The reaction network (5.7) to (5.11) can be written in condensed form using the stoichiometric matrix **T**, where the first "compound" is taken to be biomass, followed by glucose, oxygen and carbon dioxide. The order of the intracellular metabolites is taken to be ATP, NADH and NADPH. Thus,

$$\mathbf{T} = \begin{pmatrix} 1 & -(1+Y_{xc}) & 0 & Y_{xc} & -Y_{xATP} & Y_{xNADH} & -Y_{xNADPH} \\ 0 & -1 & 0 & 1 & 0 & 0 & 2 \\ 0 & -1 & 0 & 1 & 0.667 & 2 & 0 \\ 0 & 0 & -0.5 & 0 & P/O & -1 & 0 \\ 0 & 0 & 0 & 0 & -1 & 0 & 0 \end{pmatrix} \quad (5.12)$$

We now introduce the reaction rate vector **v** for the five reactions:

$$\mathbf{v} = \begin{pmatrix} \mu \\ v_{PP} \\ v_{EMP} \\ v_{OP} \\ m_{ATP} \end{pmatrix} \quad (5.13)$$

The balances for the three cofactors ATP, NADH, and NADPH can then be derived directly from Eqs. (5.2)-(5.4):

$$-Y_{xATP}\mu + 0.667 v_{EMP} + P/O\, v_{OP} - m_{ATP} = 0 \quad (5.14)$$

$$Y_{xNADH}\mu + 2 v_{EMP} - v_{OP} = 0 \quad (5.15)$$

$$-Y_{xNADPH}\mu + 2 v_{PP} = 0 \quad (5.16)$$

Using Eq. (5.1) we can also derive relationships for the four measurable rates: the specific growth rate, the specific glucose and oxygen production rates, and the specific carbon dioxide production rate, in terms of the five reaction rates:

$$\begin{pmatrix} \mu \\ r_s \\ r_o \\ r_c \end{pmatrix} = \begin{pmatrix} 1 & 0 & 0 & 0 & 0 \\ -(1+Y_{xc}) & -1 & -1 & 0 & 0 \\ 0 & 0 & 0 & -0.5 & 0 \\ Y_{xc} & 1 & 1 & 0 & 0 \end{pmatrix} \begin{pmatrix} \mu \\ v_{PP} \\ v_{EMP} \\ v_{OP} \\ m_{ATP} \end{pmatrix} = \begin{pmatrix} \mu \\ -(1+Y_{xc})\mu - v_{PP} - v_{EMP} \\ -0.5 v_{OP} \\ Y_{xc}\mu + v_{PP} + v_{EMP} \end{pmatrix} \quad (5.17)$$

Clearly the top equation is trivial, but this is due to the simplified model applied. Often many more reactions are considered for biomass synthesis and then μ becomes a function of several of the flux vector elements. Eliminating the three fluxes v_{EMP}, v_{PP}, and v_{OP} from Eqs. (5.14)-(5.16),

Biochemical Reaction Networks

the following linear relationships between measurable rates are obtained from (5.17):

$$-r_s = (a + 1 + Y_{xc} + 0.5Y_{xNADPH})\mu + b = Y_{xs}^{true}\mu + m_s \quad (5.18)$$

$$r_c = (a + Y_{xc} + 0.5Y_{xNADPH})\mu + b = Y_{xc}^{true}\mu + m_c \quad (5.19)$$

$$-r_o = (a + 0.5Y_{xNADH})\mu + b = Y_{xo}^{true}\mu + m_o \quad (5.20)$$

The two common parameters a and b are obtained as a function of the energetic parameters Y_{xATP} and m_{ATP} and the P/O ratio, according to Eqs. (5.21) and (5.22):

$$a = \frac{Y_{xATP} - Y_{xNADH}\,P/O}{0.667 + 2P/O} \quad (5.21)$$

$$b = \frac{m_{ATP}}{0.667 + 2P/O} \quad (5.22)$$

Equation (5.18) is seen to be the same as the linear rate equation (5.5) with the difference that the yields of that correlation are now obtained in terms of basic cellular energetic parameters. This is true for all parameters in the preceding linear correlations since they are coupled via the ATP, NADH, and NADPH balances. It is thus seen that the three true yield coefficients cannot have arbitrary values since they are coupled through these balances. Furthermore, the maintenance coefficients are the same. This is due to the choice of the unit C-moles per C-mole of biomass per hour for the specific rates. If other units were used for the specific rates, the maintenance coefficients would not have the same values, but they would still be simply related. This coupling of the parameters shows that there are only two degrees of freedom in the system (equivalent to defining parameters a and b in Eqs. (5.21) and (5.22)).

Equation (5.18)-(5.20) contain three parameters Y_{xNADH}, Y_{xNADPH} and Y_{xc} that are all related to the redox balance for the cell. From Table 5.2 theoretical values of Y_{xNADH} and Y_{xNADPH} are obtained and (5.7) can be used to calculate Y_{xc}. Take *S. cerevisiae* and a standard biomass with $\kappa_x = 4.20$. The redox balance for reaction (5.7) is:

$$4.20 + 2 \cdot 24.6 \cdot 15.43\,10^{-3} - 4(1 + Y_{xc}) - 2 \cdot 24.6 \cdot 8.24\,10^{-3} = 0$$
$$\Rightarrow Y_{xc} = 0.138 \text{ mole } CO_2 \text{ (C-mole biomass)}^{-1} \quad (5.23)$$

It is now for a given strain seen that the three true yield coefficients in Eqs. (5.18)-(5.20), i.e. Y_{xs}^{true}, Y_{xc}^{true} and Y_{xo}^{true}, are fixed once the parameter a has been given a value. b is the same for all three lines. Consequently (5.18)-(5.20) provides only two independent relations between the three energetic parameters Y_{xATP}, P/O and m_{ATP}. If we assume that these three parameters are fundamental parameters for the strain then a similar model as derived above can be derived for other substrates, e.g., for acetate, ethanol, glycerol or citrate, and in this model the yield coefficients will be different functions of the three energetic parameters. If the yield coefficients are experimentally determined for growth on different substrates this allows all three energetic parameters to be estimated. van Gulik and Heijnen (1995) and van Rolleghem and Heijnen (1998) applied this approach for *S. cerevisiae* and Christiansen and Nielsen (2002) applied it for analysis of *Bacillus clausii* (see Example 5.3).

Example 5.3 Energetics of *Bacillus clausii*

Due to the ability to secrete large amounts of protein, members of the genus *Bacillus* are widely used as host in the fermentation industry for production of industrial enzymes, fine chemicals, antibiotics, and insecticides. *B. clausii* is a facultative alkalophilic *Bacillus*, which is commercially used in the production of the alkaline serine protease Savinase®, an enzyme used in detergents. In order to design optimal industrial processes a detailed understanding of the bioenergetics and in particular the maintenance demands are important. Production of proteases in *Bacilli* is generally a response of poor nutritional conditions and proteases are therefore typically produced at low specific growth rates. Most industrial fermentations producing proteases are therefore operated in the fed-batch mode with very low specific growth rates in the later stages of the process. During these stages with low specific growth rates a large part of the carbon source is consumed for maintenance demands. At these conditions the growth energetics are of particular importance since the energy metabolism in the cells influences both the yield of biomass and product on substrate, but it also determines the oxygen requirements and the removal of excess heat produced. The oxygen consumption is very closely related to the energy metabolism in the cell since oxygen is used as the final electron acceptor in the respiratory chain, which plays a central role in the oxidative phosphorylation. The parameters with the most significant influence on growth energetics and oxygen consumption are: The amount of ATP needed for biomass formation (Y_{xATP}), the moles of ATP formed per mole of oxygen consumed (P/O-ratio), and the ATP needed for maintenance (m_{ATP}).

As discussed in the text it is not possible to estimate all three energetics parameters by using only data from growth on glucose, but by combining data for growth on more than one carbon source with a detailed knowledge on the biomass composition and the central carbon metabolism all the energetic parameters can be estimated. An important assumption in this approach is that the parameters are conserved properties and do not change with the carbon source used. Christiansen and Nielsen (2002) performed a study of the growth kinetics and energetics in glucose and citrate limited chemostat cultures of *B. clausii* with special focus on the oxygen consumption. The true yield coefficients of glucose, oxygen, and carbon dioxide were obtained from glucose limited chemostats. From experimental results on the amino acid composition of the biomass, the RNA content and literature data a detailed biomass composition was determined and the building block requirements for biomass formation were calculated. From this analysis the elemental composition of the biomass was found to be $CH_{1.82}O_{0.55}N_{0.16}$ and the ash content was found to be 6%. Furthermore, the ATP cost for biomass synthesis was calculated to be 16.2 mmoles ATP (g DW)$^{-1}$ (or 0.43 moles ATP (C-mole biomass)$^{-1}$) and 19.0 mmoles ATP (g DW)$^{-1}$ (or 0.73 moles ATP (C-mole biomass)$^{-1}$) for growth on glucose and citrate, respectively. The cost of biomass synthesis from precursor metabolites is obviously the same for growth on glucose and citrate, but the ATP cost for synthesis of precursor metabolites is different for the two substrates. Thus, it is energetically more expensive to synthesize pyruvate and phosphoenolpyruvate (both precursor metabolites) from citrate than from glucose. In order to account for additional drain of ATP for biomass synthesis a constant value of K_x moles of ATP per C-mole biomass is added to the above-calculated values.

Biochemical Reaction Networks

In the following we are going to consider a simplified model of Christiansen and Nielsen (2002) where enzyme production is not considered (only a small amount of free energy is going to this anyway). In the model NADH, NADPH and $FADH_2$ are lumped into one compound (NADH). In the original model both NADH and $FADH_2$ were considered as substantial amounts of $FADH_2$ are formed during growth on citrate, and the P/O ratio for this compound is different from that for NADH. Here we intend primarily to illustrate the concept of how data from growth on two different substrates can be used to estimate all three energetic parameters, and we therefore apply a less complex model.

In the model the overall stoichiometry for biomass synthesis for growth on glucose and citrate are:

$$\text{biomass} + 0.148\ CO_2 + 0.176\ NADH - 1.148\ CH_2O - (0.43 + K_x)\ ATP = 0 \tag{1}$$

$$\text{biomass} + 1.025\ CO_2 + 0.917\ NADH - 2.025\ CH_{8/6}O_{7/6} - (0.73 + K_x)\ ATP = 0 \tag{2}$$

The glucose catabolism is:

$$0.667\ ATP + 2\ NADH + CO_2 - CH_2O = 0 \tag{3}$$

and citrate catabolism is:

$$4/18\ ATP + 1.5\ NADH + CO_2 - CH_{8/6}O_{7/6} = 0 \tag{4}$$

For growth on both substrates respiration is described by:

$$P/O\ ATP - NADH - 0.50\ O_2 = 0 \tag{5}$$

Finally a maintenance reaction eq. (5.11) was considered in the model.

From the balances of co-factors and the balances for glucose and citrate in the two models the following equations can be derived in analogy with eq. (5.18):

$$-r_{glc} = \left(1.148 + \frac{(0.43 + K_x) - 0.176 P/O}{2 P/O + 0.667}\right)\mu + \frac{m_{ATP}}{2 P/O + 0.667} = Y_{xglc}^{true}\mu + m_{glc} \tag{6}$$

$$-r_{cit} = \left(2.025 + \frac{(0.73 + K_x) - 0.917 P/O}{1.5 P/O + 4/18}\right)\mu + \frac{m_{ATP}}{1.5 P/O + 4/18} = Y_{xcit}^{true}\mu + m_{cit} \tag{7}$$

From chemostat cultures Christiansen and Nielsen (2002) found the following yield coefficients and maintenance coefficient[2]:

---

[2] The true yield coefficient for citrate is not reported by Christiansen and Nielsen (2002) as they only give data for a single chemostat experiment. Their approach is therefore somewhat more complicated, and to reduce the complexity we will here use a true yield coefficient that has been calculated from their data.

- $Y_{xglc}^{true} = 1.64$ C-mole glucose (C-mole DW)$^{-1}$
- $m_{glc} = 0.032$ C-mole glucose (C-mole DW h)$^{-1}$
- $Y_{xcit}^{true} = 2.69$ C-mole citrate (C-mole DW)$^{-1}$

Using Eqs. (7) and (8) and the experimentally determined yield coefficients one obtains:

$$1.64 = 1.14 + \frac{(0.43 + K_x) - 0.176 P/O}{2P/O + 0.667} \tag{8}$$

$$2.69 = 2.01 + \frac{(0.73 + K_x) - 0.917 P/O}{1.5 P/O + 4/18} \tag{9}$$

Solving these two equations for K_x and P/O one finds values of 0.708 moles ATP (C-mole biomass)$^{-1}$ (corresponding to 26.8 moles ATP (g DW)$^{-1}$) and 0.68, respectively. Furthermore, using the estimated value for the P/O-ratio together with:

$$0.032 = \frac{m_{ATP}}{2P/O + 0.667} \tag{10}$$

we find the maintenance ATP requirements to be 2.52 mmoles ATP (g DW)$^{-1}$. As a consequence of the simpler model applied here these values deviate slightly from those reported by Christiansen and Nielsen (2002).

The estimated P/O-ratio is quite low and this points to a low thermodynamic efficiency of the oxidative phosphorylation (see also Section 4.3). The consequence of this is a very high oxygen requirement of *B. clausii*, and this is a serious problem in connection with industrial application.

When metabolic products are formed the above-described models are too simple, but a similar approach can be applied as illustrated in Example 5.4.

Example 5.4 Aerobic growth of *Saccharomyces cerevisiae*
In Example 3.5 the data of von Meyenburg were analyzed and it was found that a substantial part of the ethanol formed above D_{crit} had been stripped to the gas phase. When the data at $D = 0.30$ h^{-1} is corrected for the missing ethanol one obtains the following table of yield coefficients (all on C-mole basis).

D (h^{-1})	Y_{sx}	Y_{sc}	Y_{so}	Y_{se}
0.15	0.548	0.452	0.425	0
0.30	0.279	0.366	0.167	0.355
0.40	0.175	0.312	0.044	0.513

Biochemical Reaction Networks

These data will now be analyzed using a simple model where three reactions are considered:

$$\text{Biomass} + 0.10\ CO_2 + 0.10\ NADH - 1.10\ CH_2O - \alpha\ ATP = 0 \quad ; \quad v_1 \quad (1)$$

$$CO_2 + 2\ NADH + \beta\ ATP - CH_2O = 0 \quad ; \quad v_2 \quad (2)$$

$$CH_3O_{0.5} + 0.50\ CO_2 + 0.5\ ATP - 1.5\ CH_2O = 0 \quad ; \quad v_3 \quad (3)$$

In the first reaction glucose is converted to biomass. The second reaction is complete oxidation of glucose to carbon dioxide via the EMP pathway, TCA cycle and regeneration of NAD^+ by respiration. The last reaction is the fermentative pathway of glucose to ethanol. The biomass formula is taken to be $CH_{1.83}O_{0.56}N_{0.17}$ ($\kappa_x = 4.20$) as in Example 3.5. The stoichiometry of pathway reaction v_1 requires a few comments. Some carbon is lost to CO_2, and this carbon loss varies between 8% and 16% depending on the environmental conditions. The formation of CO_2 in connection with biomass formation is due to requirements of specific precursor metabolites, which when they are derived from glucose leads to a net formation of CO_2, e.g. ribose-5-phosphate that serves as a precursor metabolite for nucleotide biosynthesis. The stoichiometric coefficient for NADH is derived from a redox balance: $(4 \cdot 1.10 - 4.20)/2 = 0.10$.

The ATP requirement α for production of biomass will be calculated (note that α is identical to Y_{xATP}). The ATP production β per C-mole glucose converted to CO_2 via the EMP pathway and the TCA cyclus is probably not independent of D. When no ethanol is formed one might assume that the pyruvate shunt *via* acetaldehyde to acetate (in the cytosol) and to acetate incorporated as AcCoA in the mitochondria is not operative. Consequently $\beta = 2/3$ since 1/3 ATP is generated on the way to pyruvate and 1/3 GTP (energetically equivalent to ATP) is generated in the TCA cycle. Since in *S. cerevisiae* $FADH_2$ and NADH enter at the same place in the respiratory chain they will both have a theoretical P/O ratio of 2 in the production of ATP from oxidation of the reduced cofactor. Consequently reaction (2) is exact with respect to an NADH coefficient of 2 and $\beta = 2/3$ in the case of pure respiration. With increasing ethanol production rate it is conceivable that a substantial part of the pyruvate flux passes via acetate and back into the mitochondria where 1/3 ATP per glucose carbon is consumed to convert acetate to AcCoA. In the following calculations β is taken to be 0.5 at $D = 0.3\text{h}^{-1}$ and 0.36 at $D = 0.4\text{h}^{-1}$. The lowest possible value of β is 1/3 - and we shall examine the sensitivity of the result with respect to β.

Case A: $D = 0.15\ \text{h}^{-1}$, $v_3 = 0$

In this case we find:

$$\left. \begin{array}{l} -r_o = 2v_2 + 0.1v_1 \\ -r_s = 1.10v_1 + v_2 \end{array} \right\} \quad \Rightarrow \quad \begin{array}{l} v_1 = 0.952(1 - Y_{so})(-r_s) \\ v_2 = (1 - 1.048(1 - Y_{so}))(-r_s) \end{array} \quad (4)$$

where

$$Y_{so} = \frac{-r_o}{-r_s} \quad ; \quad Y_{sx} = \frac{v_1}{-r_s} = 0.952(1 - Y_{so}) \quad (5)$$

and

$$Y_{sc} = \frac{(v_2 + 0.1v_1)}{-r_s} = 1 - 0.952(1 - Y_{so}) = 1 - Y_{sx} \qquad (6)$$

Consequently the carbon balance is automatically satisfied. Also Y_{sx} is independent of the stoichiometric coefficient of CO_2 in reaction (1) as is easily seen from the two relations for $(-r_o)$ and $(-r_s)$ in (4). These results will be seen to hold, also when ethanol is produced.

From Eq. (5) Y_{sx} is calculated to be 0.548 C-mole biomass (C-mole glucose)$^{-1}$, which is exactly the same value as obtained experimentally. This is not strange since $Y_{so} = 0.425$ was calculated based on a yield coefficient of biomass $Y_{sx} = 0.548$.

The ATP balance gives:

$$-\alpha v_1 + (2v_2 + 0.1v_1)P/O + 2/3v_2 = 0 \qquad (7)$$

Inserting v_1 and v_2 from Eq. (4) gives:

$$\alpha = \frac{2.10Y_{so}P/O + 0.70(1 - 1.048(1 - Y_{so}))}{1 - Y_{so}} \qquad (8)$$

Van Gulik and Heijnen (1995) have analyzed a large number of aerobic fermentation data on yeast and they conclude that the effective P/O ratio is 1.2-1.3. If we use 1.25 then α is calculated to 2.42 mole ATP (C-mole biomass)$^{-1}$.

Case B: $D = 0.30$ h^{-1} or $D = 0.40$ h^{-1}, $v_3 \neq 0$

In this case the substrate consumption is:

$$-r_s = 1.10v_1 + v_2 + 1.5v_3 \qquad (9)$$

while the oxygen consumption is given by the upper equation in (4). Solution for v_1 and v_2 in terms of $(-r_o)$, $(-r_s)$ and $r_e = v_3$ yields:

$$v_1 = (0.952(1 - Y_{so}) - 1.429Y_{se})(-r_s) \qquad (10)$$

$$v_2 = (1 - 1.10(0.952(1 - Y_{so}) - 1.429Y_{se}) - 1.5Y_{se})(-r_s) \qquad (11)$$

which gives

$$Y_{sx} = 0.952(1 - Y_{so}) - 1.429Y_{se} \qquad (12)$$

$$Y_{sc} = \frac{0.1v_1 + v_2 + 0.5v_3}{-r_s} = 1 - Y_{sx} - Y_{se} \qquad (13)$$

As was mentioned in the case $v_3 = 0$ the carbon balance automatically closes and Y_{sx} is independent of the stoichiometric coefficient for CO_2 in reaction (1).

Biochemical Reaction Networks

For $Y_{so} = 0.167$ and $Y_{se} = 0.355$ one obtains $Y_{sx} = 0.286$ and for $Y_{so} = 0.044$ and $Y_{se} = 0.513$ one obtains $Y_{sx} = 0.177$. In both cases the calculated values of Y_{sx} are close to the experimentally determined values of the table in Example 3.5. The difference is due to round-off errors.

We shall next consider the ATP balance:

$$-\alpha v_1 + (2v_2 + 0.1v_1)\text{P/O} + \beta v_2 + 0.5v_3 = 0 \tag{14}$$

Inserting v_1 and v_2 from Eqs. (10) and (11) and $v_3 = Y_{se}(-r_s)$ one obtains:

$$\alpha = \frac{2\,\text{P/O}\,Y_{so} + \beta\left(1 - 1.048(1 - Y_{so}) + \left(0.0714 + \dfrac{0.5}{\beta}\right)Y_{se}\right)}{0.952(1 - Y_{so}) - 1.429 Y_{se}} \tag{15}$$

and consequently:

$$\text{For } D = 0.30 \text{ h}^{-1} \ (\beta = 0.50): \quad \alpha = 1.168\,\text{P/O} + 0.887 \tag{16}$$

$$\text{For } D = 0.40 \text{ h}^{-1} \ (\beta = 0.36): \quad \alpha = 0.496\,\text{P/O} + 1.517 \tag{17}$$

It hardly matters if β is 0.36 or 0.5. If the value $\beta = 0.5$ is used in (17) then $\alpha = 0.4960\,\text{P/O} + 1.545$, i.e., the influence of P/O is the same while the ATP generated by substrate level phosphorylation is about 2% higher. Likewise the stoichiometric coefficient of CO_2 in reaction (1) has only a small influence on the relation between α and P/O – and no influence at all on Y_{sx}. If a stoichiometric coefficient of 0.12 is used instead of 0.10 Eq. (16) is changed to $\alpha = 1.168\,\text{P/O} + 0.877$.

While the result of the sensitivity analysis for the influence of the CO_2 coefficient in reaction (1) and of β in reaction (2) shows that these parameters have virtually no influence on the result there is a dramatic influence of the specific growth rate $r_x = D$ on the relation between α and P/O. When D increases above the critical dilution rate, i.e., α changes from Eq. (8) to Eq. (16) or Eq. (17) the influence of the ATP generation from the oxidative phosphorylation decreases significantly – at least if the effective P/O ratio is the same as is generally accepted in the literature. Inserting P/O = 1.25 in the three expressions yields:

D (h^{-1})	0.15	0.30	0.40
α (moles ATP (C-mole biomass)$^{-1}$	2.42	2.34	2.14

In Example 5.5 we will show that Y_{xATP} for anaerobic conditions is about 1.80 moles ATP (C-mole biomass)$^{-1}$ for all dilution rates. The higher value for α found in the above table may be explained by a less effective energy generation in the oxidative phosphorylation, i.e., the P/O ratio is lower than the assumed value of 1.25. If we on the other hand assume that $Y_{xATP} = \alpha$ is the same at anaerobic and aerobic conditions, we can calculate the P/O ratio from Eq. (16) and find it to be about 0.78. This may be too low, and probably the P/O ratio is somewhere between this value and 1.25. We might also speculate that the P/O ratio is fixed at e.g. 1.5 and that $Y_{xATP} = \alpha$ is higher for aerobic growth than for anaerobic growth. This may be due to a higher cost for biomass synthesis (different macromolecular composition) and due to higher costs for maintaining membrane potentials etc. during aerobic growth.

It is interesting to observe that α decreases with D in the above table. This may as indicated above be a result of a changed macromolecular composition. However, this is unlikely as the protein content normally increases with the specific growth rate and in Section 2.1.4 we noticed that the costs for protein synthesis accounts for the major ATP costs in synthesizing a cell. The change in α with D may, however, also be interpreted differently. If α stays constant at 2.42 moles ATP (C-mole biomass)$^{-1}$ for all 3 dilution rates P/O would increase from 1.25 to about 1.82 at the highest D value. This would imply an increasing effectiveness of the oxidative phosphorylation with decreasing oxygen uptake. In reality the P/O ratio may surely vary with the operating conditions, but generally it is believed that the thermodynamic efficiency decreases with the flux through the respiratory chain.

5.3 Simple Metabolic Networks

We will now proceed with the general balance equation (5.1). As discussed earlier there are J unknowns in the balance equation, namely the elements of the flux vector $\mathbf{v}$, and there are K constraints imposed by the mass balances for the intracellular metabolites. The degrees of freedom is therefore $F=J-K$, and if exactly F rates are measured all the fluxes and the remaining $N+M-F$ rates can be calculated. Clearly one can use Gauss elimination to obtain a manual solution of the set of algebraic equations, but even for relatively simple reaction networks this becomes cumbersome, and it is much easier to use matrix manipulations. Here we start by collecting all the measured rates in the vector $\mathbf{r}_m$ and position the remaining non-measured rates in the vector $\mathbf{r}_c$. We now order the equation system such that the equations for the non-measured rates are given as the upper $N+M-F$ equations, i.e.

$$\begin{pmatrix} \mathbf{r}_c \\ \mathbf{r}_m \\ \mathbf{0} \end{pmatrix} = \begin{pmatrix} \mathbf{T}_1 \\ \mathbf{T}_2 \end{pmatrix} \mathbf{v} \qquad (5.24)$$

Here the matrix $\mathbf{T}_1$ contains the $N+M-F$ rows in $\mathbf{T}^T$ that correspond to the non-measured rates and $\mathbf{T}_2$ contains the remaining $(N+M+K)-(N+M-F)=K+F=J$ rows of $\mathbf{T}^T$. Thus, $\mathbf{T}_2$ is a square matrix, and we can calculate the elements in the flux vector $\mathbf{v}$ directly from:

$$\mathbf{v} = \mathbf{T}_2^{-1} \begin{pmatrix} \mathbf{r}_m \\ \mathbf{0} \end{pmatrix} \qquad (5.25)$$

Finally the non-measured rates can be calculated from:

$$\mathbf{r}_c = \mathbf{T}_1 \mathbf{v} \qquad (5.26)$$

A prerequisite for application of eq. (5.25) is that the matrix $\mathbf{T}_2$ is non-singular, i.e. that its determinant is different from zero. There can be three reasons for $\mathbf{T}_2$ being singular:

- *Appearance of linearly dependent columns in the stoichiometric matrix* **T**. We remember that columns in the stoichiometric matrix specify the stoichiometric coefficients for the compounds included in the model, and linearly dependent columns appear if the stoichiometric coefficients for some of the compounds are identical or linearly dependent. This rarely happens in practice, but if one included both co-factors in co-factor couples like NADH/NAD$^+$, NADPH/NADP$^+$ or ATP/ADP then the stoichiometry for one of these co-factors would be identical with that of the other – except for the sign. A simple solution to this problem is that only one of the co-factors is included for each co-factor pair. Notice that if linear dependency occurs for stoichiometric coefficients of some intracellular compounds then the linearly dependent columns will always be transferred to linearly dependent rows in T_2, as this matrix always will contain all the stoichiometric coefficients for all intracellular compounds.
- *Appearance of linearly dependent rows in the stoichiometric matrix* **T**. The rows in the stoichiometric matrix specify the stoichiometry for reactions in the model, and if for some reason one or more reaction stoichiometries are linearly dependent this will transfer to linearly dependent columns in the T_2 matrix, and it hereby becomes singular. Linearly dependent reaction stoichiometries rarely appear in simple reaction networks, but they often occur in larger reaction networks, and we will therefore return to this problem in Section 5.4.1 (see Note 5.3).
- *Non-observable system with the chosen measured rates*. In principle the measured rates can be chosen arbitrary, but in practice one may chose a set of measured rates for which the system is not observable, *i.e.* with the chosen set of measured rates the matrix T_2 is singular. It is often difficult to identify which set of measured rates results in an observable system and which do not, and if singularity occurs the only practical way is to try a different set of measured rates and check whether the matrix T_2 is non-singular. This problem is discussed further in Examples 5.4 and 5.5.

This procedure to calculate the flux vector **v** and the non-measured rates is independent of the size of the reaction network, but it only applies when exactly F rates are measured. The procedure is illustrated below with two examples on relatively simple reaction networks. In Section 5.4 we are going to analyze large reaction networks and will also discuss how the fluxes can be calculated if we measure more or fewer than F rates.

Example 5.5 Anaerobic growth of *Saccharomyces cerevisiae*
Consider anaerobic growth of *S. cerevisiae*. Here ethanol is the major metabolic product, but some glycerol is also formed as discussed in Section 3.3. Glycerol formation is primarily a result of redox balancing since NADH is formed in connection with biomass synthesis, and the only way for the cell to balance the level of NADH at anaerobic conditions is through conversion of glucose to glycerol. Glycerol is formed from dihydroxyacetone phosphate, an intermediate in the EMP pathway (see Fig. 2.4), through the following reactions:

Dihydroxyacetone phosphate $\xrightarrow{\text{NAD} \quad \text{NAD}}$ Glycerol-3-phosphate $\longrightarrow$ Glycerol

As shown in Fig. 2.6 the formation of ethanol from pyruvate in *S. cerevisiae* proceeds via acetaldehyde, which is a volatile compound, and small amounts of acetaldehyde may therefore be secreted from the cell. In a simple model for anaerobic growth of *S. cerevisiae* we can therefore use the following reaction network.

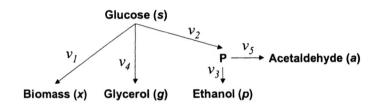

Glucose is distributed between three pathways with fluxes v_1, v_2 and v_4. The branch point P is acetaldehyde, which may either be excreted from the cell, v_5 or be reduced by v_3 to ethanol, which is immediately excreted to the medium. Carbon dioxide (*c*) is also formed, but is not shown in the network. There are five independent pathway reactions and 6 species with a net production rate different from zero (*s, x, g, e, c* and *a*). One constraint can be set up at the branch point *P*: The flux of carbon in v_2 must be distributed between excreted acetaldehyde and ethanol. This picture of the metabolism excludes the possibility of excretion of any intermediate along the pathway v_2, *i.e.* no pyruvate is excreted. Also it excludes the possibility of formation of other metabolic products – e.g. HAc or consumption of intermediates from pathway v_4 and v_2 to cell components. It is assumed that the biomass formation can be correctly described by one single pathway reaction v_1, which in terms of carbon consumption is independent of the other pathways.

It is assumed that there are no other sources or sinks of NADH or ATP than those coupled to the five pathway reactions, *i.e.* that $r_{NADH} = 0$ and $r_{ATP} = 0$. Thus there are three constraints including the trivial $v_5 = v_2 - v_3$. Since v_5 can be calculated directly from v_2 and v_3 we will only need four pathway stoichiometries.

$$\begin{aligned}
v_1 &: \quad -1.12 CH_2O + CH_{1.74}O_{0.6}N_{0.12} + 0.12 CO_2 + 0.15 NADH - 2.42 ATP = 0 \\
v_2 &: \quad -1.5 CH_2O + CH_2O_{0.5} + 0.5 CO_2 + 0.5 NADH + 0.5 ATP = 0 \\
v_3 &: \quad -CH_2O_{0.5} + CH_3O_{0.5} - 0.5 NADH = 0 \\
v_4 &: \quad -CH_2O + CH_{8/3}O - 0.333 NADH - 0.333 ATP = 0
\end{aligned} \quad (1)$$

The stoichiometry for the biomass formation reaction is given by v_1, and the stoichiometry for the following three reactions are taken from Table 3.4. We use a C-mole basis as in Chapter 3 for the black box models. Note that any of the equations can be multiplied by an arbitrary constant without affecting the final result, namely the distribution of carbon from glucose to different products. v_2 specifies the rate at which one C-mole of acetaldehyde is formed, whereas Table 3.4 gives the rate at which glucose is consumed to form $\frac{2}{3}$ C-mole acetaldehyde.

The ATP coefficient in the first reaction is an empirical quantity used when nothing is specifically stated concerning the fermentation conditions. It certainly varies significantly with the medium composition and with operating conditions such as the dilution rate. In section 5.2 we have discussed how the ATP requirement for growth can be determined from experiments (see Table 5.2 for experimental values), but in

Biochemical Reaction Networks

the present context we shall just use a value of 100 mmoles ATP $(g\ DW)^{-1}$ (corresponding to about 2.42 moles ATP $(C\ mol)^{-1}$). Notice that in the current example the amount of CO_2 formed in connection with biomass formation in the first reaction is slightly different from what was used in Example 5.4, but this is a consequence of a slightly different elemental composition of the biomass used in the present example.

The simple model can be presented in matrix form corresponding to Eq. (5.1) as:

$$\begin{pmatrix} r_s \\ r_x \\ r_a \\ r_e \\ r_c \\ r_g \\ 0 \\ 0 \end{pmatrix} = \begin{pmatrix} -1.12 & -1.5 & 0 & -1 \\ 1 & 0 & 0 & 0 \\ 0 & 1 & -1 & 0 \\ 0 & 0 & 1 & 0 \\ 0.12 & 0.5 & 0 & 0 \\ 0 & 0 & 0 & 1 \\ 0.15 & 0.5 & -0.5 & -0.333 \\ -2.42 & 0.5 & 0 & -0.333 \end{pmatrix} \begin{pmatrix} v_1 \\ v_2 \\ v_3 \\ v_4 \end{pmatrix} \qquad (2)$$

As there are four fluxes and we have two constraints, one for NADH and one for ATP (the two lower balances in Eq. (2)), the degrees of freedom $F=2$ and we therefore need to measure a minimum of 2 out of the 6 non-zero rates. The other rates can easily be calculated once the four fluxes have been determined. If more than 2 measurements are used the calculated values of v_1 to v_4 will be least squares fitted values which means that the constraints will not be exactly satisfied – but more accurate values of the fluxes will be obtained in the real situation where the measured rates are contaminated by experimental errors. The use of more than the minimum number of measured rates will be postponed until Section 5.4.

We now choose two measured non-zero rates out of the six available rates. The choice may seem arbitrary, but it will shortly be shown that only some of the 15 possible combinations of 2 out of the 6 rates will lead to solution of the problem. Let the choice be r_s and r_x, and in this case the matrix equation (5.24) becomes:

$$\begin{pmatrix} r_a \\ r_e \\ r_c \\ r_g \\ r_s \\ r_x \\ 0 \\ 0 \end{pmatrix} = \begin{pmatrix} 0 & 1 & -1 & 0 \\ 0 & 0 & 1 & 0 \\ 0.12 & 0.5 & 0 & 0 \\ 0 & 0 & 0 & 1 \\ -1.12 & -1.5 & 0 & -1 \\ 1 & 0 & 0 & 0 \\ 0.15 & 0.5 & -0.5 & -0.333 \\ -2.42 & 0.5 & 0 & -0.333 \end{pmatrix} \begin{pmatrix} v_1 \\ v_2 \\ v_3 \\ v_4 \end{pmatrix} \qquad (3)$$

And hereby we find using Eq. (5.25):

$$\begin{pmatrix} v_1 \\ v_2 \\ v_3 \\ v_4 \end{pmatrix} = \begin{pmatrix} -1.12 & -1.5 & 0 & -1 \\ 1 & 0 & 0 & 0 \\ 0.15 & 0.5 & -0.5 & -0.333 \\ -2.42 & 0.5 & 0 & -0.333 \end{pmatrix}^{-1} \begin{pmatrix} r_s \\ r_x \\ 0 \\ 0 \end{pmatrix} = \begin{pmatrix} r_x \\ -0.333 r_s + 2.047 r_x \\ 5.14 r_x \\ -0.5 r_s - 4.19 r_x \end{pmatrix} \quad (4)$$

Now all four fluxes have been determined as linear combinations of the two measured rates. Calculated values for the remaining non-zero rates are obtained by use of Eq. (5.26):

$$\begin{pmatrix} r_a \\ r_e \\ r_c \\ r_g \end{pmatrix} = \begin{pmatrix} 0 & 1 & -1 & 0 \\ 0 & 0 & 1 & 0 \\ 0.12 & 0.5 & 0 & 0 \\ 0 & 0 & 0 & 1 \end{pmatrix} \begin{pmatrix} v_1 \\ v_2 \\ v_3 \\ v_4 \end{pmatrix} = \begin{pmatrix} -0.333 r_s - 3.093 r_x \\ 5.14 r_x \\ -0.167 r_s + 1.143 r_x \\ -0.5 r_s - 4.19 r_x \end{pmatrix} \quad (5)$$

If we have measurements of r_a, r_e, r_c and r_g together with r_s and r_x we may now compare measured and calculated values of the rates. In case measured and calculated rates agree to within the experimental error it can be concluded that the simple metabolic network gives a good description of the anaerobic metabolism of yeast at the environmental conditions of the experiment.

From Eq. (5) it is seen that r_e and r_x are proportional. Consequently the two rate measurements do not form a basis on which the other four rates can be expanded. It is not immediately obvious why this combination of two rates cannot be used – while the remaining 14 pairs of r_i can be used as is easily seen from Eq. (5). Inspection of Eq. (4) explains why the system is not observable based on measurement of ethanol and biomass. Both v_1 and v_3 are proportional with r_x, and these two fluxes are therefore linearly dependent and if we lack information about one of the two other fluxes. That measurements of r_x and r_e causes problems for analysis of the system would have been difficult to identify before analysis, and it is typically only through analysis of the matrix equation for different set of measurements that it can be concluded which set of measurements ensures that the system is observable.

Finally some comments on manual checks of the correctness of the calculations

In each of the 4 reactions the carbon and redox balances close
- In a correct formulation of **T** the sum of numbers in each column should therefore give zero for the carbon containing species, *i.e.* the upper 6 rows.
- In the final result the carbon balance should close, *i.e.* $r_a + r_e + r_c + r_g = -r_s - r_x$
- In the final result the redox balance should close, *i.e.*

$$5r_a + 6r_e + 4.667 r_g + 4.18 r_x = (-1.667 r_s - 15.465 r_x) + 30.840 r_x$$
$$+ (-2.333 r_s - 19.553 r_x) + 4.18 r_x = 4(-r_s) + 0.002 r_x \approx 4(-r_s)$$

From the result in Eq. (5) the stoichiometry of a black box model for the overall growth process can be obtained. Let r_a be zero in which case:

Biochemical Reaction Networks

$$-0.333 r_s = 3.093 r_x$$

and

$$Y_{sx} = 0.108 \quad ; \quad Y_{se} = Y_{xe} Y_{sx} = 5.14 \cdot 0.108 = 0.554$$
$$Y_{sg} = 0.5 - 4.19 Y_{sx} = 0.0484 \quad ; \quad Y_{sc} = 0.167 + 1.143 Y_{sx} = 0.290$$

Thus, the black box model becomes:

$$-CH_2O + 0.554 CH_3O_{0.5} + 0.290 CO_2 + 0.108 X + 0.0484 CH_{8/3}O = 0 \tag{6}$$

Just as in Example 3.1 the stoichiometry of Eq. (6) satisfies a carbon and a redox balance, but although the same biomass composition $CH_{1.74}O_{0.6}N_{0.12}$ was used the stoichiometry is not quite the same. A little less biomass and glycerol is produced while the ethanol and carbon yields are higher. It is reasonable that Y_{sg} and Y_{sx} decrease together since conversion of glucose to ethanol is redox neutral and therefore $\frac{1}{3} r_g = \alpha r_x$ where α is the stoichiometric coefficient for NADH production in the first reaction. We see from the overall carbon balance that $Y_{xg} = 0.0484/0.108 = 0.45$, which is the ratio between $\alpha = 0.15$ and $\frac{1}{3}$ in our example. In Eq. (3.23) $Y_{sg} = 0.077/0.137 = 0.56$, which means that a portion of the glycerol formed is used as a sink for NADH produced in reactions which are not accounted for in the simple metabolic model that we use here.

Since (in the absence of acetaldehyde) the pathway to ethanol is the only one which provides ATP for glycerol production and especially for biomass formation a relatively higher yield of ethanol in Eq. (6) compared to Eq. (3.23) indicates that the ATP demand for biomass production is lower than 2.42 mol ATP (C-mol DW)$^{-1}$ assumed in the model. If the ATP coefficient in the first reaction is reduced to 1.80 the result is:

$$r_{ac} = -\tfrac{1}{3} r_s - 2.473 r_x \quad ; \quad \text{for } r_{ac} = 0 : Y_{sx} = 0.1348;$$

and consequently

$$r_e = 3.90 r_x$$

$$r_c = -\tfrac{1}{6} r_s + 0.8335 r_x$$

$$r_g = -\tfrac{1}{2} r_s - 3.26 r_x$$

The stoichiometry for $r_a = 0$ is:

$$-CH_2O + 0.526 CH_3O_{0.5} + 0.279 CO_2 + 0.135 X + 0.0606 CH_{8/3}O = 0 \tag{7}$$

The stoichiometric coefficients for ethanol, CO_2 and biomass are now close to the experimental values in Eq. (3.23). The glycerol to biomass ratio is still too low, but it is beyond the capability of the model to correct this.

Example 5.6 Heterofermentative metabolism by lactic acid bacteria

Lactic acid bacteria are often grown on complex media that are rich in amino acids (*e.g.*, yeast extract or casein peptone), providing the carbon skeletons for biosynthesis. There is, therefore, no (or very little) net consumption of redox equivalents in the anabolic pathways. This results in a conservation of redox equivalents in the conversion of the energy source (glucose or lactose) to the primary metabolites (lactate, ethanol, acetate, formate, and carbon dioxide). The catabolic pathway utilized by lactic acid bacteria, which is shown in Fig. 5.4, is therefore decoupled from growth and, as such, can be analyzed separately. Because the drain of precursor metabolites for growth is negligible for growth on a complex medium, the EMP pathway can be considered as a linear pathway with no branch points, so that all intermediates between glucose and pyruvate are eliminated. Under conditions of good growth, some species of lactic acid bacteria use only the pathway from glucose to lactate (often called homofermentative metabolism as leading to a single product). In this case the redox balance closes exactly, as the NADH formed in the conversion of glucose to pyruvate is regenerated in the conversion of pyruvate to lactate. Under conditions of extreme starvation, however, the cells strive to gain more ATP in the catabolic reactions, and channeling some pyruvate toward acetate does this. When this happens, a redox imbalance results, requiring that a fraction of the pyruvate be channeled toward ethanol where NAD^+ is regenerated. Thus, in order to obtain more ATP in the catabolism of glucose, several metabolic products are formed, and the metabolism in this case is called heterofermentative (or mixed acid fermentation).

The metabolic map shown in Fig. 5.4 is simplified, as in reality there are other reactions, e.g. an NADH oxidase, and there are also regulation that ensures that not all pathways operate at the same time, e.g. the pyruvate formate lyase, is not operating in the presence of oxygen whereas the pyruvate dehydrogenase is not operating at very low oxygen concentrations. Here we will, however, use the simplified network as it illustrates very well how the formation of different metabolic products are coupled via constrains imposed via the co-factor NADH.

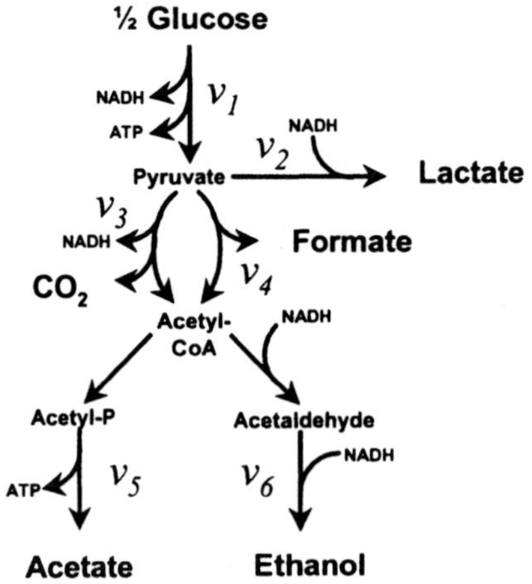

Figure 5.4 A simple model for the catabolic pathways in lactic acid bacteria. The metabolism from pyruvate is also shown in Fig. 2.6B and discussed in Section 2.1.3.3.

Biochemical Reaction Networks

Pyruvate obviously is a pathway branch point metabolite, and therefore a balance can be set up around this metabolite. In the conversion of pyruvate to metabolic products, three other pathway intermediates are involved: acetyl-CoA, acetyl-P, and acetaldehyde. Of these, only acetyl-CoA is located at a branch point and needs to be considered. Finally, conservation of the redox equivalents in the catabolic pathway yields an additional balance for NADH. One could also set up a balance for NAD^+, but this balance would be linearly dependent on the NADH balance and give no additional information, and would result in appearance of singularity in the stoichiometric matrix. Note that a balance for ATP could be set up, but because only catabolic metabolism is considered, there is no consumption of ATP and this balance therefore would not close. In summary, three pathway metabolites: pyruvate, acetyl-CoA, and NADH; one substrate: glucose (g); and five metabolic products: lactate (l), carbon dioxide (c), formate (f), acetate (a), and ethanol (e) are included in the model. Thus, Eq. (5.1) becomes:

$$\begin{pmatrix} r_g \\ r_l \\ r_c \\ r_f \\ r_a \\ r_e \\ PYR \\ AcCoA \\ NADH \end{pmatrix} = \begin{pmatrix} -0.5 & 0 & 0 & 0 & 0 & 0 \\ 0 & 1 & 0 & 0 & 0 & 0 \\ 0 & 0 & 1 & 0 & 0 & 0 \\ 0 & 0 & 0 & 1 & 0 & 0 \\ 0 & 0 & 0 & 0 & 1 & 0 \\ 0 & 0 & 0 & 0 & 0 & 1 \\ 0 & 1 & -1 & -1 & -1 & 0 & 0 \\ 0 & 0 & 0 & 1 & 1 & -1 & -1 \\ 0 & 1 & -1 & 1 & 0 & 0 & -2 \end{pmatrix} \begin{pmatrix} v_1 \\ v_2 \\ v_3 \\ v_4 \\ v_5 \\ v_6 \end{pmatrix} \quad (1)$$

Here the stoichiometry is written on a mole basis, and the flux v_1 is taken to be in mole pyruvate formed, and the stoichiometric coefficient for glucose in the first reaction is therefore –0.5. From the balance equation it is seen that all six fluxes can be directly measured from measurement of the glucose uptake rate and the formation rate of the five products, *i.e.*

$$\begin{pmatrix} r_g \\ r_l \\ r_c \\ r_f \\ r_a \\ r_e \end{pmatrix} = \begin{pmatrix} -0.5v_1 \\ v_2 \\ v_3 \\ v_4 \\ v_5 \\ v_6 \end{pmatrix} \quad (2)$$

In the balance equation (1) there are three constraints given by the pseudo-steady state assumption to the three intracellular compounds, pyruvate, AcCoA, and NADH, and with six fluxes the degrees of freedom $F=3$. Thus, if we choose three measured rates then we can calculate the three other rates in the system (and all the fluxes). If we choose rate measurements of glucose, lactate, and formate, we find by using Eq. (5.25):

$$\begin{pmatrix} v_1 \\ v_2 \\ v_3 \\ v_4 \\ v_5 \\ v_6 \end{pmatrix} = \begin{pmatrix} -0.5 & 0 & 0 & 0 & 0 & 0 \\ 0 & 1 & 0 & 0 & 0 & 0 \\ 0 & 0 & 0 & 1 & 0 & 0 \\ 1 & -1 & -1 & -1 & 0 & 0 \\ 0 & 0 & 1 & 1 & -1 & -1 \\ 1 & -1 & 1 & 0 & 0 & -2 \end{pmatrix}^{-1} \begin{pmatrix} r_g \\ r_l \\ r_f \\ 0 \\ 0 \\ 0 \end{pmatrix} = \begin{pmatrix} -2r_g \\ r_l \\ -2r_g - r_l - r_f \\ r_f \\ 0.5 r_f \\ -2r_g - r_l - 0.5 r_f \end{pmatrix} \quad (3)$$

It is observed that the formation of 2 mol of formate is accompanied by the formation of 1 mol of acetate (according to Eq. (2) the rate of acetate formation is equal to v_5). This fixed ratio between the rates of formation of two extracellular metabolites is explained by the NADH balance: When acetyl-CoA is formed together with formate, it is necessary to regenerate exactly 1 mol NAD$^+$ per mole of acetyl-CoA formed (namely, the NAD$^+$ used in the EMP pathway), and the flux from acetyl-CoA is therefore split equally between formation of ethanol (where two NAD$^+$ molecules are regenerated) and acetate. Similarly, if no formate is produced, the conversion of pyruvate to acetyl-CoA occurs solely via the pyruvate dehydrogenase route, which leads to additional NADH production. In this case, all the acetyl-CoA must be channeled toward ethanol, where the required NAD$^+$ is regenerated.

With measurements of glucose, lactate, and formate, there are no difficulties in determining all the fluxes and hereby also the non-measured rates. However, if the measurements of glucose, formate and acetate are selected instead T_2 becomes:

$$T_2 = \begin{pmatrix} -0.5 & 0 & 0 & 0 & 0 & 0 \\ 0 & 0 & 0 & 1 & 0 & 0 \\ 0 & 0 & 0 & 0 & 1 & 0 \\ 1 & -1 & -1 & -1 & 0 & 0 \\ 0 & 0 & 1 & 1 & -1 & -1 \\ 1 & -1 & 1 & 0 & 0 & -2 \end{pmatrix} \quad (4)$$

The matrix is singular and it cannot be inverted. This can be seen from the fact that subtraction of the third column from the second column gives a column that is identical with the last column. Thus, the columns 2, 3 and 6 are linearly dependent, and this results in a singularity in the matrix. In other words measurements of glucose, formate and acetate cannot be used to calculate uniquely the remaining three unkown pathway fluxes to lactate, CO_2, and ethanol. The reason, of course, is that with formate and acetate as the only measurements of metabolic products there is no information about the rate of NADH consumption (one of the rates r_1, r_c or r_e has to be measured) and the NADH balance is of little use. Since $r_a = 0.5\ r_f$ then any set of 3 measurements that includes the pair (r_a, r_f) will lead to a non-observable system. In the general case one will of course search for sets of measurements that leads to observability.

5.4 Flux Analysis in Large Metabolic Networks

When working towards an overall improvement of the yield of a given product from a certain substrate it is of great help to identify all possible routes (or pathways) between the substrate and the product and to obtain quantitative information about the relative activities of the different pathways involved in the overall conversion. Particularly in connection with *metabolic engineering* (see Chapter 1), where directed genetic changes are introduced in order to reroute the carbon fluxes towards the product of interest, it is essential to know how the different pathways operate at different growth conditions. As discussed in Section 2.1.1 the *in vivo* fluxes are the end result of many different types of regulation within the cell, and quantification of the metabolic fluxes therefore represent a detailed phenotypic characterization. Since quantification of metabolic fluxes goes hand in hand with identification of the active metabolic network, approaches to quantify fluxes have been referred to as metabolic network analysis (Christensen and Nielsen, 1999). Metabolic network analysis basically consists of two steps:
- Identification of the metabolic network structure (or pathway topology).
- Quantification of the fluxes through the branches of the metabolic network.

The extensive biochemistry literature and biochemical databases available on the web (see *e.g.* www.genome.ad.jp) provide much information relevant for identification of the metabolic network structure. Complete metabolic maps with direct links to sequenced genes and other information about the individual enzymes is typically retrieved. Thus, there are many reports on the presence of specific enzyme activities in many different species, and for most industrially important microorganisms the major metabolic routes have been identified. However, in many cases the complete metabolic network structure is not known, *i.e.* some of the pathways carrying significant fluxes have not been identified for the microorganism that is investigated. Here enzyme assays can be used to confirm the presence of specific enzymes and to determine the co-factor requirements, *e.g.* whether the enzyme uses NADH or NADPH as co-factor. Even though enzyme assays are valuable for confirming that a given pathway is present and is active, they are of limited use for a rapid screen of the totality of pathways, that are present in the studied microorganism. For this purpose isotope-labeled substrates are a powerful tool, and especially the use of ^{13}C-labelled glucose and subsequent analysis of the labeling pattern of the intracellular metabolites has proved to be very useful for identification of the metabolic network structure. This aspect is discussed further in Section 5.4.2.

When setting up the metabolic network it is important to specify a reaction (or a set of reactions) that leads to biomass formation. This reaction will specify the drain of precursor metabolites, or of building blocks, if the synthesis of these is included in the model. In some cases the stoichiometry for biomass formation has a significant influence on the analysis, and Note 5.3 shows how the so-called *biomass equation* is set up.

When the metabolic network structure has been identified the next step is to quantify the fluxes through the different branches in the network. In all cases the flux quantification relies on balancing of intracellular metabolites, just as was illustrated with several examples in Section 5.3 for simple networks. In flux analysis more detailed models are applied, but as in Section 5.3 a model with J fluxes **v** and K constraints will in practice have several degrees of freedom F=J-K, and there is an infinite number of solutions **v** to the model. In order to identify a unique solution **v** it is necessary to add more information or impose further constraints on the system. This can be done in three different ways:

- *Use of directly measurable non-zero rates.* This approach is the same as that discussed for the simple metabolic network models in Section 5.3, and when exactly F rates are measured the fluxes in the network can be calculated using eq. (5.1) – or eq. (5.25). This approach is discussed further in Section 5.4.1.
- *Use of labeled substrates.* When cells are fed with labeled substrates, e.g., glucose that is enriched for ^{13}C in the first position, then there will be a specific labeling of the intracellular metabolites. As there are different carbon transitions in the different cellular pathways, the labeling pattern of the intracellular metabolites is a function of the activity of the different pathways. Through measurements of the labeling pattern of intracellular metabolites and application of balances for the individual carbon atoms in the different biochemical reactions, additional constraints are added to the system. As discussed in Section 5.4.2 this is used to quantify the fluxes, even when only a few rates are measured.
- *Use of linear programming.* It is possible using linear programming to identify a solution (or a set of solutions) for the flux vector **v** that fulfills a specific optimization criterion, e.g., the flux vector that gives maximum growth yield. This approach is the subject of Section 5.4.3.

Note 5.3 Biomass equation in metabolic network models

Biomass formation is the result of a large number of different biochemical reactions. In Chapters 2 and 3 we looked at some of the many different reactions that are involved. Biomass synthesis starts with the formation of precursor metabolites (see Table 2.4), which are converted into building blocks (amino acids, nucleotides, lipids etc.). The building blocks are the monomers in macromolecules, which are the major constituents of biomass (see Table 2.5). The macromolecular composition of a given cell depends on the growth conditions, and on the composition, e.g. the amino acids in the proteins, of the different macromolecules. It is therefore not possible to specify a single reaction converting precursor metabolites into biomass. If this is still done one must make an assumption that the macromolecular composition is constant. There are three different ways of setting up an equation for formation of biomass with constant macromolecular composition:
- Direct synthesis from precursor metabolites
- Direct synthesis from building blocks
- Synthesis from macromolecules

In some cases one may use a combination of three approaches.

In the first approach an overall reaction is specified for conversion of precursor metabolites into biomass. Here information compiled in Table 2.4 is used together with information about the costs of ATP, NADPH and NADH to make biomass from the precursor metabolites. This identifies the stoichiometric coefficients for the different precursor metabolites involved in biomass formation. Reactions leading from the individual precursor metabolites to building blocks are not considered in this model, except perhaps for reactions leading to building blocks that are used for product formation, e.g. the synthesis of valine, cysteine and α-aminoadipic acid may have to be considered in a model for penicillin production. The synthesis of all other amino acids can be lumped into the overall biomass equation describing formation of biomass directly from precursor metabolites.

In the second approach the synthesis of most building blocks is considered in the model, and the biomass equation is described as a reaction where building blocks are converted into biomass. The stoichiometric coefficients for the building blocks are identified from knowledge of the amount of different building blocks that is needed for biomass formation. This approach typically results in a substantial increase in the model complexity, since a large number of reactions leads to the many different building blocks. Perhaps lumping of reactions that lead to many of the building blocks can be done but still the number of reactions considered in the model is large. The advantage is, however, that it is relatively easy to identify the different elements of the biomass equation.

In the last approach reactions for synthesis of the different macromolecules are included in the model, e.g. reactions for synthesis of proteins, lipids, DNA, RNA and carbohydrates. The biomass equation is described as a reaction where the macromolecules are converted into biomass, and the stoichiometric coefficients for the macromolecules are given by the macromolecular composition of the biomass. With this approach it is relatively easy to study the influence on the calculated fluxes of the macromolecular composition which directly appears in the biomass equation.

Which ever approach is used requires a substantial information about how biomass is synthesized and on the metabolic costs of the different precursor metabolites/building blocks/macromolecules. In addition the costs of ATP, NADPH and NADH for biomass formation must also be available, and this requires information about the biomass composition. In recent years this type of information has become available for many microorganisms as part of flux analysis studies. If no information is available for the investigated system one may use data from related organisms. It is already recommendable to calculate the sensitivities of the calculated fluxes to variations in the estimated (or experimentally determined) biomass equation.

5.4.1 Use of Measurable Rates

When F or more rates are measured all the fluxes can be estimated using matrix inversion as discussed in Section 5.3. Eq. (5.25) directly gives the solution for the flux vector **v** when exactly F rates are measured. This is often referred to as a *determined system*. If more than F rates are measured the system is *over-determined*. Here the matrix T_2 is not quadratic and it cannot be inverted directly. To circumvent this problem there are two possibilities:

(i) One may use a sub-set of the measured rates and calculate the fluxes using eq. (5.25) and the other rates using eq. (5.26). Through comparison of the calculated rates and the measured rates that are not used to calculate the elements of the flux vector one may check the consistency of the model (and model predicted values may be found for the measured rates if the model is believed to have the correct structure).

(ii) One may use a statistical procedure on the whole set of data to obtain good estimates for the elements in the flux vector **v** and obtain new (and better) estimates for the measured rates.

In the first case the solution is found as for a determined system. In the other case a statistical procedure similar to that described in Section 3.6 has to be applied, but there may be different approaches (see Stephanopoulos *et al.* (1998) for details). Often one will, however, simply use the least square estimate for flux vector. This estimate is found by using the pseudo-inverse of T_2 – called $T_2^\#$, i.e., the fluxes are calculated from:

$$v = T_2^\# \begin{pmatrix} r_m \\ 0 \end{pmatrix} \quad (5.27)$$

The pseudo-inverse of T_2 is given as:

$$T_2^\# = \left(T_2^T T_2\right)^{-1} T_2^T \quad (5.28)$$

The matrix $T_2^T T_2$ is always a square matrix. Furthermore, if T_2 has full rank then $T_2^T T_2$ is non-singular and the matrix can be inverted. The requirement of T_2 having full rank is synonymous with the requirement of T_2 being non-singular for the case of a determined system, and it means that there exists a $J \times J$ sub-matrix within T_2 that is non-singular. The requirements for application of eq. (5.28) are therefore the same as for analysis of the determined system, i.e., that there are no linearly dependent reaction stoichiometries and there are no metabolites with identical or linear dependent stoichiometric coefficients (such as the co-factor couple NADH/NAD$^+$). Furthermore, the set of measured rates must be chosen such that the fluxes can be calculated using the matrix equation, but this requirement is normally fulfilled for an over-determined system. Whereas it is quite simple to consider only one of the compounds in co-factor couples it is often more difficult to avoid linearly dependent reaction stoichiometries, and this is therefore discussed further in Note 5.4.

Note 5.4 Linear dependency in reaction stoichiometries.
Because most living cells are capable of utilizing a large variety of compounds as carbon, energy and nitrogen sources, many complementary pathways exist that would serve similar functions if they operated at the same time. The inclusion of all such pathways may give rise to problems when matrix inversion is applied for flux analysis. This situation usually manifests itself as a matrix singularity, whereby the non-observable pathways appear as linearly dependent reaction stoichiometries. The fluxes through these different pathways cannot be discerned by extracellular measurements alone. Here we will consider two examples:

- Glyoxylate cycle in prokaryotes
- Nitrogen assimilation via the GS-GOGAT system

In prokaryotes, the TCA cycle and all anaplerotic reactions, including the glyoxylate cycle, operate in the cytosol. Often the glyoxylate cycle is considered as a bypass of the TCA cycle because it shares a number of reactions with this cycle (see Fig. 2.5). However, the two pathways serve very different purposes: the TCA cycle has the primary purpose of oxidizing pyruvate to carbon dioxide, whereas the glyoxylate cycle has the purpose of synthesizing precursor metabolites, *e.g.*, oxaloacetate, from acetyl-CoA. Considered individually the TCA cycle and the glyoxylate shunt are not linearly dependent, but if other anaplerotic pathways, *e.g.* the pyruvate carboxylase reaction, are included, a singularity arises. This may be illustrated by writing lumped reactions for the three pathways (see Fig. 2.10 for an overview of the pathways). In all cases we use pyruvate as the starting point:

TCA cycle: - pyruvate + $3CO_2$ + 4 NADH + $FADH_2$ + GTP = 0 (1)

Glyoxylate shunt: - 2pyruvate + $2CO_2$ + oxaloacetate + 4NADH + $FADH_2$ = 0 (2)

Pyruvate carboxylase: - pyruvate - ATP - CO_2 + oxaloacetate = 0 (3)

If ATP and GTP are pooled together (which is often done in the analysis of cellular reactions), it is quite obvious that the glyoxylate shunt is a linear combination of the two other pathways, and all three pathways cannot be determined independently by flux analysis. It may be a difficult task to decide which pathway should be eliminated. Fortunately, these pathways rarely operate at the same time as their enzymes are induced differently. Information about induction and regulation of the corresponding enzymatic activities is critical in making a decision as to the exact pathway to be considered at a given set of environmental conditions. For example, expression of isocitrate lyase (the first enzyme of the glyoxylate shunt) is repressed by glucose in many microorganisms, and consequently the glyoxylate shunt is inactive for growth on glucose. In eukaryotes, the presence of the glyoxylate shunt does not give rise to a linear dependency due to compartmentation of the different reactions, *i.e.*, the TCA cycle operates in the mitochondria and the glyoxylate shunt either in the cytosol or in microbodies. In practice there are many other reactions in the network that involve intermediates of the TCA cycle and the glyoxylate shunt, and these reactions may lead to a removal of the linear dependency between these two pathways (see also Example 5.6). Even in cases where the two pathways are not linearly dependent the inclusion of both pathways in the model may lead to an ill-conditioned system, i.e., the condition number may be high (see Note 5.4).

Another example of linearly dependent reactions is the two ammonia assimilation routes: the glutamate dehydrogenase catalyzed reaction and the GS-GOGAT system (see Section 2.4.1). The stoichiometries of these two routes are

GDH: - α-ketoglutarate - NH_3 - NADPH + glutamate = 0 (4)

GS-GOGAT: - α-ketoglutarate - NH_3 - NADPH - ATP + glutamate = 0 (5)

Thus, the only difference is that ATP is used in the GS-GOGAT route (which is a high-affinity system) but not in the GDH reaction. The problem here is that an ATP balance is not easy to utilize due to lack of sufficient information about all ATP-consuming reactions. In the absence of an ATP balance to differentiate between them, the two nitrogen assimilation reactions are linearly dependent and, as such, non-observable. Because the only difference between the two routes is the consumption of ATP in the GS-GOGAT system, distinction between the two routes may not be important, and they are therefore often lumped into a single reaction in stoichiometric models.

If a singularity arises in the stoichiometric matrix, one has the following two options:
(1) Remove the linearly dependent reaction(s) from the model, invoking (or postulating) information about specific enzyme regulation and induction.
(2) Introduce additional information such as the relative flux of the two pathways. Such information may be derived from measurements of enzyme activities, *e.g.*, the relative activity of key enzymes in the two routes. However, this approach is hampered by the fact that *in vitro* enzyme activity measurements often bear little relationship to actual *in vivo* flux distributions. A more powerful technique is the use of labeled substrates, *e.g.*, ^{13}C-enriched glucose, followed by measurements of the labeling pattern of intracellular metabolites as discussed in Section 5.4.2.

The combination of a metabolic model, based only on reaction stoichiometries, and measurement of a few rates is a very simple method for estimation of intracellular fluxes, and it has been used to study many different fermentation processes (Vallino and Stephanopoulos, 1993; Vallino and Stephanopoulos, 1994a,b; Jørgensen *et al.*, 1995; van Gulik and Heijnen, 1995; Sauer *et al.*, 1996; Nissen *et al.*, 1997; Pramanik and Keasling, 1997; Pedersen *et al.*, 1999). Clearly it is valuable to quantify the fluxes through the different branches of the metabolic network considered in the model, and in Section 5.4.2 we discuss how information on fluxes may be used to guide genetic modification, resulting in strains with improved properties. The approach may, however, also be used for analysis of the metabolic network, i.e., which pathways are likely to operate. This will be illustrated in examples 5.7 and 5.8. It is important to emphasize that such analysis must always be followed up with experimental verification, but clearly the simple approach discussed in this section may be used as an efficient guide to the experimental work.

Example 5.7 Metabolic Flux Analysis of Citric Acid Fermentation by *Candida lipolytica*
Aiba and Matsuoka (1979) were probably the first to apply the concept of metabolite balancing to analyse fermentation data. They studied the yeast *Candida lipolytica* producing citric acid, and the aim of their study was not to quantify the fluxes but rather to find which pathways were active during citric acid production. For their analysis they employed the simplified metabolic network shown in Fig. 5.5. The network includes the EMP pathway, the TCA cycle, the glyoxylate shunt, pyruvate carboxylation, and formation of the major macromolecular pools, *i.e.*, proteins, carbohydrates, and lipids. At least one of the two anaplerotic routes are obviously necessary to replenish TCA cycle intermediates when citrate and isocitrate are secreted to the extracellular medium.

In the network there is a total of 16 compounds and of these 8 are intracellular metabolites for which pseudo steady state conditions apply. The compounds for which the rate of formation or consumption can be measured are:
  Glucose (glc), ammonia (N), carbon dioxide (c), citrate (cit), isocitrate (ic), protein, (prot), carbohydrates (car) and lipids (lipid).
The intracellular metabolites for which pseudo steady state applies are:
  Glucose-6-phosphate, pyruvate, acetyl-CoA, 2-oxoglutarate, succinate, malate, oxaloacetate, glyxoylate.
Notice that citrate and isocitrate do not remain in pseudo steady state: these metabolites are constantly produced. One could specify these compounds as intracellular metabolites being in pseudo steady state, but this would require including two additional reactions in the network (indicated by the broken secretion lines in Fig. 5.5).

Based on the network shown in Fig. 5.5 we can set up eq. (5.1) for the model. Mole basis is used for all stoichiometric coefficients and the protein synthesis rate is based on moles of OGT consumed. Similarly the rate of carbohydrate synthesis is based on moles of G6P consumed.

$$\begin{pmatrix} r_{gkc} \\ r_N \\ r_c \\ r_{cit} \\ r_{ic} \\ r_{prot} \\ r_{car} \\ r_{lipid} \\ 0 \\ 0 \\ 0 \\ 0 \\ 0 \\ 0 \\ 0 \\ 0 \end{pmatrix} = \begin{pmatrix} -1 & 0 & 0 & 0 & 0 & 0 & 0 & 0 & 0 & 0 & 0 & 0 & 0 & 0 & 0 \\ 0 & 0 & 0 & 0 & 0 & 0 & 0 & 0 & 0 & 0 & 0 & 0 & 0 & 0 & -1 \\ 0 & 0 & 1 & -1 & 0 & 0 & 1 & 1 & 0 & 0 & 0 & 0 & 0 & 0 & 0 \\ 0 & 0 & 0 & 0 & 1 & -1 & 0 & 0 & 0 & 0 & 0 & 0 & 0 & 0 & 0 \\ 0 & 0 & 0 & 0 & 0 & 1 & -1 & 0 & 0 & 0 & -1 & 0 & 0 & 0 & 0 \\ 0 & 0 & 0 & 0 & 0 & 0 & 0 & 0 & 0 & 0 & 0 & 0 & 0 & 0 & 1 \\ 0 & 0 & 0 & 0 & 0 & 0 & 0 & 0 & 0 & 0 & 0 & 0 & 1 & 0 & 0 \\ 0 & 0 & 0 & 0 & 0 & 0 & 0 & 0 & 0 & 0 & 0 & 0 & 0 & 1 & 0 \\ 1 & -1 & 0 & 0 & 0 & 0 & 0 & 0 & 0 & 0 & 0 & 0 & -1 & 0 & 0 \\ 0 & 2 & -1 & -1 & 0 & 0 & 0 & 0 & 0 & 0 & 0 & 0 & 0 & 0 & 0 \\ 0 & 0 & 1 & 0 & -1 & 0 & 0 & 0 & 0 & 0 & -1 & 0 & -1 & 0 \\ 0 & 0 & 0 & 0 & 0 & 0 & 1 & -1 & 0 & 0 & 0 & 0 & 0 & -1 \\ 0 & 0 & 0 & 0 & 0 & 0 & 0 & 1 & -1 & 0 & 1 & 0 & 0 & 0 & 0 \\ 0 & 0 & 0 & 0 & 0 & 0 & 0 & 0 & 1 & -1 & 0 & 1 & 0 & 0 & 0 \\ 0 & 0 & 0 & 1 & -1 & 0 & 0 & 0 & 0 & 1 & 0 & 0 & 0 & 0 & 0 \\ 0 & 0 & 0 & 0 & 0 & 0 & 0 & 0 & 0 & 0 & 1 & -1 & 0 & 0 & 0 \end{pmatrix} \begin{pmatrix} v_1 \\ v_2 \\ v_3 \\ v_4 \\ v_5 \\ v_6 \\ v_7 \\ v_8 \\ v_9 \\ v_{10} \\ v_{11} \\ v_{12} \\ v_{13} \\ v_{14} \\ v_{15} \end{pmatrix} \quad (1)$$

Notice that both the TCA cycle, the glyoxylate shunt and the pyruvate carboxylase are included in the model, but the matrix still has full rank. This is due to the reaction from 2-oxoglutarate to protein, which is an additional reaction that imposes a constraint on the flux *via* the TCA cycle.

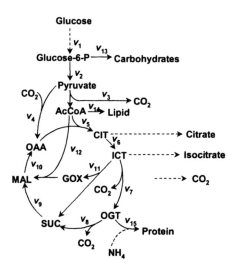

Figure 5.5 Simplified metabolic network for *Candida lipolytica*. Abbreviations: AcCoA, acetyl-CoA; OAA, oxaloacetate; CIT, citrate; ICT, isocitrate; OGT, 2-oxoglutarate; SUC, succinate; MAL, malate; GOX, glyoxylate. Broken lines represent reactions where material is transported from the cell to the environment, whereas solid lines represent intracellular reactions.

With 15 fluxes and 8 balance equations, there are 7 degrees of freedom. Thus, if 7 reaction rates are specified, the other rates can be calculated. In their analysis Aiba and Matsuoka measured six reaction rates in the network: the glucose uptake rate ($r_{glc} = -v_1$); the carbon dioxide production rate (r_c); the citric acid production rate (r_{cit}); the isocitrate production rate (r_{ict}); the protein synthesis rate (r_{prot}); and the carbohydrate synthesis rate (r_{car}). The rates r_{prot} and r_{car} were found from measurements of the protein and carbohydrate contents of the biomass (in a steady state chemostat). In addition to the six measurements, Aiba and Matsuoka imposed an extra constraint by setting one of the rates in the network equal to zero. Three different cases were examined, reflecting three different modes of operation of the network:

- **Model 1**: The glyoxylate shunt is inactive, *i.e.*, $v_{11} = 0$.
- **Model 2**: Pyruvate carboxylase is inactive, *i.e.*, $v_4 = 0$.
- **Model 3**: The TCA cycle is incomplete, *i.e.*, $v_8 = 0$.

Setting a flux to zero corresponds to removing the corresponding reaction from the model. The number of fluxes is therefore reduced to 14 and the degrees of freedom to 6. With the listed six measured rates, the system of equations is therefore exactly determined for each of the three models and can be solved to determine all the fluxes using eq. (5.25) and the unknown exchange rates using eq. (5.26). If we consider Model 1 we find:

$$\begin{pmatrix} v_1 \\ v_2 \\ v_3 \\ v_4 \\ v_5 \\ v_6 \\ v_7 \\ v_8 \\ v_9 \\ v_{10} \\ v_{12} \\ v_{13} \\ v_{14} \\ v_{15} \end{pmatrix} = \begin{pmatrix} -1 & 0 & 0 & 0 & 0 & 0 & 0 & 0 & 0 & 0 & 0 & 0 & 0 & 0 \\ 0 & 1 & 1 & -1 & 0 & 0 & 1 & 1 & 0 & 0 & 0 & 0 & 0 \\ 0 & 0 & 0 & 0 & 1 & -1 & 0 & 0 & 0 & 0 & 0 & 0 & 0 \\ 0 & 0 & 0 & 1 & 0 & 1 & -1 & 0 & 0 & 0 & 0 & 0 & 0 \\ 0 & 0 & 0 & 0 & 0 & 0 & 0 & 0 & 0 & 0 & 0 & 0 & 1 \\ 0 & 0 & 0 & 0 & 0 & 0 & 0 & 0 & 0 & 0 & 1 & 0 & 0 \\ 1 & -1 & 0 & 0 & 0 & 0 & 0 & 0 & 0 & 0 & -1 & 0 & 0 \\ 0 & 2 & -1 & -1 & 0 & 0 & 0 & 0 & 0 & 0 & 0 & 0 & 0 \\ 0 & 0 & 1 & 0 & -1 & 0 & 0 & 0 & 0 & 0 & -1 & 0 & -1 & 0 \\ 0 & 0 & 0 & 0 & 0 & 0 & 1 & -1 & 0 & 0 & 0 & 0 & -1 \\ 0 & 0 & 0 & 0 & 0 & 0 & 0 & 1 & -1 & 0 & 0 & 0 & 0 \\ 0 & 0 & 0 & 0 & 0 & 0 & 0 & 0 & 1 & -1 & 1 & 0 & 0 & 0 \\ 0 & 0 & 0 & 1 & -1 & 0 & 0 & 0 & 0 & 1 & 0 & 0 & 0 & 0 \\ 0 & 0 & 0 & 0 & 0 & 0 & 0 & 0 & 0 & 0 & -1 & 0 & 0 & 0 \end{pmatrix}^{-1} \begin{pmatrix} r_{glc} \\ r_c \\ r_{cit} \\ r_{ici} \\ r_{prot} \\ r_{car} \\ 0 \\ 0 \\ 0 \\ 0 \\ 0 \\ 0 \\ 0 \\ 0 \end{pmatrix} \quad (2)$$

or

Biochemical Reaction Networks

$$\begin{pmatrix} v_1 \\ v_2 \\ v_3 \\ v_4 \\ v_5 \\ v_6 \\ v_7 \\ v_8 \\ v_9 \\ v_{10} \\ v_{12} \\ v_{13} \\ v_{14} \\ v_{15} \end{pmatrix} = \begin{pmatrix} -1 & 0 & 0 & 0 & 0 & 0 \\ -1 & 0 & 0 & 0 & 0 & -1 \\ -2 & 0 & -0.5 & -0.5 & -0.5 & -2 \\ 0 & 0 & 0.5 & 0.5 & 0.5 & 0 \\ 1.5 & 0.5 & 1 & 1 & 0.5 & 1.5 \\ 1.5 & 0.5 & 0 & 1 & 0.5 & 1.5 \\ 1.5 & 0.5 & 0.5 & 0.5 & 1 & 1.5 \\ 1.5 & 0.5 & 0.5 & 0.5 & 0 & 1.5 \\ 1.5 & 0.5 & 0.5 & 0.5 & 0 & 1.5 \\ 1.5 & 0.5 & 0.5 & 0.5 & 0 & 1.5 \\ 0 & 0 & 0 & 0 & 0 & 0 \\ 0 & 0 & 0 & 0 & 0 & 1 \\ -3.5 & 0.5 & -1.5 & -1.5 & -1 & -3.5 \\ 0 & 0 & 0 & 0 & 1 & 0 \end{pmatrix} \begin{pmatrix} r_{glc} \\ r_c \\ r_{cit} \\ r_{ici} \\ r_{prot} \\ r_{car} \end{pmatrix} \quad (3)$$

Notice the complexity of the solution, *i.e.*, the intracellular fluxes are functions of almost all the measured rates. In such cases, manual solution of the algebraic equations is forbidding, whereas computer solution using Mathematica, Mable, or Matlab is trivial. Obviously $v_{12} = 0$ and $v_9 = v_{10}$ when the glyoxylate shunt is inactive. By calculating the unknown rates (or fluxes) at different dilution rates in a steady state chemostat, *i.e.*, for different sets of measured rates, Aiba and Matsuoka (1979) concluded that model 1 gives reasonable values for the fluxes. Furthermore, *in vitro* measurements of the activity of four different enzymes (pyruvate carboxylase, citrate synthase, isocitrate dehydrogenase, and isocitrate lyase) correlated fairly well with the calculated fluxes. When the two other models were tested, it was found that some of the fluxes were negative, *e.g.*, model 2 predicts that 2-oxoglutarate is converted to isocitrate. This is not impossible, but most of the reactions are favored thermodymically in the direction specified by the arrows in Fig. 5.5. Thus, ΔG^0 for the conversion of isocitrate to 2-oxoglutarate is -20.9 kJ (mol)$^{-1}$, and a large concentration ratio of 2-oxoglutarate to isocitrate would be required to allow this reaction to run in the opposite direction. Furthermore, there is better agreement between the measured enzyme activities and flux predictions by using model 1 than with the two other models. Aiba and Matsuoka (1979) therefore concluded that the glyoxylate shunt is inactive or operates at a very low rate in *C. lipolytica* under citric acid production conditions.

Example 5.8 Analysis of the metabolic network in *S. cerevisiae* during anaerobic growth
Nissen *et al.* (1997) constructed a network model that can be used to describe anaerobic growth of *S. cerevisiae*. The model was based on a comprehensive literature study to find which metabolic pathways are active in *S. cerevisiae*. The model contains 37 reactions and 27 intracellular metabolites. Thus, the degrees of freedom in the system are $F = 10$. In chemostat cultures operating a different dilution rates the rates of the following 13 compounds were measured: glucose, ammonia, glycerol, pyruvate, carbon dioxide, acetate, ethanol, succinate, carbohydrates, proteins, DNA, RNA and lipids. The production rate of ethanol and the uptake rate of ammonia were not used for the analysis, but these rates were used to validate the calculated fluxes, which were calculated using the other 11 measured rates together with eq. (5.27).

When the model was set up there was a number of questions concerning certain isoenzymes. Three isoenzymes of alcohol dehydrogenase (ADH) have been identified in *S. cerevisiae* (with sequencing of the genome many more putative alcohol dehydrogenases have been identified but their functions are currently unknown). The cytosolic ADH1 is constitutively expressed during anaerobic growth on glucose, and is responsible for the formation of ethanol. ADH2, which is also cytosolic, is mainly associated with growth on ethanol and is therefore not active during anaerobic growth on glucose. The function of the mitochondrial ADH3 is not known, but it has been postulated to be involved in a redox shuttle system between the mitochondria and the cytosol. Using enzyme activity assays Nissen *et al.* (1997) showed that ADH3 was active during anaerobic growth on glucose, and it was therefore included in the model.

Three isoenzymes of isocitrate dehydrogenase (IDH) have been isolated (IDH, IDP1 and IDP2). The NAD^+ dependent IDH is localized in the mitochondria and is important for the function of the TCA cycle. The function of the two $NADP^+$-dependent isoenzymes IDP1 and IDP2 localized in the mitochondria and cytosol, respectively, has not been clearly established. IDP1 is likely to be a major source for NADPH needed in the mitochondria for amino acid biosynthesis that takes place in this compartment. Some results have indicated that IDP2 is not active during growth on glucose, but it has not been clearly established.

To analyze the possible function of ADH3 and IDP2 these reactions were either included or left out of the model, and the fluxes were calculated. Table 5.3 summarizes some key fluxes calculated using the different models. During anaerobic growth on glucose the major fraction of glucose is shunted towards ethanol and most other fluxes in the network are therefore low. Some important variations are, however, clearly seen with the different models analyzed. In the reference model approximately 8% of the glucose taken up (on a C-mole basis) is shunted through the pentose phosphate pathway in order to supply the cell with ribose-5-phosphate (needed as precursor metabolite) and NADPH needed for biomass synthesis. This flux does not change if ADH3 is excluded from the model, but if IDP2 is included this flux becomes large and negative. This is due to the supply of NADPH via the IDP2 catalyzed reaction, which carries a large flux. With IPD2 it is also seen that the IDH flux becomes large and negative. Negative fluxes in the TCA cycle are not necessarily impossible, they simply imply that the flux is in the opposite direction to that specified in the model. Certain reactions in the network are, however, known to proceed only in one direction due to thermodynamic constraints, and the reaction catalyzed by glucose-6 phosphate dehydrogenase is one of these. It is therefore concluded that IDP2 cannot operate during anaerobic conditions, or its *in vivo* flux must be tightly controlled at a low level.

When ADH3 is excluded the changes in the fluxes are smaller, and the major difference is that the 2-oxoglutarate dehydrogenase catalyzed reaction carries a negative flux, i.e. succinate is converted back to 2-oxoglutarate. Even though this flux is low, it is quite well determined, and it is a result of a redox problem inside the mitochondria, i.e., there are no reactions that may oxidize NADH, and the TCA cycle reactions therefore start to operate such that they oxidize NADH. The reaction from 2-oxoglutarate to succinate is thermodynamically favored towards succinate formation, and a negative flux in this reaction is unlikely. It is concluded that ADH3 serves a very important function during anaerobic growth on glucose, namely ensuring oxidation of NADH in the mitochondria, and this is done by sending part of the acetaldehyde formed in the cytosol into the mitochondria where it is reduced to ethanol accompanied by NADH consumption. As both acetaldehyde and ethanol are easily transported across the mitochondrial membrane it is of no importance for the cell where this reaction is occurring.

Table 5.3 Fluxes through key reactions in the metabolic network during anaerobic growth of *S. cerevisiae* and using different models.[1]

		Model	
Reaction	Reference model	Including IDP2	Excluding ADH3
Glucose-6P dehydrogenase	7.5	-80	7.9
IDH	1.7	-117	1.3
IDP2	0	176	0
2-Oxoglutarate dehydrogenase	0	22	-0.3
ADH3[2]	1.0	-12	0

[1] All fluxes are normalized with respect to the glucose uptake rate that is set to 100, and the fluxes are given as C-mole (C-mole glucose)$^{-1}$. All fluxes are for a dilution rate of 0.3 h^{-1}.
[2] The flux for Adh3 is given for the direction from acetaldehyde to ethanol in which NADH is consumed.

The approach of combining a metabolic model and measurement of a few rates is attractive due to its simplicity. The approach does, however, have some pitfalls. Among these are:

- *Linearly dependent reaction stoichiometries.* This was discussed in Note 5.3, and it was indicated how the problem could be solved. However, in some cases there is no information available on regulation of the different pathways, and in fact it may be interesting to evaluate the relative activity of two different pathways that have the same overall stoichiometry. As discussed in Note 5.3 one may use enzyme activity measurements as additional constraints, but *in vitro* determined enzyme activities do rarely represent *in vivo* fluxes.
- *High sensitivity of certain fluxes to certain measured rates.* In some cases the matrix equation is ill-conditioned and this can result in very high sensitivities in the flux estimation to the measured rates. Whether the matrix equation is ill-conditioned can be checked by the *condition number* (see Note 5.4), which typically has to be less than 100. Alternatively one may check the sensitivity matrix for the model (see also Note 5.4).
- *The fluxes are determined by balances for specific co-factors.* As illustrated for the simple metabolic network models in Section 5.3 balances for the co-factors NADH, NADPH and ATP may be very useful for estimation of the fluxes, particularly as these co-factors link different parts of the metabolism together. Basically it is not a problem to include balances for co-factors in metabolic models, but a requirement for proper use of these balances is that *all* reactions involving these co-factors are included in the model. This may be problematic, since not all parts of the metabolism may be known, and one may inadvertently leave out important reactions that involve these co-factors. This may result in substantial errors in the flux estimation. Thus, particularly estimation of the flux through the pentose phosphate pathway is sensitive to whether *all* reactions involving NADPH are included in the model when the fluxes are estimated from measurable rates alone.

Introducing additional constraints on individual carbon atom transitions combined with measurements of the labeling patterns of the some metabolites when the cells are growing on specifically labeled substrates can circumvent most of the above-mentioned problems as discussed in Section 5.4.2.

Note 5.5 Sensitivity analysis of stoichiometric matrices

When metabolic fluxes are estimated using matrix inversion it is important to pay attention to the system matrix. If it is ill-conditioned even small errors in measured rates may propagate as large errors in the estimated fluxes. One way to check this is through evaluation of the condition number, which is given by:

$$\text{condition number} = \left\|\mathbf{T}_2^T\right\|\left\|\left(\mathbf{T}_2^T\right)^\#\right\| \quad (1)$$

Here $\|\ \|$ indicates any matrix norm and $\left(\mathbf{T}_2^T\right)^\#$ is the pseudo inverse of the stoichiometric matrix given by eq. (5.28) (if a determined system is analyzed the pseudo inverse becomes identical to the inverse of the matrix). If the condition number is high (above 100) the matrix is ill-conditioned. This is illustrated by a very simple equation system:

$$\mathbf{r} = \begin{pmatrix} 1 & 1 \\ 1 & 1.0001 \end{pmatrix}\begin{pmatrix} v_1 \\ v_2 \end{pmatrix} \quad (2)$$

The condition number of this matrix is 10^4, and the matrix is clearly ill-conditioned. Now let the rate vector be given by:

$$\mathbf{r} = \begin{pmatrix} 2 \\ 2 \end{pmatrix} \quad (3)$$

whereby the solution is $v_1=2$ and $v_2=0$. However, if the second element in the $\mathbf{r}$ vector is changed to 2.0001, a change of less then 0.1%, the solution is $v_1=1$ and $v_2=1$. Thus, a very small change in the measured rates results in a drastic change in the estimated flux vector, and in practice this means that even small measurement errors propagate as large errors in the estimated fluxes. If the stoichiometric matrix is ill-conditioned it is necessary to revise the model, e.g., to remove certain reactions or choosing a different set of reaction rates for the flux estimation.

A different approach to evaluate whether the fluxes are sensitive to small variations in the measured rates is through the sensitivity matrix, which is given as the inverse of $\mathbf{T}_2$:

$$\frac{\partial \mathbf{v}}{\partial \mathbf{r}_m} = -\mathbf{T}_2^{-1} \quad (4)$$

Here the partial derivate specifies the sensitivity of all the elements in the flux vector to the individual measured rates, i.e. the element in the jth row and the ith column specifies the sensitivity of the jth flux with respect to variations in the measurement of the ith rate. For the example we find the sensitivity matrix to be:

$$10^4 \begin{pmatrix} 1.0001 & -1 \\ -1 & 1 \end{pmatrix} \quad (5)$$

All the elements are very large and it is quite clear that the two fluxes are very sensitive to even small variations in the measured rates.

5.4.2 Use of Labeled Substrates

If cells are fed with a ^{13}C-labeled substrate, e.g., glucose that is enriched with ^{13}C in the first carbon position, then, when this substrate is catabolized, there will be enrichment in ^{13}C of some of the carbon atoms present in the different intermediates of the metabolism. The enrichment of the individual carbon atoms of the intermediary metabolites of a pathway will depend on the activity of the pathway. To illustrate this consider the catabolism of glucose to pyruvate by three different pathways: the Embden Meyerhof Parnas pathway (EMP pathway), the pentose phosphate pathway and the Entner Doudoroff pathway. If glucose is labeled specifically in the first position, there will be a different labeling of pyruvate depending on which pathway that operates (see Fig. 5.6).

If one measures the fractional enrichment of pyruvate (note 5.6) compared with that of glucose it is possible not only to identify which of the three pathways is operative but also to calculate the relative activity of the different pathways. Consider a simple case where we feed with 100%

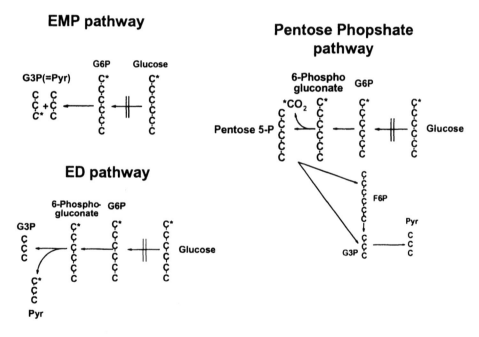

Figure 5.6 Overview of carbon transitions in three different catabolic pathways: EMP pathway, ED pathway and pentose phosphate pathway. The ^{13}C-labelled carbon in glucose is marked with *. In the EMP pathway there is labeling at the third carbon position of pyruvate. In the ED pathway there is enrichment at the first carbon position of pyruvate. In the pentose phosphate pathway there is no ^{13}C enrichment of the pyruvate since the first carbon atom of glucose escapes as carbon dioxide.

[1-^{13}C]-labeled glucose (glucose where the first carbon atom is fully labeled and the remaining carbon atoms are naturally labeled[1]), and we find that 30% of the third position carbons of pyruvate are ^{13}C enriched (specified as PYR(3)=0.3) and 10% of the first position carbons of pyruvate are ^{13}C enriched (specified as PYR(1)=0.1). Thus, a set of simple balances gives us:

$$v_{EMP}\ 0.5 + v_{ED}\ 0 + v_{PP}\ 0 = PYR(3) = 0.3 \qquad (5.29)$$

$$v_{EMP}\ 0 + v_{ED}\ 0.5 + v_{PP}\ 0 = PYR(1) = 0.1 \qquad (5.30)$$

Here the sum of the three pathway fluxes is normalized to one and $v_{EMP} + v_{ED} + v_{PP} = 1$. Clearly if the flux through the EMP pathway is 1 (corresponding to no activity of the other pathways) then PYR(3) would be 0.5. Similarly if the ED pathway is the sole pathway being active PYR(1) should be 0.5. Solving the above two equation together with the normalization equation directly gives: $v_{EMP} = 0.6$; $v_{ED} = 0.2$; $v_{PP} = 0.2$. This simple example clearly illustrates how labeling information can be used both to identify the pathway topology and to quantify the fluxes. The approach is illustrated further in Example 5.9.

Example 5.9 Identification of lysine biosynthesis
Lysine is extensively used as a feed additive, and it is produced by fermentation of *Corynebacterium glutamicum* (see also Example 3.9). It is a relatively low value added product, and it is of utmost importance to ensure a high overall yield of lysine on glucose (the typical carbon source). In bacteria lysine is derived from the precursor metabolites oxaloacetate and pyruvate, which in a series of reactions are converted into tetrahydrodipicolinate (H4D). H4D is converted to *meso*-α,ε-diaminopimelate (*meso*-DAP), and this conversion may proceed via two different routes (see Fig. 5.7). In the last step *meso*-DAP is decarboxylated to lysine. From measurement of the lysine flux v_{lys} (equal to the lysine production rate r_{lys}) it is not possible to discriminate between the activities of the two different pathways. However, the four-step pathway (the pathway at right hand side of Fig. 5.7) involves a symmetric intermediate, and the epimerase catalyzing the last step in the pathway may therefore lead to formation of *meso*-DAP with two different carbon compositions (see Fig. 5.7). Through analysis of the labeling pattern in the precursor metabolites pyruvate and oxaloacetate and the labeling pattern of lysine it is therefore possible to estimate the flux through the two different pathways. To illustrate this we set up a simple balance for the first carbon of lysine (LYS(1)):

$$LYS(1) = Y\ OAA(1) + (1-Y)/2\ OAA(1) + (1-Y)/2\ PYR(1) \qquad (1)$$

From measurements of the ^{13}C-enrichment LYS(1), OAA(1) and PYR(1) one can easily calculate Y. Notice that also other balances can be set up to estimate Y, e.g., for the sixth carbon of lysine (LYS(6)):

$$LYS(6) = Y\ PYR(2) + (1-Y)/2\ PYR(2) + (1-Y)/2\ OAA(2) \qquad (2)$$

---

[1] Notice that it is normally necessary to consider natural labeling, which is approximately 1.1% ^{13}C, but for simplicity we will neglect natural labeling here.

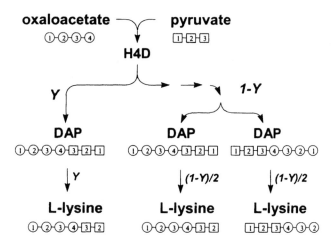

Figure 5.7 Overview of carbon transitions in two different pathways to lysine from oxaloacetate and pyruvate. The four-step succinylase pathway with the flux $(1-Y)v_{lys}$ proceeds via diaminopimelate, which is an intermediate in cell wall biosynthesis in certain bacteria. The four-step pathway contains the enzymatic sequence N-succinyl-2,6-ketopimelate synthase, N-succinylaminoketopimelate:glutamate amino transferase, N-succinyldiaminopimelate desuccinylase, and diaminopimelate epimerase. The one-step pathway with the flux Yv_{lys} involves the action of *meso*-DAP dehydrogenase.

Thus, through measurements of the ^{13}C-enrichment in several different carbon positions one has redundant information and it may be possible to obtain a robust estimate of the flux through the two different pathways. It is generally the case that measurement of the ^{13}C-enrichment combined with balances for the individual carbon atoms supplies redundant information, and that robust flux estimates are obtained.

Sonntag *et al.* (1993) used the approach of above for quantification of the fluxes through the two different pathways at different growth conditions. The ^{13}C-enrichment of secreted lysine was measured by NMR, and this information was used to quantify the flux through the two branches at different ammonia concentrations (Fig. 5.8). It was found that the relative flux through the direct pathway increased for increasing ammonia concentration and at the same time the lysine secretion was enhanced. This pointed to an important role of the direct pathway for over-production of lysine.

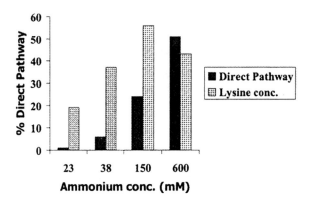

Figure 5.8 Quantification of the fluxes through the two different pathways leading to lysine shown in Fig. 5.7 for growth at different ammonia concentrations. Lysine concentration is in arbitrary units.

The approach illustrated above is very simple and straightforward. However, in practice it is difficult to identify the network topology, and the flux quantification is much more complicated. There are many more reactions involved in carbon transitions than those highlighted in Fig. 5.6, and hereby the ^{13}C-enrichment in the different carbon atoms becomes a function of many different pathways. Thus, even though none of the three catabolic pathways shown in Fig. 5.6 should result in ^{13}C-enrichment of the second carbon atom of pyruvate, there is typically some enrichment in this position. This is due to interaction with other pathways, e.g., by an active malic enzyme, but it is also due to scrambling of the labeling *via* the non-oxidative branch of the pentose phosphate pathway, where the transketolase and transaldolase catalyzed reactions are reversible and may result in many different types of carbon transitions (van Winden *et al.*, 2001). It is therefore necessary to obtain relatively detailed information about the ^{13}C-enrichment in the different carbon positions and for this purpose different analytical techniques can be applied (see Note 5.5). Furthermore, it is necessary to set up balances for each individual carbon atom in all the different reactions as will be illustrated in Example 5.10.

Note 5.6 Measurement of ^{13}C-enrichment
There are different methods for analysis of the ^{13}C-enrichment, but all methods are based on either nuclear magnetic resonance (NMR) spectroscopy or mass spectroscopy (MS). In order to shortly describe the methods it is, however, necessary to look into some definitions concerned with ^{13}C-enrichment. A molecule containing n carbon atoms has 2^n different labeling patterns, or in other words 2^n different isomers of isotopes – often referred to as *isotopomers* (see Fig. 5.9A).

Using ^{13}C-NMR it is possible to identify which of the carbon atoms that are enriched with ^{13}C, and if there are two ^{13}C positioned next to each other this will result in a split of the resonance peak into two in

Biochemical Reaction Networks

the spectrum. Combining ^{13}C-NMR with proton-NMR it is possible to resolve peaks for many of the carbon atoms, and hereby it is possible to obtain relatively detailed information about the isotopomers. However, in most cases only a fraction of the isotopomers can be identified. With proton-NMR it is possible to identify the specific resonance between protons and ^{13}C, and hereby it can be resolved whether a given carbon is ^{13}C or ^{12}C. It is, however, not possible to tell whether the ^{13}C is connected to a ^{13}C or a ^{12}C. Thus, only information about the so-called *fractional enrichment* can be obtained (see Fig. 5.9B). Information about the fractional enrichment can be used directly in balances for the individual carbon atoms as illustrated in Example 5.9. The fractional enrichment can, however, always be calculated if the distribution of isotopomers is known.

Using MS it is possible to measure the mass of the individual metabolites, since ^{13}C has a higher mass than ^{12}C. Thus, using MS it is possible to measure the so-called *mass isotopomers* (see Fig. 5.9C). It is somewhat more complicated to use this information directly for flux analysis, but as will be shown later, information about ^{13}C enrichments is rarely used directly for flux quantification. Notice that also the mass isotopomers can be calculated if the distribution of isotopomers is known.

Example 5.10 Analysis of a simple network
To illustrate a more formal approach to the use of labeled substrates for flux quantification we will consider the simple network shown below.

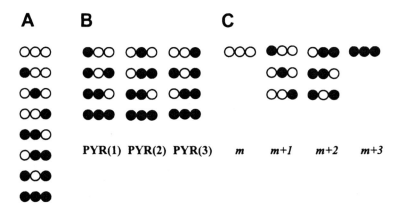

Fig. 5.9 Overview of different labeling patterns possible for a molecule with three carbon atoms, e.g., pyruvate.
A. All possible isotopomers.
B. Grouping of isotopomers to indicate the different isotopomers that are included when the fractional enrichment is measured.
C. Grouping of isotopomers into different mass isotopomers. The mass of the non-labeled metabolite is *m*.

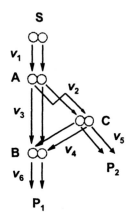

For this network we have the metabolite balance equation:

$$\begin{pmatrix} r_s \\ r_{p_1} \\ r_{p_2} \\ 0 \\ 0 \\ 0 \end{pmatrix} = \begin{pmatrix} -1 & 0 & 0 & 0 & 0 & 0 \\ 0 & 0 & 0 & 0 & 0 & 1 \\ 0 & 0 & 0 & 0 & 1 & 0 \\ 1 & -1 & -1 & 0 & 0 & 0 \\ 0 & 0 & 1 & 1 & 0 & -1 \\ 0 & 1 & 0 & -1 & -1 & 0 \end{pmatrix} \begin{pmatrix} v_1 \\ v_2 \\ v_3 \\ v_4 \\ v_5 \\ v_6 \end{pmatrix} \qquad (1)$$

Clearly three fluxes can be measured directly and with three degrees of freedom we can in principle estimate the fluxes using matrix inversion. However, the matrix does not have full rank (addition of the second and fourth column gives the third column), and we therefore need additional information. We now set v_1 to 100 and then express the other fluxes as function of two of the fluxes (v_3 and v_4), i.e.

$$\begin{pmatrix} v_1 \\ v_2 \\ v_3 \\ v_4 \\ v_5 \\ v_6 \end{pmatrix} = \begin{pmatrix} 100 \\ 100 - v_3 \\ v_3 \\ v_4 \\ 100 - v_3 - v_4 \\ v_3 + v_4 \end{pmatrix} \qquad (2)$$

The two fluxes v_3 and v_4 will be termed free fluxes, and if we can obtain information about these from balances around the individual carbon atoms then we can calculate all the fluxes in the network using eq. (2). With the carbon transitions specified in the network we can set up balances for the individual carbon atoms. Thus, for the first reaction we have:

$$-100S_1 = -(v_2 + v_3)A_1 \qquad (3)$$

$$-100S_2 = -(v_2 + v_3)A_2 \qquad (4)$$

where S_1 indicates the enrichment of the first carbon atom in the substrate and S_2 indicates the enrichment of the second carbon atom in the substrate (similarly for A_1 and A_2). The balances simply state that enriched carbon entering the first position in the metabolite pool A balance the enriched carbon in the first position leaving this metabolite pool. Similar to (3)-(4) balances can be written for all metabolites, and in matrix formulation we get:

$$\begin{pmatrix} -(v_2+v_3) & 0 & 0 & 0 & 0 & 0 \\ 0 & -(v_2+v_3) & 0 & 0 & 0 & 0 \\ v_3 & 0 & -v_6 & 0 & v_4 & 0 \\ 0 & v_3 & 0 & -v_6 & 0 & v_4 \\ 0 & v_2 & 0 & 0 & -(v_4+v_5) & 0 \\ v_2 & 0 & 0 & 0 & 0 & -(v_4+v_5) \end{pmatrix} \begin{pmatrix} A_1 \\ A_2 \\ B_1 \\ B_2 \\ C_1 \\ C_2 \end{pmatrix} = \begin{pmatrix} -100S_1 \\ -100S_2 \\ 0 \\ 0 \\ 0 \\ 0 \end{pmatrix} \qquad (5)$$

Equation (5) can be used to calculate the fluxes, but they cannot be obtained directly from the matrix equation which is non-linear due to the occurrence of products of fluxes and labeling. However, the matrix has full rank, and if the fluxes are given the ^{13}C enrichment can be calculated by a simple matrix inversion. This is the basis for the iterative method of Figure 5.10.

As discussed in Example 5.10 estimation of the fluxes is not straightforward and it requires a robust estimation routine. One approach to estimate the fluxes is illustrated in Fig. 5.10, but other approaches have been described in the literature (see Wiechert (2001) for a recent review). Instead of using the fractional labeling of the metabolites one may also directly calculate the NMR spectra that would arise for a given isotopomer distribution and then compare the calculated NMR spectra with the experimentally determined spectra. Schmidt et al. (1999) used this approach for estimation of the fluxes in *E. coli* based on experimental NMR data. A requirement for any estimation procedure is a mathematical model describing the carbon transitions and including all relevant biochemical reactions in the network. The carbon transitions for most biochemical reactions are described in biochemical textbooks, and procedures to implement these carbon transitions in a formalized way have been developed (Zupke and Stephanopoulos (1995); Schmidt et al. (1997); Wiechert et al. (1997)). Although there are good procedures for estimation of the fluxes, it is still important to specify the correct model structure, and in many cases it is necessary to evaluate different model structures in order to identify a proper network that fits the experimental data. In these cases the network identification and flux quantification goes hand in hand.

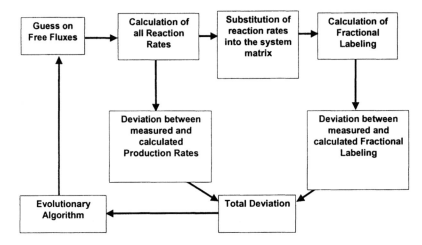

Figure 5.10 A procedure for estimating metabolic fluxes using ^{13}C-labeled data. The procedure starts with a guess of the free fluxes, and then all the fluxes are calculated using matrix inversion (in analogy with eq. (2) in Example 5.10). The calculated fluxes can be compared directly with measured rates and they can be used to calculate the fractional labeling of all the intracellular metabolites. The calculated fractional labeling can then be compared with experimental data on the fractional labeling, and a total deviation between calculated fluxes and fractional labeling of metabolites can be found. Through the use of an optimization algorithm it is possible to propose a new set of fluxes, and hereby iterate until the total deviation is minimized, resulting in a good estimate of the fluxes.

When a proper model has been identified it is possible to estimate the fluxes through the different branches of the central carbon metabolism. In the literature one may find several examples of the use of ^{13}C-labeled data to estimate the metabolic fluxes (see e.g. Pedersen *et al.* (2000); Christensen and Nielsen (2000); Dauner *et al.* (2001)). Thus Fig. 5.11 shows a flux map for the central carbon metabolism in *S. cerevisiae* during aerobic growth. Metabolic flux maps of the type shown in Fig. 5.11 contain useful information about the *contribution* of various pathways to the overall metabolic processes of substrate utilization and product formation. However, the real value of such metabolic flux maps lies in the *flux differences* that are observed when flux maps obtained with different strains or under different conditions are compared with one another. In Fig. 5.11 the fluxes are compared for growth at glucose repressed conditions (batch culture) and at glucose de-repressed conditions (chemostat culture at low dilution rate). Through such comparisons the impact of genetic and environmental perturbations can be fully assessed, and the importance of specific pathways, or reactions within a given pathway can be accurately described. Hereby new insight into the function of the different pathways may be obtained and this can be used to guide metabolic engineering (Vallino and Stephanopoulos, 1991).

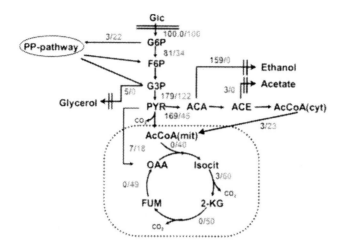

Fig. 5.11 Metabolic fluxes in *S. cerevisiae* estimated using ^{13}C-labeled glucose. For each step the flux is shown as a/b. The yeast was grown at high and low specific glucose uptake rates, respectively, and a is the flux at high specific glucose uptake rate while b is the flux at low specific glucose uptake rate. At high specific glucose uptake rates there is ethanol formation due to the Crabtree effect, and respiration is repressed resulting in almost no flux through the TCA cycle. Due to the Crabtree effect the yield of biomass on glucose is low and there is consequently a low requirement for NADPH and precursors of the pentose phosphate pathway. Thus, the flux through this pathway is low. [The data are taken from Gombert *et al.* (2001)].

5.4.2 Use of Linear Programming

In a metabolic model the elements of the flux vector can in principle take any value. Balancing of all the intracellular metabolites does, however, impose a set of constraints on the flux vector **v**. Even with the constraints imposed by the metabolite balancing there are in practice several degrees of freedom in the system, and infinitely many solutions for the flux vector **v** result. The total solution space for the flux vector **v** that satisfies the constraints imposed by the metabolite balances is given by the *null space* of the stoichiometric matrix **T**. Thus, the null space defines the possible solution space for the flux vector. Furthermore, the shape of the null space is given by the stoichiometry of all the reactions used in the metabolic network model, and the network model represented by the stoichiometric matrix **T** defines all possible modes of operation of the network. Above we discussed how experimentally determined rates or measurements of ^{13}C-enrichments can be used to further constrain the flux vector. This corresponds to the situation where some of the fluxes are measured (or additional constraints are added and this results in a reduction in the dimension of the null space), and hereby one may find a unique solution, that is given as the intercept of the flux vector with one of the edges of the null space.

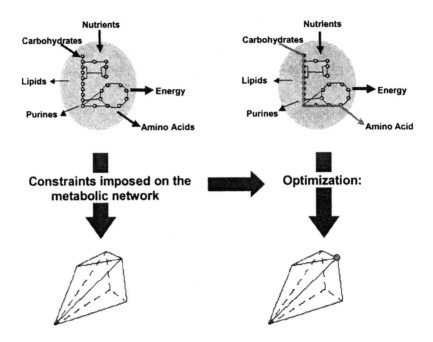

Figure 5.12 Illustration of the concept behind the use of linear programming to constrain the operation of metabolic networks. A given network model allows an infinite number of solutions represented by the null space at the bottom. When an optimization criterion is applied, it is possible to identify a flux vector that specifies how the network should operate to fulfill the optimization criterion. [Adapted from I. Famili, J. Förster, J. Nielsen and B. Ø. Palsson (2002) Systems properties of a reconstructed genome-scale metabolic network for *Saccharomyces cerevisiae*.].

Another possibility to constrain the flux vector is to impose an objective function on the system and apply linear programming to find the optimal solution within the underdetermined system (see Fig. 5.12). The solution to the optimization problem will give a flux vector that represents the set of fluxes that leads to the optimum. A typical optimality criterion is to optimize the specific growth rate, but clearly the concept can also be used to identify the operation of the network that gives the maximum yield of a given metabolite (see discussion later). The optimality criterion may not always give a unique solution for the flux vector, as there may be several modes of operation of the network that result in the same overall optimal operation. Thus, one may not necessarily obtain robust estimates of the fluxes in the network using linear programming, but clearly the advantage is that one may use metabolic models for prediction of how a metabolic network may operate whereas the approaches discussed earlier only allows analysis based on a given set of experimental data. In Note 5.6 we will discuss the concepts of linear programming and in Example 5.11 it is illustrated how a simple model for *E. coli* can be analyzed using linear programming.

Note 5.7 The concept of linear programming
To illustrate the concept of linear programming consider a very simple system with only two variables, x and y related through a constraint (which can be interpreted as a metabolic model):

$$2x + y = 4 \qquad (1)$$

The constraint specifies that the solution lies on the straight line AB (see Fig. 5.13), which in this case represents the null space of the system. In addition to eq. (1), another representation common in linear programming problems is that all variables (in this case x and y) are required to be nonnegative, i.e.

$$x \geq 0 \quad \text{and} \quad y \geq 0 \qquad (2)$$

The elements of the flux vector **v** may be either positive or negative, and in order to apply linear programming it is therefore necessary to consider reactions in both directions. Thus, for all reactions that may operate as reversible reactions in the network both a forward and a backward reaction has to be included in the model. Hereby it is possible to apply the constraint that all fluxes (or all elements of the flux vector **v**) must be positive. With the constraints $x \geq 0$ and $y \geq 0$ the solutions of the preceding problem are restricted to those lying in the positive quadrant of (x,y) space.

The next step is to identify a unique solution by applying the requirement that *the solution maximizes or minimizes a certain cost (or objective) function*. For example, if we want to maximize the objective function $2x + 3y$, then the problem is to find a solution that lies on the line AB and at the same time maximizes this function. A family of cost lines is shown in Fig. 5.13. The line that yields the largest value for the objective function and at the same time satisfies the constraint is given by $2x + 3y = 12$. It intersects with the line AB at $x = 0$ and $y = 4$.

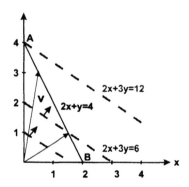

Figure 5.13 The concept of linear programming. The solution to the problem is constrained to be on the line AB in the first quadrant, and some solutions indicated by the vector **v** are indicated. The objective function consists of a family of lines (of which three are shown as broken lines). The line that maximizes the objective function intercepts the constraint line AB at $(x,y) = (0,4)$, which represents the solution for this optimization problem.

In this simple example one feasible solution is obtained, but one may encounter situations where there is 1) no feasible solution, 2) infinitely many solutions or 3) solutions where one or more elements of the flux vector is infinitely large. Case 1) is found if the objective function is $x = {}^-6$, case 2) when the objective function is $2x+y$, where infinitely many pairs (x,y) give the value 4 for the objective function.
In practice the dimensions are very large, and the network may operate in many different ways and still fulfill the optimum criterion. If e.g. the two pathways to lysine shown in Fig. 5.7 are included, the model cannot discriminate between them, and for overall operation of the cell it does not matter which pathway is chosen. The flux through each of the two pathways may therefore take any arbitrary value, whereas the sum of the two fluxes may indeed be very important for the overall operation of the network. This illustrates the situation where a solution is found, but the resulting flux vector $\mathbf{v}$ is not unique.

Consequently the use of linear programming to metabolic networks can be specified as:

$$\mathbf{r} = \mathbf{T}^T \mathbf{v} \quad ; \quad Z = \mathbf{cv} \quad ; \quad \max(Z) \tag{3}$$

where the stoichiometric matrix is formulated such that all elements in the flux vector are non-negative, i.e., for reversible reactions the stoichiometry is given both for the forward and the reverse reaction. Z is the objective function. It is a linear combination of the elements in the flux vector, possibly with weight factors for some of the fluxes, i.e. some of the fluxes may have a larger influence on the objective function than others (many elements in the weight factor vector $\mathbf{c}$ may be zero). The objective function may be the specific growth rate of the cells, in which case the operation of the network that gives the maximum specific growth rate is identified.

Finding the optimal solution to the problem of eq. (3) may be complicated due to the potentially large number of feasible solutions that satisfy all the constraints. However, the solution is always located at the end (or corner) of the set of feasible solutions (a result of the linearity of both the constraints and the objective functions), just like the situation for the simple example of Fig. 5.13. It is therefore (in principle) possible to find the solution by, first, enumerating all solution corners and, second, evaluating the objective function at each point of the space of feasible solutions. This approach is, however, practically infeasible due to the very large number of corners of the cone, even for relatively small network models. Instead the so-called *simplex* method is often used. Applied to find a minimum the method works as follows:: First locate a corner of the feasible set and then proceed from corner to corner along the edges of the feasible solutions. At a given corner there are several edges to choose from, some leading away from the optimal solution and others leading gradually toward it. In the simplex method one moves along the edge that is guaranteed to decrease the objective function. That edge leads to a new corner with a lower value, with no possibility of returning to any corner that yields a higher value for the objective function. Eventually, a specific corner is reached, from which all edges lead to higher or identical objective function values, and this corner represents the optimal solution.

Example 5.11 The use of linear programming for analysis of *E. coli* metabolism
Optimal flux distributions have been obtained by linear programming for several metabolic models. In this example, we will consider the analysis of *E. coli* metabolism as described by Varma *et al.* (1993a,b). Based on a detailed stoichiometric model for *E. coli* metabolism, the authors analyzed glucose

catabolism and the biochemical production capabilities of this organism. The model consisted of 107 metabolites (including substrates and metabolic products) and 95 reversible reactions, and because each reaction has to be included in both a forward and backward direction this gives a stoichiometric matrix of 107 x 190, *i.e.*, there are 83 degrees of freedom. None of the fluxes were measured, but one flux (the specific glucose uptake rate) was used to scale the fluxes. The only experimental data used for the flux calculations were the following:

- Metabolic demands for growth were used to calculate the drain of precursor metabolites for biomass synthesis. In practice, the drain of precursor metabolites is included as stoichiometric coefficients in a reaction leading to biomass synthesis.
- ATP requirements for maintenance. By fitting the specific glucose uptake rate at different specific growth rates to experimental data, a requirement of 23 mmoles ATP (g DW)$^{-1}$ for growth-associated maintenance and 5.87 mmoles ATP (g DW h)$^{-1}$ for non-growth-associated maintenance was estimated, and these values are used in the calculations.
- The specific oxygen uptake rate was given an upper bound of 20 mmoles (g DW h)$^{-1}$.

The objective function was the specific growth rate, which in all cases was maximized. Thus, if a specific glucose uptake rate is given, the model determines the corresponding maximum specific growth rate along with all the fluxes in the network.

To evaluate the model the production rates of metabolites formed by *E. coli* were plotted as functions of the specific growth rate, and the results are shown in Fig. 5.14. It is observed that at low specific glucose uptake rates (corresponding to low specific growth rates), no byproducts are formed, *i.e.*, growth is by pure respiration. As a result, in this regime the specific oxygen uptake rate increases linearly with the specific growth rate. When the specific glucose uptake rate approaches 8 mmol (g DW h)$^{-1}$ (corresponding to a specific growth rate of about 0.9 h^{-1}) the oxygen requirement for complete oxidation of glucose to carbon dioxide exceeds the specified maximum of 20 mmol (g DW h)$^{-1}$. The cells therefore shift to a mixed metabolism with both respiration and fermentation. The first fermentative metabolite that is excreted is acetate, and at higher glycolytic flux (or higher specific growth rate) formate is also excreted. Finally, at very high glycolytic flux ethanol is excreted. It is interesting that the model predicts the right sequence of excretion of metabolites, *i.e.*, first acetate, then formate, and finally ethanol, which is consistent with experimental findings. This could indicate that when respiration operates at its maximum rate the cells are constrained by the supply of ATP (notice that more ATP is gained by the formation of acetate than by formation of formate and ethanol, see Fig. 2.6). When the glycolytic flux increases further oxidation of NADH starts to become a problem for the cell, and it will therefore first try to reduce the formation of NADH in connection with acetyl-CoA synthesis, namely by by-passing the pyruvate dehydrogenase complex (see Fig. 2.6) through the use of pyruvate-formate lyase, and when this is not sufficient start to convert acetyl-CoA to ethanol which is accompanied by a net consumption of NADH.

An advantage of linear programming is that additional information can be obtained by calculating the sensitivity λ of the objective function Z with respect to the system variables. These sensitivities are sometimes called *shadow prices*. For the *E. coli* metabolic model discussed above (maximization of μ)

$$\lambda_i = \frac{\partial \mu}{\partial r_i} \qquad (1)$$

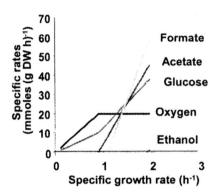

Fig. 5.14 The specific glucose and oxygen uptake rates and the specific rates of formation of acetate, formate, and ethanol as functions of the specific growth rate calculated by maximizing the specific growth rate of the *E. coli* metabolic model using linear programming.

where r_i is the net specific rate of consumption of the *i*th compound, *e.g.*, oxygen, acetate, or an amino acid. Thus, the shadow price reflects the effects that changes in individual metabolic rates have on the specific growth rate. Table 5.4 gives some of the shadow prices calculated by the metabolic model for different specific oxygen uptake rates (r_o). At low r_o the shadow price for oxygen is positive, *i.e.*, by increasing r_o the specific growth rate can be increased. When r_o is at its maximum, the specific growth rate cannot be increased further by increasing r_o and the shadow price is therefore zero. At anaerobic conditions ($r_o = 0$), the shadow price for NADH is negative due to the inability of the cell to oxidize this cofactor, *i.e.*, if NADH production increases the specific growth rate will decrease. This indicates that cofactor regeneration can limit growth at anaerobic conditions. For ATP it is observed that the shadow price is high at anaerobic conditions and decreases with increasing r_o. Thus, supply of ATP for growth becomes less limiting for cells growing aerobically compared with anaerobic growth. The positive shadow price for ATP supports the above statement, that the cells are ATP limited and when respiration reaches its upper limit the cells will start to seek other routes for ATP formation – namely by formation of acetate. The shadow price for acetate is zero when the specific oxygen uptake rate is below 20 mmoles (g DW h)$^{-1}$. The likely reason is that the ATP yield in connection with acetate formation is much less than the ATP yield in respiration, and the cells will therefore prefer to increase respiration. It is interesting to notice that the shadow price for ethanol is positive even when the specific oxygen uptake rate is below 20 mmoles (g DW h)$^{-1}$. This could be a consequence of oxidation of NADH being a problem for the cell since formation of ethanol involves a route from acetyl-CoA where NADH is oxidized. However, the energy metabolism is likely to dominate at these conditions, and the cell will therefore prefer to oxidize NADH in the oxidative respiration, without activating the route to ethanol.

Table 5.4 Shadow prices at different specific oxygen uptake rates [in mmol (g DW h)$^{-1}$] calculated using the metabolic model of Varma et al. (1993a)[a]

Metabolite	Shadow price [g DW (mmol metabolite)$^{-1}$]		
	0	12	20
Oxygen	0.0399	0.0283	0
ATP	0.0109	0.0106	0.0049
NADH	-0.0054	0	0.0065
Acetate	0	0	0.0242
Ethanol	0	0.0106	0.0422
Lactate	0.0054	0.0106	0.0422

[a] The specific glucose uptake rate is 10 mmol (g DW h)$^{-1}$.

The model was also applied to investigate the potential of *E. coli* to produce specific compounds, *e.g.*, amino acids (Varma et al., 1993b). This potential can be determined from the magnitude of the shadow prices, since these provide a measure of the trade off in biomass growth for the production of a specific compound. It was found that the shadow prices are low for amino acids like glycine, alanine, and aspartate, which have simple biosynthetic routes, whereas the shadow prices are high for the aromatic amino acids phenylalanine, tryptophan, and tyrosine, which have complex biosynthetic routes. Thus, if the product is the end-result of a long and complex biosynthetic route, the requirement of cell metabolism is large and the effect on biomass growth is therefore severe. On the contrary, the cells can at a small cost produce a compound with a simple metabolic route, and the effect on biomass growth is therefore small.

Application of linear programming has been used for flux analysis of many cellular systems, and recently it has been used to analyze genome-scale metabolic models, i.e., metabolic models that are reconstructed from genomic information, possibly combined with information from biochemical reaction databases (Covert et al., 2001). Genome-scale models contain a large number of reactions. Using these models the role of specific genes in the overall metabolism can be evaluated, and operation of the network during growth on different media can be analyzed in great detail (see e.g. Edwards and Palsson (2000) and Edwards et al. (2001)).

The concept of linear programming is well suited to evaluate the potential of a given metabolic network for producing a specific compound, and the maximum possible yield can easily be calculated. Calculation of maximum theoretical yields provides a benchmark for real processes. Furthermore, the influence of the network structure on the overall yield can be analyzed by inserting or deleting reactions from the network. Burgaard and Maranas (2001) provided a formalized approach based on *mixed linear integer programming* (MILP) to this type of analysis.

As discussed in Note 5.6 a problem with linear programming is that the flux vector **v** may not be uniquely defined by the optimization process, and consequently there may be many different pathway routes that lead to the maximum overall yield of the product of interest, and in practice only a few of these may be of interest. It is problematic to use linear programming to identify different solutions for the flux vector **v** that gives the optimum yield, and alternative approaches are therefore interesting.

One alternative approach to analyze the capabilities of a metabolic network is *elementary flux modes* (Schuster *et al.*, 2000), which is shortly described in Note 5.7. The elementary flux modes completely characterize the metabolic network and any operation of the network as a linear combination of the elementary flux modes. Even though this approach offers the possibility to obtain all possible modes of operation of the network, and hereby also several different modes that gives a high yield of the product of interest, its practical limitations lies in the fact that even for relatively small networks there is a huge number of elementary flux modes, and when more reactions are added to the network the number of elementary flux modes increases drastically. Schilling *et al.* (2000) handled the problem using so-called *extreme pathway analysis*, where the extreme pathways are represented by the edges of the cone in Fig. 5.12 and any mode of operation of the network can be expressed as a linear combination of these extreme pathways. With the introduction of large (and almost complete) metabolic networks it is of course of significant interest to evaluate the metabolic capabilities, not only for biotech applications but also in order to gain further insight into the systemic properties of large networks. Development of new algorithms for analysis of metabolic networks is therefore a key theme in Systems Biology, and it is currently a hot topic of research.

Note 5.8 Elementary flux modes
The elementary flux modes of a metabolic network is a unique representation of the system, and it basically represents a new definition to the concept of metabolic pathways substantially different from the traditionally classified pathways, such as glycolysis, pentose phosphate pathway, TCA cycle etc. Schuster and co-workers (Schuster *et al.*, 1999; Schuster *et al.*, 2000) developed the concept of elementary flux modes, which allows identification of a unique set of reaction paths that span all possible modes of operation of the network. Thus, any operation of the network can be represented as a linear combination of the elementary flux modes. To illustrate, consider the very simple pathway structure below.

$$S \xrightarrow{v_1} I \xrightarrow{v_2} A$$
$$\downarrow v_3$$
$$B$$

Pathway fluxes
$v_1=2$ and $v_2=v_3=1$

$$S \xrightarrow{v_1} I \xrightarrow{v_2} A \quad + \quad S \xrightarrow{v_1} I$$
$$\downarrow v_3$$
$$B$$

Pathway 1   Pathway 2
fluxes: $v_1=1, v_2=1$ fluxes: $v_1=1, v_3=1$

This structure may be represented by two elementary flux modes, pathway 1 and pathway 2, and the mode of operation of the combined pathway (indicated by the fluxes) may be represented as the sum of the two elementary modes.

The concept of elementary flux modes is similar to identification of the extreme pathways (Schilling *et al.*, 2000), and clearly any flux vector (corresponding to a given operation of the network) can also be specified as a linear combination of the extreme pathways. However, the elementary flux modes give a more detailed representation of the network structure than the extreme pathways. Schuster and co-workers recently released the software "Separator", which enables a systematic division of metabolic networks into subsystems, and investigated the metabolic network of *Mycoplasma pneumoniae* (Schuster *et al.*, 2002).

Elementary flux mode analysis has successfully been applied for the improvement of product yield in aromatic amino acid production (Liao *et al.*, 1996). Furthermore, it may be used to analyze the function of specific reactions in metabolic networks (Förster *et al.*, 2002). However, a major problem with elementary flux modes is that the number of modes increases drastically with the number of reactions in the model. Thus, for a relatively simple metabolic network of *S. cerevisiae* consisting of 45 reactions (16 reversible reactions and 29 irreversible reactions), 42 internal metabolites and 7 external metabolites, the number of elementary flux modes was found to be 192 (Förster *et al.*, 2002). If one additional isoenzyme is included in the model the number of elementary flux modes increases to 307, and if one further reaction is included the number of elementary flux modes increases to 1117. Clearly for genome-based metabolic models there will be a very large number of elementary flux modes, and this limits the practical use of this approach for analysis of such large networks.

PROBLEMS

Problem 5.1 Determination of stoichiometric coefficients in the biomass pathway reaction for *S. cerevisiae*.

Assume that the metabolic pathway model corresponding to Example 3.8 can be approximated by the following 3 pathway reactions:

v_1 : $-(1 + \alpha) CH_2O + \alpha\, CO_2 + \beta\, NADH + CH_{1.61}O_{0.52}N_{0.15} - \gamma\, ATP = 0$

v_2 : $-1.5\, CH_2O + 0.5\, CO_2 + CH_3O_{0.5} + 0.5\, ATP = 0$

v_3 : $- CH_2O + CH_{8/3}O - \frac{1}{3} ATP - \frac{1}{3} NADH = 0$

Determine the value of α, β and γ in the first reaction as functions of Y_{xg} and Y_{sx}.

(Answer: $\alpha = (1/6)\, Y_{xg} + 0.03$ and $\gamma = (1/3)\, Y_{xs} - (13/18)\, Y_{xg} - 0.343$)

Consider the set of experimentally obtained yield coefficients in eq (3.23). Determine numerical values of α and γ for this data set.

The yield coefficients of eq. (3.23) were obtained for a slightly different composition of the biomass. Determine α and γ with the biomass composition used in eq. (3.23) and discuss the sensitivity of the result with respect to changes in κ_x. Also relate the present results with the discussion in Example 5.5 in which exactly the same data were used in the end. In Example 5.5 one reached the conclusion that γ is more likely to be 1.8 than the value found here. What is the reason for this different conclusion?

Problem 5.2 Analysis of a PFL deletion mutant of *Lactococcus lactis*.

We shall study a *Lactococcus lactis* deletion mutant in which the gene encoding pyruvate formate lyase (PFL) has been knocked out. The metabolic capability of the strain is exactly as shown in Fig. 2.4 and 2.6B except that HCOOH cannot be produced. The biomass composition is $X = CH_2O_{0.5}N_{0.25}$ and biomass is produced according to

v_2 : $-1.1\ CH_2O + 0.1\ CO_2 + X + 0.075\ NADH - 2.48\ ATP = 0$

The other network reactions are:

v_1:   glucose $\rightarrow$ pyruvate
v_3:   pyruvate$\rightarrow$AcCo A (by PDH)
v_4:   pyruvate$\rightarrow$lactate
v_5:   AcCo A$\rightarrow$ethanol (e)
v_6:   AcCo A$\rightarrow$acetate (HAc).

    a. Assume that the net rates of production of pyruvate, AcCoA, NADH and ATP are zero. Calculate the six fluxes as well as the remaining non-zero production rates in terms of the measured rates $r_x = \mu$ and $r_e = r_{ethanol}$. List the three pairs of two rate measurements for which the system is non-observable. Give a physiological reason for the negative value of $r_{HAc} = v_6$.

    b. Repeat the calculations of question a., but assume that $r_{pyruvate} = r_{PYR} \neq 0$.

    c. Why does this not help to make r_{HAc} positive?

    d. Finally consider a *pdh(-)* strain. Here the functioning of PFL is unimpaired while PDH does not function. Repeat the calculation of question a. for this strain. Explain why r_{HAc} is positive.

Comment : A *pfl(-)* strain will not grow anaerobically unless it is fed with a little HAc besides glucose (which is of course the main carbon and energy substrate.). This has been proved experimentally by Henriksen and Nilsson (2001) who also found that the *pfl(-)* strain fed properly will have an almost two-fold increase of r_{Hlac} compared to the wild type strain. Why is that so?

Problem 5.3 The maximum theoretical yield of lysine on glucose

In Problem 4.2 it was shown that a growth yield of 6/7 for L-lysine on glucose (Example 3.9) could not be excluded for thermodynamic reasons. In the present problem it will be shown that the explanation for the lower theoretical yield (3/4) – see references to Problem 4.2 – is to be found in the need for NADPH and not NADH in the synthesis of lysine from glucose. This will help the reader to appreciate that the crude definition of redox equivalents as "H_2" cannot always be used.

a. Go through the biosynthesis pathway in *Corynebacterium glutamicum* of L-lysine from 1 mole of pyruvate and one mole of phosphoenolpyruvate. For each step write down the chemical formula for the intermediate and the amount of redox units harbored in the compound. Show that the net consumption rate of CO_2 and of 2-oxoglutarate is zero, and that 4 NADPH is consumed while 2 NADH is produced for each mole of L-lysine produced from glucose. Prove that this leads to a theoretical yield of ¾ C-mole lysine (C-mole glucose)$^{-1}$, and that 1/6 O_2 must be used for each C-mole lysine produced.

b. Can any of the NADPH requiring enzymes in the pathway from glucose to lysine possibly be exchanged with an enzyme that uses NADH as co-factor? Or could an NADH producing enzyme be exchanged with an NADPH producing isoenzyme? What would be the possible thermodynamic consequences of exchanging the glutamate dehydrogenase encoded by GDH1 (using NADPH as cofactor) with the GDH2 encoded isoenzyme (using NADH to produce glutamic acid from 2- oxoglutarate)?

To solve this problem use the following references

1 "Biochemical Pathways" ed. G. Michal, Spektrum Verlag, Berlin, 1999
2 "ExPASy-Molecular Biology Server, Subsection Biological Pathways"
    This reference is found at: **http:// www.expasy.ch**

Problem 5.4 Production of solvents (acetone and butanol) by fermentation.

Reardon *et al.* (1987) and Papoutsakis (1984) have both studied anaerobic fermentations of butyric acid bacteria using metabolic flux analysis. We shall base our analysis of the primary metabolism of *Clostridium acetobutylicum* on the simplified pathway scheme in Fig. 3 of Reardon *et al.* (1987), but the corresponding pathway scheme, Fig. 1 of Papoutsakis (1984), differs only in some minor details and could also have been used. The figure below shows the branchpoints and reactions.

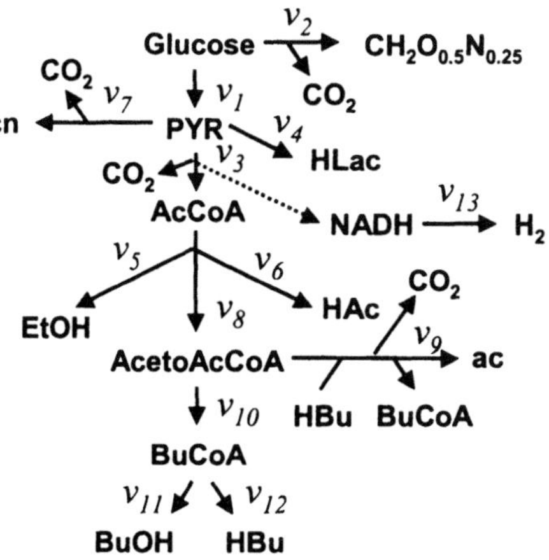

Production of biomass by reaction 2 follows the stoichiometry of (1) in Problem 5.2. Note that everything until AcCoA is the same as in Problem 5.2 except that acetoin (CH_3CHOH-CO-CH_3) is formed in reaction 7 by condensation of two pyruvate molecules followed by decarboxylation (no NADH or ATP involved). Entry into the solvent/acid pathways starts with Acetoacetyl-CoA (CH_3COCH_2CO-S-CoA) formed by condensation of Acetyl-CoA (CH_3CO-S-CoA) in reaction 8. Acetoacetyl-CoA may either decarboxylate to acetone (CH_3COCH_3) in reaction 9 or be reduced (reaction 10) to buturyl-CoA (BuCoA) ($CH_3CH_2CH_2CO$-S-CoA). Reaction 9 requires ATP since 1 C-mole of BuCoA is formed from butyrate for every C-mole of Acetoacetyl-CoA converted. BuCoA can be either reduced to butanol in reaction 11 or hydrolyzed to butyrate in reaction 12 (similar to reaction 6, and again one ATP is produced per molecule of CoA, i.e., one-fourth ATP per C-mole of BuCoA). In the decarboxylation of pyruvate to Acetyl-CoA (reaction 3) in *C. acetobutylicum*, pyrovate-ferredoxin oxido reductase acts as electron receptor ($Fd_{ox} \rightarrow Fd_{red}$).

There are two ways of reoxidation of Fd_{red}: Either NAD^+ is reduced to NADH (the feature used in Problem 5.2) or $2H^+$ can be reduced to form H_2. This last reaction is included as a separate net reaction: NADH → $NAD^+ + H_2$ (reaction 13) to account for the considerable amounts of free H_2 produced, especially at high pH (the acid forming process). At low pH, less H_2 and more reduced solvents (e.g., butanol) are produced (the solvent forming process).

a. For each reaction v_1 to v_{13}, write up the stoichiometry and include the ATP/NADH produced or consumed. For all carbon-containing compounds the stoichiometry should be on the basis of 1 C-mole consumed. Reactions 7, 8, 9 and 12 do not involve any NADH/NAD^+ conversion. 0.25 ATP is liberated per C-mole converted in reaction 12, and 0.25 ATP is consumed per C-mole converted in reaction 9. Reactions 7, 8, 10, 11 and 13 do not involve ATP/ADP conversion.

Assume that $r_{pyr} = r_{AcCoA} = r_{AcetoAcCoA} = r_{BuCoA} = r_{NADH} = r_{ATP}$ are all zero. Determine the minimum number of measurements needed to observe the system.

b. Show that the system is observable if r_s, r_x, r_e, r_{acn}, r_{HBu}, r_{ac}, and r_b are the measured rates. Calculate the remaining nonzero rates as linear combinations of the measured rates.

c. If v_8 is set to zero, none of the metabolites that are progenies of AcetoAcCoA are formed. If, furthermore, $v_7 = 0$ the whole pathway is the one considered in Problem 5.2. Show that the expressions for r_{HAc}, r_{Hlac}, and r_c are the same in the degenerate version of Problem 5.2 if $r_{H2} = 0$.

d. Reardon *et al.* (1987), Fig. 4, gives some results from a batch fermentation carried out at pH = 6. Between 10 and 15 h after the start of the batch the glucose concentration s and the concentrations of HAc, HBu, and cells are all fitted well by linear functions of fermentation time. Virtually no acetoin, lactate, ethanol, acetone, or butanol is produced. CO_2 and H_2 production were not measured. The following concentration changes can be read from the figure (with at least 5% uncertainty) for the 5-h period:

Glucose	-117 mM (= -21.1 g/L)
HBu	76 mM (= 6.69 g/L)
HAc	55 mM (= 3.30 g/L)
Biomass	3.3 g dry weight/L

Calculate from the answer to b the amount of HAc that should be produced when the other rates are as stated above. Note that the result fits remarkably well with the experimentally determined r_{HAc}. Next calculate the amounts of H_2 and CO_2 liberated between 10 and 15 h of fermentation time. Check that $r_{HLac} \approx 0$ to within the accuracy of the data.

e. Between 15 and 20 hr of fermentation time, biomass growth appears to stop [a decrease in x is even indicated in Reardon et al. (1987), Fig. 4]. Reardon et al. obtained the following changes in concentrations:

Glucose	-93 mM
HBu	50 mM
HAc	35 mM
Acetoin	4.5 mM
Acetone	8 mM
Butanol	12 mM
Ethanol	2 mM

Do these data fit the metabolic model of b? If not, give a reasonable physiological explanation for the lack of fit of, e.g., the carbon balance. Why is no extra biomass formed?

Reardon et al. (1987) notice a sharp increase in culture fluorescence after 15 h. They tentatively explain this by an increase in NADH concentration in the cells. If this is true, does it point to other flaws in the model of b after 15 h?

f. Papoutsakis (1984) cites some very old (1930) data from van der Lek for a typical "solvent type fermentation" of *C. acetobutylicum*. The biomass production is not stated, but for 100 moles (18 kg) of glucose fermented the following amounts (in moles) of metabolic products are found:

HBu	4.3
HAc	14
H_2	139
CO_2	221
Acetoin	6.3
Acetone	22.4
Butanol	56
Ethanol	9.3

Use the calculated expression in b for r_{HAc} to eliminate r_x from the expressions for r_{H2} and r_c in b and check whether the model in b gives a satisfactory prediction of r_{H2} and r_c.

g. In the pathway diagram two reactions used by Reardon et al. (1987) in their Fig. 3 have been left out. These reactions are:

$$-BuCoA - 0.5 \, HAc + 0.5 \, AcCoA + HBu = 0 \qquad (v_{14})$$

$$-AcetoAcCoA - 0.5 \, HAc + 0.5 \, AcCoA + 0.75 \, ac - 0.25 \, CO_2 - 0.25 \, ATP = 0 \quad (v_{15})$$

Reaction 14 is a parallel reaction to reaction 12, but without release of ATP. Reaction 15 is a parallel reaction to reaction 9 where the CoA is transferred to acetate rather than to butyrate. Expand the stoichiometric calculation in b with these two reactions. Obviously, two more rates have to be measured – but can you solve for the remaining rates? Discuss the result, referring to the paper by Reardon *et al.* to see what they have done. Recalculate the result of b with reaction 15 instead of reaction 9.

Final note: The present problem is very suitable for studies of pathway analysis in real industrial processes. Both Papoutsakis (1984) and Reardon *et al.* (1987) have many suggestions for further variations of the problem – e.g. abandoning the assumption of $r_{NADH} = 0$ or of $r_{ATP} = 0$. The papers should be consulted for these additional possibilities. In recent years a number of studies, especially in the group of Terry Papoutsakis has shown that both yield (of solvents) and productivity can be improved by metabolic engineering of the pathway. Hereby it may become economically viable to resurrect the old (1st World War) process of making solvents by fermentation of house hold waste.

Problem 5.5 Production of propane 1,3-diol (3G) by fermentation

Zeng and Biebl (2002) have recently reviewed the possibilities of making 3G by a fermentation route. The diol can be produced by several chemical routes from either ethylene (Shell) or from acrolein (Degussa), but it may be cheaper to produce it by fermentation using either glycerol or the much cheaper glucose as substrate. An enormous increase in demand for 3G is envisaged, especially since it can condense with terephtalic acid to form an excellent polyester (Sorona^{TM} from DuPont).

3 G is a natural metabolic product in anaerobic cultivation of *Klebsiella pneumonica* on glycerol:

$$\text{Glycerol} \rightarrow \text{3 Hydroxy propanal} \rightarrow \text{1,3 Propane diol} \qquad (1)$$

The first step (a dehydration) is catalysed by glycerol dehydratase that requires the vitamin B_{12} as cofactor. Step 2 is catalysed by 1,3 propane diol dehydrogenase, and NADH is used as cofactor

When *Klebsiella* grows on glycerol the two steps (Fig. 2.4) from DHAP to glycerol are reversed. One ATP and one NAD$^+$ is used per glycerol molecule. DHAP enters the EMP pathway and *via* pyruvate it is metabolised to all the end products shown on Fig. 2.6A. (actually a pathway from pyruvate to 2,3 butane diol *via* α-Acetolactate and acetoin can also be followed, but this will not be considered here.)

   a. Determine the maximum theoretical yield of 3G on glycerol by anaerobic fermentation with *Klebsiella*. The analysis should be based on the constraints $r_{NADH} = r_{ATP} = 0$. Which end products of the mixed acids from pyruvate metabolism are desirable and which should if possible be avoided?

   b. *Klebsiella* is not the optimal production organism. In a series of brilliant metabolic engineering studies (covered by many patents) DuPont and Genencor have succeeded to insert the pathway from DHAP to glycerol (from *S. cerevisiae*) and from glycerol to 3G (from *Klebsiella*) into *E.coli*. Hereby a well researched organism is used as host for production of 3G and it has been shown that a very high titer of 129 g L^{-1} 3G can be obtained by fed-batch fermentation of the engineered *E.coli* (Emptage *et al.*, 2001).

We shall study the DuPont/Genencor process in the following. Three pathways are considered from glucose:

v_1 :   glucose→ biomass ($CH_{1.8}O_{0.5}N_{0.2}$) by :
             $-1.1\ CH_2O + 0.1\ CO_2 + X + 0.1\ NADH - 2.42\ ATP = 0$
v_2 :   glucose → 3G
v_3 :   glucose→ CO_2 + 2NADH (*via* the EMP pathway and the TCA cycle)
v_4 :   NADH + ½ O_2 → P/O ATP

The rate of glucose consumption is $-r_s = v_2 + v_3 + 1.1\ v_1$.

Assume that no glycerol is excreted and that Y_{so} is 0.045 mol O_2 (C-mol glucose)$^{-1}$. Determine the stoichiometry (the "black box" model) for the overall reaction based on the assumption that $r_{NADH} = r_{ATP} = 0$, and P/O = 2. What is the lowest value of P/O for which biomass can be produced?

c. Assume that (the very small) Y_{sx} can be set to zero. Show that it is now possible to calculate both Y_{sp} and Y_{so} based on knowledge of the P/O ratio.

(Answer: $Y_{so} = \dfrac{1}{12P/O+3}$ and $Y_{sp} = \dfrac{6P/O+1}{8P/O+2}$)

In this way the effect of using a more realistic value than 2 for P/O can be studied.

d. Again assume that $Y_{sx} = 0$ and that some glycerol is excreted from the cells. Glycerol is of course an unwanted byproduct and we wish to study the influence of Y_{sg} on the product yield. Show that

Y_{sp} = C-mole 3G (C-mole glucose)$^{-1}$ = $\dfrac{6P/O+1-(2+7P/O)Y_{sg}}{8P/O+2}$

$Y_{so} = \dfrac{1+0.5Y_{sg}}{12P/O+3}$

Calculate the reduction of Y_{sp} compared with what is found in question c. when $Y_{sg} = 0.05$ and 0.1.

e. Return to the stoichiometry of question b. with P/O = 2. In parallel with the glucose used to feed this carbon-, redox- and ATP balanced set of pathways that determines production of 3 G some glucose may be used to produce biomass and CO_2 by pure respiration, and for maintenance of cells by combustion of glucose to CO_2 alone. Write a carbon-, redox- and ATP- balanced metabolic scheme for purely respiratory biomass production (overall reaction 2), and a carbon-and redox balanced stoichiometry for the maintenance reaction (overall reaction 3).
Show that any experimentally obtained stoichiometry containing only X, 3G and CO_2 can be written as a linear combination of the three overall stoichiometries.

Use P/O = 2 in overall reaction 2 and show that the experimentally observed stoichiometry:

$$- CH_2O + 0.543\ P + 0.120\ X + 0.337\ CO_2 - 0.150\ O_2 = 0$$

is obtained by combining overall reactions (1) to (3) in the ratio 0.7714: 0.1738: 0.0548.

The maintenance demand is significant, but not alarming. There is probably a kinetic bottleneck in the pathway v_2 from glucose to 3 G which causes overflow to (useless) biomass formation. This needs to be looked at as a part of the physiological design of the final production strain.
The solution to this question is helpful in understanding the (mathematically) complex methods of section 5.4.2.

REFERENCES

Aiba, S., Matsuoka, M. (1979) Identification of metabolic model: Citrate production from glucose by *Candida lipolytica*. *Biotechnology and Bioengineering*. **21**, 1373-1386.

Albers, E., Larsson, C., Liden, G., Niklasson, C., Gustafsson, L. (1996) Influence of the nitrogen source on *Saccharomyces cerevisiae* anaerobic growth and product formation. *Appl. Environ. Microbiol.* **62**, 3187-3195

Albers, E., Liden, G., Larsson, C., Gustafsson, L. (1998) Anaerobic redox balance and nitrogen metabolism in *Saccharomyces cerevisiae*. *Rec. Res. Devel. Microbiol.* **2**, 253-279

Bauchop, T., Esden, S. R. (1960) The growth of microorganisms in relation to their energy supply. *J. Gen. Microb.* **23**, 35-43.

Benthin, S. (1992) Growth and Product Formation of *Lactococcus cremoris*, Ph.D. thesis, Department of Biotechnology, Technical University of Denmark, Lyngby

Benthin, S., Nielsen, J., Villadsen, J. (1991) Characterization and application of precise and robust flow-injection analysers for on-line measurements during fermentations. *Anal. Chim. Acta* **247**, 45-50.

Benthin, S., Schulze, U., Nielsen, J., Vlladsen, J. (1994) Growth energetics of *Lactococcus cremoris* FD1 during energy-, carbon- and nitrogen-limitation in steady state and transient cultures. *Chem. Eng. Sci.* **49**, 589-609.

Brown, W. V., Collins, E. B. (1977) End product and fermentation balances for lactis Streptococci grown aerobically on low concentrations of glucose. *Appl. Environ. Microbiol.* **59**, 3206-3211.

Christensen, B., Nielsen, J. (1999) Metabolic network analysis – powerful tool in metabolic engineering. Adv. Biochem. Eng./Biotechnol. **66**, 209-231

Christensen, B., J. Nielsen, J. (2000) Metabolic network analysis on *Penicillium chrysogenum* using 13C-labelled glucose. *Biotechnol. Bioeng.*, **68**, 652-659

Christiansen, T., Nielsen, J. (2002) Growth energetics of an alkaline serine protease producing strain of *Bacillus clausii* during continuous cultivation. *Bioproc. Biosystems Eng.*, in press

Emptage, M., Haynie, S., Laffend, L., Pucci, J., Whited, G. (2001). Process for the biotechnological production of 1,3 propane diol with high titer. US patent application No. PCT/US 00/22874; cited in WO (World Intellectual Property Organization) 01/12833

Gombert, A. K., dos Santos, M. M., Christensen, B., Nielsen, J. (2001) Network identification and flux quantification in the central metabolism of *Saccharomyces cerevisiae* at different conditions of glucose repression. *J. Bacteriol.* **183**, 1441-1451

van Gulik, W. M., Heijnen, J. J. (1995) A metabolic network stoichiometry analysis of microbial growth and product formation. *Biotechnol. Bioeng.* **48**, 681-698

Henriksen, C.M., and Nilsson, D. (2001) Redirection of pyruvate catabolism in *Lactococcus lactis* by selection of mutants with additional growth requirements. *Appl. Microbiol. Biotechnol.* **56**, 767-775

Hempfling, W. P., Mainzer, S. E. (1975) Effects of varying the carbon source limiting growth on yield and maintenance characteristics of *Escherichia coli* in continuous culture. *J. Bacteriol.* **123**, 1076-1087.

Herbert, D. (1959). Some principles of continuous culture. *Recent Prog. Microb.* **7**, 381-396.

Ingraham, J. L., Maaløe, O., & Neidhardt, F. C. (1983) Growth of the Bacterial Cell. Sunderland: Sinnauer Associated.

Jørgensen, H. S., Nielsen, J., Villadsen, J., Mølgaard, H. (1995) Metabolic flux distributions in *Penicillium chrysogenum* during fed-batch cultivations. *Biotechnol. Bioeng.* **46**, 117-131

Liao, J. C., Hou, S.-Y., Chao, Y.-P. (1996) Pathway analysis, engineering, and physiological considerations for redirecting central metabolism. *Biotechnol. Bioeng.* **52**, 129-140

Luedeking, R.& Piret, E. L. (1959a) A kinetic study of the lactic acid fermentation. Batch process at controlled pH. *J. Biochem. Microb. Technol. Eng.* **1**, 393-412

Luedeking, R., Piret, E. L. (1959b) Transient and steady state in continuous fermentation. Theory and experiment. *J. Biochem. Microb. Technol. Eng.* **1**, 431-459

Major, N. C., Bull, A. T. (1985) Lactic acid productivity of a continuous culture of *Lactobacillus delbrueckii*. *Biotechnol. Letters.* **7**, 401-405.

Nielsen, J. (1997) Physiological Engineering Aspects of *Penicillium chrysogenum*. Singapore: World Scientific Publishing Co.

Nissen, T.L., Schulze, U., Nielsen, J., Villadsen, J. (1997) Flux distributions in anaerobic, glucose limited continuous cultures of *Saccharomyces cerevisiae*. *Microbiology* **143**, 203-218

Otto, R., Sonnenberg, A. S. M., Veldkamp, H., Konings, W. N. (1980) Generation of an electrochemical proton gradient in *Streptococcus cremoris* by lactate efflux. *Proc. Nat. Acad. Sci.* **77**, 5502-5506.

Oura, E. (1983) Biomass from carbohydrates. In *Biotechnology*, H. Dellweg, ed., Vol. 3, 3-42. VCH Verlag, Weinheim, Germany.

Papoutsakis, E. T. (1984). Equations and calculations for fermentations of butyric acid bacteria. *Biotechnol. Bioeng.* **26**, 174-187

Pedersen, H., Carlsen, M., Nielsen, J. (1999). Identification of enzymes and quantification of metabolic fluxes in the wild type and in a recombinant strain of *Aspergillus oryzae* strain. *Appl. Environ. Microbiol.* **65**, 11-19

Pirt, S. J. (1965). The maintenance energy of bacteria in growing cultures. *Proc. Royal Soc. London. Series B* **163**, 224-231.

Pramanik, J., Keasling, J. D. (1997) Stoichiometric model of *Escherichia coli* metabolism: Incorporation of growth-rate dependent biomass composition and mechanistic energy requirements. *Biotechnol. Bioeng.* **56**, 398-421

Reardon, K. F., Scheper, T. H., and Bailey, J. E. (1987). Metabolic pathway rates and culture fluorescence in batch fermentations of *Clostridium acetobutylicum*. *Biotechnol. Prog.* **3**, 153-167

Sauer, U., Hatzimanikatis, V., Hohmann, H. P., Manneberg, M., van Loon, A. P. G. M., Bailey, J. E. (1996) Physiology and metabolic fluxes of wild-type and riboflavin-producing *Bacillus subtilis*. *Appl. Environ. Microbiol.* **62**, 3687-3696

Schmidt, K., Carlsen, M., Nielsen, J., Villadsen, J. (1997) Modelling isotopmer distributions in biochemical networks Using isotopomer mapping matrices. *Biotechnol. Bioeng.* **55**, 831-840

Schmidt, K., Nielsen, J., Villadsen, J. (1999). Quantitative analysis of metabolic fluxes in *E. coli*, using 2 dimensional NMR spectroscopy and complete isotopomer models. *J. Biotechnol.* **71**, 175-190

Schuster, S., Dandekar, T., Fell, D. A. (1999) Detection of elementary flux modes in biochemical networks: a promising tool for pathway analysis and metabolic engineering. *TIBTECH* **17**, 53-60

Schuster, S., Fell, D. A., Dandekar, T. 2000. A general definition of metabolic pathways useful for systematic organization and analysis of complex metabolic networks. *Nature Biotechnol.* **18**, 326-332

Sonntag, K., Eggeling, L., de Graaf, A. A., Sahm, H. (1993) Flux partitioning in the split pathway of lysine synthesis in *Corynebacterium glutamicum* – Quantification by ^{13}C- and ^{1}H-NMR spectroscopy. *Eur. J. Biochem.* **213**, 1325-1331

Stephanopoulos, G., Aristodou, A. & Nielsen, J. (1998). Metabolic Engineering. Principles and Methodologies. Academic Press, San Diego

Stouthamer, A. H. (1979) The search for correlation between theoretical and experimental growth yields. In *International Review of Biochemistry: Microbial Biochemistry*, Vol. 21, pp. 1-47. Edited by J. R. Quayle. Baltimore: University Park Press

Stouthamer, A. H., Bettenhaussen, C. (1973) Utilization of energy for growth and maintenance in continuous and batch cultures of microorganisms. *Biochim. Biophys.Acta* **301**, 53-70.

Theobald, U., Mailinger, W., Baltes, M., Rizzi, M. & Reuss, M. (1997). In vivo analysis of metabolic dynamics in *Saccharomyces cerevisiae*: I. Experimental observations. *Biotechnol. Bioeng.* **55**, 305-316

Vallino, J. J., Stephanopoulos, G. (1993) Metabolic flux distributions in *Corynebacterium glutamicum* during growth and lysine overproduction. *Biotechnol. Bioeng*, **41**, 633-646

Vallino, J. J., Stephanopoulos, G. (1994a) Carbon flux distributions at the pyruvate branch point in *Corynebacterium glutamicum* during lysine overproduction. *Biotechnol. Prog.* **10**, 320-326

Vallino, J. J.; Stephanopoulos, G. (1994b) Carbon flux distributions at the glucose-6-phosphate branch point in *Corynebacterium glutamicum* during lysine overproduction. *Biotechnol. Prog.* **10**, 327-334

Varma, A., Boesch, B.W., Palsson, B.O. (1993a) Stoichiometric interpretation of *Escherichia coli* glucose metabolism under various oxygenation rates. *Appl. Environ. Microbiol.* **59**, 2465-2473

Varma, A., Boesch, B. W., Palsson, B. O. (1993b). Biochemical production capabilities of *Escherichia coli*. *Biotechnology and Bioengineering* **42**, 59-73.

Verduyn, C., Postma, E., Scheffers, W. A., van Dijken, J. P. (1990) Energetics of *Saccharomyces cerevisiae* in anaerobic glucose limited chemostat cultures. *J.Gen. Microbiol.* **136**, 405-412.

de Vries, W., Kapteijn, W. M. C., van der Beek, E. G., Stouthamer, A. H. (1970) Molar growth yields and fermentation balances of *Lactobacillus casei* L3 in batch cultures and in continuous cultures. *J. Gen. Microbiol.* **63**, 333-345.

Wiechert, W. (2001) ^{13}C Metabolic Flux Analysis. Metabolic Eng. **3**, 195-206

Wiechert, W., Siefke, C., de Graaf, A. A., Marx, A. (1997) Bidirectional reaction steps in metabolic networks. Part II: Flux estimation and statistical analysis. *Biotechnol. Bioeng.* **55**:118-135

van Winden, W., Verheijen, P., Heijnen, S. (2001) Possible pitfalls of flux calculations based on ^{13}C-labeling. *Metabolic Eng.* **3**, 151-162

Zeng; An-Ping, and Biebl, H. (2002). Bulk chemicals from biotechnology : The case of 1,3 propane diol production and the New Trends. In *Advances in Biochemical Engineering Biotechnology* **74**, 239-259

Zupke, C., Stephanopoulos, G. (1995) Intracellular flux analysis in hybridomas using mass balances and *in vivo* ^{13}C NMR. *Biotechnol. Bioeng.* **45**, 292-303

6

Enzyme Kinetics and Metabolic Control Analysis

There is little doubt that "Enzymes" was the word emblazoned over the gate through which most chemists, physicists and chemical engineers entered the realm of bio-reactions.

Chemical reactions catalyzed by enzymes are not much different from dehydrogenation reactions catalyzed by noble metal catalysts. There is only one reaction to be considered rather than the myriad of reactions which take place in the metabolic network of cells, and the stoichiometry of the reaction is known. As we have seen in Chapters 3 and 5 the study of stoichiometry and of the steady state flux distribution in a metabolic network is a difficult task, and it must necessarily be done before any meaningful investigation of the kinetics of cellular reactions can be started. Enzymes are usually highly specific catalysts which allow only one form of the substrate (reactant) to be treated – e.g. the L- but not the D-form of an enantiomer. Structural studies of the enzymes reveal how the substrate docks on the catalyst, where the cofactor sneaks in and how the product is detached. Poisoning (or "inhibition") of enzymes follows the same rules as have been developed for the inhibition (or reversible poisoning) of inorganic catalysts, and one can explain why overloading of the enzyme with substrate, or blocking of the enzyme activity by a reaction product or with a foreign molecule can reduce the rate of conversion of the substrate. In short: the enzymatic reaction can if desired be treated by mechanistic models to obtain the true kinetics of the reaction. This is in sharp contrast to the wholly empirical approach that has to be used for cell reactions and where the outcome of the modeling exercise frequently has a very limited predictive power.

Quite apart from the pedagogical qualities of enzyme reactions which open up new avenues of studies for readers of classical textbooks on reaction engineering (e.g. Levenspiel (1999)) the subject has of course enormous significance for modern biotechnology.

Enzymes (rather than living cells) are used in a multitude of processes, in industry as well as in the daily household, and enzymatically based processes are replacing many classical processes due to their mild reaction conditions and environmentally friendly outcome. Large companies such as Novozymes and Genencor, the two dominant players on the enzyme market, are entirely devoted to the production of enzymes (by fermentation processes) for their various applications. The market (approximately 1.5 billion USD in 2000) is expanding, and new applications of enzymes are constantly being explored as can be seen in the yearly reports of the enzyme producers.

The main theme of our textbook is the cellular reaction for which the kinetics is treated in Chapter 7. It is impossible to treat the kinetics of cellular reactions solely on the assumption of known kinetics for all the enzyme reactions in the cell – in itself a never to be satisfied assumption. The cell is much more "than a bag of enzymes", and *in vitro* studies of enzyme kinetics tell nothing about the influence of the control structure of the cell which is destroyed when the enzymes are investigated outside their natural habitat.

Still, the apparent analogies between enzyme kinetics and cell kinetics have often been used to home in on the right structure of a kinetic expression for the conversion of substrates by a living cell. This alone is a good reason for giving a short treatment of enzyme kinetics. Furthermore, enzyme-based assays are used routinely to study the state of the cell at a certain set of environmental conditions. Some of these assays are not easily understood without prior knowledge of the behavior of the enzymatic reaction on which the assay is based.

Finally, when taking the step from analysis of the steady state flux distribution in a metabolic network to an analysis of the rate of production that can be supported by the network, one needs to include a kinetic analysis of the different enzymatic reactions in the network. The outcome of this analysis is a prediction of how genetic manipulations of the network can lead to a change in the flux distribution and perhaps to a higher yield and a higher production rate of the desired product in a given branch of the network.

Consequently, enzyme kinetics will be treated with a view of analyzing the rates of production in different branches of a metabolic network. The analysis is based on the sensitivity of each enzymatic reaction in a particular branch to changes in the activity of each of the enzymes in the pathway, and to changes in the metabolite levels at the different steps.

The activity of enzymes is a function of a number of environmental factors, e.g. temperature and pH. Many enzyme preparations are sold as particles of immobilized enzymes in a support. Here mass transport and chemical reaction both play a role for the overall reaction rate as is well known for normal chemical reactions. It therefore becomes important to analyze when mass transport is likely to influence the bio-reaction mediated by the enzyme. Both of these topics will, however, be treated in the context of cellular reactions (Chapters 7 and 10) which are modeled by expressions similar to the mechanistic models of enzyme reactions, but with an extra factor, the biomass concentration x, which is responsible for the autocatalytic nature of the cellular kinetics. When yeast cells are embedded in an alginate matrix and the biomass concentration is assumed to be constant there is in fact no difference between the behavior of this system and an immobilized glucose isomerase or penicillin deacylase embedded in a glutaraldehyde matrix.

6.1 Michaelis-Menten and Analogous Enzyme Kinetics

Very often the rate of an enzymatic reaction when pictured as a function of the substrate concentration s takes the form of Fig. 6.1.

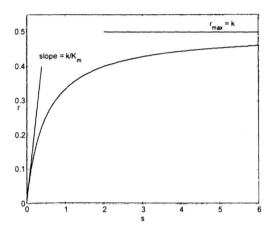

Figure 6.1 Schematic representation of the rate of an enzymatic reaction. s is the substrate concentration (typical unit: g substrate (L medium)$^{-1}$) and r is the rate of conversion of s.

The shape of Fig. 1 with an initial near proportionality between r on s and a final, maximum rate of conversion is well correlated by the following expression:

$$r = \frac{ks}{s + K_m} = \frac{k_0 e_0 s}{s + K_m} \qquad (6.1)$$

The substrate concentration s is in units of g (or mole) substrate(L medium)$^{-1}$. The reaction rate r can be specified in units of either g substrate(L medium h)$^{-1}$ or in units of g substrate (g enzyme preparation h)$^{-1}$. The last unit corresponds best with the standard used for catalytic reactions with solid catalysts : g converted(g catalyst h)$^{-1}$, and the use of the letter r for the rate would correspond to that used for specific rates elsewhere in the text. But the whole literature on enzyme kinetics is based on the first unit for r, and we shall follow this standard nomenclature. Consequently k is in units of g substrate(L medium h)$^{-1}$, e_0 : g enzyme preparation (L medium)$^{-1}$ and k_0 : g substrate(g enzyme preparation h)$^{-1}$. r/e_0 is consequently the equivalent of the specific rates used elsewhere in the book. In practical investigations of enzymatic reactions only k is determined, and it signifies the maximum of r, or r_{max}. k_0 and e_0 can not be determined separately, but splitting k into two factors does show that an enzyme with a low activity per unit mass of enzyme will lead to a low rate of conversion of the substrate. In the following derivation of some classical enzyme kinetic models it will be convenient to describe r by the second expression in eq (6.1).

In 1925 Briggs and Haldane derived the form of (6.1) by mechanistic modeling. They considered that the enzyme could exist in two forms E = free enzyme, and ES = an enzyme complex with the substrate. The conversion of substrate S to product P proceeds in two steps

$$E + S \underset{k_{-1}}{\overset{k_1}{\rightleftarrows}} ES \xrightarrow{k_2} E + P \qquad (6.2)$$

The first reaction is reversible, the second irreversible. Both reactions are assumed to be elementary reactions with rate proportional to the concentration of reactants. Furthermore Briggs and Haldane assumed that the concentration (es) of ES was constant in time, i.e. ES is in a "pseudo-steady state".

$$\frac{d(es)}{dt} = 0 = k_1 e \cdot s - k_{-1}(es) - k_2(es) \qquad (6.3)$$

The total enzyme "concentration" e_0 was assumed to be constant and representing the "activity" in eq. (6.1)

$$e_0 = e + (es) \qquad (6.4)$$

From (6.3)

$$(es) = \frac{k_1 e_0 s}{k_1 s + (k_{-1} + k_2)} = \frac{e_0 s}{s + \frac{k_{-1} + k_2}{k_1}} \qquad (6.5)$$

The rate of the reaction is determined by the decomposition of ES (i.e. the first reaction is much faster in both directions than the second), and if ES decomposes by a first order reaction then

$$r = k_2 (es) = \frac{k_2 e_0 s}{s + \frac{k_{-1} + k_2}{k_1}} = \frac{k_2 e_0 s}{s + K_m} \qquad (6.6)$$

If the first reaction is infinitely fast and the substrate concentration is much higher than the enzyme concentration e_0 then the concentration of the enzyme-substrate complex is

$$(es) = \frac{e \cdot s}{K_{eq}} \qquad (6.7)$$

where K_{eq} is the equilibrium constant for the dissociation of ES to E and S. Inserting (6.4) in (6.7) and again assuming first order decomposition of ES by the second, rate determining reaction:

$$r = \frac{k_2 e_0 s}{s + K_{eq}} \tag{6.8}$$

Derivation of eq (6.8) by Michaelis and Menten in 1913 signified virtually the beginning of quantitative enzymology. Their equilibrium assumption for step 1 of (6.2) was relaxed by Briggs and Haldane (1925), who used a similarly speculative hypothesis of pseudo-steady state for ES. Both models – which typically for mechanistically based models can assign a definite physical meaning to the parameters – lead to eq. (6.1) where K_m is called the Michaelis constant in honor of the recognized "father" of enzymology[1]. The picture of K_m as an equilibrium constant for a dissociation reaction according to (6.8) is illustrative: a substrate that is easily captured by the enzyme (the enzyme has a high affinity for the substrate) will have a small K_m value. If k_2 is large – i.e. if the rate of the second reaction in (6.2) is high then the rate of conversion of the substrate by the enzyme is high, also at low values of s.

Note 6.1 Assumptions in the mechanistic models for enzyme kinetics
The simplicity of the derivation of the two mechanistically based models for the rate of an enzyme reaction could lead the reader to believe that here, for once, there is a trustworthy piece of modeling in biotechnology. Indeed, the rate expression (6.1) is very robust – much more so than the apparently equivalent Monod model for cell kinetics, eq. (7.16). It allows for a significant extrapolation from the data that are used to determine the two parameters k and K_m, and the parameters are clearly related to stringently defined parameters of elementary reactions. Still, it may be useful to give a few comments to illustrate the effect of the approximations that lurk behind both the Michaelis Menten and the Briggs Haldane derivation of eq. (6.1). First of all the comments given below may help the user to avoid mistakes in an experimental set up which is aimed to calculate the two kinetic parameters.

First one can easily see that the derivation of eq. (6.6) leads to the same result as (6.8) if, indeed the first step of (6.2) is an equilibrium reaction. Then both k_{-1} and k_1 in (6.2) are infinitely large and the ratio k_{-1}/k_1 = K_m is equal to the equilibrium constant K_{eq} of (6.7) and (6.8). Also the assumption of (6.3) that (es) is constant is true. The amount of ES consumed by the second step of (6.2) is immediately replenished by the fast equilibrium reaction in step 1.

But (6.1) is also derived by (6.2) to (6.6) without the apparently unnecessary assumption of a fast equilibrium – the value of K_m must only be interpreted differently. How trustworthy is the assumption of quasi-stationarity of (es)? Clearly if all three rate constants k_1, k_{-1} and k_2 are small it may take a very long time until (es) becomes constant. In fact eq. (6.3) can easily be integrated (by separation of variables) for a constant s, and (es) can therefore be found as a function of time. When es(t) is inserted in the overall rate expression (6.6) one obtains

$$r = \frac{k_2 e_0 s}{s + K_m} (1 - \exp[-(k_1 s + k_{-1} + k_2)t]) \tag{1}$$

It is well established that k_{-1}, k_2 and $k_1 s$ are very large, 100 to 1000 s^{-1} for normal enzymes, and the

---

[1] Menten had a sad fate, discriminated as a Jew and ending her life in London in the early 1930's

exponential function vanishes within milliseconds, before s has started to decrease from its initial value.

A much more serious criticism of both the Michaelis Menten and the Briggs Haldane mechanism lies in the assumed irreversibility of the step 2 in (6.2). Since enzymes like other catalysts must in principle enhance the rate of both the forward reaction and the reverse reaction (but certainly not to the same degree – the product might not be able to dock on the enzyme) one must consider

$$E + S \underset{k_{-1}}{\overset{k_1}{\rightleftarrows}} ES \underset{k_{-2}}{\overset{k_2}{\rightleftarrows}} E + P \qquad (2)$$

rather than (6.2). Assuming pseudo- stationarity for ES yields

$$(es) = \frac{k_1 s + k_{-2} p}{k_{-1} + k_2 + k_1 s + k_{-2} p} e_0 \qquad (3)$$

$$r = k_2 (es) - k_{-2} p \cdot (e_0 - (es)) = \frac{k_1 k_2 s - k_{-1} k_{-2} p}{k_{-1} + k_2 + k_1 s + k_{-2} p} e_0 \qquad (4)$$

where $p = 0$ yields eq. (6.6).

An enzymatic assay is made with no P initially, and until p is built up equation (6.1) describes the rate of substrate consumption quite accurately. The overall reaction $S \rightarrow P$ may also be thermodynamically favored which means that almost all S can be converted to P without any influence of the reverse reaction ($k_{-2}k_{-1} \approx 0$), and as we have seen in Chapter 4 an enzymatic reaction in a pathway can convert S to P also when the reaction has a small negative or even a small positive ΔG^0, if only P is sucked away from the equilibrium by further reactions. Consequently, when using models for enzyme kinetics in assays or in the analysis of pathway reactions the simple form (6.1) is usually adequate.

This is not necessarily so when studying the enzymatic conversion of substrate in an industrial bioreactor. When glucose is converted to fructose in a commercial plug flow reactor using immobilized glucose isomerase it is desired to approach the equilibrium conversion (about 50% at 40-60°C) quite closely to obtain an adequate sweetening of the sugar solution. Calculation of the necessary amount of catalyst will be wildly wrong if (6.1) rather than (4) is used (Gram et al., 1990)

Finally, one may raise the objection to (6.2) that most enzymatic reactions need a second substrate to proceed. The cooperation of cofactors such as $NAD^+/NADPH$, $NADP^+/NADPH$, ATP/ADP is seen in many of the pathway reactions of Chapter 2. It is tacitly assumed that these cofactors are regenerated by other cellular reactions and that their level is fixed through constraints on redox charge or energy charge. But when planning to use a genetically engineered strain in which the balances between cofactors has been artificially changed (an $NADP^+$ dependent enzyme may have been exchanged with an NAD^+ dependent variant in a high flux pathway) the rate of the enzyme reaction may be significantly changed.

Enzyme Kinetics and Metabolic Control Analysis

6.2 More Complicated Enzyme Kinetics

Equation (6.1) derived by either eq. (6.3) or (6.7) has an almost exact analogue in Langmuirs treatment during World War 1 of catalytic surfaces on which a reactant can adsorb on a finite number of active sites and react in the adsorbed state, from which the product can finally be desorbed. The presence of other species which could inhibit the overall reaction either reversibly (the activity of the catalyst returns when the inhibitor is removed from the feed) or irreversibly (it could be the permanent damage caused by sulfides or by sintering of the catalyst during a temperature excursion) have been exhaustively studied both by chemists and by biochemists, each in their own field.

Some of the most common inhibitory effects contributing to a reduction of the enzymatic activity either through an action on k or on K_m in (6.1) will be discussed in Section (6.2.1). In Section 6.2.2 we shall discuss certain rate expressions in which the dependence on s is strong in a certain s-interval and small outside this interval. Enzymes exhibiting this type of behavior are important for the regulation of cell metabolism. The kinetics developed in Sections 6.2.1 and 6.2.2 below both have apparent analogues in expressions for cell kinetics, but there they are used as pure data fitters without any mechanistic foundation at all.

6.2.1 Variants of Michaelis-Menten Kinetics

Although enzymes are usually very specific catalysts there are important cases of interactions between enzymes and substrate-analogues that practically impede the desired reaction. The membrane bound transport enzymes for hexoses are typical examples. A specific example is the membrane bound mannose-PTS system of lactic bacteria (Fig. 2.3C) that can transfer a number of sugars from the medium to the cell where it arrives in phosphorylated form. Despite its name the uptake of mannose by the enzyme is almost completely inhibited by the presence of glucose, and also by the presence of glucose-analogues that are not even metabolized by the lactic bacteria (Benthin et al, 1993)

The free enzyme E is clearly removed from eq (6.2) by a competing reaction

$$E + S_1 \underset{k_{-3}}{\overset{k_3}{\rightleftarrows}} ES_1 \tag{6.9}$$

Following the derivation of the Michaelis Menten expression (6.8) one obtains

$$r = \frac{k_2 e_0 s}{s + K_{eq}(1 + \frac{s_1}{K_{eq1}})} \tag{6.10}$$

When S_1 is tightly bound to the enzyme form E, i.e. when the dissociation constant for ES_1 in (6.9) is small then the apparent affinity of E for S is small. The result is that r decreases unless s is large enough to make the denominator constant small compared to s. *Competitive inhibition* – the deposition of active enzyme in a "dead-end "compound ES_1- can be alleviated by increasing the substrate concentration, but this may not always be possible, e.g. in an enzymatic assay.

The competing substrate for E could be any foreign chemical or it could be the product P of the reaction. Here a negative influence of P on the rate is found also when the overall reaction is completely irreversible. When the assay is run at different levels of P one may detect a possible product inhibition.

If both E and ES can react with the foreign substance S_1 to form inactive complexes ES_1 and ESS_1 respectively, and if also ES_1 can react with S to form ESS_1 then the following reaction network is found

$$
\begin{array}{ccc}
 & S & \\
E & \rightleftarrows \ ES & \rightarrow E + P \\
S_1 \updownarrow & \updownarrow S_1 & \\
ES_1 & \rightleftarrows \ ESS_1 & \\
 & S &
\end{array}
\qquad (6.11)
$$

Here "the dead end complex" ES_1 can be "activated" through conversion to ESS_1 which again can form the active enzyme complex ES. This type of indiscriminate inhibition where S_1 binds randomly to both primary forms of the enzyme does not lower the over all affinity of the enzyme for S, but leads to a general decrease of the rate of the enzymatic reaction through a smaller apparent rate constant.

Assume that ES_1 and ESS_1 have the same dissociation constant K_{eq1} to respectively E and ES and also that the substrate S binds equally well to E and to ES_1. Then in analogy with (6.7)

$$e = \frac{(es)}{s} K_{eq} \ ; \ (ess_1) = \frac{(es)s_1}{K_{eq1}} \ ; \ (es_1) = \frac{(es)s_1 K_{eq}}{K_{eq1}\, s} \qquad (6.12)$$

$$(es) = \frac{e_0}{\dfrac{K_{eq}}{s}\left(1 + \dfrac{s_1}{K_{eq1}}\right) + \left(1 + \dfrac{s_1}{K_{eq1}}\right)} \qquad (6.13)$$

$$r = \frac{k_2 e_0 s}{K_{eq} + s}\left(1 + \frac{s_1}{K_{eq1}}\right)^{-1} \qquad (6.14)$$

which is the same as (6.8) except for the reduced rate constant. Here an increase of the concentration of S will not alleviate the inhibitory effect of S_1. If the nature of S_1 is known it is possible to study the inhibition process by repeating the experiment at different S_1 levels, but sadly one does not always know what the inhibitory substance is, and the so called *non-competitive* enzyme inhibition may be difficult to detect in assays which probably do not contain all the inhibitory substances that are present in the cell.

If the inhibitor attacks ES but not E the inhibitor is said to be *un-competitive* (for lack of a better word).

With the enzyme balance

$$e_0 = (es) + \frac{K_{eq}}{s}(es) + \frac{(es)s_1}{K_{eq1}} \qquad (6.15)$$

$$r = \frac{k_2 e_0 s}{K_{eq} + \left(1 + \frac{s_1}{K_{eq1}}\right)s} \qquad (6.16)$$

and both the Michaelis constant and the rate constant decrease by a factor $\left(1 + \frac{s_1}{K_{eq1}}\right)$.

In either (6.10), (6.14) or (6.16) the inhibitor S_1 could be the substrate S in which case we have *substrate inhibition*. Unless the "dead-end" complex ES_1 is interpreted as an enzyme complex between S and another form of E neither (6.10) nor (6.14) are of much help in interpretation of substrate inhibition. It may, however well be that S complexes with ES to form ESS which is now the "dead-end" complex. This is the mechanism of (6.16), and this rate expression (with s used instead of s_1) will also appear in Chapter 7 as a typical rate expression for substrate inhibited cellular kinetics (Eq (7.21)).

Fig. 6.2 illustrates the difference (and similarity) between the different types of inhibition. In all four cases $1/r$ is shown as a function of $1/s$ which will give a linear plot of (6.1), (6.10), (6.14) and (6.16). If in (6.16) $s_1 = s$ then $1/r$ is a hyperbolic function of $1/s$ with a vertical and an inclined asymptote. Plots such as Fig. 6.2 (and similar linear plots found in textbooks on enzyme kinetics) are excellent in a first phase of an experimental investigation of enzyme kinetics. They clearly show what type of inhibition is present (if the nature of S_1 is known!). When the experiment is carried out at different levels of S_1 the parameters of the inhibition and thereafter the "true" kinetic parameters k and K_m of (6.1) can be obtained, possibly by extrapolation to the limit $s_1 = 0$ of data obtained from samples spiked with known concentrations of S_1

It is, however, not really good to determine kinetic parameters from any type of linear plot of the variables r and s in a non-linear relation between the two. In Fig. 6.2 the error bars on $1/r$

increase dramatically for increasing 1/s, since it is likely that r is determined experimentally with a given absolute standard deviation irrespective of the value of s (which can always be determined with a fixed relative standard error).

Once the correct type of inhibition has been found all parameters in the rate expression should be determined by non-linear Least-Squares regression. The correlation matrix between the parameters should be shown together with the estimate of the parameter vector and the standard deviation of the parameters.

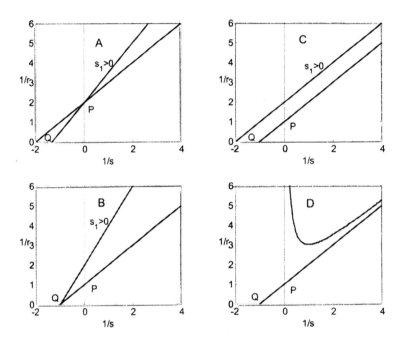

Figure 6.2 Lineweaver-Burk (double reciprocal plots) of the relation between the rate r and the substrate concentration s. In all four plots the intersection P with the ordinate axis is k^{-1} in eq (6.1) for inhibitor concentration $s_1 = 0$ and the apparent value of k^{-1} according to (6.10), (6.14), and (6.16) respectively. |Q| is the corresponding value of K_m^{-1} in (6.1) for $s_1 = 0$, and of the apparent value of K_m^{-1} for s_1 greater than zero. The slope of the lines is equal to $K_{m,apparent} / k_{apparent}$.
  A. Competitive inhibition by s_1, eq(6.10).
  B. Non-competitive inhibition by s_1, eq (6.14).
  C. Un-competitive inhibition by s_1, eq (6.16)
  D. Substrate inhibition by (6.16). Values of k^{-1} and of K_m^{-1} are obtained from the ordinate at P and the abscissa at Q. Slope of the inclined asymptote is K_m/k.

Enzyme Kinetics and Metabolic Control Analysis

Table 6.1 Enzymatic rate data r (g substrate converted $L^{-1} h^{-1}$) at four levels of s and p.

	Product concentration (mg L^{-1})			
S (g L^{-1})	3	9	27	81
0.1	0.073	0.058	0.036	0.017
0.4	0.128	0.102	0.064	0.030
1.6	0.158	0.126	0.079	0.037
6.4	0.168	0.134	0.084	0.039

The same advice is also valuable when parameters in expressions for cell kinetics are determined from (hopefully) a large set of data representing an adequate variation of all the variables of the model.

Example 6.1 Analysis of enzymatic reaction data.
In an enzymatic reaction substrate S is converted irreversibly to product P. It is suspected that the product inhibits the reaction, and consequently the rate r of the reaction is determined for four values of s at each of four levels of the product concentration p. The 16 rate measurements are collected in Table 6.1. s is in g L^{-1} and p in mg L^{-1}.

Fig. 6.3 is a Linewaever-Burk plot of the data. For increasing level of product concentration p the slope of the lines increases (from A to D), and all four lines intersect the abscissa axis at more or less the same point. Clearly a non-competitive product inhibition is indicated. The value of K_{eq} in eq (6.14) is determined from the average of the four intersection abscissas: K_{eq} = 0.142 gL^{-1}. Next the slopes of the four regression lines are plotted against p:

$$\text{Slope} = \frac{K_{eq}}{k}\left(1 + \frac{p}{K_{eq1}}\right) \qquad (1)$$

Fig. 6.4 is a plot of the slope vs. the product concentration. From the slope and the intersection with the ordinate axis, and using the previously determined value for K_{eq} the following values are determined for $r_{max} = k_2 e_0$ in eq. (6.14) and for K_{eq1}:

$$r_{max} = 0.212 \text{ g substrate } L^{-1} h^{-1} \text{ and } K_{eq1} = 18.45 \text{ mg } L^{-1} \qquad (2)$$

The data in Table 6.1 were generated with K_{eq} = 0.136, K_{eq1} = 20.92 and r_{max} = 0.196. The data are remarkably accurate with a standard deviation of 1 mg L^{-1} for s and 1 μ gL^{-1} for p. Still the regressed parameters are about 10% from their true values. This is an indication of the difficulties that are encountered in experimental studies of enzyme kinetics: the parameters are strongly correlated and the parameter estimation can easily be swamped by even small experimental errors. In the present case more data should have been obtained for small s values to obtain a better value for K_{eq}, while the values of p were judiciously chosen to obtain a good spread of points on Fig. 6.4. The piece-wise determination of parameters illustrated in Fig. 6.3 and 6.4 is a nice, intuitive approach, but a full non-linear regression of the 16 rate data is probably better, once the structure of the model has been revealed by the graphical approach. Using a non-linear least squares regression one obtains:

$$K_{eq} = 0.139 \text{ gL}^{-1}, \quad K_{eq1} = 20.97 \text{ mgL}^{-1} \text{ and } r_{max} = 0.197 \text{ gL}^{-1}\text{h}^{-1} \qquad (3)$$

With a correlation matrix

$$M = \begin{pmatrix} 1 & 0.4366 & -0.7459 \\ 0.4366 & 1 & -0.0012 \\ -0.7459 & -0.0012 & 1 \end{pmatrix} \qquad (4)$$

The parameter values in (3) are of course very satisfactory, but the correlation between the parameters is considerable.

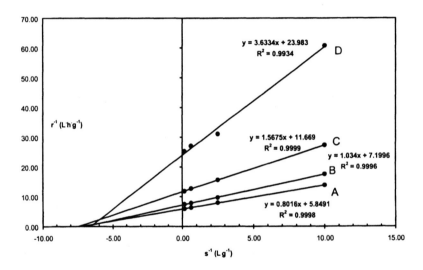

Figure 6.3 Reciprocal plot of r vs. s for a case of competitive inhibition.

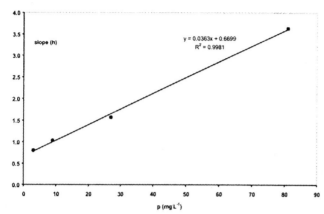

Figure 6.4 The slope of the lines on Fig. 6.3 as a function of inhibitor concentration.

6.2.2 Cooperativity and Allosteric Enzymes

We have seen already in Chapter 2 that the living cell is able to control the concentration of metabolic intermediates in the pathways within narrow limits. On the other hand the enzymatic reaction in a pathway should be able to react quickly to a sudden environmental change in the input of substrate to the pathway. In particular the rate of conversion of the substrate by a pathway enzyme should be sensitive to changes in the substrate concentration around the "normal" level that prevails when the cell is operating at steady state.

The downwards convex hyperbolic shape of Fig. 6.1 shows that Michaelis Menten kinetics is unable to give a high sensitivity of r to changes in s except when the "normal" substrate level is close to zero. One might guess that the saturation constant K (or K_m in eq (6.1)) of an enzyme in a microorganism has by evolution settled on a value which would give $r/r_{max} = 0.5$ around the s-value which is "normally" offered to the enzyme. If suddenly the enzyme is exposed to a pulse of substrate 5 times higher than the normal level $s = K$ then the enzyme kinetics given by eq (6.1) would only increase the rate of the reaction from $r/r_{max} = 0.5$ to 5/6. It would take too long for the enzyme to bring the substrate level back to $s = K$, and in the mean time other pathway reactions may be inhibited (or even repressed) by the high level of substrate waiting to be consumed by the enzyme.

In 1910 Hill proposed an empirical expression for the uptake of oxygen by the protein hemoglobin. He found that the rate of oxygen uptake when pictured against the oxygen partial pressure had a sigmoidal shape, and the data could be fitted to

$$r = \frac{r_{max}\, s^n}{s^n + K^n} \qquad (6.17)$$

The maximum uptake rate r_{max} and the denominator constant K (typical unit mg substrate/mg enzyme) are similar to the two parameters of the Michaelis Menten kinetics, and $r/r_{max} = 0.5$ when $s = K$. The new parameter n is greater than one when the substrate is cooperatively bound to the enzyme.

Eq (6.17) has the property that dr/ds is zero for $s = 0$ whenever n is greater than one, and r/r_{max} increases rapidly from about zero to nearly one in a rather narrow s-interval around K. To be more specific:

$$\text{The point } P \text{ of inflexion of } r(s) \text{ is at} \quad s = s_{inflex} = \left(\frac{n-1}{n+1}\right)^{\frac{1}{n}} K \qquad (6.18)$$

$$\text{The value of } r/r_{max} \text{ is } (n-1)/(2n) \text{ at } s = s_{inflex} \qquad (6.19)$$

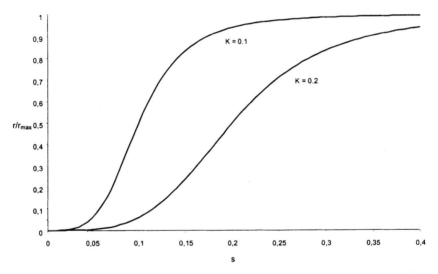

Figure 6.5 The Hill equation (6.17) for $n = 4$ and $K = 0.1$ and 0.2 (arbitrary units).

Enzyme Kinetics and Metabolic Control Analysis

$$\text{The slope of } r/r_{max} \text{ at } P \text{ is } \frac{(n-1)(n+1)}{4ns_{inflex}} \qquad (6.20)$$

Independently of the value of K an increase of r/r_{max} from 0.5 at $s = K$ to 0.9 will be mediated by the same relative change in s, e.g for $n = 4$ by changing s from K to 1.7 K. A corresponding change in r/r_{max} would require an increase of s from K to 9 K if Michelis Menten kinetics had described the functionality between r and s.

Fig. 6.5 shows r/r_{max} for $n = 4$ in eq (6.17) and for two values of K. It is easily seen that the formulas (6.18) to (6.20) can be used to calculate the major features of the two curves.

If desired a linear plot can be constructed to determine n and K from experimental data. Reordering (6.17) gives

$$\frac{r_{max}}{r} - 1 = \frac{s^n + K^n}{s^n} - 1 = \frac{K^n}{s^n}, \quad or \quad \ln\frac{r}{r_{max} - r} = n \ln s - n \ln K \qquad (6.21)$$

The linearity is most pronounced around $s = K$ and deviations are likely to occur both at very small s and when s is much larger than K. The reason is that eq. (6.17) is only an approximation to more complete models of cooperativity as will be discussed below.

Naturally a lot of speculation has been devoted to find a mechanistic explanation for the Hill equation. The simplest mechanism was already proposed by Hill himself: The enzyme is thought to consist of n subunits, and each subunit must bind one substrate molecule before the enzyme can function. Later research has shown that haemoglobin is an aggregate of fixed size with four binding sites for oxygen, but Hill's experiments showed that the binding data corresponded to $n = 2.7$. Apparently the enzyme does not abide by the musketeers oath ("one for all, and all for one") but the binding of the first substrate molecule facilitates the binding of the next molecule, and so on. This is why the binding is called cooperative. As shown in modern models of cooperativity both the numerator and the denominator of (6.17) should consist of a sum of powers of s up to s^n, and the approximation (6.17) is only accurate when the dominant power is s^n in both the numerator and the denominator. The derivation of models for cooperativity is complicated, and the reader is referred to textbooks on enzymology, e.g. Cornish-Bowden (1995) and Fell (1997) for an adequate treatment of the subject.

As described above the cooperation may involve only the enzyme and the substrate, or it could involve cooperation between enzyme, substrate and an effector (either an inhibitor or an activator) of the enzyme. This last interaction is of great importance for the regulation of metabolic pathways. Thus, in lactic acid bacteria fructose 1,6 bisphosphate is an effector of both lactate dehydrogenase (LDH) that converts pyruvate to lactic acid and of pyruvate formate lyase (PFL) that directs pyruvate into the mixed acids pathways, (see Fig. 2.6). A high glucose flux through the EMP pathway gives a high level of fructose 1,6 diphosphate. This activates the LDH and strongly inhibits PFL. At low glucose flux (either because the concentration of glucose in the medium is low or because a less easily fermented sugar such as lactose is used as feed) the inhibition of PFL is immediately lifted (Melchiorsen et al., 2001)

Both in the relatively simple case of substrate cooperativity and in the case of cooperativity between enzyme, substrate and effector the relationship between r and s (with the concentration s_1 of the effector as a parameter) has the shape of Fig. 6.5. Enzymes which exhibit the property of cooperativity are collectively named *allosteric enzymes*.

Note 6.2 Competition of two substrates for the same enzyme
When two substrates S and S_1 can be consumed simultaneously on the same site of an enzyme – e.g. when mannose and glucose compete for the same membrane bound PTS transport protein in lactic acid bacteria – it becomes of interest to calculate the relative uptake rate of the two substrates.

Let the rate of conversion of S and S_1 be r and r_1 respectively, and assume a competitive inhibition of r by S_1 and of r_1 by S. In both cases the inhibition is described by eq. (6.10) where the Michaelis constant is denoted K_m and K_{m1} for r and r_1 respectively.

$$r = \frac{r_{max}\, s}{s + K_m\left(1 + \dfrac{s_1}{K_{eq}}\right)} \quad ; \quad r_1 = \frac{r_{max1}\, s_1}{s_1 + K_{m1}\left(1 + \dfrac{s}{K_{eq1}}\right)} \tag{1}$$

In r the equilibrium constant K_{eq} for the capture of E by S_1 to form the "dead-end" complex ES_1 is equal to the Michaelis constant K_{m1} used in r_1 to describe the first step in the conversion of S_1 by the enzymatic reaction. Similarly, in r_1 K_{eq1} is equal to K_m. Consequently the two expressions in (1) have the same denominator and (1) can be rearranged to:

$$r = \frac{\dfrac{r_{max}\, s}{K_m}}{\dfrac{s}{K_m} + 1 + \dfrac{s_1}{K_{m1}}} \quad\quad r_1 = \frac{\dfrac{r_{max1}\, s_1}{K_{m1}}}{\dfrac{s_1}{K_{m1}} + 1 + \dfrac{s}{K_m}} \tag{2}$$

The ratio of the two rates of conversion is consequently

$$\frac{r}{r_1} = \frac{\left(\dfrac{r_{max}}{K_m}\right)}{\left(\dfrac{r_{max1}}{K_{m1}}\right)} \frac{s}{s_1} \tag{3}$$

and the value of the *specificity constant* r_{max}/K_m determines the competitive edge of one substrate relative to the other.

Enzyme Kinetics and Metabolic Control Analysis

Example 6.2 Determination of NADH in cell extract using a cyclic enzyme assay

One of the most important applications of enzymes in studies of cell physiology is as a tool to determine the intracellular concentration of metabolites. As has often been pointed out in Chapters 3 and 5 the intracellular concentrations of NADH and NAD^+ and in particular of the ratio $NADH/NAD^+$ is of crucial importance for the cellular behavior.

In lactic bacteria the intracellular concentration of NADH is miniscule – of the order of 1 µmole (g DW)$^{-1}$ – and a very large quantity of cell extract must be used to obtain a reasonable accuracy in the assay. This is where cyclic assays can come in very handy. For the case of NADH determination Fig. 6.6 shows that the redox reaction between NADH and NAD^+ is coupled to two other redox reactions, the oxidation of ethanol to acetaldehyde catalyzed by the enzyme alcohol dehydrogenase (ADH), and the reduction by NADH of an indicator, MTT for which the reduced form absorbs light at 570 nm. As long as there is ethanol and MTT_{ox} present the absorbance signal will continue to increase. Since both substrates are available in surplus during the whole experiment, and since NADH is continuously being regenerated by the oxidation of ethanol its concentration stays constant. As the reaction of NADH with MTT_{ox} is first order in the NADH concentration the slope of the absorbance versus time curve is proportional to the amount of NADH initially present in the sample.

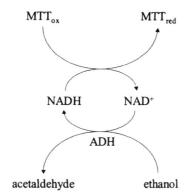

Figure 6.6 A cyclic assay in which the net reaction is: Ethanol + MTT_{ox} → Acetaldehyde + MTT_{red}

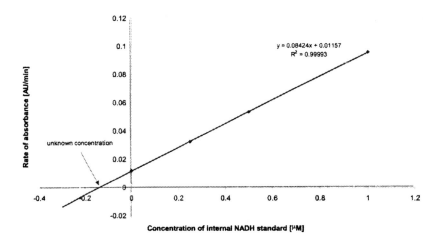

Figure 6.7 Determination of NADH in a cell extract from cultivation of *Lactococcus lactis*

Applied to a cell extract from a lactic bacteria culture where the NAD^+ concentration is much higher than the NADH concentration completely wrong results would be obtained if the NAD^+ content of the sample is not quantitatively removed before adding the ADH enzyme to start the experiment. Nordkvist (2001) has shown that treatment of a sample which originally contains 1 μM NADH and 10 μM NAD^+ for 10 minutes at 56° C and pH = 12.5 completely destroys the NAD^+: The initial slope of the absorbance signal vs. time is exactly the same as in a sample with only 1 μM NADH.

The assay must be calibrated by measurements of the initial slope of the absorbance signal with known amounts of NADH added. But the cell extract sample has to undergo a harsh treatment with alkali at 56°C to remove NAD^+, and therefore a number of cell extracts are spiked with known amounts of NADH and incubated at 56°C for 10 minutes. Ethanol, and MTT_{ox} are added in excess, and adding a sufficient amount of ADH starts the experiment.

Fig. 6.7 shows that the initial slope of the absorbance vs. time signal is a linear function of the added NADH (μM). The intersection with the abscissa axis represents the amount of NADH in the assay that came from the original cell extract.

From Fig. 6.7 one determines the concentration of NADH in the original cell sample withdrawn from the reactor to be 0.156 μM. The biomass concentration in the reactor was 0.684 g L^{-1}, and consequently the intracellular NADH concentration is 0.228 μmoles (g DW)$^{-1}$, or with a cell density of 0.59 g mL^{-1} cell the molarity of NADH in the cell is calculated to 0.134 mM. In a series of 5 experiments based on samples withdrawn from the same exponentially growing anaerobic lactic bacteria culture the mean NADH concentration was found to be 0.172 mM with a standard deviation of 0.029 mM, or 17%.

6.3 Metabolic Control Analysis

We have now seen how the kinetics of an enzyme reaction influences the rate of conversion of a substrate to a product of the enzymatic reaction. But this is not really what we are interested in when the objective is to design a pathway for optimal conversion of a substrate S to the final product P of the pathway. It does not help much if a number of enzymes in the pathway operate at high efficiency if just one single enzymatic reaction is slow. The steady state flux through the pathway from S to P is determined by the bottleneck of the pathway, just as the flow of water by gravity from one reservoir to another is determined by the piece of pipeline that offers the highest resistance to the flow.

In the analysis of a sequence of chemical reactions one often talks about the "rate limiting" step. The concepts of *Metabolic Control Analysis* (MCA) do, however, tell us that it is meaningless to talk about "rate limiting" steps since flux control may well shift from one enzyme to another enzyme depending on the environmental conditions. As will be illustrated in the following description of MCA it is much more fruitful to look on the whole pathway rather than on the individual enzymatic reactions, and that the task is to change the architecture of the pathway such that the resistance to the flow through different steps is more or less the same when the pathway is used to produce a valuable product at relevant environmental conditions. To quantify the degree of flux control exerted by the different enzymes in a pathway and to compare different pathway architectures the so called *Flux Control Coefficients* C^J_i provide a good tool. Basically C^J_i denotes how much the flux J through the pathway will change when the reaction rate of the i'th step of the pathway is increased by improving the catalytic activity of the i'th enzyme. Clearly C^J_i is a systemic property since we look at the change in the outcome of all I reactions in the pathway when we change just one of the steps in the sequence of reactions.

To illustrate the concepts of MCA it will be fruitful to consider a simple example that can be treated analytically. In the example all the topics of MCA are discussed, but the heavy matrix algebra that must necessarily be used to analyze a real pathway is avoided.

Consider therefore a two-step pathway:

$$S \xrightarrow{r_1} S_1 \xrightarrow{r_2} P \tag{6.22}$$

For simplicity the reverse reactions are neglected. The concentrations s of S and p of P are fixed, and we wish to calculate the steady state flux J of e.g. carbon from S to P via the intracellular species S_1 whose concentration will vary not only with s but also with the architecture of the pathway, i.e. with the parameters of the two enzymatic reactions r_1 and r_2.

Since the steady state is considered the two rates r_1 and r_2 must be equal and also equal to the steady state flux J through the pathway.

We shall consider two different expressions for r_1:

$$r_1 = \frac{k_1 s}{s + K_1} \tag{6.23a}$$

$$r_1 = \frac{k_1 s}{s + K_1\left(1 + \dfrac{s_1}{K_{eq}}\right)} \tag{6.23b}$$

In (6.23a) the simple Michaelis Menten kinetics is used, and in (6.23b) the competitive inhibition scheme (6.10) where the intermediate S_1 forms a dead-end complex with the free form of the enzyme that catalyzes r_1.

The second reaction r_2 of (6.22) is assumed to follow simple Michaelis Menten kinetics:

$$r_2 = \frac{k_2 s_1}{s_1 + K_2} \tag{6.24}$$

In order to reduce the number of parameters somewhat the following dimensionless variables and parameters are introduced:

$$\begin{aligned} x = \frac{k_1}{k_2} \quad,\quad y = \frac{s_1}{s} \quad,\quad J' = \frac{J}{k_2} \\ a = \frac{K_{eq}}{s} \quad,\quad b = \frac{K_1}{s} \quad,\quad c = \frac{K_2}{s} \end{aligned} \tag{6.25}$$

$$r_1 = \frac{k_1}{1+b} \quad \text{(6.23a), and} \quad r_1 = \frac{k_1}{1 + b\left(1 + \dfrac{y}{a}\right)} \quad \text{(6.23b)}$$

$$J' = \frac{J}{k_2} = \frac{y}{y+c} \tag{6.26}$$

Enzyme Kinetics and Metabolic Control Analysis

When r_1 given by (6.23a) is equated with r_2 one obtains:

$$\frac{x}{1+b} = \frac{y}{y+c} \quad \text{or} \quad y = \frac{cx}{1+b-x} \qquad (6.27)$$

We note that, as expected $J = k_2 \dfrac{y}{y+c} = k_2 \dfrac{x}{1+b} = \dfrac{k_1}{1+b}$ can be calculated both as r_1 and as r_2.

When (6.23b) is used for r_1 then y is determined from

$$\frac{k_1}{1+b\left(1+\dfrac{y}{a}\right)} = \frac{k_2 y}{y+c} \quad \text{or}$$

$$2y = -\left(\frac{a}{b}+a-\frac{a}{b}x\right)+\left(\left(\frac{a}{b}+a-\frac{a}{b}x\right)^2+4\frac{ac}{b}x\right)^{\frac{1}{2}} \qquad (6.28)$$

Fig. 6.8 shows $J(x)$ for the two expressions (6.23a and 6.23b) for r_1. There is an important difference between the two expressions. With (6.23a) J is proportional to x, and J is completely independent of the parameters of r_2. With (6.23b) $J(x)$ has the typical downwards concave shape of the hyperbolic functions which determine the kinetics for individual enzymes. An increase in k_1 is profitable for small x (i.e. small k_1) but has little effect for large x, since dJ/dx becomes zero as J increases towards 1.

For small x a power series expansion of $y(x)$ in (6.28) yields

$$y \approx \frac{cx}{b+1} \quad \text{and} \quad J \approx \frac{x}{1+b} \qquad (6.29)$$

Consequently, for small values of k_1 the flux J depends only on the kinetic parameters k_1 and K_1 of r_1. The value of s_1 never becomes large enough to give any influence of the denominator term s_1/K_{eq} in (6.23b). For larger k_1 values s_1 increases, and the parameters K_2 and K_{eq} start to influence the flux J through the pathway.

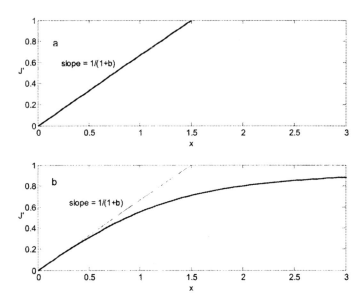

Figure 6.8 Dimensionless flux $J' = J/k_2$ through a two step pathway (6.22) as a function of the ratio $r_{max1}/r_{max2} = k_1/k_2 = x$. In both cases r_2 is given by (6.24), and the parameters of (6.25) are $a=1$, $b=c=0.5$.
a. r_1 given by (6.23a).
b. r_1 given by (6.23b).

We can conclude the following:
- If both rate constants are multiplied with the same factor f the flux J through the pathway also increases by a factor f. The shape of J' does not change, since J' is only a function of x. For large values of k_1 the flux J becomes proportional to k_2.
- If s is doubled the value of all parameters a, b and c will decrease by a factor 2, and $y(x)$ as determined from (6.28) will have a different shape. $J'(x)$ will increase faster with x, but the initial slope $(b + 1)^{-1}$ will only increase by a factor 2 if b is much greater than 1. Thus J' does not react proportionally to a change in s.

These observations originate in the form of r_1. In (6.23a) the parameters of r_2 have no influence at all on the flux. With (6.23b) the control of the flux is distributed between the two reactions, and it moves towards the second reaction when k_1 increases, since a larger value of k_1 leads to a higher value of y in (6.28), and this influences r_1 negatively through the denominator term with y. The smaller the value of c the faster is step two of the sequence, and S_1 is removed fast enough to avoid a build up of S_1 waiting to be processed by step two.

It now appears that the question of flux control by either step 1 or 2 in the sequence depends on

Enzyme Kinetics and Metabolic Control Analysis

the slope of $J'(x)$. This is where the flux control coefficients are introduced.

Define

$$C_1^J = \frac{\partial J}{\partial e_1} \frac{e_1}{J} \quad \text{and} \quad C_2^J = \frac{\partial J}{\partial e_2} \frac{e_2}{J} \tag{6.30}$$

In (6.30) the enzyme concentrations e_1 and e_2 are used instead of the rate constants k_1 and k_2. It would be more correct to use the enzyme activities determined through the rate constants than the enzyme concentrations (or enzyme "dosages") since the activity of an enzyme is not necessarily proportional with the enzyme concentration. Here we will, however, assume that there is proportionality between enzyme concentration and enzyme activity. Flux and enzyme concentrations are measured in units that are not commensurable, and consequently the sensitivity of the flux with respect to changes in e_i is scaled by e_i/J in order to give dimensionless flux control coefficients.

In the example we assume that the rate constants are proportional to the enzyme concentrations and we obtain the following for the two cases of r_1.

$$C_1^J = \frac{\partial J}{\partial k_1} \frac{k_1}{J} = (1+b)^{-1} \frac{k_1}{\left(\frac{k_1}{1+b}\right)} = 1$$

$$C_2^J = \frac{\partial J}{\partial k_2} \frac{k_2}{J} = 0 \tag{6.26a}$$

since k_1 (or x) is the independent variable while k_2 is a constant.

To obtain the result for (6.23b) the following "chain-rule" for differentiation is used:

$$C_1^J = \frac{\partial J}{\partial k_1} \frac{k_1}{J} = \frac{\partial J'}{\partial x} \frac{x}{J'} = \frac{\partial J'}{\partial y} \frac{dy}{dx} \frac{x}{J'}$$

$$C_2^J = \frac{\partial J}{\partial k_2} \frac{k_2}{J} = \left(k_2 \frac{\partial J'}{\partial k_2} + J'\right) \frac{1}{J'} = \left(k_2 \frac{\partial J'}{\partial x} \frac{dx}{dk_2} + J'\right) \frac{1}{J'} = \left(-x \frac{\partial J'}{\partial x} + J'\right) \frac{1}{J'} \tag{6.31}$$

which means that also in general $C_1^J + C_2^J = 1$, just as in the simple case (6.23a).

Inserting

$$\frac{\partial J'}{\partial y} = \frac{c}{(c+y)^2} = \frac{c J'}{y(c+y)}$$

in (6.31) together with $y(x)$ and dy/dx from (6.27) one obtains C_1^J as an explicit function of x. We shall see in Example 6.3 that there is an easier way to calculate the flux control coefficients, but from the calculations above we can already draw the main conclusion that C^J_1 decreases while C^J_2 increases when x increases. Since $J(x)$ is a monotonous function of x so must also $\ln J(x)$ be a monotonous function of $\ln x$ and consequently

$$C^J_1 = \frac{\partial J}{\partial k_1}\frac{k_1}{J} = \frac{\partial J'}{\partial x}\frac{x}{J'} = \frac{\partial \ln J'}{\partial \ln x}$$

must decrease monotonously to zero when x (or k_1) increases to infinity. Since $C^J_1 + C^J_2 = 1$ the value of C^J_2 must correspondingly increase towards 1.

The conclusion is that the control shifts from the first reaction step to the second for increasing activity of the first enzyme. The rate of change varies with the environmental conditions as was seen for the case of increasing s. For large s the concentration of the first enzyme (represented by the parameter k_1) must be relatively large to change the relative size of C^J_1 and C^J_2 since the competitive inhibition of the first enzyme by S_1 is less pronounced when s is large.

Having introduced the flux control coefficients and explained their significance by means of a simple example we can now turn to a general unbranched pathway with I reaction steps:

$$S \leftrightarrows S_1 \leftrightarrows S_2 \cdots\cdots S_{I-1} \leftrightarrows P \qquad (6.32)$$

There are $I - 1$ intermediates $S_1\ldots\ldots S_{I-1}$ and I pathway reactions mediated by the enzymes $E_1\ldots\ldots E_I$. In the steady state the rates of all the pathway reactions r_i are equal and equal to the flux J through the pathway. We wish to calculate the flux J of substrate of known concentration that can be processed by the pathway, and we wish to investigate how the pathway architecture can be improved to allow a larger flux to be processed.

The flux control coefficients C^J_i are defined by

$$C^J_i = \frac{e_i}{J}\frac{\partial J}{\partial e_i} \qquad (6.33)$$

In Note 6.3 it is shown that irrespective of the kinetics of the individual enzyme catalyzed reactions the sum of the flux control coefficients is 1 for the unbranched pathway.

$$\sum_{i=1}^{I} C^J_i = 1 \qquad (6.34)$$

This result is named the *flux control summation theorem*

Note 6.3 Proof of the flux-control summation theorem
If in a thought experiment we increase all the enzyme activities simultaneously by the same fractional amount a, i.e..

Enzyme Kinetics and Metabolic Control Analysis

$$\frac{de_i}{e_i} = a \; ; \qquad i = 1,\ldots,I \tag{1}$$

then the fractional change in all the reaction rates will also be a and the level of the intermediates is therefore kept constant. Consequently, the fractional change in the overall flux is given by

$$\frac{dJ}{J} = a \tag{2}$$

where J is a function of the level of all enzyme activities in the pathway and dJ is consequently the sum of all the individual fractional changes in the flux when the level of each of the individual enzymes is changed. Thus

$$\frac{dJ}{J} = \sum_{i=1}^{I} \frac{1}{J}\left(\frac{dJ}{de_i}\right)_{e_j \; j=1,\ldots,I \text{ and } j \neq i} de_i \tag{3}$$

or when the definition of the control coefficients in Eq. (6.33) is introduced:

$$\frac{dJ}{J} = \sum_{i=1}^{I} C_i^J \left(\frac{de_i}{e_i}\right) \tag{4}$$

By inserting Eqs. (1) and (2), it is easily seen that the flux-control summation theorem in Eq. (6.34) is derived.

Similar to the flux-control coefficients we define *Concentration-Control Coefficients* by

$$C_{ij} \equiv \frac{e_i}{s_j} \frac{\partial s_j}{\partial e_i} \; ; \qquad i = 1,\ldots,I \text{ and } j = 1,\ldots,I-1 \tag{6.35}$$

These coefficients specify the relative change in the level of the jth intermediate when the level of the ith enzyme is changed. Since the level of the intermediates is not changed when all the enzyme considerations are changed by the same factor a (see Note 6.3), we have

$$\frac{ds_j}{s_j} = \sum_{i=1}^{I}\left(\frac{1}{s_j}\frac{\partial s_j}{\partial e_i}\right)_{e_j \, ; \; j=1,\ldots,I, \; j \neq i} de_i = \sum_{i=1}^{I} \frac{e_i}{s_j}\frac{\partial s_j}{\partial e_i}\frac{de_i}{e_i} = a\sum_{i=1}^{I} C_{ij}$$

or

$$\sum_{i=1}^{I} C_{ij} = 0; \quad j = 1,\ldots,I-1 \tag{6.36}$$

This implies that for each of the $I - 1$ intermediates, at least one of the enzymes must exert a negative control, i.e., when the level of one of the enzymes increases then the metabolite concentration decreases.

The control coefficients specify the influence of the enzyme level on the overall flux, but it is also valuable to examine the sensitivity of the net rate of individual enzymatic reactions to variations in the size of each of the $I - 1$ metabolite pools. Thus we define *elasticity coefficients* for the i th enzyme by

$$\varepsilon_{ji} = \frac{s_j}{r_i}\frac{\partial r_i}{\partial s_j}; \quad i = 1,\ldots,I \text{ and } j = 1,\ldots,I-1 \tag{6.37}$$

where r_i is the net rate of the ith enzymatic reaction. A negative value of ε_{ji} implies that when the level of the jth intermediate is increased the net rate of the ith reaction decreases, and *vice versa*. For a simple, reversible enzymatic reaction (see Example 6.4) the elasticity coefficient $\varepsilon_{i,i}$ is negative, whereas the elasticity coefficient $\varepsilon_{i-1,i}$ will be positive. Furthermore, if the jth intermediate does not influence the rate of the ith enzymatic reaction (the rate of this reaction may be independent of the concentration of s_j), then $\varepsilon_{ji} = 0$. The elasticity coefficients are connected to the flux control coefficients through the *flux-control connectivity theorem*:

$$\sum_{i=1}^{I} C_i^J \varepsilon_{ji} = 0; \quad j = 1,\ldots,I-1 \tag{6.38}$$

for which the proof is given in Note 6.4. From Eq. (6.38) it is seen that enzymes that have high elasticities tend to have low flux-control coefficients. The increase or decrease of an internal reaction rate r_i with an increase in the internal metabolite concentration s_j is not in itself of much interest; the object of MCA is to identify the enzyme an increase in the activity of which will maximize the increase of the productivity of the whole pathway, i.e., maximize the change in J. Thus C_i^J is of prime importance. But experimentally it may be much more difficult to vary the activity level of the enzymes than it is to study the influence of metabolite concentrations on the rates of the individual intracellular reactions r_i for a fixed set of enzyme activities.

Note 6.4 Proof of the flux-control connectivity theorem
The differential change in the ith reaction caused by a differential change in the jth intermediate s_j is

$$dr_i = \left(\frac{\partial r_i}{\partial s_j}\right)_{s_k, k \neq j} ds_j \tag{1}$$

or

$$\frac{dr_i}{r_i} = \varepsilon_{ji} \frac{ds_j}{s_j}; \quad i = 1,...,I \quad \text{and} \quad j = 1,...,I-1 \qquad (2)$$

This fractional change in the rate of the i th reaction can also be accomplished by changing the level of the ith enzyme, i.e.,

$$\frac{dr_i}{r_i} = \frac{de_i}{e_i} \qquad (3)$$

If we balance the change in the reaction rate due to the differential change of the intermediate by a change in the enzyme level, we find from Eqs. (2) and (3)

$$\frac{de_i}{e_i} = \varepsilon_{ji} \frac{ds_j}{s_j}; \quad i = 1,...,I \quad \text{and} \quad j = 1,...,I-1 \qquad (4)$$

If we change the internal rates without changing the total flux J through the pathway, then by Eq. (4):

$$\frac{dJ}{J} = \sum_{i=1}^{I} C_i^J \frac{de_i}{e_i} \qquad (5)$$

By combination of Eq. (5) with Eq. (4) we get

$$\sum_{i=1}^{I} C_i^J \left(\frac{ds_j}{s_j} \right) = \frac{ds_j}{s_j} \sum_{i=1}^{I} C_i^J \varepsilon_{ji} = 0; \quad j = 1,...,I-1 \qquad (6)$$

and since we imposed a fractional change on each of the $I - 1$ intermediates, i.e., $ds_j \neq 0$, we obtain the flux-control connectivity theorem, Eq. (6.38).

In addition to the flux-control connectivity theorem, Westerhoff and Chen (1984) introduced two other connectivity theorems:

$$\sum_{i=1}^{I} C_{ij} \varepsilon_{ji} = -1; \quad j = 1,...,I-1 \qquad (6.39)$$

$$\sum_{i=1}^{I} C_{ij} \varepsilon_{ki} = 0; \quad j = 1,...,I-1 \quad \text{and} \quad k \neq j \qquad (6.40)$$

In matrix notation we can summarize all the theorems given in Eqs. (6.34), (6.36), and (6.38)-(6.40) by

$$\begin{pmatrix} 1 & 1 & \cdots & 1 \\ \varepsilon_{11} & \varepsilon_{12} & \cdots & \varepsilon_{1I} \\ . & . & \cdots & . \\ . & . & \cdots & . \\ \varepsilon_{I-1,1} & \varepsilon_{I-1,2} & \cdots & \varepsilon_{I-1,I} \end{pmatrix} \begin{pmatrix} C_1^J & -C_{11} & \cdots & -C_{1,I-1} \\ C_2^J & -C_{21} & \cdots & -C_{2,I-1} \\ . & . & \cdots & . \\ . & . & \cdots & . \\ C_I^J & -C_{I,1} & \cdots & -C_{I,I-1} \end{pmatrix} = \begin{pmatrix} 1 & 0 & \cdots & 0 \\ 0 & 1 & \cdots & 0 \\ . & . & \cdots & . \\ . & . & \cdots & . \\ 0 & 0 & \cdots & 1 \end{pmatrix} \quad (6.41)$$

or

$$\mathbf{EC}^* = \mathbf{I} \quad (6.42)$$

where **I** is the unity matrix (dimension $I \times I$). For a nonsingular **E**, the control coefficients are obtained as

$$\mathbf{C}^* = \mathbf{E}^{-1} \quad (6.43)$$

and the control coefficients are elements of $\mathbf{C}^*$ determined as:

$$C_i^J = C_{i1}^* \text{ and } C_{i,j-1} = -C_{ij}^*, \quad j > 1 \quad (6.44)$$

Example 6.3 Flux control coefficients from elasticities in a simple example.
In the introductory example case (6.23b) one calculates

$$\varepsilon_{11} = \frac{\partial r_1}{\partial s_1} \frac{s_1}{r_1} = -\frac{k_1 s}{\left(s + K_1\left(1 + \frac{s_1}{K_{eq}}\right)\right)^2} \frac{K_1}{K_{eq}} \frac{s_1}{r_1} = \frac{-\frac{K_1}{K_{eq}} s_1}{\left(s + K_1\left(1 + \frac{s_1}{K_{eq}}\right)\right)} = -\frac{\frac{b}{a} y}{\left(1 + b\left(1 + \frac{y}{a}\right)\right)} \quad (1)$$

$$\varepsilon_{12} = \frac{\partial r_2}{\partial s_1} \frac{s_1}{r_2} = \frac{K_2}{s_1 + K_2} = \frac{c}{y + c} \quad (2)$$

Using (6.41) one obtains :

Enzyme Kinetics and Metabolic Control Analysis

$$\begin{pmatrix} C_1^J \\ C_2^J \end{pmatrix} = \begin{pmatrix} 1 & 1 \\ \varepsilon_{11} & \varepsilon_{12} \end{pmatrix}^{-1} \begin{pmatrix} 1 \\ 0 \end{pmatrix} = \frac{1}{\varepsilon_{12} - \varepsilon_{11}} \begin{pmatrix} \varepsilon_{12} \\ -\varepsilon_{11} \end{pmatrix} \qquad (3)$$

Obviously $C_1^J + C_2^J = 1$, but more specifically

$$C_1^J = \frac{\varepsilon_{12}}{\varepsilon_{12} - \varepsilon_{11}} = \frac{1}{1 + \dfrac{\dfrac{b}{ac} y(y+c)}{1 + b\left(1 + \dfrac{y}{a}\right)}} = \frac{1}{1 + \dfrac{by^2}{acx}} \qquad (4)$$

Fig. 6.9 results by calculation of C_1^J as a function of x using (4) and eq. (6.28) to find y for a given value of x. The use of elasticities to calculate C_1^J obviously bypasses a lot of the algebra associated with a direct calculation of C_1^J from J by (6.31). The results are easily checked with values obtained by numerical differentiation of $J'(x)$ in Fig. 6.8b since by (6.31)

$$C_1^J = \frac{\partial J'}{\partial x} \frac{x}{J'} \qquad (5)$$

In general the enzymatic reaction rates are given as $r_i = N/D$. Consequently eq. (6) may be useful when calculating elasticities, either manually or by a symbolic computer software program.

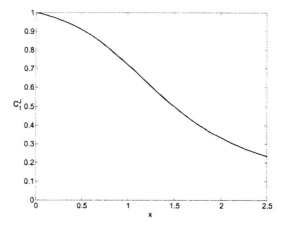

Figure 6.9 Flux coefficient C_1^J calculated from elasticities.

$$\varepsilon_{ji} = \frac{\left(D\dfrac{\partial N}{\partial s_j} - N\dfrac{\partial D}{\partial s_j}\right)}{D^2} \cdot \frac{s_j D}{N} = s_j\left(\frac{1}{N}\dfrac{\partial N}{\partial s_j} - \frac{1}{D}\dfrac{\partial D}{\partial s_j}\right) \qquad (6)$$

Equation (6) is used in Example 6.4 to calculate elasticities.

Example 6.4. Illustration of metabolic control analysis

We now consider the linear pathway in Eq (6.32) for three intermediates. Consequently, there are four enzymatic reactions, i.e., $I = 4$. Feedback inhibition of the second enzyme, e_2, by the last intermediate, s_3, is included. All the reactions are taken to be reversible, and the feedback inhibition is taken to be non-competitive. Thus we may write the net rates for the four enzymatic reactions as

$$r_1 = k_1 e_1 \frac{s}{s + K_1\left(1 + \dfrac{s_1}{K_1^*}\right)} \qquad (1)$$

$$r_2 = k_2 e_2 \frac{s_1}{s_1 + K_2\left(1 + \dfrac{s_2}{K_2^*}\right)} \cdot \frac{1}{1 + \dfrac{s_3}{K_{inhib}}} \qquad (2)$$

$$r_3 = k_3 e_3 \frac{s_2}{s_2 + K_3\left(1 + \dfrac{s_3}{K_3^*}\right)} \qquad (3)$$

$$r_4 = k_4 e_4 \frac{s_3}{s_3 + K_4\left(1 + \dfrac{p}{K_4^*}\right)} \qquad (4)$$

The rate of the reverse reaction is included in a somewhat unusual fashion, in the inhibition term in the denominator, which increases with increasing conversion. Close to equilibrium, the rate expressions in Eqs. (1)-(4) cannot possibly hold. For the purpose of illustrating the methodology the realism of the kinetic expressions is, however, not of overriding importance.

$$\mathbf{E} = \begin{pmatrix} 1 & 1 & 1 & 1 \\ -\dfrac{s_1 K_1/K_1^*}{a_1} & \dfrac{K_2(1+s_2/K_2^*)}{a_2} & 0 & 0 \\ 0 & -\dfrac{s_2 K_2/K_2^*}{a_2} & \dfrac{K_3(1+s_3/K_3^*)}{a_3} & 0 \\ 0 & -\dfrac{s_3}{K_{inhib}+s_3} & -\dfrac{s_3 K_3/K_3^*}{a_3} & \dfrac{K_4(1+p/K_4^*)}{a_4} \end{pmatrix} \quad (5)$$

where

$$\begin{pmatrix} a_1 \\ a_2 \\ a_3 \\ a_4 \end{pmatrix} = \begin{pmatrix} s + K_1(1+s_1/K_1^*) \\ s_1 + K_2(1+s_2/K_2^*) \\ s_2 + K_3(1+s_3/K_3^*) \\ s_3 + K_4(1+s_4/K_4^*) \end{pmatrix} \quad (6)$$

From Eq. (5) it is seen that the elasticity coefficients for the ith enzyme with respect to its substrate -i.e., the $(i-1)$th intermediate - is positive, whereas its elasticity coefficient with respect to its product (i.e., the ith intermediate) is negative.

If the kinetic parameters are known together with steady-state levels of the intermediates, the numerical values of the elasticity coefficients can be calculated, but normally it is difficult to obtain *in vivo* experimental values for the parameters, and it is likely that the *in vitro*-determined parameters do not represent the true situation. Here we will, however, assume that at a particular steady state [$s, p, (s_1, s_2, s_3(s,p))$] we have determined the elasticity coefficients to be given by

$$\mathbf{E} = \begin{pmatrix} 1 & 1 & 1 & 1 \\ -0.9 & 0.5 & 0 & 0 \\ 0 & -0.2 & 0.7 & 0 \\ 0 & -1.0 & -0.5 & 0.5 \end{pmatrix} \quad (7)$$

in which case the control coefficient matrix is obtained from Eqs. (6.41 and 6.44):

$$\mathbf{C} = \begin{pmatrix} 0.14 & 0.96 & 0.39 & 0.27 \\ 0.24 & -0.27 & 0.69 & 0.48 \\ 0.07 & -0.08 & -1.23 & 0.14 \\ 0.55 & -0.61 & 0.15 & -0.89 \end{pmatrix} \quad (8)$$

The desired flux-control coefficients C_i^J are in the first column of $\mathbf{C}^*$. The last enzyme in the pathway has a flux-control coefficient significantly higher than $1/I = \frac{1}{4} = 0.25$, and it is therefore the rate controlling step. The rate of conversion of s_3 is too small. This in itself reduces the total flux through the sequence, and furthermore a buildup of a high level of the metabolite s_3 impedes the second reaction, the conversion of s_1 to s_2. Thus an increase in the level of the enzyme E_2 that is under feedback control will not necessarily be the best remedial action.

From Eq. (8), it is observed that the concentration-control coefficients for the ith metabolite with respect to the ith enzyme is positive, but with respect to the $(i + 1)$th enzyme it is negative. If the concentration of the ith enzyme increases, the concentration of the ith intermediate (product of the reaction) increases, whereas the concentration of the $(i - 1)$th intermediate (the substrate) decreases. The last row of $\mathbf{C}^*$ contains in its last three columns the concentration-control coefficients pertinent to the last enzyme. When the activity of this enzyme is increased, the level of the last intermediate s_3 and of the first intermediate s_1 sharply decrease while the level of the second intermediate increases. This is easily understood; Both s_3 and s_1 are substrates that are more rapidly consumed when the level of E_4 increases, s_3 directly and s_1 indirectly when the inhibition control of E_2 is relieved. Notice that the last three columns of $\mathbf{C}^*$ sum to zero, as they should according to Eq. (6.36), and that the first row contains only positive control coefficients: The total flux as well as the concentration level of all intermediates increases when the activity of the first enzyme is increased.

The effect of the feedback inhibition can be illustrated by setting $\varepsilon_{32} = 0$ and calculating the control matrix for this situation. The result is given in Eq. (9):

$$\mathbf{C} = \begin{pmatrix} 0.27 & 0.82 & 0.75 & 0.52 \\ 0.47 & -0.52 & 1.34 & 0.94 \\ 0.13 & -0.15 & -1.05 & 0.27 \\ 0.13 & -0.15 & -1.04 & -1.73 \end{pmatrix} \qquad (9)$$

By comparison of the control coefficients with those in Eq. (8), where feedback inhibition is present, it is observed that the rate control is now at the second enzyme in the pathway. Thus this enzyme is a potential rate-controlling enzyme for the true system with feedback inhibition, since if the inhibition is removed or reduced in strength the second reaction is controlling the overall flux. When the feedback inhibition is lifted, all three concentration-control coefficients in the last row become negative; Increasing the activity of the last enzyme in a straight sequence lowers the level of all intermediates.

We now consider a microorganism in which the pathway described above (with feedback inhibition) is active. The product of the pathway is a desired product, e.g., an antibiotic, and we want to design a new strain – using MCA – in which an increased flux through the pathway is possible. Thus we want to decrease the rate control, which can be done, e.g., by inserting a gene coding for an enzyme that also catalyzes the conversion of s_1 to s_2 but has different elasticity coefficients, i.e., kinetic parameters compared with the native enzyme – e.g. a higher value of K_{inhib} (less inhibition) and/or a lower value of K_2/K_2^*.

Assume that the search leads to a strain that has an enzyme similar to E_4 but with other elasticity coefficients. The gene for this slightly different enzyme is cloned into the chosen production strain. Take the elasticity coefficients for the second pathway reaction to be

$$\begin{pmatrix} \varepsilon_{11} \\ \varepsilon_{12} \\ \varepsilon_{13} \end{pmatrix} = \begin{pmatrix} 1.0 \\ -0.6 \\ -0.3 \end{pmatrix} \quad (10)$$

i.e., the feedback inhibition by s_3 is weaker and the ratio K_2/K_2^* is slightly lower. With the slightly different enzyme in the production strain instead of the native enzyme, the control matrix is found to be

$$C = \begin{pmatrix} 0.25 & 0.83 & 0.72 & 0.50 \\ 0.23 & -0.25 & 0.65 & 0.45 \\ 0.19 & -0.22 & -0.88 & 0.39 \\ 0.33 & -0.36 & -0.49 & -1.34 \end{pmatrix} \quad (11)$$

Thus with the new enzyme the flux control of the last enzyme has been reduced, and the control coefficients are all close to $1/I = 0.25$. Consequently, none of the enzymes dominate the flux control in the pathway.

MCA suggests a systematic way of improving the overall performance of a metabolic pathway and points to specific experiments that may assist the protein engineering work. What has not been included in the above discussion is the role of the operating conditions on the performance of the cell. All intracellular metabolite concentrations are calculable once s, p, and the kinetics are given. The elasticity coefficients are complex functions of s_i and before any enzyme modification (i.e., changes of kinetic parameters) is attempted the current reaction network must be optimized to give the maximum flux by simulations at different s and p values.

When flux control is lifted, all the control coefficients are of approximately the same size, but this does not necessarily indicate that the flux through the pathway is significantly higher than in the native strain. This has to be checked separately. However, if one obtains a modified strain in which all the control coefficients have approximately the same value, the flux may be increased by amplifying the level of all the enzymes, e.g., by inserting a stronger promotor upstream of the genes coding for the four enzymes.

We have until now considered only linear pathways, since these are much easier to analyze than branched pathways. The MCA can, however, easily be extended to branched pathways. If the total flux J branches into two fluxes J_1 and J_2, then [see, e.g., Westerhoff and Kell (1987)]:

$$J_2 \sum_{\text{branch 1}} C_i^J - J_1 \sum_{\text{branch 2}} C_i^J = 0 \quad (6.45)$$

An attempt to increase the flux through the second branch by decreasing the control coefficients in that branch relative to the rest of the control coefficients in the total pathway often fails because of the interbranch interactions described by Stephanopoulos and Vallino (1991). With a branched pathway the number of intermediates becomes $I - 2$ when the total number of enzymes is I, and the **E** matrix in Eq. (6.42) is therefore not quadratic. But Eq. (6.45) represents an additional constraint (extra row), which allows calculation of all the control coefficients from the elasticity coefficients, for branched pathways also. In branched pathways negative flux-control coefficients can appear since an increase of the activity of an enzyme in one branch could well have a negative influence on the flow in the other branch.

MCA is a powerful tool for qualitative studies of metabolic pathways. A serious drawback of the method is, however, the requirement that either the elasticity coefficients or the control coefficients have to be measured. This is not an easy task, and practical applications are as yet rarely seen in the literature. In some cases one may learn something of the enzyme kinetics from *in vitro* experiments, and this may be used to calculate the elasticity coefficients as illustrated in Example 6.4 and in an appealingly simple study by Delgado *et al.* (1993) for two glycolytic enzymes. It is, however, unlikely that the *in vivo* enzyme kinetics is the same as that determined from *in vitro* experiments, and conclusions based on *in vitro* experiments must therefore be regarded as tentative only.

The most direct way of obtaining the control coefficients is to examine the effect of variations of the enzyme activities on the pathway flux. The enzyme level may be varied genetically, e.g., by insertion of stronger promoters or additional copies of the genes, as illustrated by Flint *et al.* (1981), who examined the arginine pathway of the filamentous fungus *Neurospora crassa*. From measurements of fluxes in different mutants expressing different levels of the individual enzymes, the control coefficients could be found directly. This approach is, however, very time-consuming and it is not generally applicable since different mutants of the applied strain are not normally available.

Another limitation of MCA is that it is based on a steady state of the pathway, and strictly speaking the analysis is valid only close to a given operating point (s,p). However, the purpose of metabolic engineering is to design a strain with a completely different flux distribution than the parental strain, and the analysis of the parental strain based on MCA is therefore not likely to hold for the new strain. Despite these drawbacks MCA is still a useful tool for examination of metabolic pathways, but it should be used in conjunction with other methods. Both Stephanopolous et al. (1998) and Fell (1997) give detailed accounts of the theoretical as well as the experimental foundation of MCA. One of the original contributors to the development of MCA has proposed (Small and Kacser, 1993) that as long as the pathway flux J is a downwards convex function of the activity of the ith enzyme then the control coefficient can be obtained experimentally also by measuring the change in flux mediated by an integral change in e_i. This important result will be proved and used in Example 6.5.

The drawbacks of the steady-state assumptions applied in MCA are circumvented in a method described by Liao and Lightfoot (1988). It is a generalization of a classical analysis by Wei and Prater (1962) of *characteristic reaction paths* to identify rate-controlling steps using *time-scaling separation*. In an extension of these techniques, Delgado and Liao (1991) illustrated the use of measurements of the pathway intermediates in a transient state after a perturbation is applied to the cells. The method is based on a linearization of all the reaction rates in the pathway, i.e.,

Enzyme Kinetics and Metabolic Control Analysis

$$r_i = \sum_{j=1}^{I-1} a_{ji} s_j + b_i \qquad i = 1,\ldots,I \tag{6.46}$$

The coefficients a_{ij} and b_i change from one steady state to the next, but as pointed out by Small and Kacser (1993) it is not certain that the flux-control coefficients change dramatically. By using the flux-control connectivity theorem [Eq. (6.38)] it is possible (see Note 6.5) to derive Eq. (6.47), which correlates the control coefficients with the changes in internal reaction rates immediately after a perturbation in the system. Thus, by measuring the transient responses of the pathway intermediates after, e.g., changing the concentration s of the substrate, one can calculate the variations in the fluxes and thereby the control coefficients:

$$\sum_{i=1}^{I} C_i^J \Delta r_i = 0 \tag{6.47}$$

In two further papers, Delgado and Liao (1992a,b) have reshaped their original method into a practical procedure for obtaining control coefficients from measurements of transient metabolite concentrations. We will end our introduction to MCA with an extensive example (Example 6.5) that will deal with these methods and discuss their drawbacks.

Note 6.5 Derivation of Eq. (6.47)
After linearization of the reaction rates, we find from the definition of the elasticity coefficients

$$\varepsilon_{ji} = \frac{s_j}{r_i} a_{ji}; \qquad i = 1,\ldots,I \quad \text{and} \quad j = 1,\ldots,I-1 \tag{1}$$

and consequently by Eq. (6.38):

$$\sum_{i=1}^{I} C_i^J a_{ji} \frac{s_j}{r_i} = \sum_{i=1}^{I} C_i^J a_{ji} = 0; \qquad j = 1,\ldots,I-1 \tag{2}$$

since $s_j \neq 0$ and all reaction rates r_i are equal at any steady state. We now make a perturbation of the system and observe each $\Delta s_j = s_j(t) - s_j(t=0)$. Multiply the last part of Eq. (2) by Δs_j:

$$\sum_{i=1}^{I} C_i^J a_{ji} \Delta s_j = 0; \qquad j = 1,\ldots,I-1 \tag{3}$$

and add the $I - 1$ equations:

$$\sum_{j=1}^{I-1}\sum_{i=1}^{I} C_i^J a_{ji} \Delta s_j = 0$$

$$\Downarrow \qquad (4)$$

$$\sum_{i=1}^{I} C_i^J \sum_{j=1}^{I-1} a_{ji} \Delta s_j = 0$$

Since the change in the *i*th internal rate (or flux) is found from Eq. (6.46) to be given by

$$\Delta r_i = \sum_{j=1}^{I-1} a_{ji} \Delta s_j ; \quad i = 1,...,I \qquad (5)$$

Eq. (6.47) is obtained by inserting Eq. (5) in Eq. (4).

Example 6.5 Approximate calculation and experimental determination of flux control coefficients. Consider the two-step pathway (6.23b) and (6.24). If K_1, K_2 and K_{eq} are constant there is only one variable left to characterize the architecture of the pathway. This could be k_1 or it could be $x = k_1/k_2$ since with only two control coefficients one of them can always be calculated from the flux control summation theorem. Consequently a study of $J(x)$ or of $J'(x) = J(x)/k_2$ will show how the control is distributed between the two reaction steps when the enzyme concentration e_1 (considered to be proportional to k_1) is varied relative to e_2.

There is one environmental variable and this is the concentration s of the substrate to the first reaction in the pathway. In our first treatment of the problem s was held constant, and in (6.25) the variable s_1 and the parameters could be normalized by this constant. If we wish to study the influence of s on the flux and on the flux control distribution we must normalize s_1 with a reference value s^0 of the exterior substrate concentration.

Consequently we have the following dimensionless variables:

$$x = k_1/k_2, \quad s' = s/s^0 \quad \text{and} \quad y = s_1/s^0 \qquad (1)$$

and the parameters :

$$a = K_{eq}/s^0, \quad b = K_1/s^0 \quad \text{and} \quad c = K_2/s^0 \qquad (2)$$

The variable y is, however, a function of the other two variables x and s' since eq (6.28) now reads:

$$2y = -\left(a + \frac{a}{b}s' - \frac{a}{b}s'x\right) + \sqrt{\left(a + \frac{a}{b}s' - \frac{a}{b}s'x\right)^2 + \frac{4ac}{b}s'x} \qquad (3)$$

The rates r_1 and r_2 of the two reaction steps are normalized by k_2:

Enzyme Kinetics and Metabolic Control Analysis

$$R_1 = r_1/k_2 = \frac{s'x}{s'+b\left(1+\frac{y}{a}\right)} \quad \text{and} \quad R_2 = r_2/k_2 = \frac{y}{y+c} \tag{4}$$

We now expand R_1 and R_2 from the point

$$P = (x^0, s' = 1, y^0 = s_1^0/s^0) \tag{5}$$

$$R_1 = R_1^0 + \frac{\partial R_1}{\partial x}dx + \frac{\partial R_1}{\partial y}dy + \frac{\partial R_1}{\partial s'}ds' = R_1^0 + \frac{\partial R_1}{\partial x}dx + \frac{\partial R_1}{\partial y}\left(\frac{\partial y}{\partial x}dx + \frac{\partial y}{\partial s'}ds'\right) + \frac{\partial R_1}{\partial s'}ds' \tag{6}$$

$$R_2 = R_2^0 + \frac{\partial R_2}{\partial y}dy = \frac{\partial R_2}{\partial y}\left(\frac{\partial y}{\partial x}dx + \frac{\partial y}{\partial s'}ds'\right) \tag{7}$$

At P the dimensionless rates are:

$$R_1^0 = \frac{1 \cdot x^0}{\left(1+b\left(1+\frac{y^0}{a}\right)\right)} = R_2^0 = \frac{y^0}{y^0+c} = J'(x^0, s^0) \tag{8}$$

As point P we choose $x^0 = 0.5$, while s^0 is chosen such that a, b and c have the values given to them in Fig. 6.8, i.e. $a = 1$ and $b = c = 0.5$. Using (3) (or (6.28)) and (4)

$$y^0 = 0.22474 \quad \text{and} \quad R_1^0 = R_2^0 = J^0 = 0.3101 \tag{9}$$

The partial derivatives evaluated at P are:

$$2\frac{\partial y}{\partial x} = \frac{a}{b} + \frac{\frac{2ac}{b} - \frac{a}{b}\left(a+\frac{a}{b}-\frac{a}{b}x^0\right)}{2y^0 + \frac{a}{b} + a - \frac{a}{b}x^0} \tag{10}$$

$$2\frac{\partial y}{\partial s'} = -\left(\frac{a}{b}-\frac{a}{b}x^0\right) + \frac{\left(\frac{a}{b}-\frac{a}{b}x^0\right)\left(a+\frac{a}{b}-\frac{a}{b}x^0\right)+\frac{2ac}{b}x^0}{2y^0+a+\frac{a}{b}-\frac{a}{b}x^0} \tag{11}$$

$$\frac{\partial R_1}{\partial x} = \frac{1}{1+b\left(1+\frac{y^0}{a}\right)} \quad ; \quad \frac{\partial R_1}{\partial y} = \frac{-\frac{b}{a}x^0}{\left(1+b\left(1+\frac{y^0}{a}\right)\right)^2} \quad ; \quad \frac{\partial R_1}{\partial s'} = \frac{b\left(1+\frac{y^0}{a}\right)x^0}{\left(1+b\left(1+\frac{y^0}{a}\right)\right)^2} \quad (12)$$

$$\frac{\partial R_2}{\partial y} = \frac{c}{(y^0+c)^2} \quad (13)$$

Evaluation of the derivatives at $x^0 = 0.5$ and $s' = 1$ yields:

$$\begin{aligned} R_1 &= 0.3101 + 0.6202\, dx - 0.09616\,(0.59175\, dx + 0.11237\, ds') + 0.11778\, ds' \\ &= 0.3101 + 0.5633\, dx + 0.10697\, ds' \end{aligned} \quad (14)$$

$$\begin{aligned} R_2 &= 0.3101 + 0.9519\,(0.59175\, dx + 0.11237\, ds') \\ &= 0.3101 + 0.5633\, dx + 0.10697\, ds' \end{aligned} \quad (15)$$

Equations (14) and (15) can be used for an approximate calculation of the steady state flux $J'(x,s')$ at any point $Q = (x,s')$.

The flux control coefficient $C_1^J(x^0,1)$ is immediately obtained from (14) or (15):

$$C_1^J(0.5,1) = \frac{\partial R_1}{\partial x} \cdot \frac{x^0}{R_1^0} = 0.5633 \cdot \frac{0.5}{0.3101} = 0.9082 \quad (16)$$

This value is of course exact since (16) is exact for $dx \to 0$ and $ds' \to 0$.

Extrapolation of (15) to $x = 1$ (i.e $dx = 0.5$) for constant s' gives $J'(1,1) = 0.59175$. This value is too high since $J'(x)$ is downwards concave, and extrapolation along the tangent of the curve will predict too high values of J'. The true value of J'(1,1) is found by solution of (3) for y and insertion of y into (4). $J'_{exact}(1,1) = 0.55278$.

Similarly the control coefficients at (1,1) can be estimated by the method of example (6.3). The elasticities are:

Enzyme Kinetics and Metabolic Control Analysis

$$\varepsilon_{11} = \frac{-\frac{b}{a}y}{s' + b\left(1 + \frac{y}{a}\right)}, \quad \varepsilon_{12} = \frac{c}{c+y} \quad \text{and} \quad C_1^J = \frac{\varepsilon_{12}}{\varepsilon_{12} - \varepsilon_{11}} = \frac{1}{1 + \frac{\frac{b}{ac}y(y+c)}{s' + b\left(1 + \frac{y}{a}\right)}} \quad (17)$$

Solving (3) for $s' = 1$ and $x = 1$ yields $y = 0.6180$ and $C_{1\,\text{exact}}^J(1,1) = 0.7236$. An approximate value for $C_1^J(1,1)$ is obtained by inserting into (17) an approximate value for $y(1,1)$ calculated as $y = y^0 + dy$ from (14) or (15):

$$y(1,1) = y^0 + 0.59175\, dx = 0.22474 + 0.59175 \cdot 0.5 = 0.52062$$

$$C_{1\,\text{approx}}^J(1,1) = 0.7681.$$

This is a fair approximation of the true value $C_{1\,\text{exact}}^J(1,1) = 0.7236$.

Another approximate method for calculation of C_1^J is based on the shape of $J'(x)$. If $J'(x)$ had been given by

$$J'(x) = \frac{x}{x+K} \quad (18)$$

Then for two different enzyme concentrations represented by x^0 and x:

$$\frac{x(J' - J'(x^0))}{J'(x - x^0)} = \frac{x}{J'} \frac{K(x - x^0)}{(x - x^0)(x + K)(x^0 + K)} = \frac{KJ'(x^0)}{x^0} \quad (19)$$

But

$$C_1^J(x_0) \equiv \left.\frac{\partial J'}{\partial x}\right|_{x^0} \frac{x^0}{J'(x^0)} = \frac{K}{(x^0 + K)^2} \frac{x^0}{J'(x^0)} = \frac{KJ'(x^0)}{x^0} \quad (20)$$

This means that an exact value for the control coefficient $C_1^J(x^0)$ can be found simply by measuring the flux through the pathway for two known enzyme concentrations.

In our case and based on experimental determination of the flux at $x = 0.5$ and $x = 1$:

$$C_1^J(0.5) \approx \frac{1 \cdot (0.55278 - 0.31010)}{0.55278 \cdot 0.5} \approx 0.878 \quad (21)$$

This result is not far from the true value 0.9082, and in reality we have no way of obtaining the true value of C^J since this would demand that we either knew the exact relationship between the rates and the variable x or the linear approximations (14),(15).

As mentioned earlier formula (19) is a consequence of a remarkable paper by Small and Kacser (1993) who introduced the concept of flux control coefficients obtained by *large deviations*. Since for each enzyme, even in long pathways, the relationship between flux J and enzyme activity usually has an approximately hyperbolic form the large deviation approach is eminently practical in an experimental study. An integral change in flux is measured in a series of experiments where, one at a time, each enzyme dosage is integrally changed from a given initial level, e.g. that in the wild type strain. Formula (19) is afterwards used to obtain approximate values of C_i^J.

The other coefficient 0.10697 in (14) and (15) is the *response coefficient*, i.e. the response of the pathway flux to a change in the environmental variable s. If s is changed from s^0 to $2 s^0$ at constant $x = 0.5$ then $J'(x^0, s')$ increases by 0.1070 according to (15), while an exact calculation from (3) with $s' = 2$ and $x = 0.5$ gives $y = 0.30278$ and $\Delta J' = 0.0671$.

From (15) $y_{approx}(0.5,2) = 0.22474 + 0.11237 = 0.33711$, and using (17) for the exact and the approximate values of y yields:

$$C^J_{I\,exact}(0.5,2) = 0.9160 \quad \text{and} \quad C^J_{I\,approx}(0.5,2) = 0.9044.$$

We shall finally revert to the method of Delgado and Liao (1992) which as shown in eq. (6.47) promises to deliver the flux control coefficients based on the linearized rates, i.e. from the elasticities ε_{11} and ε_{12} at the point of linearization.

Let us therefore consider R_1 and R_2 at $P = (0.5,1)$, and perturb y from its value y^0 at the steady state. From (14) or (15):

$$\begin{aligned} R_1 &\approx J'(x^0) - 0.09616\,(y(t) - y^0) = J'(x^0) + \varepsilon_{11}(y - y^0) \\ R_2 &\approx J'(x^0) + 0.9519\,(y(t) - y^0) = J'(x^0) + \varepsilon_{12}(y - y^0) \end{aligned} \quad (22)$$

$$\begin{aligned} \frac{dy}{dt} &= R_1 - R_2 = (\varepsilon_{11} - \varepsilon_{12})(y(t) - y^0) \\ \frac{d\left(\dfrac{p}{s^0}\right)}{dt} &= R_2 = J'(x^0) + \varepsilon_{12}(y(t) - y^0) \end{aligned} \quad (23)$$

The solution of (21) and (22) is:

Enzyme Kinetics and Metabolic Control Analysis

$$y - y(t=0) = y - y(0) = -(y(0) - y^0)(1 - \exp(-(\varepsilon_{12} - \varepsilon_{11})t)) \quad (24)$$

$$\frac{1}{s^0}(p - p(0)) = tJ'(x^0) + \frac{\varepsilon_{12}}{\varepsilon_{12} - \varepsilon_{11}}(y(0) - y^0)(1 - \exp(-(\varepsilon_{12} - \varepsilon_{11})t))$$

But from (17):

$$C_1^J = \frac{\varepsilon_{12}}{\varepsilon_{12} - \varepsilon_{11}}$$

Consequently if the first equation in (24) is multiplied by C_1^J and added to the second equation one obtains:

$$C_1^J(y - y(0)) + \frac{1}{s^0}(p - p(0)) = J'(x^0)t \quad (25)$$

In a series of experiments the internal metabolite concentration $y(t)$ and $p(t)/s^0$ are measured at $t = t_1, t_2, t_3, \ldots t_N$, and by linear regression the two coefficients α_1 and α_2 of (26) are determined:

$$\alpha_1(y - y(0)) + \alpha_2\left(\frac{p}{s^0} - \frac{p(0)}{s^0}\right) = t \quad (26)$$

The flux control coefficients are now constructed from α_1 and α_2 using

$$\begin{bmatrix} C_1^J & C_2^J \end{bmatrix} = \begin{bmatrix} \alpha_1 & \alpha_2 \end{bmatrix} \mathbf{A} J'(x^0) = \begin{bmatrix} \alpha_1 & \alpha_2 \end{bmatrix} \begin{bmatrix} 1 & -1 \\ 0 & 1 \end{bmatrix} J'(x^0) \quad (27)$$

The stoichiometric matrix $\mathbf{A}$ has metabolite 1 (here y) in the first row and metabolite 2 (here p/s^0) in the second row. The steady state flux $J'(x^0,1)$ is simply found as α_2^{-1}. Fig. 6.10 shows $y(t)$ and $p/s^0(t)$ for fixed $s = s^0$ in an *in silico* experiment with no error in the two dependent variables. There are 12 t values, $t = 0.5, 1, 1.5, \ldots 6$. The flux control coefficient and the steady state flux are of course exactly retrieved.

In reality there are experimental errors in all the concentration measurements. Fig. 6.11 shows another *in silico* experiment where the exact values of the dependent variables are randomly overlaid by 2 % noise taken from a Gauss distribution with mean equal to the exact concentration values. Now C_1^J is calculated to 0.8671 and $J' = 0.3131$.

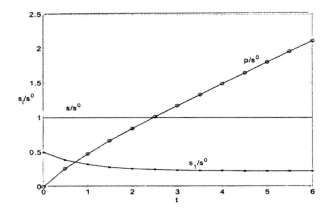

Figure 6.10 Result of a pulse addition of the internal metabolite at t = 0. p/s^0 is a measure of the flux out of the system integrated in time. The slope of p/s^0 is $J_n^{'}$ (= 0.3101) for large t.

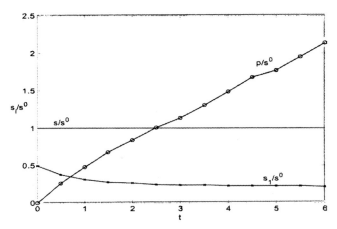

Figure 6.11 Results of the same experiment as Fig. 6.10, but with randomly added experimental error.

When 10 independent experiments are made, each with 2% randomly distributed noise the average values are:

$$C_1^J = 0.8933 \pm 0.085 \quad \text{and} \quad J' = 0.3105 \pm 0.0052 \tag{28}$$

The flux is well determined but (considering the necessary work to perform the real experiments) the flux control coefficient is not very accurate. One may easily imagine that with more enzymes in the pathway the calculation of the flux control coefficients would be swamped by the experimental errors. This is indeed what Ehlde and Zacchi (1996) point out in a critical study of the accuracy of the transient method.

Enzyme Kinetics and Metabolic Control Analysis

The general formula corresponding to (27) for I enzymes in the linear pathway is:

$$(C_1^J, C_2^J, C_3^J, \ldots C_I^J) = (\alpha_1, \alpha_2, \alpha_3, \ldots \alpha_I) \mathbf{A} \, \mathbf{J}'(x^0) \qquad (29)$$

Besides the sad issue of the influence of experimental errors on the calculated flux control coefficients it should also be noted that (29) holds *only* when each reaction R_i in the pathway is independent of the concentration of the substrate which feeds the pathway and of the concentration of the final product of the pathway. This is very inconvenient since experimental determination of $s(t)$ and of $p(t)$ is much easier than determination of $s_i(t)$, $I = 1, 2, 3 \ldots I - 1$, and we would like to start the transient experiment by adding a pulse of S (e.g. glucose). It is inconceivable that R_1 should in general be independent of s – perhaps with the exception that the first enzyme could be saturated with its substrate.

The transient method can, however be reformulated into a procedure that has a much better relation to experimental practice – and as a bonus appears to give much better estimates of the flux control coefficients.

Consider in a thought experiment a continuous, stirred tank reactor that is fed with S while P is withdrawn with the effluent from the reactor. S is converted to P via an intermediate S_1 that cannot leave the reactor – we might imagine that S_1 is only present in the immobilized enzyme particles that remain in the reactor. There is no P in the feed to the reactor. Mass balances for S, S_1 and P are:

$$\frac{ds}{dt} = -r_1 + D(s_f - s) \; ; \qquad \frac{ds^1}{dt} = r_1 - r_2 \; ; \qquad \frac{dp}{dt} = r_2 - Dp \qquad (30)$$

With the kinetics given in (4) and linearized around the point $(k_1=0.5, k_2=1, s=s^0)$ one obtains (14)-(15):

$$\frac{ds}{dt} = r_1^0 - \varepsilon_{11}(s_1 - s_1^0) - \varepsilon_{01}(s - s^0) + D(s_f - s^0) - D(s - s^0)$$

$$\frac{ds_1}{dt} = r_1^0 + \varepsilon_{11}(s_1 - s_1^0) + \varepsilon_{01}(s - s^0) - r_2^0 - \varepsilon_{12}(s_1 - s_1^0) \qquad (31)$$

$$\frac{dp}{dt} = r_2^0 + \varepsilon_{12}(s_1 - s_1^0) - Dp^0 - D(p - p^0)$$

In (31) ε_{01} is the elasticity of r_1 with respect to changes in s. s_1^0 and p^0 are steady state values of S_1 and P, respectively at the point of expansion. In the steady state $r_1^0 = r_2^0$, and the terms $D(s_f - s^0)$ and Dp^0 just balance these steady state rates. Consequently after scaling the equations with s^0 one obtains the following differential equations for the deviations from the steady state values of s, s_1 and p.

$$\frac{d(s'-1)}{dt} = -\varepsilon_{11}(y - y^0) - \varepsilon_{01}(s' - 1)$$

$$\frac{d(y-y^0)}{dt} = \varepsilon_{11}(y-y^0) + \varepsilon_{01}(s'-1) - \varepsilon_{12}(y-y^0) \qquad (32)$$

$$\frac{d(p'-p'^0)}{dt} = \varepsilon_{12}(y-y^0) - D(p'-p'^0) \; ; \; p' = p/s^0$$

In the following we shall repeat the perturbation experiment of eq (22) to (25). At constant $s' = 1$ y is perturbed from its steady state value y^0 and the transient of y and of p' back to their steady state values is followed. The solution of the two last equations of (32) for initial values y (t=0) = y(0), p'(0) = p'^0 is :

$$y(t) - y(0) = -(y(0) - y^0)(1 - \exp((\varepsilon_{11} - \varepsilon_{12})t))$$

$$p'(t) - p'^0 = \frac{\varepsilon_{12}(y(0) - y^0)}{\varepsilon_{12} - \varepsilon_{11} - D}(\exp(-Dt) - \exp((\varepsilon_{11} - \varepsilon_{12})t)) \qquad (33)$$

When the first equation, multiplied by $\alpha = \varepsilon_{12}/(\varepsilon_{12}-\varepsilon_{11}-D)$ is added to the second equation the terms in $\exp((\varepsilon_{11} - \varepsilon_{12})t)$ cancel and

$$\alpha(y(0) - y) + (p' - p'^0) = \alpha(y(0) - y^0)(1 - \exp(-Dt)) \qquad (34)$$

Except for the factor α the right hand side of (34) is a known function $f(t)$ of time. Except for α the left hand side of (34) is also a known function of time. Consequently α is determined by regression of

$$\alpha = \frac{-(p'(0) - p')}{(y(0) - y) - f(t)} \qquad (35)$$

using the experimentally determined values of the deviation variables $(y - y(0))$ and $(p' - p'^0)$ at a set of t-values. Knowing α for two experimental runs with different values of D permits the elasticities ε_{11} and ε_{12} to be determined by simple algebraic calculations. Finally the control coefficient C_1^J is determined using eq (17). Note that the same steady state value s^0 can be maintained for the two different D values by a simultaneous change of s_f to keep r_1^0 constant in the two experiments.

In silico experiments show that this variant of the Delgado and Liao transient method is much more robust with respect to the influence of experimental errors than the original method. A series of 10 experiments were run (t values the same as in Fig. 6.10 and 6.11) with respectively $D = 1$ and $D = 0.1$ (same unit for (time^{-1}) as used in k_1 and k_2). C_1^J was determined to 0.911 ± 0.015 when the data was overlaid with random noise 10% relative to the mean, and this is much better than the result obtained previously with only 2 % relative noise.

In this numerical experiment we have kept s at s^0 throughout the experiment. If s is allowed to deviate from s^0, i.e. if we make a simulation of the real pulse experiment, then the elasticity ε_{01} (= 0.11778 in (14)) can also be found, but we must of course now measure all three deviation variables as functions of time during the transient. If $\varepsilon_{01} = 0$ then measurement of $s' - 1$ contains no information of value for determination of the other elasticities since the first mass balance is now uncoupled from the other two balances. This topic is taken up in Problem 6.3.

PROBLEMS

Problem 6.1 Simulation of an enzymatic reaction.

Consider the enzyme kinetics discussed in detail in Section 6.3, and specifically in Example 6.5. The rate constants of r_1 and r_2 (k_1 and k_2) are respectively 0.5 and 1 g substrate S converted (L reactor h)$^{-1}$. s^0 is 1 g L^{-1}. The parameters a, b and c have the values 1, 0.5 and 0.5 at $s^0 = 1$ g L^{-1}.

A continuous, stirred tank reactor is fed with S in a concentration s_f. $p_f = 0$. Make a simulation study in which the productivity of P is determined for different values of dilution rate D and of s_f.

Problem 6.2 Analysis of an enzymatic two-step reaction sequence with a non-competitive inhibition in the first step.

The two-step reaction sequence with a competitive inhibition of the first step has been exhaustively analyzed in Section 6.3. Make the same analysis when the first step is non-competitively inhibited by the intermediate S_1. Try as far as possible to work with dimensionless groups of parameters in order to make the analysis of more general value. Give the dimensionless parameter groups specific numerical values in order to make a simulation study of the results.

Problem 6.3 The Delgado and Liao method applied to deviation variables

a. Prove that if $\varepsilon_{01} = 0$ no information concerning ε_{11} and ε_{12} can be obtained by measurement of $s(t)$ in the experimental set-up described in eq (30) of example 6.5.

b. For $\varepsilon_{01} \neq 0$ derive formulas similar to eq (35) in Example 6.5 for determination of the elasticities by measurement of the three deviation variables in a transient experiment – possibly at several values of D. The analysis may be somewhat simplified if you work on the system matrix alone without at first bothering to determine the actual solutions to the mass balances.

c. Make an *in silico* experiment (as in Example 6.5) with random relative errors on the concentration measurements to determine how robust the method is for determination of the flux control coefficients

Problem 6.4 Flux control coefficients from *real* experiments.
Nielsen and Jørgensen (1995) and Delgado *et al.* (1993) are some recent references to studies of the flux control in short pathways. There are other references that you might wish to retrieve by a literature search. For the two references (and for other references that you might find) compare the experimental results with calculations of the kind shown in this chapter. Is the quality of the experiments good enough to merit a quantitative treatment of the results? Any other comments to the papers?

REFERENCES

Benthin, S., Nielsen, J., and Villadsen, J. (1993). Transport of sugars *via* two anomer-specific sites on the mannose-phosphotransferase system in *Lactococcus cremoris*: *in vivo* study of mechanism, kinetics and adaptation. *Biotechnol. Bioeng.* **42**, 440-448.

Cornish-Bowden,A. (1995). "Fundamentals of Enzyme Kinetics" , Portland Press, London (UK).

Delgado, J.P .and Liao, J.C. (1991). Identifying rate-controlling enzymes in metabolic pathways without kinetic parameters. *Biotecnol.Prog.* **7**, 15-20.

Delgado, J.P. and Liao, J.C. (1992a) . Determination of flux control coefficients using transient metabolite concentrations. *Biochem. J.* **282**, 919-927.

Delgado, J.P. and Liao, J.C. (1992b). Metabolic control analysis using transient metabolite concentrations. Determination of metabolite concentration control coefficients. *Biochem. J* **282**, 919-927.

Delgado, J.P., Meruane,J., and Liao, J.C. (1993). Experimental determination of flux control distribution in biochemical systems : *In vitro* model to analyze metabolite concentrations. *Biotechnol. Bioeng.***41**, 1121-1128.

Ehlde, M., and Zacchi, G. (1996). Influence of experimental errors on the determination of flux control coefficients from transient metabolite measurements. *Biochem. J* **313**,721-727.

Fell, D. (1997). "Understanding the Control of Metabolism". Portland Press, London (UK).

Flint, H.J., Tateson, R.W., Barthelmess, I.B., Porteous, D.J., Donachie, W.D., and Kacser, H. (1981). Control of the flux in the arginine pathway of *Neurospora crassa. Biochem. J*.**200**, 231-246.

Gram, J., de Bang, M., and Villadsen, J. (1990). An automated glucose- isomerase reactor system with on line flow injection analyzers for monitoring of pH, glucose- and fructose concentrations. *Chem. Eng. Sci*, **45**, 1031-1042.

Levenspiel, O. (1999). "Chemical Reaction Engineering" 3.rd edition. Wiley, New York.

Liao, J.C., and Lightfoot , E.M. (1988). Characteristic reaction paths of biochemical reaction systems with time scale separation. *Biotechnol. Bioeng.* **31**, 847-854.

Melchiorsen, C.R., Jensen, N.B.S., Christensen, B., Jokumsen, K.V., and Villadsen, J. (2001). Dynamics of pyruvate metabolism in *Lactococcus lactis. Biotechnol. Bioeng.* **74**, 271-279.

Nordkvist, M. (2001). "Physiology of Lactic Acid Bacteria". M.Sc. thesis , Biocentrum, DTU, Denmark.

Nielsen, J., Jørgensen, H. S. (1995) Metabolic control analysis of the penicillin biosynthetic pathway in a high yielding strain of *Penicillium chrysogenum, Biotechnol. Prog.* **11**:299-305

Small, J.R. and Kacser, H. (1993). Response of metabolic systems to large changes in enzyme activities and effectors. *Eur. J. Biochem.* **213**, 613-640.

Stephanopolous, G., and Vallino, J.J. (1991) Network rigidity and metabolic engineering in metabolite overproduction. *Science* **252**, 1675-1681.

Stephanopolous, G., Aristidou, A.A., and Nielsen, J. (1998). "Metabolic Engineering", Academic Press, San Diego (USA).

Wei, J., and Prater, C.D. (1962). "The Structure and Analysis of Complex Reaction Systems", in *Advances in Catalysis* 13 203-392

Westerhoff, H.V. and Chen, Y.-D.(1984). How do enzyme activities control metabolite concentrations. An additional theorem in the theory of metabolic control. *Eur. J. Biochem.* **142**, 425-430.

Westerhoff, H.V., and Kell, D.B. (1987). Matrix method for determining steps most rate- limiting to metabolic fluxes in biotechnological processes *Biotechnol .Bioeng* **30** 101-107.

7

Modeling of Growth Kinetics

In chapters 3 and 5 we have discussed how the two important design parameters *yield* and *productivity* can be derived from experimental data, *e.g.* from measurements of the substrate consumption and the product formation. Furthermore, we have shown how measured steady state rates (or fluxes) in and out of the cell can be used to calculate the fluxes through the different branches of the metabolic network functioning in a given cell. However, we have not yet established a quantitative relation between the fluxes and the variables (concentrations etc.) that characterize the environment of the cell, and we have also not considered how the fluxes change with changes in the operating conditions, *e.g.* the response to a change in medium composition or temperature. In order to do this it is necessary to define kinetic expressions for the key reactions and processes considered in the model – or in other terms to set up a mathematical model that can simulate the studied process (see Note 7.1). Setting up kinetic expressions is normally referred to as *kinetic modeling*, and this involves defining verbally or mathematically expressed correlations between rates and reactant/product concentrations that, inserted in mass balances, permits a prediction of the degree of conversion of substrates and the yield of individual products at other operating conditions. Conceptually this is a great step forward compared to the methodology applied in chapters 3 and 5. Thus, if the rate expressions are correctly set up, it is possible to express the course of a fermentation experiment based on initial values (or input) for the components of the state vector, e.g., concentration of substrates. This leads to simulations that finally may result in an optimal design of the equipment or an optimal mode of operation for a given system. Independent of the model structure, the process of defining a quantitative description of a fermentation process often involves an iterative process where the model is continuously revised when new process information is obtained. However, it is always *important to clearly define the aim of the model*, i.e., what the model is going to be used for. The model structure and complexity should relate to this.

Models are used by all researchers in life sciences when results from individual experiments are interpreted and when results from several different experiments are compared with the aim of setting up a model that may lead to deeper insight into the biological system. Biologists constantly use models *e.g.* when experiments on gene regulation and expression are to be interpreted, and these models are very important when the inherent message from often quite complicated experiments is to be extracted. Most of these biological models are qualitative *only*, and they do not allow quantitative analysis. Often these *verbal models* can be quite easily transferred to quantitative models, but a major obstacle in applying quantitative models is estimation of the parameters of the model. In order to do this it is necessary to have precise measurements of the different variables of the system, and preferably at very different experimental conditions. Thus, in order to obtain a quantitative description of the biological system considered (or model simulations) no effort should be spared to obtain reliable data. Ideally carbon and nitrogen balances should close to within 99%, and the concept of elemental balancing described in Chapter 3 therefore represents a natural first step in evaluating experimental data that are going to be used for quantitative modeling of a given system. A failure to close the mass balances will lead to inaccuracies in the estimated fluxes in and out of the cell and in general make it impossible to develop meaningful kinetics. The last ten years have witnessed a revolution in experimental techniques applied to life sciences. This has made it possible to measure more variables and at a higher precision, resulting in a far more detailed modeling of cellular processes. Furthermore, the availability of powerful computers permits complex numerical problems to be solved with a reasonable computational time. At the present even complex mathematical models for biological processes can therefore be handled and experimentally verified. Such detailed (or mechanistic) models are often of little use in the design of a bioprocess, whereas they serve a purpose in fundamental research on biological phenomena. In this presentation we will focus on models that are useful for design of bioprocesses, but we will also shortly treat mathematical models that can give a better insight into biological processes at the molecular level.

Note 7.1 Mathematical models.

A mathematical model is a set of *relationships* between the *variables* in the system being studied, and generally it can predict the output variables, "the state of the system", from the input variables. These relationships are normally expressed in the form of mathematical equations, but they may also be specified as logic expressions (or cause/effect relationships) that are used in the operation of a process. The variables include as inputs any property that is of importance for the process, *e.g.* the agitation rate in the bioreactor, the feed rate to the bioreactor, pH of the medium, the temperature of the medium, the concentration of substrates in the feed. Output variables are concentrations in the reactor of substrates, metabolic products, the biomass concentration, and the state of the biomass – often represented by a set of key intracellular components. In order to set up a mathematical model it is necessary to specify a control volume wherein all the variables of interest are taken to be uniform, *i.e.* there is no variation in their values throughout the control volume. For fermentation processes the *control volume* is typically given by the whole bioreactor, but for large bioreactors the medium may be inhomogeneous due to mixing problems and here it is necessary to divide the bioreactor into several control volumes (see Chapter 11). When the control volume is the whole bioreactor it may either be of constant volume or it may change with time depending on the operation of the bioprocess. When the control volume has been defined a set of balance equations can be specified for the variables of interest. These balance equations specify how material is flowing in and out of the control volume and how material is converted within the control volume. The conversion of material within the control volume is specified by so-called rate equations (or kinetic expressions), and together with the mass balances these specify the complete model.

7.1 Model Structure and Complexity

Setting up a mathematical model involves several steps as illustrated in Fig. 7.1. A very important first step is to specify the model complexity. This depends on the aim of the study, *i.e.* what the model is going to be used for. Specification of the model complexity involves defining the number of reactions to be considered in the model, and specification of the stoichiometry for these reactions. When the model complexity has been specified, rates of the cellular reactions considered in the model are described in terms of a specified set of output variables that may be concentrations in the reactor of substrates, metabolic products and certain biomass components. Every reactant – substrate or product – which is suspected to influence the rate is included, but in simple models mass balances for many of the output variables can be expressed in terms of other output variables using assumptions such as constant yield coefficients. The expressions that relate the rates to the output variables are normally referred to as *kinetic expressions*, since they specify the kinetics of the reactions considered in the model in terms of a selected set of output variables. This is an important step in the overall *modeling cycle* and in many cases different kinetic expressions have to be examined before a satisfactory model is obtained. The next step in the modeling process is to combine the kinetics of the cellular reactions with a model for the reactor in which the cellular process occurs. Such a model specifies how the concentrations of substrates, biomass, and metabolic products change with time, and what flows in and out of the bioreactor. When no spatial variation occur in the reactor we speak of models for *ideal bioreactors*, and these models, to be treated in Chapter 9, are normally represented in terms of simple mass balances over the whole reactor. More detailed reactor models may also be applied, if inhomogeneity of the medium is likely to play a role (see Chapter 11). The combination of mass balances, including kinetic rate equations, and the reactor model makes up a complete mathematical description of the fermentation process, and this model can be used to simulate how output variables depend on the set of input variables. The steady state model consists of a set of algebraic equations that relate the output variables to the input variables. Dynamic – or transient models consist of a set of differential equations. For an ideal bioreactor the output variables are found as functions of time for any given set of input variables and the initial values of the output variables. However, before this can be done it is necessary to assign values to the parameters of the model. In order to do so one most compare model simulations with experimental data, and hereby estimate a parameter set that gives the best fit of the model to the experimental data. This is referred to as *parameter estimation*. The evaluation of the fit of the model to the experimental data can be done by simple visual inspection of the fit, but generally it is preferable to use a more rational procedure, *e.g.* by minimizing the sum of squared errors between the model and the experimental data. If the model simulations are considered to represent the experimental data sufficiently well the model is accepted, whereas if the fit is poor even for the set of parameters that gives the best fit it is necessary to revise the kinetic model and go through the modeling cycle again.

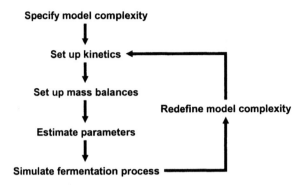

Figure 7.1 Different steps in quantitative description of fermentation processes.

Biological processes are *per se* extremely complex, and in Chapter 2 we illustrated that cell growth and metabolite formation is the result of a very large number of cellular reactions and events such as gene expression, translation of mRNA into functional proteins, further processing of proteins into functional enzymes or structural proteins, and long sequences of biochemical reactions leading to building blocks needed for synthesis of cellular components. It is quite clear that a complete description of all these reactions and events cannot be attempted in a mathematical model – at least with our current knowledge of the underlying molecular events and with the currently available computer power. When in a fermentation process the cell population is characterized by a stochastic distribution of properties such as cell activity and cellular composition further complexity is added to the model. In setting up fermentation models lumping of cellular reactions and events is therefore always done, but the level of detail considered in the model, *i.e.* the degree of lumping, depends on the aim of the modeling.

Fermentation models can roughly be divided into four groups depending on the level of detail in the model (Fig. 7.2). The simplest description is the so-called *unstructured models* where the biomass is described by a single variable (often the total biomass concentration) and where no segregation in the cell population is considered, *i.e.* the cell population is assumed to be completely homogeneous. These models can be extended to a *segregated* population model, where the individual cells in the population are described by a single variable, *e.g.* the cell mass or cell age, but often it is also relevant to add further structure to the model when segregation in the cell population is considered. In the so-called *structured models* the biomass is described with more than one variable, *i.e.* structure in the biomass is considered. Structure may be anything from diving cell mass into a few compartments to a microscopic view of the cell as made up of individual enzymes and macromolecular pools.

Modeling of Growth Kinetics

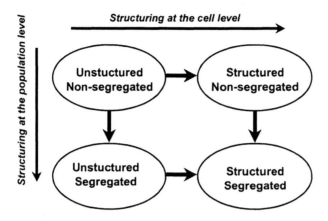

Figure 7.2 Different types of model complexity, with increasing complexity from the upper left corner to the lower right corner. With structuring at the cell level, specific intracellular events or reactions are considered in the model, and the biomass is structured into two or more variables. With structuring at the population level, segregation of the population is considered, *i.e.* it is recognized that not all the cells in the population are identical.

From the above it is clear that a very important element in mathematical modeling of fermentation processes is to define the model structure (or specifying the complexity of the model). A general rule can be stated as: *As simple as possible but not simpler*. This rule implies that the basic mechanisms should always be included and that the model structure depends on the aim of the model (see Note 7.2). Thus, if the aim is to simulate the biomass concentration in a fermentation process, a simple unstructured model (Section 7.3) may be sufficient. Even though these models are completely empirical, they are valuable for simple design problems and for extracting key kinetic parameters of the growth kinetics. If on the other hand the aim is to simulate dynamic growth conditions one may turn to simple structured models (Section 7.4), *e.g.* the compartment models, which are also useful for illustrating the concept of structured modeling. However, if the aim is to analyze a given system in further detail it is necessary to include far more structure in the model, and in this case one often describes only selected processes within the cell, *e.g.* a certain pathway or gene transcription from a certain promoter. Similarly if the aim is to investigate the interaction between different cellular processes, *e.g.* the influence of plasmid copy number on chromosomal DNA replication, a single cell model has to be applied. A short introduction to these more detailed (or mechanistic) models is given in Section 7.5. Finally, if the aim is to look at population distributions, which in some cases may have an influence on growth or production kinetics, either a segregated or a morphologically structured model has to be applied. In Section 7.6 we will consider simple cellular segregation – here referred to as morphologically structured models, whereas population based models is the topic of Chapter 8.

Note 7.2 Model complexity
A simple illustration of difference in model complexity is the quantitative description of the fractional saturation y of a protein at a ligand concentration c_l. This may be described either by the Hill equation (see Chapter 6):

$$y = \frac{c_l^h}{c_l^h + K} \quad (1)$$

where h and K are empirical parameters, or by the equation of Monod *et al.* (1963):

$$y = \frac{\left(La\left(1 + \frac{ac_l}{K_R}\right)^3 + \left(1 + \frac{c_l}{K_R}\right)^3\right)\frac{c_l}{K_R}}{L\left(1 + \frac{ac_l}{K_R}\right)^4 + \left(1 + \frac{c_l}{K_R}\right)^4} \quad (2)$$

where L, a, and K_R are parameters. Both equations address the same experimental problem, but whereas eq. (1) is completely empirical with h and K as fitted parameters eq. (2) is taken from a truly mechanistic model for enzymatic reactions where the parameters have a direct physical interpretation. If the aim of the modeling is to understand the underlying mechanism of the process eq. (1) can obviously not be applied since the kinetic parameters are completely empirical and give no (or little) information about the ligand binding to the protein. In this case eq. (2) should be applied, since by estimating the kinetic parameter one obtains valuable information about the system, and the parameters can be directly interpreted. If, on the other hand, the aim of the modeling is to simulate the ligand binding to the protein eq. (1) may be as good as eq. (2) - one may even prefer eq. (1) since it is more simple in structure and has fewer parameters, and it actually often gives a better fit to experimental data than eq. (2). Thus, the answer to which model one should prefer depends on the aim of the modeling exercise. In the list of unstructured models (Table 7.2) the expression (1) appears as the Moser kinetics, but here the mechanistic foundation is completely absent.

7.2 A General Structure for Kinetic Models

In order to describe kinetic growth models it is useful to apply a general framework that allows uniform presentation of the models. Hereby the model construction process is facilitated and it is also easy to compare different model structures. In this section we will present such a general framework. Although it may look like a rather theoretical presentation of cellular growth, it will facilitate our subsequent discussion significantly.

7.2.1 Specification of Reaction Stoichiometries

As shown in Fig. 7.1 model construction start with defining the stoichiometry of the reactions to be considered in the model. In order to start this process we consider the general model

Modeling of Growth Kinetics

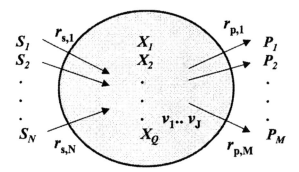

Figure 7.3 A general representation of reactions involved in cellular growth. N substrates enter the cell and are converted into M metabolic products and Q biomass constituents. The conversions involves J cellular reactions for which the rates are given by $v_1...v_J$. Rates of substrate uptake ($r_{s,1},...,r_{s,N}$) and product formation ($r_{p,1},...,r_{p,M}$) are also shown. In addition to Fig. 5.3 intracellular structure in the biomass is considered.

illustrated in Fig. 7.3, where N substrates are taken up by the cells and converted into M metabolic products and Q biomass constituents. The conversions are carried out in J reactions. The number of reactions and processes involved in cellular growth is very large, and as mentioned above the reactions are therefore typically lumped reactions, e.g., conversion of glucose into cellular protein. Often the lumped reactions are completely empirical, e.g., conversion of glucose into "active biomass".

In order to describe the stoichiometry of the reactions we introduce stoichiometric coefficients for all the components in the system. Thus, stoichiometric coefficients for the substrates are termed α, the stoichiometric coefficients for the metabolic products are termed β and the stoichiometric coefficients for the biomass constituents are termed γ. The numerical value of these stoichiometric coefficients are analogous to the yield coefficients of the black box model introduced in Chapter 3, but since we will use the yield coefficients to specify overall conversion yields it is necessary to distinguish between the stoichiometric coefficients that appear in independent reactions and the yield coefficients. Clearly the stoichiometric coefficients are identical to the yield coefficients if all the reactions are lumped into a single reaction, as is done in the black box kinetic models to be described in Section 7.3.1.

Since there are many different compounds involved in many different reactions it is necessary to use two indices on the stoichiometric coefficients to indicate the reaction number and the compound, e.g., α_{ji} is the stoichiometric coefficient for the ith substrate in the jth reaction. Thus, in the general framework for growth models, we introduce stoichiometric coefficients for all substrates, metabolic products and biomass constituents in each of the J reactions, but many of the stoichiometric coefficients will be zero, since only a few compounds participate in any given reactions.

As shown in Fig. 7.3 the substrates are S_i, the metabolic products P_i and the biomass constituents X_i. With these definitions, the stoichiometry for the jth cellular reaction can be specified as

$$\sum_{i=1}^{N} \alpha_{ji} S_i + \sum_{i=1}^{M} \beta_{ji} P_i + \sum_{i=1}^{Q} \gamma_{ji} X_i = 0 \tag{7.1}$$

In a growth model there will be a stoichiometric equation like (7.1) for each of the J cellular reactions, and it is therefore convenient to write the stoichiometry for all J cellular reactions in a compact form using matrix notation:

$$\mathbf{AS} + \mathbf{BP} + \mathbf{\Gamma X} = \mathbf{0} \tag{7.2}$$

where the matrices $\mathbf{A}$, $\mathbf{B}$ and $\mathbf{\Gamma}$ are stoichiometric matrices containing stoichiometric coefficients in the J reactions for the substrates, metabolic products and biomass constituents, respectively. In the matrices $\mathbf{A}$, $\mathbf{B}$ and $\mathbf{\Gamma}$ rows represent reactions and columns compounds. Thus, the element in the jth row and the ith column of $\mathbf{A}$ specifies the stoichiometric coefficient for the ith substrate in the jth reaction. The stoichiometric coefficients may be positive, negative, or zero. Typically they are negative for substrates (and other compounds consumed in a given reaction) and positive for metabolic products (and compounds formed in a given reaction). With a stoichiometric formulation of the general type of (7.2), a large number of the stoichiometric coefficients become zero, and one may find it cumbersome to specify stoichiometric coefficients for all compounds and all reactions considered in the model. However, the advantage is that the general matrix formulation facilitates much of the subsequent analysis because it can be done in the compact matrix symbolism that at a later stage may be advantageous for computer simulations. When the stoichiometry has been formulated in matrix notation, it is very easy to spot the participation of a given compound in the various reactions. One just has to look at the column for this compound in the appropriate matrix.

7.2.2 Reaction Rates

The stoichiometry specified in Section 7.2.1 defines the relative amounts of the compounds produced or consumed in each of the J intracellular reactions, but does not allow one to calculate the rates or the relative amounts at which metabolic products are secreted in the medium. This can be done by introducing the rates of the individual reactions and further coupling them to determine the overall rates of product secretion. The rate of a given reaction (or process) considered in the model is now defined by the forward reaction rate (or velocity) v, which specifies that a compound with a stoichiometric coefficient β is formed at the rate βv in this particular reaction. Normally, one of the stoichiometric coefficients in each reaction is arbitrarily set to be 1, whereby the forward reaction rate becomes equal to the consumption or production rate of this compound in the particular reaction. For cellular reactions we often use the biomass as reference to define the so-called *specific rates*, usually with the unit g (g DW h)$^{-1}$. We now collect the forward reaction rates of the J reactions considered in the model in the rate vector $\mathbf{v}$. Thus, $\beta_{ji} v_j$ specifies the specific rate of formation of the ith metabolic product in the jth reaction. Because the stoichiometric coefficients for the substrates, *i.e.*, the elements of $\mathbf{A}$, are generally

Modeling of Growth Kinetics

negative, the specific conversion rate of the ith substrate in the jth reaction is given by $-\alpha_{ji}v_j$. When we want to calculate the overall production or consumption of a compound, we have to sum the contributions from the different reactions. We can therefore write the net specific *consumption* rate for the ith substrate as the sum of its rate of consumption in all J reactions:

$$(r_{s,i})_{consumption} = -\sum_{j=1}^{J} \alpha_{ji} v_j \tag{7.3}$$

and similarly for the net specific rate of *formation* of the ith metabolic product:

$$r_{p,i} = \sum_{j=1}^{J} \beta_{ji} v_j \tag{7.4}$$

Equations (7.3) and (7.4) specify important relationships between what can be directly measured, namely, the specific uptake rates of substrates and the specific formation rate of products on one hand and the rates of the reactions considered in the model on the other. Notice that in Chapter 5 we used the term *fluxes* for these intracellular reactions, but there we also considered well defined chemical reactions, whereas in growth models the reactions are typically lumped reactions of empirical nature.

Similar to eqs. (7.3) and (7.4) we find for the biomass constituents:

$$r_{X,i} = \sum_{j=1}^{J} \gamma_{ji} v_j \tag{7.5}$$

These rates are not as easy to determine experimentally as the specific substrate consumption rates and the specific product formation rates, but they are related to the specific growth rate of the biomass. Thus, the rates in Eq. (7.5) are the net specific formation rates of the individual biomass constituents, and the specific growth rate is therefore given as the sum of net formation rates of all the biomass constituents:

$$\mu = \sum_{i=1}^{Q} r_{X,i} \tag{7.6}$$

The summation equations (7.3) - (7.5) can be formulated in matrix notation as

$$(\mathbf{r}_s)_{consumption} = -\mathbf{A}^T \mathbf{v} \tag{7.7}$$

$$\mathbf{r}_p = \mathbf{B}^T \mathbf{v} \tag{7.8}$$

$$\mathbf{r}_X = \mathbf{\Gamma}^T \mathbf{v} \tag{7.9}$$

where the specific rate vector $(\mathbf{r}_s)_{consumption}$ contains the N specific substrate uptake rates, $\mathbf{r}_p$ the M specific product formation rates, and $\mathbf{r}_X$ the net, specific formation rates of the Q biomass constituents.

7.2.3 Dynamic Mass Balances

In Section 3.1 we specified mass balances at steady state conditions for substrates, metabolic products and biomass – equations (3.1)-(3.3). In these mass balances the specific rates derived above can be inserted and hereby steady state conditions can be calculated. In many cases fermentation processes are, however, operated at non-steady-state conditions, i.e., at dynamic growth conditions. At these conditions the equations are based on the general mass balance, which for the case of a substrate takes the form:

$$\frac{d(\mathbf{c}V)}{dt} = \mathbf{r}(\mathbf{c})xV + v_{feed}\mathbf{c}_f - v_{exit}\mathbf{c} \tag{7.10}$$

$\mathbf{c}$ is the vector of medium concentrations of N substrates, M metabolic products and Q biomass components. v_{feed} is the volumetric flow rate to the bioreactor, v_{exit} is the volumetric flow rate out of the reactor. Different parts of the production rate vector $\mathbf{r}$ are taken from (7.7)-(7.9). In analogy with the steady state balances presented in Section 3.1 transfer of substrates and metabolic products from the gas to the liquid phase can if desired be included in the dynamic mass balance (7.10).

The mass balances for the biomass and the Q biomass constituents X_i that together make up the biomass require special attention. The specific rate of formation of biomass is the sum of contributions from each of the J reactions as specified in eq. (7.6), which also can be written as:

$$r_x = \mu = \sum_{i=1}^{K} \Gamma_i^T \mathbf{v} \tag{7.11}$$

where Γ_i is the ith column of Γ. Consequently, with a sterile feed into a bioreactor the mass balance for biomass is:

$$\frac{d(xV)}{dt} = r_x xV - v_{exit}x = q_x V - v_{exit}x \tag{7.12}$$

The concentrations of biomass constituents in the cell are normally expressed in units of g (g DW)$^{-1}$ and consequently the sum of all the concentrations equals 1. Furthermore, a mass balance for component i in a cell of mass m is:

$$\frac{d(mX_i)}{dt} = \Gamma_i^T \mathbf{v} m \tag{7.13}$$

or

$$\frac{dX_i}{dt} = \Gamma_i^T \mathbf{v} - \frac{1}{m}\frac{dm}{dt}X_i = \Gamma_i^T \mathbf{v} - \mu X_i \tag{7.14}$$

It should be noted that (7.14) is independent of the type of reactor operation that is used to produce the biomass. Eq. (7.14) shows that the rate of formation by all the J reactions of a biomass constituent must at least be μX_i to preserve the same fraction of the constituent i in the biomass. This is a consequence of expansion of the biomass upon growth and this results in dilution of all biomass components. As the concentrations of all biomass constituents sum to 1, Eq. (7.11) can easily be derived from Eq. (7.14).

7.3 Unstructured Growth Kinetics

In unstructured models, all cellular components are pooled into a single biomass component represented by the total biomass concentration x. Initially unstructured models were based on a single reaction describing the overall conversion of substrate into biomass, and typically the kinetics of this overall reaction was represented by the specific growth rate of the biomass μ. Many different kinetic expressions have been proposed for the specific growth rate of the biomass, and we will start our discussion of unstructured kinetic models with a description of these simple models, and then move on to models considering more than one reaction and also look into the effect of temperature and pH. Many of the unstructured models described in the following look similar to the mechanistic based models for enzyme kinetics described in Chapter 6, but for description of overall cell growth they can only be regarded as (often very useful) date fitters.

7.3.1 The Black Box Model

The simplest mathematical presentation of cell growth is the so-called *black box* model of Section 3.3, where all the cellular reactions are lumped into a single overall reaction. This implies that the yield of biomass on the substrate (as well as the yield of all other compounds consumed and produced by the cells) is constant. Consequently the specific substrate uptake rate is proportional with the specific growth rate of the biomass:

$$-r_s = Y_{xs}\mu \tag{7.15}$$

Similar relations describe the specific uptake rate of other substrates, *e.g.* uptake of oxygen, and the formation rate of metabolic products. Thus, in the black box model kinetic modeling reduces to a description of the specific growth rate as function of the variables in the system. In the simplest model it is assumed that there is only one limiting substrate, typically the carbon source (which is often glucose), and no influence of other substrates on r_s. Hence the specific growth rate is a function of the concentration of this substrate only. The consumption of all other substrates and the production of metabolic products and biomass is found by mass balances using the yield coefficients of the black box model. A very general observation for cell growth on a single limiting substrate is that at low substrate concentrations (s) the specific growth rate μ is proportional with s, but for increasing values of s the specific growth rate approaches an upper limit. This verbal model can be described with many different mathematical models of which the Monod (1942) model is the simplest, and most frequently used.

$$\text{Monod model}: \quad \mu = \mu_{max} \frac{s}{s + K_s} \tag{7.16}$$

Fig. 7.4 shows the time course of a batch fermentation and a cross plot of μ versus s. After a *lag phase* of duration about 1 hour there is a long *exponential growth phase* where the biomass concentration increases exponentially with time. At the end, where the substrate is used up, growth stops quite abruptly. This is the typical behavior when the growth follows Monod kinetics: The lag-phase is a period during which the biomass composition changes and the specific growth rate increases. Here the assumption of the Monod model – or any other unstructured model that assumes the growth to be an independent of the biomass composition – is not satisfied. Since the batch cultivation starts with an initial substrate concentration $s = s_0$ much higher than K_s the specific growth rate μ, which is the slope of the curve $\ln(x)$ versus time, is almost constant for most of the fermentation period. When finally s decreases into the range of K_s the specific growth rate declines and eventually growth stops when the substrate concentration becomes close to zero.

In the Monod model K_s is the value of the limiting substrate concentration at which the specific growth rate is half its maximum value. Roughly speaking, it divides the plot of μ versus s into a low-substrate-concentration range where the specific growth rate is strongly (almost linearly) dependent on s, and a high-substrate-concentration range where μ is more or less independent of s. This is illustrated in Fig. 7.4, where μ is plotted versus the limiting substrate s for the batch fermentation data. The cross plot of μ versus s has the typical shape of Eq. (7.16). Fig. 6.1 is of course identical to Fig. 7.4B.

When glucose is the limiting substrate the value of K_s is normally in the micromolar range (corresponding to the milligram per liter range), and it is therefore experimentally difficult to determine. Some K_s values reported in the literature are compiled in Table 7.1. It should be stressed that the K_s value in the Monod model does not represent the saturation constant for substrate uptake but only an overall saturation constant for the whole growth process. However, since the substrate uptake is often involved in the control of substrate metabolism, the value of K_s is also often in the range of the K_m values of the substrate uptake system of the cells.

Modeling of Growth Kinetics

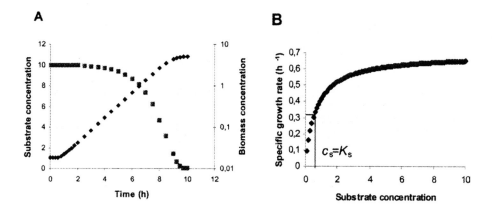

Figure 7.4 Typical profile of biomass and substrate concentrations during a batch culture.
A. The time profile of the biomass concentration (♦) and the concentration of the limiting substrate (■).
B. A cross plot of the specific growth rate (given by $1/x \, dx/dt$) versus the substrate concentration. Corresponding values of μ and s are taken from Fig. 7.4A. The value of K_s is indicated.

The satisfactory fit of the Monod model to many experimental data should as mentioned in Chapter 6 never be misconstrued to mean that Eq. (7.16) is a mechanism of fermentation processes. The Langmuir rate expression of heterogeneous catalysis and the Michaelis-Menten rate expression in enzymatic catalysis are formally identical to Eq. (7.16), but in both cases the denominator constant has a direct physical interpretation (the equilibrium constant for dissociation of a catalytic site-reactant complex), whereas K_s in Eq. (7.16) is no more than an empirical parameter used to fit the average substrate influence on all cellular reactions, pooled into the single reaction in which

Table 7.1 Compilation of K_s values for different microbial cells growing on different sugars.

Species	Substrate	K_s (mg L^{-1})
Aerobacter aerogenes	Glucose	8
Escherichia coli	Glucose	4
Klebsiella aerogenes	Glucose	9
	Glycerol	9
Klebsiella oxytoca	Glucose	10
	Arabinose	50
	Fructose	10
Lactococcus cremoris	Glucose	2
	Lactose	10
	Fructose	3
Saccharomyces cerevisiae	Glucose	180
Penicillium chrysogenum	Glucose	4
Aspergillus oryzae	Glucose	5

substrate is converted to biomass. In fact, one may argue that the two features which make the Monod model work so well in fitting experimental data are deeply rooted in any naturally occurring conversion process: The size of the cell machinery which converts substrate must have an upper value, and all chemical reactions will end up as first-order processes when the reactant concentration tends to zero.

Example 7.1 Steady-state chemostat described by the Monod model with sterile feed
For continuous steady-state operation the mass balances for biomass and a single substrate reduce to:

$$0 = r_s x + D(s_f - s) \tag{1}$$

$$0 = (\mu - D)x \tag{2}$$

Equation (2) immediately gives the key relationship for steady state continuous bioreactors:

$$\mu = D \tag{3}$$

which holds irrespective of the functional relationship between μ and s, p and x. When the specific growth rate is given by the Monod model, then:

$$D = \frac{\mu_{max} s}{s + K_s} \tag{4}$$

or solved for s:

$$s = \frac{DK_s}{\mu_{max} - D} \tag{5}$$

Finally the biomass concentration and the concentration of metabolic products in the reactor (and in the effluent) is given by the total mass balances for the bioreactor and using the constant stoichiometric coefficients that are inherent in the black box model:

$$x = Y_{sx}(s_f - s) \tag{6}$$

$$p = Y_{sp}(s_f - s) \tag{7}$$

for $x_f = 0$ and $p_f = 0$. Note in Eq. (5) that the substrate concentration in the reactor is independent of the substrate feed concentration. This is true for any functional relationship $\mu = \mu(s)$. If μ depends on the concentration of one of the metabolic products then s depends also on s_f. The biomass concentration always depends on the substrate concentration in the feed (s_f), and the higher the feed concentration the higher will be the biomass concentration, at least for simple kinetic models.

The right hand side of Eq. (4) is limited from above by:

Modeling of Growth Kinetics

$$D_{max} = \mu_{max} \frac{s_f}{s_f + K_s} \tag{8}$$

If we try to operate the stirred tank continuous reactor with $D > D_{max}$ the rate of removal of biomass will be higher than the maximum specific rate of biomass production. No steady state will be possible except the trivial $x = 0$ that is also a solution to Eq. (2).

Since from Table 7.1 K_s is often in the ppm range the value of s in the reactor is usually orders of magnitude lower than s_f except when D approaches D_{max} and s sharply increases towards s_f. Consequently x is approximately equal to $Y_{sx} s_f$ until D approaches D_{max} that according to Eq. (8) with $K_s \ll s_f$ is very close to μ_{max} for most organisms fed with glucose.

In Fig. 7.5, experimental data are shown for growth of *Aerobacter aerogenes* in a chemostat with glycerol as the limiting substrate. The biomass concentration is observed to be approximately constant for dilution rates between 0.4 and 0.95 h^{-1}. The glycerol concentration is very low for dilution rates below 0.95 h^{-1}, and when the dilution rate approaches $D_{max} = 1.0$ h^{-1} the glycerol concentration increases rapidly to $s = s_f = 10.0$ g L^{-1}. The Monod model, i.e., Eqs. (6) and (7), with the following parameters listed:
- $\mu_{max} = 1.0$ h^{-1}
- $K_s = 0.01$ g glycerol L^{-1}
- $Y_{sx} = 0.55$ g DW (g glycerol)$^{-1}$

describes the glycerol concentration quite well in the whole dilution-rate range, and also the approximately constant biomass concentration in the range 0.4-0.95 h^{-1}. However, for low dilution rates the model predicts too high biomass concentrations. This is explained by maintenance metabolism of the glycerol (see Section 5.2.1), which is not included in the Monod model. When substrates not used for maintenance processes are limiting, e.g., the nitrogen source, the Monod model normally describes the biomass concentration in the whole dilution range quite well (see Example 7.2).

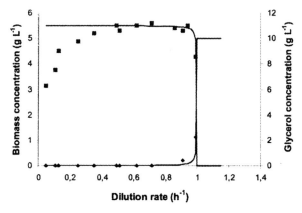

Figure 7.5 Growth of *A. aerogenes* in a chemostat with glycerol as the limiting substrate. The lines are model calculations using the Monod model. The data are taken from Herbert (1959).

It is not to be expected that the empirical Monod model can be used to fit all kinds of fermentation data. Many authors have tried to improve on the Monod model, and some of these embellished models are listed in Table 7.2. Except for the Moser model, all the models contain two adjustable parameters. The logistic law equation contains a constant K_x that must obviously depend on the substrate feed concentration. It cannot either be true that μ is independent of the substrate level in the reactor. The Contois model includes inhibition of the specific growth rate by the biomass. For very high biomass concentration the biotic phase may take up a substantial part of the total reactor volume, and the uptake of substrates could presumably be hampered just by the presence of the biomass. It is, however, difficult to imagine how cells by their mere presence should inhibit their own growth, and probably the ability of the Contois kinetics to fit experimental data is explained by some unaccounted-for effect of substrates or of the metabolic products, which may be toxic. One may have endless discussions concerning the pros and cons of empirical rate expressions such as the Monod model or its relatives in Table 7.1. They may serve a useful purpose as data fitters and as control models in industrial fermentations, but their value for research purposes is very limited since they reveal next to nothing about the possible mechanisms behind the observed phenomena.

All the unstructured models presented above assume that there is only one limiting substrate, but often more than one substrate concentration influences the specific growth rate. In these situations complex interactions can occur, which are difficult to model with unstructured models unless many adjustable parameters are admitted. Tsao and Hansen (1975) proposed a general, multiparameter, unstructured model for growth on multiple substrates

$$\mu = \left(1 + \sum_i \frac{s_{e,i}}{s_{e,i} + K_{e,i}}\right) \prod_j \frac{\mu_{max,j} s_j}{s_j + K_{s,j}} \quad (7.17)$$

Table 7.2 Compilation of different unstructured, kinetic models.

Name	Kinetic expression
Teissier	$\mu = \mu_{max}\left(1 - e^{-c_s/K_s}\right)$
Moser	$\mu = \mu_{max} \dfrac{s^n}{s^n + K_s}$
Contois	$\mu = \mu_{max} \dfrac{s}{s + K_s x}$
Blackman	$\mu = \begin{cases} \mu_{max} \dfrac{s}{2K_s} & ; \ s \leq 2K_s \\ \mu_{max} & ; \ s \geq 2K_s \end{cases}$
Logistic law	$\mu = \mu_{max}\left(1 - \dfrac{x}{K_x}\right)$

$s_{e,i}$ are the concentrations of growth-enhancing substrates, and s_j are the concentration of substrates that are essential for growth. The presence of growth-enhancing substrates results in an increased specific growth rate, whereas the essential substrates must be present for growth to take place. Application of Eq. (7.17) may be quite successful, but in the presence of many substrates there will of course be a large number of parameters, and the fit of the model to the experimental data may be coincidental. A special case of Eq. (7.18) is growth in the presence of two essential substrates, i.e.,

$$\mu = \frac{\mu_{max,1} \mu_{max,2} s_1 s_2}{(s_1 + K_{s,1})(s_2 + K_{s,2})} \qquad (7.18)$$

Equation (7.18) may, for example, be used as a model for the growth of methanotrophic bacteria on the two substrates O_2 and CH_4 – an important process leading to single-cell protein (see Example 3.4). If the concentrations of both substrates are at levels where the specific growth rate for each substrate reaches 90% of its maximum value, i.e., $s_i = 9 K_i$, then the total rate of growth is limited to 81% of the maximum possible value. This is hardly reasonable, and Roels (1983) has therefore proposed two alternatives to Eq. (7.18), both of which may be generalized for application to more than two limiting substrates:

$$\frac{\mu}{\mu_{max}} = \min\left(\frac{s_1}{s_1 + K_1}, \frac{s_2}{s_2 + K_2}\right) \qquad (7.19)$$

$$\frac{\mu_{max}}{\mu} = 1 + \frac{1}{2}\left(\frac{K_{s,1}}{s_1} + \frac{K_{s,2}}{s_2}\right) \qquad (7.20)$$

Both models will give growth at 90% of the maximum value in the situation mentioned above.

Growth on two or more substrates that may substitute for each other, e.g., glucose and lactose, cannot be described by any of the unstructured models described above. Consider, e.g., growth of *E. coli* on glucose and lactose. Glucose is metabolized first since it is the "best" carbon source, and metabolism on lactose will only start after glucose is used up. The bacterium needs one of the sugars to grow, but in the presence of glucose there is not even a growth-enhancing effect of lactose. Application of Eq. (7.17) to this example of multiple substrates for glycolysis will clearly not be feasible. To describe the so-called diauxic growth it is necessary to apply a structured model, as illustrated in Section 7.4. Baltzis and Fredrickson (1988) offer a systematic procedure for treatment of growth limitation by two different substrates that are either complementary (such as O_2 and CH_4 in the SCP process) or substitutable. See also Egli (1991) for some interesting experimental results with dual limitation by C and N.

In some cases, growth is inhibited either by high concentrations of the limiting substrate or by the presence of a metabolic product. In order to account for these possibilities, the Monod kinetics is often extended with additional terms. Thus, for inhibition by high concentrations of the limiting substrate,

$$\mu = \mu_{max} \frac{s}{s^2/K_i + s + K_s} \qquad (7.21)$$

and for inhibition by a metabolic product

$$\mu = \mu_{max} \frac{s}{s+K_s} \frac{1}{1+p/K_i} \qquad (7.22)$$

or

$$\mu = \mu_{max} \frac{s}{s+K_s}\left(1 - \frac{p}{p_{max}}\right) \qquad (7.23)$$

Equations (7.21)-(7.23) may be useful models for including product or substrate inhibition in a simple way, and often these expressions are also applied in connection with structured models. Extension of the Monod model with additional terms or factors should, however, be done with some restraint since the result may be a model with a large number of parameters but of little value outside the range in which the experiments were made.

The discussion of unstructured kinetic models applied to a black box stoichiometry model can be summed up as follows:
1. In the black box stoichiometric model the yield coefficients Y_{ji} are assumed to be constant. Consequently only one rate expression, namely that for the limiting substrate, needs to be set up. All other rates can be derived from total mass balances.
2. For a single limiting substrate Eq. (7.16) or its relatives Eqs. (7.21)-(7.23) will usually give an adequate representation of the reaction rate. The parameters of the kinetic model are determined from experiments in a continuous steady state bioreactor.
3. Picking the correct limiting substrate can be difficult since the feed concentration of the many different nutrients will determine which is the limiting one. Experiments with different levels of feed concentrations of substrates will help to resolve the question. An indication that a growth enhancing substrate runs out during a batch fermentation is that the specific growth rate decreases, i.e., in a plot of log(x) versus time the curve bends over, although there is still much left of the substrate that was thought to be limiting. The value of K_s is usually in the ppm range as seen in Table 7.1 and K_s values in the g L^{-1} range sometimes seen in publications result from misinterpretation of data as described above. Another reason for obtaining a large K_s value can be that product inhibition sets in. Here of course one has to use models with structure as (7.22) or (7.23) rather than the simple Monod model (7.16).
4. None of the unstructured models gives a sensible description of data resulting from fast transients in a continuous stirred tank reactor. All the models assume that a change in limiting substrate concentration causes an immediate change in rate – such as is the case in simple gas phase catalytic reactions. With the growth being a result of a large number of biochemical reactions this is clearly not the case, and the slow start up of batch growth

Modeling of Growth Kinetics

– called the lag phase, is one example of this. Whereas a change in substrate concentration in the reactor takes place within seconds, both in pulse addition of substrate and by changing the feed rate, the change in biomass composition resulting from up – or down regulation of genes has a time constant of one to several hours. Consequently application of simple unstructured kinetic models results in underestimation of time constants for dynamic changes by several orders of magnitude.

7.3.2 Multiple Reaction Models

In the black box model all the yield coefficients are taken to be constant. This implies that all the cellular reactions are lumped into a single overall growth reaction where substrate is converted to biomass. A requirement for this assumption is that there is a constant distribution of fluxes through all the different cellular pathways at different growth conditions. In the present section we shall start to put some biochemical structure into kinetic models, moving from "unstructured, non-segregated" towards "structured non-segregated" on Fig. 7.2. The truly structured models will be the topic of Sections 7.4 to 7.6, and the models treated below are basically unstructured in the sense that the influence of the biomass is still expressed solely through the biomass concentration x. However, biochemical information will be included in a semi-quantitative way, and it will be acknowledged that the overall kinetics is the result of several overall reactions operating in the cell.

The first and obvious example is the substrate consumption for maintenance of Section 5.2.1, a process that runs independently of the growth process. From eq. (5.8) we find:

$$-r_s = Y_{xs}^{true}\mu + m_s \qquad (7.24)$$

Here Y_{xs}^{true} is referred to as the *true yield coefficient* and m_s as the *maintenance coefficient*. The extra substrate consumption is accompanied by synthesis of extra metabolic products and

$$r_p = Y_{xp}^{true}\mu + m_p \qquad (7.25)$$

Equations (7.24) and (7.25) can be used together with any black-box model for $\mu(s,p)$ described in Section 7.3.1. One simply expands the model with a constant term for substrate consumption and product formation. This may in principle give rise to a conflict since the black box rate expressions are zero for $s = 0$ and in some cases it might be necessary to specify m_s as a function of s.

With the introduction of the linear correlations the yield coefficients can obviously not be constants. Thus for the biomass yield on the substrate:

$$Y_{sx} = \frac{\mu}{Y_{xs}^{true}\mu + m_s} \qquad (7.26)$$

which for a constant m_s shows that Y_{sx} decreases to zero for $\mu \to 0$ where an increasing fraction of the substrate is used to meet the maintenance requirements of the cell. For large specific growth rates Y_{sx} approaches $(Y_{xs}^{true})^{-1} = Y_{sx}^{true}$. This corresponds to the situation where the maintenance substrate consumption becomes negligible compared with the substrate consumption for biomass growth, and the model reduces to the black box model of Section

7.3.1. Despite its simple structure the linear rate equation (7.24) of Pirt is found to mimic the results of many fermentation processes quite well. Table 7.3 compiles true yield coefficients and maintenance coefficients for various microbial species with glucose and oxygen as substrates. This table is a complement to Table 5.1 that gives values for the corresponding energetic parameters Y_{xATP} and m_{ATP}.

Example 7.2 Steady-state chemostat described by the Monod model including maintenance
We now reconsider the experimental data from Example 7.1, but include maintenance requirements. The steady-state balances are

$$\left(Y_{xs}^{true} \mu_{max} \frac{s}{s+K_s} + m_s \right) x = D(s_f - s) \quad (1)$$

$$\mu_{max} \frac{s}{s+K_s} x = Dx \quad (2)$$

Table 7.3 A compilation of "true" yield and maintenance coefficients for different microbial species growing at aerobic growth conditions. Data are shown for growth on glucose or glycerol.[#]

Species	Substrate	Y_{xs}^{true} (g (g DW)$^{-1}$)	Y_{xo}^{true} (mmoles (g DW)$^{-1}$)	m_s (g (g DW h)$^{-1}$)	m_o (mmoles (g DW h)$^{-1}$)
Aspergillus awamori	Glucose	1.92	68	0.016	0.62
Aspergillus nidulans		1.67	111	0.020	0.54
Aspergillus niger					
Aspergillus oryzae					
Bacillus clausii		1.82	24	0.043	1.49
Candida utilis		2.00	84	0.031	0.87
Escherichia coli		2.27	70	0.057	0.45
Klebsiella aerogenes		2.27	74	0.063	0.99
Penicillium chrysogenum		2.17	70	0.021	0.87
Saccharomyces cerevisiae		1.85	47	0.015	0.62
Aerobacter aerogenes	Glycerol	1.79	63	0.089	2.52
Bacillus megatarium		1.67	90	-	0.70
Klebsiella aerogenes		2.13	58	0.074	2.52

[#] The yield coefficients are all for growth with no formation of metabolites, i.e., carbon dioxide is the only metabolic product. The coefficients are specified both for the listed limiting substrate and for oxygen. NH_3 is the nitrogen source.

Modeling of Growth Kinetics

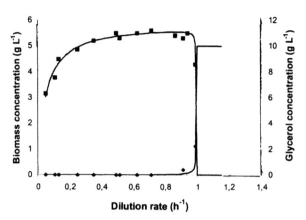

Figure 7.6 Growth of *A. aerogenes* in a chemostat with glycerol as the limiting substrate. The lines are model calculations using the Monod model including maintenance. The data are taken from Herbert (1959).

With Eq. (2), the steady-state concentration of the limiting substrate is still given by Eq. (5) in Example 7.1, since in both cases $\mu = D$ at steady state. However, the steady-state biomass concentration is different from that found when the simple Monod model is used. Combination of the two balances yield

$$x = \frac{D}{Y_{xs}^{true} D + m_s} (s_f - s) \qquad (3)$$

By comparison with Eq. (7.26), it is observed that Eq. (3) is an analogue of Eq. (6) in Example 7.1. At low dilution rates the term m_s is significant and the biomass concentration becomes smaller than if maintenance is neglected. This is illustrated in Fig. 7.6, where it is observed that the model correctly describes the biomass concentration in the whole dilution-rate range. The model parameters in the revised model are:
- $\mu_{max} = 1.0 \text{ h}^{-1}$
- $K_s = 0.01 \text{ g glycerol L}^{-1}$
- $m_s = 0.08 \text{ g glycerol (g DW h)}^{-1}$
- $Y_{xs}^{true} = 1.82 \text{ g glycerol (g DW)}^{-1}$

With these values the yield coefficient Y_{sx} is calculated to 0.53 and 0.56 at dilution rates of 0.4 h^{-1} and 0.8 h^{-1}, respectively. This is close to the constant value of the yield coefficient for the simple Monod model used in Example 7.1.

Including maintenance is in principle the same as considering two reactions in the model: a reaction where substrate is converted to biomass and a reaction where substrate is used for cellular maintenance. Extending the number of reactions, but still describing the biomass with a single variable, may allow modeling of more complex phenomena as illustrated in Examples 7.3 and 7.4.

Example 7.3 An unstructured model describing the growth of *S. cerevisiae*

S. cerevisiae is an industrially important microorganism used for production of baker's yeast and ethanol and today also for production of recombinant proteins. Sonnleitner and Kappeli (1986) proposed a simple *kinetic model* for the growth of this organism, and in this example we are going to discuss this model. However, before the model is discussed some fundamental aspects of the fermentation physiology of *S. cerevisiae* will have to be reviewed.

As already discussed in Example 3.5 and further in Example 5.4 aerobic growth of *S. cerevisiae* on glucose involves a mixed metabolism, with both respiration and fermentation being active. At high glucose uptake rates there is a limitation in the respiratory pathway, which results in an overflow metabolism towards ethanol. The exact location of the limitation has not been identified, but it is probably at the pyruvate node (Pronk *et al.*, 1996). The glucose uptake rate at which fermentative metabolism is initiated is often referred to as the critical glucose uptake rate, and the critical glucose uptake rate is found to depend on the oxygen concentration. Thus, at low dissolved oxygen concentrations the critical glucose uptake rate is lower than at high dissolved oxygen concentrations (and clearly at anaerobic conditions there is only fermentative metabolism corresponding to the critical glucose uptake rate being zero). The influence of dissolved oxygen concentration on the critical glucose uptake rate is often referred to as the Pasteur effect. A simple verbal model for the mixed metabolism is given in Fig. 7.7. When *S. cerevisiae* is grown in an aerobic, glucose-limited chemostat two distinct growth regimes are observed (see Fig. 7.8): (1) at low dilution rates all (or most) of the glucose is converted to biomass and carbon dioxide, and (2) at high dilution rates ethanol is formed in addition to biomass and carbon dioxide. As observed in Fig. 7.8, the shift to ethanol formation results in a dramatic decrease in the biomass yield from glucose.

In Fig. 7.8 the specific oxygen uptake rate is observed to increase with the dilution rate up to D_{crit}, whereafter it is approximately constant, while the specific carbon dioxide formation rapidly increases. Sometimes there is a more distinct decrease in the specific oxygen uptake rate above the critical dilution rate, but this depends on the strain and the operating conditions. The shift in metabolism at D_{crit} is normally referred to as the Crabtree effect, and it is a consequence of a bottleneck in the oxidation of pyruvate and repression of the oxidative system by high glucose concentrations. The bottleneck in the oxidation of pyruvate is illustrated in Fig. 7.7, where the flux through the glycolysis (indicated by the arrow) has to be smaller than the oxidative capacity of the cell (indicated by the bottleneck) if ethanol formation is to be avoided. If the glucose flux is larger than permitted by the bottleneck, the excess glucose is metabolized by the fermentative metabolism, and ethanol is formed.

Modeling of Growth Kinetics

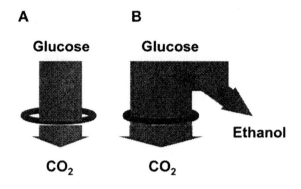

Figure 7.7 Illustration of the "bottle-neck" of the oxidative metabolism in *S. cerevisiae*.
A. Excess oxidative capacity, and therefore completely oxidative metabolism.
B. Limited oxidative capacity. The excess carbon flux is converted to fermentative products.

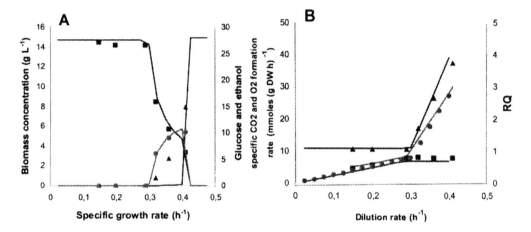

Figure 7.8 Chemostat culture of *S. cerevisiae*. The experimental data are from Rieger *et al.* (1983). The lines are model simulations with the parameters specified in Table 7.4.
A. The concentrations of glucose (▲), biomass (■) and ethanol (●) as function of the dilution rate. All concentrations are given as g L^{-1}.
B. The specific carbon dioxide formation rate (●) and the specific oxygen uptake rate (■), both in mmoles (g DW h)$^{-1}$. The respiratory quotient RQ is also shown (▲).

In the model of Sonnleitner and Käppeli (1986) the verbal model of Fig. 7.7 is applied with two reactions: (1) oxidative glucose metabolism and (2) fermentative glucose metabolism. In the model two substrates, glucose (*S*) and oxygen (O_2), and two metabolic products, ethanol (*P*) and carbon dioxide (CO_2) are considered. In the original model formulation ethanol may also serve as a substrate when glucose is not present, but here we will consider a simplified version of the model where ethanol uptake is not considered. The stoichiometry for the two reactions considered in the model is:

1. Oxidative glucose metabolism:

$$\gamma_1 X + \beta_{11} CO_2 - S - \alpha_{12} O_2 = 0 \tag{1}$$

2. Fermentative glucose metabolism:

$$\gamma_2 X + \beta_{21} CO_2 + \beta_{22} P - S = 0 \tag{2}$$

The nitrogen source (ammonia) and water are not included in the model.

When glucose is consumed solely by respiratory metabolism Eq. (1) is identical with the black box description of the cell growth, and all stoichiometric coefficients can therefore be interpreted as the yield coefficient Y_{si}. These stoichiometric coefficients can therefore easily be experimentally determined. At anaerobic conditions the metabolism is exclusively fermentative, and Eq. (2) therefore represents a black box description of cell growth at these conditions. The stoichiometric coefficients can therefore be found as shown in Section 3.3. Fermentative metabolism has a much lower ATP yield than respiratory metabolism, and the yield of biomass in the fermentative metabolism is therefore much lower than in the respiratory metabolism.

In order to describe biomass growth when both (1) and (2) contribute to growth the rates of the two reactions has to be specified. In order to do this Sonnleitner and Käppeli first specified the total glucose uptake rate (equal to the breadth of the arrow in Fig. 7.7) as:

$$-r_s = k_s \frac{s}{s + K_s} \tag{3}$$

and the maximum possible rate of respiration – or the maximum specific oxygen uptake rate (equal to the opening of the circle in Fig. 7.7) as:

$$-r_{o,\max} = k_o \frac{s_o}{s_o + K_o} \tag{4}$$

The maximum possible rate of respiration is a function of the dissolved oxygen concentration s_o, but normally s_o is much larger than K_o and the maximum rate of respiration becomes equal to k_o. According to the verbal model of Fig. 7.7 the rate of the respiratory metabolism is the minimum of either the glucose uptake rate or the maximum possible rate of respiration, *i.e.*

$$v_{res} = \min\left(r_s, \frac{r_{o,\max}}{\alpha_{12}}\right) \qquad (5)$$

Here the maximum possible rate of respiration is scaled with the stoichiometric coefficient for oxygen in eq. (1) to give the right comparison of the rates. Thus, at low glucose uptake rate, the rate of the oxidative metabolism is determined by the glucose uptake rate, whereas at high glucose uptake rate, it is determined by the maximum possible rate of respiration. If the glucose uptake rate is larger than $r_{o,\max}/\alpha_{12}$ the excess glucose will be metabolized by the fermentative metabolism given by eq. (2), i.e.

$$v_{fer} = \left|-r_s\right| - \left|\frac{r_{o,\max}}{\alpha_{12}}\right| \quad ; \quad \left|-r_s\right| > \left|\frac{r_{o,\max}}{\alpha_{12}}\right| \qquad (6)$$

Clearly reaction (2) acts as overflow reaction for excess glucose, and it is not active if $r_s < r_{o,\max}/\alpha_{12}$.

Biomass is formed by both (1) and (2), and the specific growth rate is:

$$\mu = \gamma_1 v_{res} + \gamma_2 v_{fer} \qquad (7)$$

Hereby the yield coefficient for biomass on glucose becomes:

$$Y_{sx} = \frac{\mu}{-r_s} = \frac{\gamma_1 v_{res} + \gamma_2 v_{fer}}{v_{res} + v_{fer}} \qquad (8)$$

At low glucose uptake rate where the metabolism is purely respiratory the yield coefficient is γ_1, whereas with mixed metabolism (high specific glucose uptake rates) the yield coefficient decreases. In Baker's yeast production where the yield of biomass on glucose has to be maximized it is obviously desirable to reduce fermentative metabolism by controlling the glucose uptake rate below the critical value (given by $r_{res,\max}/\alpha_{12}$).

With the rate of the two reactions specified it is also possible to find for the specific rate of carbon dioxide production:

$$r_{CO_2} = \beta_{11} v_{res} + \beta_{21} v_{fer} \qquad (9)$$

and the specific rate of ethanol production:

$$r_p = \beta_{22} v_{fer} \qquad (10)$$

The respiratory quotient (RQ) is often used to evaluate the metabolic state of baker's yeast, and with the Sonnleitner and Käppeli (1986) model we find:

$$RQ = \left|\frac{r_{CO_2}}{r_{O_2}}\right| = \frac{\beta_{11} v_{res} + \beta_{11} v_{fer}}{\alpha_{12} v_{res}} \qquad (11)$$

If there is no fermentative metabolism RQ becomes equal to β_{11}/α_{12}, which is close to 1, whereas if there is fermentative metabolism RQ increases above 1. Since it is relatively easy to measure the carbon dioxide production rate and the oxygen uptake rate by head space gas analysis, the RQ can be evaluated almost continuously, and a value above 1 will indicate some fermentative metabolism. This can be used in the control of e.g. the feed of glucose to the reactor, i.e., if the feed is too fast the glucose uptake rate will be above the critical value and there will be fermentative metabolism. The feed has to be reduced in order to avoid this.

With the specific rates given for all the major substrates, metabolic products, and biomass, it is possible to simulate the concentration of the variables in a bioreactor (see Fig. 7.8) with the model parameters listed in Table 7.4. The model predicts that when the glucose uptake rate increases above the critical value $r_{o,max}/\alpha_{12}$ (corresponding to a certain value of the specific growth rate), ethanol is formed by the cells. In a steady state chemostat this is seen as the presence of ethanol in the medium at specific growth rates above this critical value. When the critical value is exceeded and ethanol is produced the biomass yield drops dramatically, and this results in a rapid decrease in the biomass concentrations.

The Sonnleitner and Käppeli (1986) model is an excellent example of how mechanistic concepts can be incorporated into an unstructured model to give a fairly simple and in many situations adequate description of the complex growth of *S. cerevisiae*. With its limited structure the model does, however, give a poor description of transient operating conditions, e.g., the lag phase between growth on glucose and the subsequent growth on ethanol in a batch fermentation can not be predicted. It would not be difficult to include intracellular structure to describe, e.g., the level of the oxidative machinery, but this was not the target of Sonnleitner and Käppeli. Any basically sound model can be made to fit new experiments when more structure is added.

Example 7.4 Extension of the Sonnleitner and Käppeli model to describe protein production
As mentioned in Chapter 2 *S. cerevisiae* is used to produce heterologous proteins, and in particular it is used to produce human insulin. Strong glycolytic promoters often drive the production of heterologous proteins, and the productivity of the protein is therefore closely associated with the biomass production. Carlsen *et al.* (1997) studied the kinetics of proteinase A production by *S. cerevisiae*, and used a recombinant system very similar to that used for industrial insulin production (proteinase A was taken as a model protein for insulin).

Table 7.4 Model parameters in the Sonnleitner and Käppeli model.

Stoichiometric coefficients		Kinetic parameters	
α_{12}	12.4 mmoles/g of glucose	k_s	3.50 g glucose (g DW h)$^{-1}$
β_{11}	13.4 mmoles/g of glucose	k_o	8.00 mmoles (g DW h)$^{-1}$
γ_1	0.49 g/g of glucose	K_s	0.1 g L^{-1}
β_{21}	10.5 mmoles/g of glucose	K_o	0.1 mg L^{-1}
β_{22}	0.48 g/g of glucose		
γ_2	0.05 g/g of glucose		

Modeling of Growth Kinetics

From analysis of the production kinetics in chemostat cultures they found that the specific proteinase A production could be describe as:

$$r_{prot} = \beta_{13} v_{res} + \beta_{23} v_{fer} \tag{1}$$

where v_{res} and v_{fer} are given by eq. (5) and (6) in Example 7.3, respectively. When this production kinetics was applied it was found that the model could be fitted quite well to experimental data. Using the model they predicted the volumetric productivity of proteinase A as a function of the specific growth rate, and found the results shown in Fig. 7.9. It is interesting to see that the volumetric productivity, which is the quantity of interest in connection with heterolous protein production, has a very distinct maximum at the critical dilution rate. Thus, in order to ensure optimal productivity it is necessary to operate at (or very close) to the critical dilution rate. Even small changes in the dilution rate results in a significant decrease in the volumetric productivity. The reason for the very distinct optimum is that the protein production is related to biomass production, which also decreases drastically around the critical dilution rate.

7.3.3 The Influence of Temperature and pH

The reaction temperature and the pH of the growth medium are other process variables with a bearing on growth kinetics. It is normally desired to keep both of these variables constant (and at their optimal values) throughout the fermentation process – hence they are often called *culture parameters* to distinguish them from other variables such as reactant concentrations, stirring rate, oxygen supply rate, etc., which can change dramatically from the start to the end of a fermentation. The influence of temperature T and pH on individual cell processes can be very different, and since the growth process is the result of many enzymatic processes the influence of both temperature and pH on the overall bioreaction is quite complex.

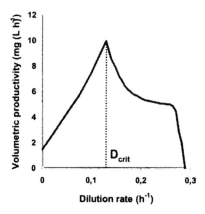

Figure 7.9 The volumetric production of protease A by a recombinant strain of *S. cerevisiae* at different dilution rates.

Large industrial bioreactors present greater control problems than laboratory reactors, but the economic consequences of unscheduled excursions in pH and temperature during a large-scale bioreaction are significant, and the investment in expensive multilevel pH and T control adds up to a substantial part of the total reactor investment. The control algorithms are sometimes quite complex since the optimum pH and T may change during the process – e.g., from an initial biomass growth phase to a production phase in which a secondary metabolite is produced.

The influence of temperature on the maximum specific growth rate of a microorganism is similar to that observed for the activity of an enzyme: An increase with increasing temperature up to a certain point where protein denaturation starts, and a rapid decrease beyond this temperature. For temperatures below the onset of protein denaturation the maximum specific growth rate increases in much the same way as for a normal chemical rate constant:

$$\mu_{max} = A \exp\left(-\frac{E_g}{RT}\right) \tag{7.27}$$

where A is a constant and E_g is the activation energy of the growth process. Assuming that the proteins are temperature-denatured by a reversible chemical reaction with free energy change ΔG_d and that denatured proteins are inactive, one may propose (Roels, 1983) an expression for μ_{max} that is closely related to the Hougen-Watson expression for catalyst activity in classical reaction engineering:

$$\mu_{max} = \frac{A \exp(-E_g / RT)}{1 + B \exp(-\Delta G_d / RT)} \tag{7.28}$$

Figure 7.10 is a typical Arrhenius plot (reciprocal absolute temperature on the abscissa and log μ on the ordinate) for *E. coli*. The linear portion of the curve between approximately 21 and 37.5 °C is well represented by Eq. (7.27), while the sharp bend and rapid decrease of the specific growth rate for $T > 39$ °C shows the influence of the denominator term in Eq. (7.28). Table 7.4 lists the parameters found by fitting the model in Eq. (7.28) to the data in Fig. 7.10. The results of the model calculations are shown as lines on the figure. Esener *et al.* (1981a) also applied Eq. (7.28) to describe the influence of the temperature on the maximum specific growth rate of *Klebsiella pneumoniae*, and the resulting parameters are also included in Table 7.5. It is observed that in the low temperature range the influence of the temperature is stronger for *K. pneumoniae* than for *E. coli*, i.e., E_g is larger for *K. pneumoniae* than for *E. coli*. On the other hand, denaturation of the proteins is much more temperature-sensitive in *E. coli* than in *K. pneumoniae*. Figure 7.10 also illustrates the general observation that the maximum specific growth rate is always lower for growth on a minimal medium compared with growth on a complex medium. The parameter A is smaller for growth on the glucose-minimal medium than for growth on the glucose-rich medium, and A is therefore not a characteristic parameter for the individual strain but rather a function of, e.g., the medium composition. E_g is the same for the two media, and it may therefore be a characteristic parameter for a given strain.

$$e \leftrightarrow e^- + H^+ \leftrightarrow e^{2-} + 2H^+ \tag{7.29}$$

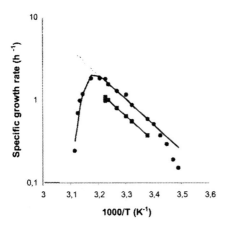

Figure 7.10 The influence of temperature on the maximum specific growth rate of *E. coli* B/r. The circles represent growth on a glucose-rich medium, and the squares represent growth on a glucose-minimal medium. The lines are calculated using the model in Eq. (7.28) with the parameters listed in Table 7.5. The data are taken from Herendeen *et al.* (1979).

The model presented above for the temperature influence on the maximum specific growth rate has a reasonable physical interpretation, and it may with some confidence also be used to express the temperature dependence of rate constants in structured models for cellular kinetics. One important aspect not considered is the influence of temperature on maintenance processes, which are normally very temperature dependent. An expression similar to Eq. (7.28) can be used, but the activation energy of the maintenance processes is likely to be different from that of the growth process. Thus the relative rate of the two processes may vary with the temperature.

The influence of pH on cellular activity is determined by the sensitivity of the individual enzymes to changes in the pH. Enzymes are normally active only within a certain pH interval, and the total enzyme activity of the cell is therefore a complex function of the environmental pH. As an example we shall consider the influence of pH on a single enzyme, which is taken to represent the cell activity. The enzyme is assumed to exist in three forms:

Table 7.5 Model parameters in Eq. (7.28) for *K. pneumoniae* and *E. coli*.[#]

Parameter	*K. pneumoniae*	*E. coli* (rich)	*E. coli* (min)	
E_g	86.40	58	58	kJ mole^{-1}
ΔG_d	287.78	550	-	kJ mole^{-1}
A	5.69 10^{14}	1.0 10^{10}	6.3 10^9	h^{-1}
B	1.38 10^{48}	3.0 10^{90}	-	-

[#]For *E. coli* the parameters are specified both for growth on a glucose-rich medium and a glucose-minimal medium.

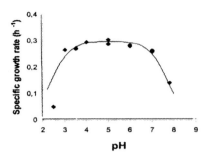

Fig. 7.11 The influence of pH on the maximum specific growth rate of the filamentous fungus *Aspergillus oryzae*. The line is simulated using eq. (7.31) with $K_1=4\cdot 10^{-3}$, $K_2=2\cdot 10^{-8}$, and $ke_{tot}=0.3$ h^{-1}.

where e^- is taken to be the active form of the enzyme while the two other forms are assumed to be completely inactive. With K_1 and K_2 being the dissociation constants for e and e^- respectively. The fraction of active enzyme e^- is calculated to be:

$$\frac{e^-}{e_{tot}} = \frac{1}{1+[H^+]/K_1 + K_2/[H^+]} \quad (7.30)$$

and the enzyme activity is taken to be $k=k_e e^-$. If the cell acticity is determined by the activity of the enzyme considered above the maximum specific growth rate will be:

$$\mu_{max} = \frac{ke_{tot}}{1+[H^+]/K_1 + K_2/[H^+]} \quad (7.31)$$

Although the dependence of cell activity on pH cannot possibly be explained by this simple model it is, however, found that Eq. (7.31) gives an adequate fit for many microorganisms, as illustrated in Fig. 7.11, where data are shown for *Aspergillus oryzae*. From the profiles of the maximum specific growth rate as a function of pH it is observed that the optimum is relatively broad. Thus the cell activity does not change much when one moves one pH unit away from the optimum, but larger deviations from the optimum pH leads to drastic reductions in the microbial activity.

Microbial cells have a remarkable ability to maintain the intracellular pH at a constant level even with large variations in the pH of the extracellular medium, but only at the expense of a significant increase in the maintenance demands, since Gibbs free energy has to be used for maintaining the proton gradient across the cell membrane. This ability of the microbial cells to function at suboptimal environmental pH may, however, be destroyed by the presence of *uncoupling agents*, e.g., organic acids (see Example 2.1). At a low extracellular pH, organic acids present in the medium may be transported by passive diffusion into the cell, where they are dissociated at the higher intracellular pH. The protons have to be pumped back to the extracellular medium, and this requires a considerable amount of ATP, which would otherwise be utilized for growth or other useful purposes. At high concentrations of the undissociated acids, the cell machinery breaks down with fatal outcome for the cell.

7.4 Simple Structured Models

As discussed in the introduction to this chapter, the unstructured models are quite adequate for steady-state conditions, and as shown in Example 7.3 even complex growth patterns can be incorporated. Unstructured models do, however, give a poor description of dynamic experiments where the biomass composition changes and the activity of the biomass therefore varies. The poor fit of unstructured models to dynamic changes in fermentations has been confirmed by many different fermentation studies, see e.g. Esener *et al.* (1981b) who illustrated that the Monod model (including maintenance) could not fit biomass concentration data both in the exponential growth phase and in the transition phase from exponential growth to a phase with slower growth in a fed-batch fermentation of *K. pneumoniae*. It must be realized that unstructured models are primarily data fitters for a restricted set of data, and they can rarely be used at significantly different experimental conditions. Simple structured models are, in one sense or the other, improvements on the unstructured models since some basic mechanisms of the cellular behavior are at least qualitatively incorporated. Thus, the structured model may have some predictive strength. It may describe the growth process at different operating conditions with the same set of parameters, and it can therefore be applied for, e.g. optimization of the process.

In simple structured models, biomass components are lumped into a few key variables – the vector $\mathbf{X}$ of Section 7.2 – that, hopefully, are representative of the cell behavior. Hereby the microbial activity becomes not only a function of the abiotic variables, which may change with very small time constants, but also of the cellular composition. The microbial activity therefore depends of the history of the cells, i.e., on the environmental conditions that the culture has experienced in the past. In simple structured models the cellular components included in the model represent pools of different enzymes, metabolites, or other cellular components. The cellular reactions considered in these models are therefore empirical since they do not represent the conversion between true components. Similarly, the kinetics for the individual reactions is normally described with empirical expressions, of a form that appears to fit the experimental data with a small number of parameters. Thus, Monod-type expressions are often used since they summarize some fundamental features of most cellular reactions, i.e., being first order at low substrate concentration and zero order at high substrate concentration. Despite their empirical nature, simple structured models are normally based on well-established cell mechanisms, and they are able to simulate certain features of experiments quite well. In the following we will consider two different types of simple structured models: Compartment Models and Cybernetic Models.

7.4.1 Compartment Models

In compartment models the biomass is divided into a few compartments or macromolecular pools. These compartments must be chosen with care, and cell components with similar function should be placed in the same compartment (e.g., all membrane material and otherwise rather inactive components in one compartment, and all active material in another compartment). If some thought is put into this crude structuring process one may regard individual, true cell components, which are not accounted for in the model, as being either in a frozen state or in pseudo steady state (very long or very short relaxation times compared with the time constants for the change in environment).

With the central role of the protein synthesizing system (PSS; see Section 2.1.4) in cellular metabolism, this is often used as a key component in simple structured models. Besides a few enzymes, the PSS consists of ribosomes (Ingraham et al., 1983), which contain approximately 60% ribosomal RNA and 40% ribosomal protein. Since the ribosomal RNA makes up more than 80% of the total stable RNA in the cell (see Note 7.3), the level of the ribosomes is easily identified through measurements of the RNA concentration in the biomass. As seen in Fig. 7.12, the RNA content of *E. coli* increases approximately as a linear function of the specific growth rate at steady-state conditions, and a similar observation is made for other microorganisms (see Fig. 7.12). Thus the level of the PSS is well correlated with the specific growth rate, and there is no doubt that X_{PSS} (or X_{RNA}) is a good representative of the state of activity of the cell. As seen from the chemostat experiments in Fig. 7.13, where the RNA concentration is measured in *Lactococcus cremoris* together with the biomass concentration during the transient from one steady state to another steady state, the linear relation between X_{RNA} and μ seems to hold throughout the transient.

Most simple structured models are based on a division of the cell into an active and an inactive part, where the PSS is always included in the active part of the cell (Nielsen and Villadsen, 1992). In some models the DNA content of the cell, X_{DNA}, is also taken to be part of the active cell compartment, but even though DNA is an essential cell component, from a mechanistic viewpoint X_{DNA} *per se* has virtually nothing to do with the growth rate of cell components (except as a possible determinant of RNA synthesis rate). This is discussed further in Note 7.4.

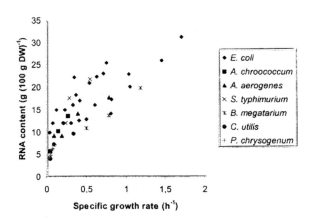

Figure 7.12 Measurements of steady-state RNA content in various microorganisms as a function of the specific growth rate (equal to the dilution rate). The data are compiled from various literature sources.

Modeling of Growth Kinetics

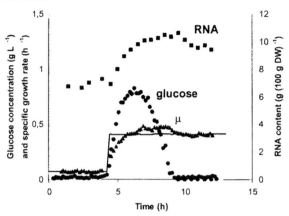

Figure 7.13 Measurements of the RNA content of *L. cremoris* and of the specific growth rate in a glucose-limited chemostat. The chemostat is at steady state at the dilution rate of 0.075 h^{-1} during the first 4.3 h of monitoring, whereafter D is changed to 0.409 h^{-1}. μ is calculated from the biomass concentration profile (not shown), and it is noted that the μ profile closely mimics the RNA profile with an overshoot before it settles down at the new D value. The glucose concentration profile is also shown. At $D = 0.075$ h^{-1}, $s = 5.5$ mg L^{-1} and after a large excursion (to 0.75 g L^{-1}) the glucose concentration reaches a new steady-state value of 10 mg L^{-1}. Note the dramatic dependence of the RNA content of cells on glucose concentration in the low concentration range $s \sim 5$-10 mg L^{-1}. The data are taken from Benthin *et al.* (1991).

Note 7.3 Stable and unstable RNA
In microbial cells there are three types of RNA:
1. Ribosomal RNA (rRNA)
2. Transfer RNA (tRNA)
3. Messenger RNA (mRNA)

Both rRNA and tRNA are fairly stable, whereas mRNA has a typical half-life of 1-2 min [1.3 min is an average value for *E. coli* (Ingraham *et al.*, 1983)]. For an *E. coli* cell at a specific growth rate of 1.04 h^{-1} the relative content of the three types of RNA is 5% mRNA, 18% tRNA, and 77% rRNA (see Table 2.5), and the relative content of stable RNA (tRNA plus rRNA) is therefore 95%. With the extremely low half-life of mRNA, this will normally not be included in the RNA measured by standard techniques [see, e.g., Benthin *et al.* (1991)], and the data specified for RNA measurements therefore represents the sum of tRNA and rRNA. With the values specified above, it is found that rRNA accounts for approximately 80% of the total measured RNA, and measurements of RNA can therefore be considered a good estimate for the ribosome level in the cell.

Whereas the content of stable RNA increases with the specific growth rate (see Fig. 7.12), the tRNA content decreases with the specific growth rate. Thus the relative content of tRNA is higher than 18% at specific growth rates below 1.04 h^{-1}. However, tRNA comprise a small fraction of the total RNA and even at very low specific growth rates it is less than 25% of the total RNA (Ingraham *et al.*, 1983). It is therefore still reasonable to consider the stable RNA content of the cell to be proportional with the ribosome level.

Note 7.4. What should be positioned in the active compartment in a simple structured model?
When the cell is divided into an active and an inactive part, it is necessary to define which biomass components should be considered active and which should be considered inactive. Most of the biomass components are involved in the overall growth process, and the only truly inactive components are structural material such as membranes or the cell wall. Thus for most microorganisms the inactive part of the cell is small. There are, however, a larger number of components that do not participate directly in the growth process, and furthermore the concentration of a number of components does not influence the overall growth kinetics. An example is DNA, which is of course essential for growth, but it is not an active component since it does not synthesize new material by itself. It is rather the polymerases that synthesize new DNA and mRNA which are active components, but these are present in very low concentrations and do not play a key role in regulation of DNA transcription and gene expression. The compartment model concept is based on an assumption of a constant composition of the individual macromolecules within each compartment. With only two compartments, i.e., an active and an inactive one, one has to make a very rough division of the components, and this partition can therefore not be expected to hold under all operating conditions. However, one should attempt to group the various macromolecules so that all components that decrease or increase with, e.g., dilution rate in approximately the same fashion are grouped in the same compartment.

A conceptual problem with the division of the cell into two parts is that many active components are placed in the inactive compartment. The tRNA is found to have a constant ratio to DNA at different specific growth rates and should therefore be included in the same compartment as the "inactive" DNA. However, tRNAs are strongly involved in protein synthesis and they are therefore active compounds. Similarly, many non-PSS proteins that are involved in the primary metabolism, e.g. enzymes involved in the EMP pathway and TCA cycle, are obviously active even though they are positioned in the inactive compartment. The words *active* and *inactive* should therefore not be interpreted in a strict sense, but rather according to the following: the size of the active compartment increases with and is linearly correlated with the specific growth rate, whereas the size of the inactive compartment decreases with the specific growth rate.

Ramkrishna *et al.* (1967) presented one of the first structured models. In their model biomass was divided into two compartments: a G compartment composed of nucleic acids and a D compartment with enzymatic proteins. Hence the whole biomass is considered to be active since both genetic material and proteins are involved in the growth process. Both G and D compartments can be degraded to inactive material by the action of a non-specified *inhibitor*. These degradation processes are especially important for description of cell death at the end of batch fermentations, but since this is only of small relevance for normal operating conditions we shall leave these empirical degradation reactions out of our discussion. The stoichiometry of the processes by which G and D compartment are synthesized is

$$X_D + 2X_G - \alpha_{11}s - X_D - X_G = 0 \quad ; \quad v_1 = k_1 \frac{X_G}{X_G + K_{GG}} \frac{s}{s + K_{sG}} X_D \qquad (7.32)$$

$$2X_D + X_G - \alpha_{21}s - X_D - X_G = 0 \quad ; \quad v_2 = k_2 \frac{X_G}{X_G + K_{GD}} \frac{s}{s + K_{sD}} X_D \qquad (7.33)$$

Proteins and genetic material participate in the doubling of genetic material and external substrate is consumed [Eq. (7.32)]. Enzymatic proteins are formed in the reaction in Eq. (7.33) with active participation of genetic material, and again external substrate is consumed. The two stoichiometric coefficients α_{11} and α_{21} are model parameters to be fitted from experiments. As for many of the original structured models the original model was (as pointed out by Fredrickson (1976)) not formulated in terms of intracellular concentrations X_G and X_D (grams per gram dry weight), but if one reinterprets the variables in the model as being intracellular variables, one arrives at the kinetic expressions in Eqs. (7.32) and (7.33). In this model the specific growth rate is the sum of v_1 and v_2, and it is clearly seen that besides a dependence of the substrate concentration the specific growth rate is a function of the biomass composition. Depending of the model parameters it is possible to describe an increase in the specific growth rate when X_D increases.

In another simple structured model, Williams (1967) also divides the biomass into two compartments. The interpretation of the compartments is, however, different from that used by Ramkrishna et al. (1967). Williams includes small metabolites and ribosomes in one compartment, which we designate as the A (or active) compartment. All macromolecules such as protein and DNA are pooled in another compartment, the G compartment. Synthesis of the two compartments is described by

$$\gamma_{11} X_A - s = 0 \quad ; \quad v_1 = k_1 \frac{s}{s + K_s} \tag{7.34}$$

$$\gamma_{22} X_G - X_A = 0 \quad ; \quad v_2 = k_2 X_A X_G \tag{7.35}$$

Thus the A compartment is formed directly from s, whereas the G compartment is formed from building blocks present in A. With the definition of the compartments, A is certainly an active part of the cell, and the rate of formation of G must depend on the size of A. The role of the G compartment is less clear. Most of the G compartment is probably inactive, but a constant fraction of G may contain enzymes necessary for growth of G. This must be the reason why Williams postulated that the rate of growth of X_G is proportional to both X_A and X_G, as in Eq. (7.35). In Eq. (7.34), a Monod-type dependence of the substrate is used for the rate of formation of A, whereas Williams used a first-order dependence of s. Furthermore, the original model was not formulated in terms of intrinsic variables, and the interpretation of the Williams model given above includes the corrections pointed out by Fredrickson (1976) and by Roels and Kossen (1978).

The Williams model has been used by Roels and co-workers to describe different fermentation processes (Roels and Kossen, 1978; Esener et al., 1981b,c; Esener et al., 1982; Harder and Roels, 1982; Jöbses et al., 1985), and their application of the model is discussed in Example 7.5. Conversion of the G compartment to the A compartment, as described in Eq. (7.36), was included in the modifications of Roels and co-workers. The argument for the applied stoichiometry is that G is degraded to small molecules that are included in the A compartment without any loss. It is reasonable to include some kind of turnover of cellular material, but it is less justified to let the normally very stable structural material degrade to the active material present in A.

$$X_A - X_G = 0 \quad ; \quad v_3 = k_3 X_G \tag{7.36}$$

In analogy with the models of Roels and co-workers (Nielsen *et al.*, 1991a,b) also proposed a simple two compartment model for simulation of the growth of lactic acid bacteria that is analogous to the Williams model and a simplified version of this model is discussed in Example 7.6.

Example 7.5 Analysis of the model of Williams
With the formulation of the Williams (1967) model given by Eqs. (7.34)-(7.36), the stoichiometric matrices become

$$\mathbf{A} = \begin{pmatrix} -1 \\ 0 \\ 0 \end{pmatrix} \quad ; \quad \Gamma = \begin{pmatrix} \gamma_{11} & 0 \\ -1 & \gamma_{22} \\ 1 & -1 \end{pmatrix} \tag{1}$$

Using Eq. (7.11), the specific growth rate is calculated:

$$\mu = \begin{pmatrix} \gamma_{11} & -1 & 1 \end{pmatrix} \begin{pmatrix} v_1 \\ v_2 \\ v_3 \end{pmatrix} + \begin{pmatrix} 0 & \gamma_{22} & -1 \end{pmatrix} \begin{pmatrix} v_1 \\ v_2 \\ v_3 \end{pmatrix} = \gamma_{11} v_1 - (1 - \gamma_{22}) v_2 \tag{2}$$

or

$$\mu = \gamma_{11} k_1 \frac{s}{s + K_s} - (1 - \gamma_{22}) k_2 X_A X_G \tag{3}$$

The third reaction does not contribute to the specific growth rate since its sole function is to recycle biomass. Using Eq. (7.14) we find the mass balance for the active compartment:

$$\frac{dX_A}{dt} = \gamma_{11} k_1 \frac{s}{s + K_s} - k_2 X_A X_G + k_3 X_G - \mu X_A \tag{4}$$

and since $X_G = 1 - X_A$, we obtain the following relation for steady-state chemostat operation

$$X_A = \frac{1}{\gamma_{22} k_2} (D + k_3) \tag{5}$$

Clearly the model predicts that X_A increases with D, as it should according to Fig. 7.12, since the A compartment contains the PSS. If the parameters in the correlation are estimated from the experimental data, k_3 is found to be 0.5 h^{-1}, which as pointed out by Esener *et al.* (1982), is far too high to be reasonable. The rather arbitrarily postulated kinetics is probably the reason for the inaccurate result. From Eq. (5) it follows that if degradation of the G compartment is not included (i.e., $k_3 = 0$), the model predicts $X_A = 0$ when $D = 0$ h^{-1}. This does not correspond with the experimental data of Fig. 7.12; i.e., even resting cells ($\mu = 0$) must have ribosomes on stand-by if they are to develop into actively growing cells. Thus the

Modeling of Growth Kinetics

degradation reaction of G compartment to active A compartment is taken to be the mechanism that ensures that some A compartment is present at $\mu = 0$. Despite the inconsistencies of the Williams model, it gives a far better description of a fed-batch fermentation with *Klebsiella pneumoniae* than a traditional Monod model including maintenance (Esener *et al.*, 1981b,c; Esener *et al.*, 1982), and the number of parameters is not higher than that which can be estimated from steady-state growth measurements.

Jöbses *et al.* (1985) also applied the modified Williams model for analysis of fermentations with *Zymomonas mobilis*. The experimental data for the substrate uptake at different specific growth rates in a steady-state chemostat indicate a nonlinear increase in the volumetric glucose uptake rate q_s (grams of glucose per liter per hour) with D. This cannot be described by a traditional unstructured model, but the two-compartment model fits the experimental data well.

It is normally claimed that the Williams model is able to describe a lag phase initially in a batch fermentation (Bailey and Ollis, 1986; Roels and Kossen, 1978; Williams, 1967). Inserting $X_G = 1 - X_A$ in Eq. (3) we find:

$$\frac{d\mu}{dX_A} = -(1 - \gamma_{22})k_2(1 - 2X_A) \qquad (6)$$

and since $\gamma_2 \leq 1$ we find for high substrate concentrations (i.e., $s \gg K_s$), that X_A must be > 0.5 if the specific growth rate is to increase with X_A. In model simulations presented by Williams, X_A increases from 0 to 0.75 at the beginning of a batch fermentation, and the specific growth rate therefore decreases until X_A becomes larger than 0.5. This is, however, not easily seen from the presented model simulations since the biomass concentration is shown in a linear plot rather than a semi-logarithmic plot. Since $X_A > 0.5$ is not biologically reasonable (Esener *et al.* (1982) uses a maximum of 0.3 for X_A) the conclusion is that in reality the model does not predict that the specific growth rate increases with X_A but rather that μ decreases when the culture is inoculated with cells having a low X_A, i.e., resting cells. This illustrates a general problem with compartment models. Through the introduction of structure into the biomass it may be possible to describe the specific growth rate as a function of the cellular composition and hereby describe some phenomena quite well, but the compartments (or variables) used in the model may not necessarily relate to any biological variables.

Example 7.6 Two compartment model for lactic acid bacteria

Nielsen *et al.* (1991a,b) presented a two compartment model for the lactic acid bacterium *Lactococcus cremoris*. The model is a progeny of the model of Williams with a similar definition of the two compartments:
- Active (A) compartment contains the PSS and small building blocks
- Structural and genetic (G) compartment contains the rest of the cell material

The model considers both glucose and a complex nitrogen source (peptone and yeast extract), but in the following presentation we discuss the model with only one limiting substrate (glucose). The model considers two reactions for which the stoichiometry is:

$$\gamma_{11} X_A - s = 0 \qquad (1)$$

$$\gamma_{22} X_G - X_A = 0 \qquad (2)$$

As for the Williams model the first reaction represents conversion of glucose into small building blocks in the A compartment and these are further converted into ribosomes. The stoichiometric coefficient Y_{11} can be considered as a yield coefficient since metabolic products (lactic acid, carbon dioxide etc.) are not included in the stoichiometry. In the second reaction building blocks present in the A compartment are converted into macromolecular components of the G compartment. In this process some by-products may be formed and the stoichiometric coefficient Y_{22} is therefore slightly smaller than 1. It is assumed that the formation of macromolecules is the rate-controlling process in the formation of both the A and G compartments, and the kinetics of the two reactions therefore has the same form, *i.e.*

$$v_i = k_i \frac{s}{s + K_{s,i}} X_A \quad ; \quad i = 1,2 \tag{3}$$

v_i is taken both to be a function of the glucose concentration (the carbon and energy source) and of the concentration of the active compartment (the catalyst for formation of biomass). From Eq. (7.11) the specific growth rate for the biomass is found to be:

$$\mu = \begin{pmatrix} \gamma_{11} & -1 \end{pmatrix} \begin{pmatrix} v_1 \\ v_2 \end{pmatrix} + \begin{pmatrix} 0 & \gamma_{22} \end{pmatrix} \begin{pmatrix} v_1 \\ v_2 \end{pmatrix} = \gamma_{11} v_1 - (1 - \gamma_{22}) v_2 \tag{4}$$

or with the kinetic expressions for v_1 and v_2 inserted:

$$\mu = \left(\gamma_{11} k_1 \frac{s}{s + K_{s,1}} - (1 - \gamma_{22}) k_2 \frac{s}{s + K_{s,2}} \right) X_A \tag{5}$$

Thus the specific growth rate is proportional to the size of the active compartment. The substrate concentration s influences the specific growth rate both directly and indirectly by also determining the size of the active compartment. The influence of the substrate concentration on the synthesis of the active compartment can be evaluated through the ratio v_1/v_2:

$$\frac{v_1}{v_2} = \frac{k_1}{k_2} \frac{s + K_{s,2}}{s + K_{s,1}} \tag{6}$$

If $K_{s,1}$ is larger than $K_{s,2}$ the formation of X_A is favored at high substrate concentration, and it is hereby possible to explain the increase in the active compartment with the specific growth rate. Consequently, when the substrate concentration increases rapidly there are two effects on the specific growth rate:
- A fast increase in the specific growth rate, which results from an increase of $s/(s+K_{s,1})$ with s. The time constant for increase of s in the medium is small compared to the time constant for the growth process (see e.g. Fig. 7.13).
- A slow increase in the specific growth rate, which is a result of a slow build up of the active part of the cell, *i.e.* additional cellular synthesis machinery has to be formed in order for the cells to grow faster.

This is illustrated in Fig. 7.14, which shows the biomass concentration in two independent wash out experiments. In both cases the dilution rate was shifted to a value (0.99 h^{-1}) above the critical dilution rate (0.55 h^{-1}), but in one experiment the dilution rate before the shift was low (0.1 h^{-1}) and in the other experiment it was high (0.5 h^{-1}). The wash out profile is seen to be very different, with a much faster wash out when the shift is from a low dilution rate. When the dilution rate is changed to 0.99 h^{-1} the glucose concentration increases rapidly to a value much higher than $K_{s,1}$ and $K_{s,2}$, and this allows growth at the maximum rate. However, when the cells have been grown at a low dilution rate the size of the active compartment is not sufficiently large to allow rapid growth, and X_A therefore has to be built up before the maximum specific growth rate is attained. On the other hand if the cells have been grown at a high dilution rate X_A is already near its maximum value and the cells immediately attain their maximum specific growth rate. It is observed that the model is able to correctly describe the two experiments (all parameters were estimated from steady state experiments), and the model correctly incorporates information about the previous history of the cells.

The model also included formation of lactic acid and the kinetics was described with a rate equation similar to Eq. (3). Thus, the lactic acid formation increases when the activity of the cells increases, and hereby it is ensured that there is a close coupling between formation of this primary metabolite and growth of the cells.

It is interesting to note that even though the model does not include a specific maintenance reaction, it can actually describe a decrease in the yield coefficient of biomass on glucose at low specific growth rates. The yield coefficient is given by:

$$Y_{sx} = \gamma_{21}\left(1 - (1-\gamma_{22})\frac{k_2}{k_1}\frac{s+K_{s,1}}{s+K_{s,2}}\right) \quad (7)$$

Since $K_{s,1}$ is larger than $K_{s,2}$ the last term in the parenthesis decreases for increasing specific growth rate,

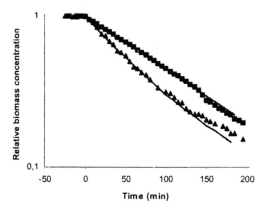

Figure 7.14 Measurement of the biomass concentration in two transient experiments in a glucose limited chemostat cultures with *L. cremoris*. The dilution rate was shifted from an initial value of 0.10 h^{-1} (▲) or 0.50 h^{-1} (■) to 0.99 h^{-1}, respectively. The biomass concentration is normalized by the steady state biomass concentration before the step change, which was made at time zero. The lines are model simulations. The data are taken from Nielsen *et al.* (1991b).

and the yield coefficient will therefore also increase for increasing substrate concentration.

The present two-compartment model, which solves some of the conceptual problems with the Williams model, has also been extended for description of recombinant *E. coli* fermentations (Nielsen *et al.*, 1991c; Strudsholm *et al.*, 1992).

Many other simple structured models are presented in the literature (see Harder and Roels (1982) and Nielsen and Villadsen (1992) for reviews). Most of these are similar in structure to the models described above, but the same ideas may be formulated very differently by different authors. Thus Powell (1967) introduced a class of structured models which he called "bottleneck models." They are based on an assumption of one cellular element being the bottleneck for cellular growth. These models are formally identical with the two-compartment models where the active compartment is the bottleneck for growth, and Powell also infers that the bottleneck may be the PSS. Other models are based on an extension of unstructured models, where some or all of the rate constants are described as functions of the environmental conditions, e.g.,

$$\frac{dk_i}{dt} = \frac{1}{\tau_i}\left(k_i^{max} f_i - k_i\right) \qquad (7.37)$$

where τ_i is a characteristic time for adaptation to new environmental conditions, k_i^{max} is a parameter, and f_i is a function of the substrates, which determines the "target value" of the parameter k_i. Sweere *et al.* (1988) used this concept to improve on the capability of the Sonnleitner and Käppeli model (see Example 7.3) to describe dynamic growth conditions, and the revised model fits experimental data for batch fermentation better than the original model. However, the approach is empirical, and the parameters in Eq. (7.37) cannot be estimated from steady-state experiments alone (the adaptation time τ_i has to be estimated from a transient experiment).

7.4.2 Cybernetic Models

Most academic fermentation studies are made with only a single limiting substrate, but in industrial processes several different components of a complex substrate may become rate-controlling in various parts of the fermentation. Modeling the parallel uptake of substrate components that serve different purposes in the microorganism, e.g., an energy source like glucose and a nitrogen source like ammonia, can be done in a rather simple fashion as illustrated with Eq. (7.18)-(7.20). Modeling the sequential uptake of different substrates that serve the same purpose in the microorganism, e.g., glucose and lactose, is, however, much more difficult. Sequential uptake of substrates in batch fermentations normally results in different exponential growth phases separated by lag phases where synthesis of enzymes necessary for metabolism of the next substrate is carried out. This is referred to as *diauxic growth* with two substrates (triauxic growth with three substrates). The sequential uptake of substrates is a consequence of complex regulatory structures in the cell. Thus, there is typically glucose repression (or more precise carbon catabolite repression) on the utilization of other carbon sources like galactose, lactose, maltose, sucrose, ethanol, acetate and carbohydrate polymers. The reason is that glucose is the preferred substrate for most organisms as it supports rapid growth, and the microorganism will therefore attempt to use this carbon and energy source first. Some of the characteristics of microbial growth on sequentially metabolized substrates are compiled in Table 7.6. They capture in a simple fashion the overall consequences for the sequential utilization of carbon and energy sources as a result of the complex regulatory mechanisms behind carbon catabolite repression.

It is possible to set up detailed mechanistic models describing the regulatory phenomena behind glucose repression, and hereby describe how enzymes involved in utilization of other carbon sources are synthesized. In Section 7.5.1 we will illustrate this with a model for the lac-operon in *E. coli*, i.e., the genes involved in lactose utilization. However, the complexity of these models is substantial, and if the aim is to simply simulate biomass growth on multiple substrates a simpler modeling concept is to be preferred. Ramkrishna and co-workers have developed an ingenious modeling concept, which is especially suited for description of growth on multiple carbon and energy sources. Their cybernetic modeling approach is based on the hypothesis that while the detailed modeling of regulatory processes is complicated, it may be possible to interpret the functioning of the cell as being guided by a strategy of optimal allocation of resources (Ramkrishna, 1982; Ramkrishna *et al.*, 1984). Since there is no direct way of confirming whether microorganisms really optimize the allocation of their resources towards growth, the cybernetic models should be accepted on the same basis as other simple structured models: They provide a good modeling framework that can be used to simulate growth on multiple substrates and hereby be used for design of fermentation processes.

The basic idea of the cybernetic model is that one *key enzyme* plays a bottleneck role in growth on a

Table 7.6 Characteristics of microbial growth on truly substitutable substrates (Ramkrishna *et al.*, 1987).

1. Given multiple substrates, microorganisms prefer to utilize the substrate on which they can grow the fastest, commonly resulting in a sequential utilization of the substrates in a batch culture.
2. Sequential utilization may turn into simultaneous utilization when another substrate, e.g., the nitrogen source, becomes limiting.
3. Even during simultaneous utilization of multiple substrates, the specific growth rate is never higher than that which can be obtained with growth on any substrate alone.
4. If during growth on a "slower" substrate, a "faster" substrate is added to the medium, the growth on the slower substrate quickly stops.
5. In continuous cultures, multiple substrates are consumed simultaneously at low dilution rates, and the faster substrate is preferentially consumed at high dilution rates.

particular substrate, and this key enzyme must be synthesized before growth can occur on that substrate. In reality this key enzyme may represent several enzymes, e.g., lactose permease and β-galactosidase in case of lactose metabolism. The reaction scheme for the growth process on each substrate can therefore be summarized by the following three reactions:

$$\gamma_i X - s_i = 0 \quad ; \quad v_i = r_i^* w_i = k_i \frac{s_i}{s_i + K_i} X_{E_i} w_i \tag{7.38}$$

$$X_{E_i} - X = 0 \quad ; \quad v_{E_i} = k_{E,i} \frac{s_i}{s_i + K_i^E} u_i \tag{7.39}$$

$$X - X_{E_i} = 0 \quad ; \quad v_{\text{deg},i} = k_{\text{deg},i} X_{E_i} \tag{7.40}$$

The index i indicates the substrate, and the model may consider N different substrates. The first set of reactions (7.38) is the formation of all biomass components except the enzyme E_i used for uptake of the ith substrate, and the second set of reactions (7.39) is the formation of the particular enzyme. The enzymes are synthesized from the general biomass compartment. Finally, the third set of reactions represents degradation of enzymes. In the model the "potential" substrate uptake rate is given by r_i^*, i.e., the substrate uptake rate that occurs when there is no limitation in the enzyme system involved in substrate uptake. The kinetics for the potential substrate uptake is described by Monod-kinetics. Similarly the formation of the key enzymes is given by Monod-type expressions with respect to the substrate. Degradation of the enzyme is described by a first-order reaction.

The production of the ith enzyme cannot proceed without some critical cellular resources, which must be suitably allocated for different enzyme synthesis reactions. This feature is included in the kinetics through the cybernetic variable u_i, which may be regarded as the fractional allocation of resources for the synthesis of the ith enzyme, and it can be interpreted as a controller of the enzyme production. The kinetics for substrate assimilation (and hereby biomass growth) is determined by another cybernetic variable w_i. This variable ensures that the growth takes place primarily on the best-suited substrate, and it may be interpreted as a control mechanism at the enzyme level. It is doubtful whether there are control mechanisms that work directly on the transport enzyme, but with the complex interactions between different intracellular pathways it is reasonable to include this control function in the model.

Several different models for the cybernetic variables have been described (Dhurjati et al., 1985; Kompala et al., 1984; Kompala et al., 1986), but based on an examination of the various models Kompala et al. (1986) conclude that the best model is obtained when Eqs. (7.41) and (7.42) are used for the two cybernetic variables.

$$u_i = \frac{r_i^*}{\sum_{j=1}^{N} r_j^*} \tag{7.41}$$

Modeling of Growth Kinetics

$$w_i = \frac{r_i^*}{\max_j(r_j^*)} \quad (7.42)$$

The definition of the cybernetic variable u_i in Eq. (7.41) is based on the so-called matching law model, which specifies that the total return from allocation of resources to different alternatives is maximized when the fractional allocation equals the fractional return. Thus in the cybernetic model the resources are allocated for synthesis of that enzyme which gives the highest specific growth rate (or highest μ). Originally the cybernetic variable w_i was also defined according to the matching law model, but the definition in Eq. (7.42) is superior to the double matching law concept – especially for description of simultaneous metabolism of two equally good substrates (here the double matching concept predicts a specific growth rate that is only half of that obtained on each substrate).

With the definition of the cybernetic models in Eqs. (7.38)-(7.42), the cybernetic model can handle both diauxic and triauxic batch fermentations (Kompala et al., 1986). A major strength of the cybernetic models is that all the parameters can be estimated on the basis of experiments on the individual substrates, and thereafter the model does a good job in fitting experiments with mixed substrates (see Fig. 7.15). Considering the large amount of experimental data presented by Kompala et al. (1986) and the quantitatively correct description of many experiments, it must be concluded that despite their rather empirical nature cybernetic models are well suited to description of growth on truly substitutable substrates.

According to the model of Kompala et al. (1986), the biomass in a chemostat that has been subjected to feed for a long period with only one carbohydrate should contain transport enzymes only for uptake of this particular substrate. All other enzyme systems would have been degraded or diluted to virtually zero by the growth of the biotic phase. Thus a pulse of another carbohydrate added to the chemostat would not be consumed, but this is contradicted by experimental observations. To account for this weakness in the cybernetic model, Turner and Ramkrishna (1988) introduced a term for constitutive enzyme synthesis with the rate $k_{con,i}$ in the Kompala et al. (1986) model. The mass balance for the ith enzyme is then given by Eq. (7.43).

$$\frac{dX_{E_i}}{dt} = k_{E,i} \frac{s_i}{s_i + K_i^E} u_i - k_{\deg,i} X_{E_i} + k_{con,i} - \mu X_{E_i} \quad (7.43)$$

Now the microorganism is always allowed a latent capability to metabolize – at least at a low rate – substrate different from that on which it is accustomed to grow. As a final note it can be mentioned that the cybernetic modeling concept has been extended to describe also other phenomena prevailing during growth of microorganisms, but these extensions will not be discussed here [see e.g. Varner and Ramkrishna (1999)]. As is the case for compartment models of Section 7.4.1 the cybernetic models have now been developed to a stage where the complexity of the model and its many empirical parameters obscures the main object of the exercise, namely to provide a reasonable verbal model to explain certain observations. The simple, semi-mechanistic model

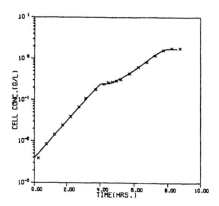

Figure 7.15 Cybernetic modeling of diauxic growth of *K. oxytoca* on glucose and xylose (inocolum precultured on glucose). The data points are measurements, and the line is model simulations. [Reprinted by permission from D. S. Kompala, D. Ramkrishna, and G. T. Tsao (1986), "Investigation of bacterial growth on mixed substrates: Experimental evaluation of cybernetic models," *Biotechnol. Bioeng.* **28**, 1044-1055].

concept was a substantial step forward in the 1980's but future efforts in model building must be spent on truly mechanistic models such as those described in the following section.

7.5 Mechanistic Models

With the many complex regulatory structures that operate in the living cell (see Fig. 2.2) it may be desirable to evaluate the relative importance of different cell processes. This can only be done through quantitative analysis, and here mathematical models are well suited as they allow integration of information from different experiments into an overall description of the system. As yet it is impossible to describe all the individual processes involved in cell growth – not all processes are identified and even the most powerful computers could not handle the large and complex models involved. Still there is a move towards construction of complete cellular models in modern biology. As mentioned in Chapter 1 this is referred to as *systems biology*, and in Section 7.5.2 we are going to consider some of the first attempts to describe cell growth using very detailed models. In many cases it is, however, interesting to study specific processes in the cell separately, and here models where mechanistic information is included must be used. These models are normally set up at the molecular level just as the models for enzyme kinetics in Chapter 6. They include a certain amount of information about the prevailing mechanisms – and they are therefore referred to as *mechanistic models*. Among the best studied processes in the cell at the molecular level is regulation of gene transcription, which clearly plays a very important role in overall control of cell function (but as discussed in Section 2.1.1 not necessarily a dominant role). Before we discuss single cell models we will therefore consider genetically structured models as an example of detailed mechanistic models.

7.5.1 Genetically Structured Models

Gene transcription models aim at quantifying gene transcription based on knowledge of the promoter function. Among the best-studied promoters is the so-called lac-promoter of *E. coli*, which takes part in regulating expression of genes that are involved in lactose uptake, i.e., lactose permease and β-galactosidase. Understanding the regulation of this promoter represented a major breakthrough in molecular biology, and the history of how the mechanisms were unraveled is an excellent introduction to modern molecular biology (Müller-Hill, 1996). The lac-promoter is also of substantial industrial interest as it is often used to drive expression of heterologous genes encoding recombinant proteins in *E. coli*.

The three genes coding for enzymes necessary for lactose metabolism in *E. coli* are coordinated in a so-called operon, and gene expression is coordinately controlled by two regulatory sites positioned upstream of the genes (see Fig. 7.16):
- Control at the operator by a repressor protein.
- Carbon catabolite repression at the promotor.

The repressor protein X_r has two binding sites – one site that specifically ensures binding to the operator (X_o) and one site which may bind lactose (S_{lac}). When lactose binds to the repressor protein, its conformation changes so that its affinity for binding to the operator is significantly reduced. Thus lactose prevents the repressor protein from binding to the operator, and transcription of the genes by RNA polymerase is therefore allowed. Consequently, lactose serves as an inducer of transcription; i.e., expression of the three genes *lacZ*, *lacY*, and *lacA* is not possible unless lactose (or another inducer, e.g. isopropyl-β-D-thiogalactoside, abbreviated IPTG) is present.[1] The binding of the repressor protein to lactose and the operator may be described by reactions (7.44), where $n = 4$ is the number of binding sites for lactose on the repressor protein.

$$X_r + nS_{lac} \overset{K_1}{\leftrightarrow} X_r nS_{lac} \tag{7.44a}$$

$$X_o + X_r \overset{K_2}{\leftrightarrow} X_o X_r \tag{7.44b}$$

$$X_o X_r + nS_{lac} \overset{K_3}{\leftrightarrow} X_o X_r nS_{lac} \tag{7.44c}$$

$$X_o + X_r nS_{lac} \overset{K_4}{\leftrightarrow} X_o X_r nS_{lac} \tag{7.44d}$$

---

[1] More precisely, the regulatory molecule is a lactose isomer, allolactose, that acts as an inducer. This species is formed *in situ* by β-galactosidase, which transforms lactose to allolactose. Thus, β-galactosidase, which is actually under the control of the *lac* operon, also has to be expressed constitutively at a basal level as a primer of the induction sequence.

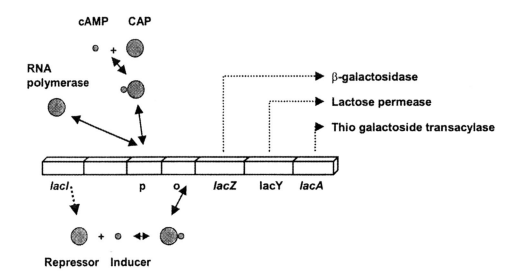

Figure 7.16 The lac-operon of *E. coli*. The operon includes a gene (*lacI*) for the repressor protein (r), promotor (p), operator (o), and the three genes *lacZ*, *lacY*, and *lacA*, which code for different enzymes: *lacZ* codes for β-galactosidase, *lacY* for lactose permease, and *lacA* for thio galactoside transacylase. In its free form the repressor protein may bind to the operator; when it complexes with the inducer (i), conformational changes of the repressor protein prevent binding to the operator. cAMP complexes with CAP and the complex may bind to the promoter, whereby the RNA polymerase may start the transcription from the promotor.

The model in Eq. (7.44) gives a simplified description of the true system since there may be different binding affinities for the repressor protein depending on how much lactose is bound to the protein (see Problem 7.3). With the concentration of the species (indicated with squared brackets) being in moles per gram dry weight, the equilibrium constants K_i, $i = 1, 2, 3, 4$ are given by:

$$K_1 = \frac{[X_r nS_{lac}]}{[X_r][S_{lac}]^n} \quad (7.45)$$

$$K_2 = \frac{[X_o X_r]}{[X_o][X_r]} \quad (7.46)$$

$$K_3 = \frac{[X_o X_r nS_{lac}]}{[X_o X_r][S_{lac}]^n} \quad (7.47)$$

Modeling of Growth Kinetics

$$K_4 = \frac{[X_o X_r n S_{lac}]}{[X_r n S_{lac}][X_o]} \quad (7.48)$$

Application of Eqs. (7.45)-(7.48) is based on the following assumptions (Harder and Roels, 1982):
- A macroscopic description can be used to express the influence of the reacting species on the kinetics, i.e., the concentrations of the different components are used. However, microorganisms only contain a few (1-4) copies of one type of operator per cell, and the number of repressor proteins per cell is also low (10-20). For such small entities the meaning of concentrations and of thermodynamic equilibrium is disputable, and it may be more correct to apply a stochastic modeling approach.
- As in Michaelis-Menten kinetics for enzymes (Section 6.1) all reactions in (7.44) are assumed to be equilibrium reactions. This is reasonable since the relaxation times for the equilibria are much smaller than for most other cellular reactions.

Balances for the repressor, operator, and inducer are

$$[X_r]_t = [X_r] + [X_r n S_{lac}] + [X_o X_r] + [X_o X_r n S_{lac}] \quad (7.49)$$

$$[X_o]_t = [X_o] + [X_o X_r] + [X_o X_r n S_{lac}] \quad (7.50)$$

$$[S_{lac}]_t = [S_{lac}] + n[X_r n S_{lac}] + n[X_o X_r n S_{lac}] \quad (7.51)$$

where the index t refers to the total concentration. In wild-type *E. coli* there are 10-20 times more repressor molecules than there are operators, and in this case the last two terms in Eq. (7.49) can be neglected. Furthermore, with the weak binding of the inducer-repressor complex to the operator, $[X_o X_r n S_{lac}]$ can be neglected in Eq. (7.50). Finally, Eq. (7.51) can be simplified by assuming that the intracellular concentration of inducer molecules is in sufficient excess over repressor molecules, and consequently that $[X_r n S_{lac}] + n[X_o X_r n S_{lac}] \ll [S_{lac}]$. With these simplifications the fraction of repressor free operators[2] is found (see Note 7.5) to be

$$Q_t = \frac{[X_o]}{[X_o]_t} = \frac{1 + K_1 [S_{lac}]_t^n}{1 + K_1 [S_{lac}]_t^n + K_2 [X_r]_t} \quad (7.52)$$

Since the transcription of the three genes in the operon is likely to be determined by the fraction of repressor-free operators eq. (7.52) is valuable for description of the synthesis of the enzymes necessary for lactose metabolism in a structured model that aims at describing diauxic growth on glucose and lactose. The inducer concentration S_{lac} is likely to be correlated with the extracellular lactose concentration, whereas the total content of repressor protein can be assumed to be constant.

---

[2] Since the total number of operators of a given type in the cell is very small, it does not make much sense to talk about the fraction of repressor-free operators. However, in a description of enzyme synthesis one may use Eq. (7.52) as an expression for the probability that the operator is repressor free.

Note 7.5 Derivation of Eq. (7.52)
With the assumptions specified for the derivation of Eq. (7.52), we have

$$[X_r]_t \approx [X_r] + [X_r nS_{lac}] \tag{1}$$

$$[X_o]_t \approx [X_o] + [X_o X_r] \tag{2}$$

$$[S_{lac}]_t \approx [S_{lac}] \tag{3}$$

Using Eqs. (2) and (7.46) the fraction of repressor-free operators is

$$Q_1 = \frac{[X_o]}{[X_o]_t} = \frac{1}{1 + K_2[X_r]} \tag{4}$$

After multiplication by $1 + K_1[S_{lac}]^n$ in both the nominator and the denominator, we find

$$Q_1 = \frac{1 + K_1[S_{lac}]^n}{1 + K_1[S_{lac}]^n + K_2([X_r] + K_1[X_r][S_{lac}]^n)} \tag{5}$$

Using Eqs. (7.45), (1), and (3), we obtain the expression in Eq. (7.52) for the fraction of repressor-free operators.

Small molecules that influence the transcription of genes are called effectors, and in the lac-operon the effector (lactose) is an *inducer*. In other operons there may, however, be a negative type of control, and here the effector is called an *anti-inducer*. With an inducer the binding affinity to the operator of the free repressor is much larger than that of the inducer-repressor complex, i.e., $K_2 \gg K_4$, whereas with anti-inducer it is the other way round. For an anti-inducer, the fraction of repressor-free operators can be found from an expression similar to Eq. (7.52) (see Harder and Roels (1982) and Problem 7.3).

The other control mechanism in the lac-operon is the so-called carbon catabolite repression, which ensures that no enzymes necessary for lactose metabolism are synthesized as long as a preferred substrate is available, e.g., glucose. The mechanisms behind carbon catabolite repression are not completely known, but it is known that the binding of the RNA polymerase to the promotor is strongly enhanced when a complex of cAMP and a protein called CAP (catabolite activator protein) is bound to the promotor. When the concentration of glucose (or another energy source) in the extracellular medium is high, the intracellular cAMP level is found to be low. Consequently, the level of the cAMP-CAP complex is low at high glucose concentrations. The site of the binding of the cAMP-CAP complex has been located in several operons that are under carbon catabolite repression, and binding of the complex to DNA has been found to promote helix destabilization downstream. This in turn facilitates the binding of the RNA polymerase and hereby stimulates gene expression. The carbon catabolite repression can be described by the following equilibria:

Modeling of Growth Kinetics

$$X_{CAP} + mX_{cAMP} \overset{K_5}{\leftrightarrow} X_{CAP}mX_{cAMP} \qquad (7.53a)$$

$$X_p + X_{CAP}mX_{cAMP} \overset{K_6}{\leftrightarrow} X_pX_{CAP}mX_{cAMP} \qquad (7.53b)$$

where m is a stoichiometric coefficient. Equilibrium between CAP and the promotor is not considered, since this binding coeffieient is taken to be very small. Again we apply an assumption of a pseudo steady state and assume that the concentrations of the individual components can be used. Thus the association constants are

$$K_5 = \frac{[X_{CAP}mX_{cAMP}]}{[X_{CAP}][X_{cAMP}]^m} \qquad (7.54)$$

$$K_6 = \frac{[X_pX_{CAP}mX_{cAMP}]}{[X_p][X_{CAP}mX_{cAMP}]} \qquad (7.55)$$

and the total balances for CAP and promotor are

$$[X_{CAP}]_t = [X_{CAP}] + [X_{CAP}mX_{cAMP}] + [X_pX_{CAP}X_{cAMP}] \qquad (7.56)$$

$$[X_p]_t = [X_p] + [X_pX_{CAP}mX_{cAMP}] \qquad (7.57)$$

We can now derive an expression for the fraction of promotors being activated:

$$Q_2 = \frac{[X_pX_{CAP}mX_{cAMP}]}{[X_p]_t} = \frac{K_5K_6[X_{cAMP}]^m[X_{CAP}]}{1 + K_5K_6[X_{cAMP}]^m[X_{CAP}]} \qquad (7.58)$$

The quantity Q_2 of Eq. (7.58) is used to model the repression effect of glucose, just as Q_1 in Eq. (7.52) is used to describe the induction of lactose on gene expression and hereby synthesis of enzymes necessary for lactose metabolism. However, in order to apply Eq. (7.58), one needs to know the intracellular level of CAP (which in a simple model may be assumed to be constant) and also the level of cAMP. Harder and Roels (1982) suggest the following empirical correlation between X_{cAMP} and the extracellular glucose concentration s_{glc}:

$$X_{cAMP} = \frac{K}{K + s_{glc}} \qquad (7.59)$$

With Eq. (7.59) the genetically structured model is linked up to the glucose concentration in the medium, and the genetically structured model may be used to describe diauxic growth as discussed in Example 7.7. Eq. (7.59) is a totally empirical description of all the different processes involved in determining the cAMP level in the cell at different glucose concentrations. This illustrates a general problem when a genetically structured model is combined with overall models for cell function: Certain mechanisms may be described in great detail – in this case the gene expression, whereas other processes are described by completely empirical expressions. Hereby the performance of the overall model is largely determined by the performance of the empirical expressions in the model, and it may be adequate to apply a simpler model for the gene expression. The real strength of the genetically structured models is, however, not its linkage to the overall growth model, but rather the possibility offered to analyze the influence of specific model parameters on the process. Thus, using the above model the importance of the different equilibrium constants, which are related to the binding affinities e.g. of the repressor to the operator, can be studied in detail. This can be done by comparison with experimental data for the mRNA level, preferably at conditions where the overall cell activity is the same in all experiments.

Example 7.7 A model for diauxic growth
Based on Eqs. (7.52) and (7.58), Harder and Roels (1982) developed a structured model for diauxic growth. It describes the synthesis of mRNA encoding for the three enzymes necessary for lactose metabolism and also for translation of the mRNA into proteins (which are collected in one compartment called X_E). The residual biomass, including building blocks for mRNA and enzyme synthesis, is pooled into one compartment X which constitutes almost all of the cell mass, i.e., $X \approx 1$.

Synthesis of mRNA is described by

$$-X + X_{mRNA} = 0 \; ; \; v_1 = k_1 f(\mu) Q_1 Q_2 \tag{1}$$

where $f(\mu)$ is a linear function of the specific growth rate. The function $f(\mu)$ is used to describe the way the activity of the cell (e.g., expression of genes) increases with the specific growth rate. The expression is completely empirical, but one could combine the Harder and Roels model with one of the two-compartment models of Section 7.4.1 and replace $f(\mu)$ in (1) with the concentration of the active compartment. In Eq. (1) the functions Q_1 and Q_2 of Eqs. (7.52) and (7.58) both appear as factors. The fraction of repressor-free operators and the fraction of activated promoters must both be high to obtain a rapid mRNA synthesis.

The rate of synthesis of enzymes necessary for lactose metabolism is assumed to be first order in the mRNA concentration, i.e.,

$$-X + X_E = 0 \; ; \; v_2 = k_2 X_{mRNA} \tag{2}$$

The half-life of mRNA is short due to rapid degradation by an assumed first-order process:

$$-X_{mRNA} + X = 0 \; ; \; v_3 = k_3 X_{mRNA} \tag{3}$$

Similarly, degradation of the lactose-metabolizing enzymes is included as one first-order process:

$$-X_E + X = 0 \; ; \; v_4 = k_4 X_E \tag{4}$$

With these four reactions, the mass balances for X_{mRNA} and X_E are

$$\frac{dX_{mRNA}}{dt} = k_1 Q_1 Q_2 f(\mu) - k_3 X_{mRNA} - \mu X_{mRNA} \tag{5}$$

$$\frac{dX_E}{dt} = k_2 X_{mRNA} - k_4 X_E - \mu X_E \tag{6}$$

The formation of residual biomass from either glucose or lactose is described with Monod-type kinetics, but for the metabolism of lactose a dependence of X_E is included. Thus

$$-s_{glc} + \gamma_{glc} X = 0 \; ; \; r_{glc} = k_{glc} \frac{s_{glc}}{s_{glc} + K_{glc}} \tag{7}$$

$$-s_{lac} + \gamma_{lac} X = 0 \; ; \; r_{lac} = k_{lac} \frac{s_{lac}}{s_{lac} + K_{lac}} X_E \tag{8}$$

Since the reactions in Eqs. (1)-(4) do not contribute to a net formation of new biomass, the specific growth rate is found to be

$$\mu = \gamma_{glc} r_{glc} + \gamma_{lac} r_{lac} \tag{9}$$

For a batch fermentation where the glucose and the lactose concentrations are both initially high, $[S_{lac}]_t$ is high and, according to Eq. (7.52), Q_1 is therefore high. However, since the glucose concentration is also high, X_{cAMP} is low, according to Eq. (7.59) and Q_2 is therefore low. Thus the rate of synthesis of mRNA is small, and with the rapid turnover of mRNA (k_3 high) the intracellular concentration of mRNA tends to be very low. This again results in a low level of X_E, and consequently the last term in Eq. (9) is negligible, i.e., only the metabolism of glucose contributes to the formation of cell mass. When later in the batch fermentation the glucose concentration decreases, X_{cAMP} and thus Q_2 increases, and as a result the rate of formation of mRNA becomes sufficiently high to ensure an increasing level of mRNA. The result is *de novo* synthesis of enzymes, and this will lead to a larger and larger contribution of lactose metabolism to the total formation of residual biomass. Finally, when the glucose concentration is zero, the cells grow only on lactose. Since some time is needed for synthesis of the enzymes needed for metabolism of lactose, the specific growth rate may be low in a period where both s_{glc} and X_E are low. Thus the model may predict a lag phase between growth on glucose and on lactose.

Equations (7.52) and (7.58) are true mechanistic elements of the model. Unfortunately the number of adjustable parameters is quite large, and the model is still empirical due to Eq. (7.59). It is, however, an excellent example of how known mechanisms can be included in structured models, and it may be possible to find values for some of the binding coefficients in the literature (see Note 7.6).

As mentioned earlier the lac-promoter is often applied as a tool for expression of heterologous genes in connection with industrial production of recombinant proteins. Most industrial enzymes exhibit a complex regulation with induction and carbon catabolite repression, and also here genetically structured models can be used to gain insight into the expression of the gene encoding the enzyme of interest. A more general expression for synthesis of mRNA encoding a given protein than that used in Example 7.7 is:

$$r_{mRNA} = k_m \eta_{tr} X_g = k_m Q_1 Q_2 Q_3 X_g \qquad (7.60)$$

where k_m is the overall transcription rate constant, η_{tr} is the overall transcription efficiency, and X_g is the copy number of the gene to be transcribed (could be given as number of genes per g DW). The overall transcription efficiency is given as the product of Q_1, Q_2, and Q_3. The factors Q_1 and Q_2 represent, respectively, the fraction of repressor free operators and the fraction of activated promoters, i.e., those that may bind RNA polymerase. These factors are not necessarily identical with those derived for the lac-operon above, e.g., if the control mechanism involves an anti-inducer Q_2 is not given by Eq. (7.58) (see Problem 7.3). The factor Q_3 is the fraction of promoters that form complexes with the RNA polymerase, i.e., it is a function of the cellular content of RNA polymerases.

The overall transcription rate constant k_m is a function of the environmental conditions, and Lee and Bailey (1984c) specified it as a function of the specific growth rate (see Note 7.6). With mRNA being very unstable, it is necessary to include degradation of mRNA in the model. This is normally done as a first-order process as illustrated in Example 7.7:

$$r_{mRNA,deg} = k_{m,deg} X_{mRNA} \qquad (7.61)$$

Translation of the mRNA to form the desired protein is generally described by Eq. (7.62).

$$r_p = k_p \xi X_{mRNA} \qquad (7.62)$$

where k_p is the overall translation rate constant (see Note 7.6) and ξ is the translation efficiency (this is often set to unity). Similar to the degradation of mRNA, a turnover of protein is often included as a first-order process:

$$r_{p,deg} = k_{p,deg} X_p \qquad (7.63)$$

The above model for protein synthesis is generally applicable, and the parameter values have been identified for many different systems (see Note 7.6). The model is, however, often simplified in order to keep its complexity at a reasonably low level.

Note 7.6. Mechanistic parameters in the protein synthesis model
Lee and Bailey (1984c) used the above model for an analysis of the influence of the specific growth rate on the productivity of recombinant *E. coli*. Because of the mechanistic nature of the model, each of the parameters has a physical meaning, and here we will illustrate how Lee and Bailey calculated the model parameters.

The overall transcription rate constant and the overall translation rate constant are given by:

Modeling of Growth Kinetics

$$k_m = a_m k_{me} N_p \tag{1}$$

$$k_p = a_p k_{pe} N_r \tag{2}$$

where a_m and a_p are conversion factors, k_{me} and k_{pe} represent the mRNA chain elongation rate per active RNA polymerase and the polypeptide chain elongation rate per active ribosome, respectively. N_p and N_r are the number of active RNA polymerase molecules per gene and the number of active ribosomes per mRNA, respectively. The rate of elongation of mRNA chains per active RNA polymerase (k_{me}) is about 2400 nucleotides min^{-1}, and this value does not vary significantly with the specific growth rate. The polypeptide chain elongation rate per active ribosome (k_{pe}) is about 1200 amino acids min^{-1} when $\mu \gg \ln(2)$, whereas it is proportional to μ for $\mu \ll \ln(2)$. Thus

$$k_{pe} = 1200a \tag{3}$$

where

$$a = \begin{cases} 1 & ; \mu > \ln(2) \\ \mu/\ln(2) & ; \mu < \ln(2) \end{cases} \tag{4}$$

N_p is estimated from the size of the gene and the intermolecular distances between transcribing RNA polymerase molecules (d_p). Similarly, N_r is found from the size of the mRNA and the intermolecular distance between translating ribosomes (d_r). The intermolecular distances depend on the cellular activity, and they are correlated with the specific growth rate (in h^{-1}):

$$d_p = 233\mu^{-2} + 78 \quad \text{nucleotides} \tag{5}$$

$$d_r = 82.5\mu^{-1} + 145 \quad \text{nucleotides} \tag{6}$$

where the intermolecular distances are specified as the number of nucleotides between each RNA polymerase and the next (or a translating ribosome).

Assume that there are z deoxyribonucleotides in the gene. These are transcribed into z ribonucleotides, and at a specific growth rate $\mu = \ln(2)$ h^{-1} (corresponding to a doubling time of 1 h) we obtain

$$k_m = \left(\frac{1 \text{ mRNA molecule}}{z \text{ ribonucleotides}}\right)\left(\frac{2400 \text{ ribonucleotides}}{\text{active RNA polymerase} \cdot \text{min}}\right)\left(\frac{1 \text{ active RNA polymerase}}{563 \text{ deoxyribonucleotides}}\right)$$
$$\left(\frac{z \text{ deoxyribonucleotides}}{\text{gene}}\right) = 4.26 \text{ mRNA molecules per gene per min} \tag{7}$$

Similarly, if y amino acid molecules are used to synthesize one protein molecule the overall protein translation rate constant is determined by

$$k_p = \left(\frac{1 \text{ protein molecule}}{y \text{ ribonucleotides}}\right)\left(\frac{1200 \text{ amino acids}}{\text{active ribosomes} \cdot \text{min}}\right)\left(\frac{1 \text{ active ribosome}}{264 \text{ ribonucleotides}}\right)$$
$$\left(\frac{3y \text{ deoxyribonucleotides}}{\text{mRNA}}\right) = 13.8 \text{ proteins per mRNA per min} \qquad (8)$$

In Eq. (8) for each amino acid incorporated in the protein three ribonucleotides on the mRNA have to be translated. The transcription and translation constants calculated above can be used to estimate reasonable values of the parameters in other, less mechanistic models. Furthermore, it is illustrated how the parameters in a very mechanistic model can be calculated from information in the biochemistry literature.

The rate of degradation of mRNA, $k_{m,deg}$ is on the order of 0.53 min^{-1} and fairly constant for different mRNAs. The rate of degradation of protein, $k_{p,deg}$ is different from protein to protein but it is a much slower process—for most proteins the rate constant is below 0.1 h^{-1}.

For recombinant microorganisms the cellular content of the gene to be expressed (normally called the *gene dosage*) is not necessarily constant. If the gene is inserted directly in the chromosome the gene dosage is approximately independent of the operating conditions. However, in bacteria and yeast the inserted gene is often present in so-called plasmids, which are circular non-chromosomal DNA. The plasmids are replicated independently of the chromosomal DNA, and the ratio of the plasmid number to the chromosome number (often called the *plasmid copy number*) may therefore vary with the operating conditions. The concentration of the gene X_g in Eq. (7.60) should therefore be replaced with the concentration of plasmid copy number when recombinant bacteria are considered. The plasmid is normally designed with a certain replication control mechanism, and in some cases one uses a replication control mechanism that permits induction of rapid plasmid replication, e.g., by the addition of chemical components or changing the temperature.

With the detailed knowledge of recombinant *E. coli*, it has been possible to set up truly mechanistic models for this organism. The largest contribution to the modeling of recombinant *E. coli* has been made in a series of papers from the group of Jay Bailey. Thus Lee and Bailey (1984a-e) describe very detailed modeling of both plasmid replication and protein synthesis. In Lee and Bailey (1984a-c) a mechanistic model for replication of the plasmid in *E. coli* is described. The plasmid copy number was found to vary with the specific growth rate. Replication control of the plasmid involves both a repressor and an initiator (which are both proteins). In their model formation of the repressor and the initiator is described by transcription of the genes followed by translation of the mRNA using kinetic expressions similar to Eqs. (7.60) and (7.62). The repressor affects the transcription efficiency of the genes coding for both the repressor and the initiator, whereas the initiator is necessary for formation of a so-called replication complex. The plasmid replication is initiated when the replication complex increases above a certain threshold value, and once plasmid replication is initiated it is assumed that the replication is almost instantaneous, a reasonable assumption considering the small size of the plasmid. The influence of the specific growth rate is included through the overall transcription and translation constants, as discussed in Note 7.6. The model correctly describes a decreasing plasmid content with increasing specific growth rate, and model simulations reveal that the primary reason for the higher copy number at the lower specific growth rate is reduced synthesis of the repressor protein. In Lee and Bailey (1984d,e) the gene transcription efficiency η is examined for the recombinant protein when the lac promotor is included in the plasmids (often the promotors of operons for which the control mechanisms are well known are used in plasmids, since thereby the transcription of the gene can be controlled). A similar model describes the lac-operon in the chromosome, but the binding of RNA polymerase to the promotor is included. This is important since in the recombinant strain the promotors in the plasmids and in the chromosome compete for the available RNA polymerases. The overall transcription efficiency is given as the product of Q_1, Q_2, and Q_3 (see Equation 7.60). In a study of the effects of multicopy plasmids containing the lac promotor, Lee and Bailey (1984d,e) found that Q_1 increases with the plasmid copy number and that both Q_2 and Q_3 decrease with the plasmid copy number. The overall effect is a decreasing gene-expression efficiency with increasing plasmid copy number, and the overall transcription rate of the cloned gene is therefore not increasing linearly with the plasmid copy number, as has also been experimentally verified (Seo and Bailey, 1985). The decrease in Q_3 with the plasmid copy number is explained by an increasing competition for the available RNA polymerases. Lee and Bailey (1984e) suggest that the empirical expression in Eq. (7.64) may be used in simple structured models to account for this effect:

$$Q_3 = \left(1 - \frac{X_p}{X_{p,\max}}\right)^n \qquad (7.64)$$

The modeling work of Lee and Bailey has been used to study host-plasmid interactions and to explain experimental observations, which are seldom obvious due to the many interactions present in recombinant microorganisms.

7.5.2 Single Cell Models

In single-cell models, characteristic features of the individual cells are considered (e.g., the cell geometry) and particular events during the cell cycle may be studied. All the models that we have described so far are based on the assumption of an equal distribution of cellular material to the daughter cells upon cell division, and the intracellular concentration of a component is therefore not affected by cell division. Furthermore, no special events in the cell cycle have been included in the model, and cell age is consequently of no importance. Therefore the overall kinetics is completely defined by the composition of each cell (the state vector **X**). The single-cell models are used to study microbial behavior at the cellular level. The advantages of single-cell models are (Shuler and Domach, 1982):

1. It is possible to account explicitly for cell geometry and thereby examine its potential effects on nutrient transport.
2. Temporal events during the cell cycle can be included in the model.
3. Spatial arrangements of intracellular components can be considered.
4. Biochemical pathway models and metabolic control models can be included with ease.

To set up a single-cell model one must have a detailed knowledge of the microorganism, and good single-cell models are therefore developed only for well examined microorganisms: *E. coli*, *Bacillus subtilis*, and *S. cerevisiae*. Among the most comprehensive single-cell models constructed is the so-called Cornell model for *E. coli*, which was developed by Mike Shuler and co-workers at Cornell University. The original model by Shuler *et al.* (1979) contained 14 components, and it formed the basis for all later versions of this model – particularly a more detailed model with 20 intracellular components (Shuler and Domach, 1982). The additional components were introduced in order to describe the incorporation of ammonium ions into amino acids, to allow more accurate estimates of cellular energy expenditures, and to allow a more complete simulation of the systems which control transcription and translation of the genes (Shuler and Domach, 1982). The model correctly predicts an increase in cell volume with increasing specific growth rate during both glucose- and ammonia-limited growth (Domach *et al.*, 1984), a decrease in the glycogen content with increasing specific growth rate during ammonia-limited growth (Shuler and Domach, 1982), and many other observations made with *E. coli*.

Peretti and Bailey (1986) revised the Cornell model by introducing a more refined description of the protein and RNA synthesis, including initiation of translation and distribution of RNA polymerase along with initiation of DNA and chromosomal replication. Changes and additions to the Cornell model are motivated by a desire to expand the range of applications of single-cell models. In particular it is desired to study the effect of plasmid insertion into a host cell and the expression of any plasmid genes, as in Peretti and Bailey (1987), who describe host-plasmid interactions in *E. coli*. Their extension of the model can be applied to study the effect of copy number, promoter strength, and ribosome-binding-site strength on the metabolic activity of the host cell and on the plasmid gene expression.

With the rapid development in experimental techniques it becomes feasible to extend the rather empirical single cell model described above to "complete" models of living cells. In these models all known molecular mechanisms are incorporated, and the interaction between the different components in the system can be analyzed at the quantitative level (Endy and Brent, 2000). This approach is currently referred to as systems biology, and even though the ultimate goal is to describe all processes in the cell, it is currently only possible to describe a few of the key processes. However, eventually models of different cellular processes, e.g., gene transcription, specific biochemical pathways, cell cycle control, mating type shifts, may be assembled into an overall model that may reveal the relative importance of the many different processes operating in a living cell. In the quest for a complete mathematical description of cellular function interaction between model construction and evaluation is important, since a final model will be the result of several iterations where the model is continuously revised as soon as new biological data become available.

7.6 Morphologically Structured Models

In sections 7.3-7.5 we specified the growth kinetics assuming that all the cells in a culture have the same metabolism; i.e., the cell population is assumed to be completely homogeneous, and a non-segregated model for cellular performance resulted. For some microbial systems differentiation of the cells in the culture does, however, play an important role in the overall performance of the culture, and both growth kinetics and productivity are effected by the presence of more than one cell type in the culture. In Chapter 8 we are going to consider complete segregation, and instead of considering a finite number of distinguishable cell types the culture is going to be characterized by a continuous distribution of an important cell property, e.g. the cell age. Obviously the model for a culture with only a few distinguishable cell types (e.g. cells that produce a desired protein and cells that have lost this property) is much simpler than a model that has to take a continuous distribution of a property into account. We shall refer to the crudely segregated models as *morphologically structured models*. These models are particularly relevant for description of the growth of filamentous fungi, where cellular differentiation takes place in connection with hyphal extension, but they also find application for description of other cellular systems, e.g., cultures with bacteria containing unstable plasmids and to explain why yeast cultures sometimes exhibit oscillatory behavior in several variables.

In the morphologically structured models the cells are divided into a finite number Q of cell states **Z** (or morphological forms), and conversion between the different cell states is determined by a sequence of empirical *metamorphosis reactions*. Ideally these metamorphosis reactions can be described as a set of intracellular reactions, but the mechanisms behind most morphological conversions are largely unknown. Thus, it is not known why filamentous fungi differentiate into cells with a completely different phenotype than that of their origin. It is therefore not possible to set up detailed mechanistic models describing these changes in morphology, and empirical metamorphosis reactions have to be used. The stoichiometry of the metamorphosis reaction where the jth form is converted to the ith form is given by very simple relations:

$$Z_i - Z_j = 0 \tag{7.65}$$

Z_q will be used in the following to describe both the qth morphological form itself and the fraction of cell mass that is Z_q (g qth morphological form (g DW)$^{-1}$). In the metamorphosis reaction one morphological form is spontaneously converted to another form. This is of course an extreme simplification since the conversion between morphological forms is the sum of many small changes in the intracellular composition of the cell. Clearly there may be many different metamorphosis reactions, and the stoichiometry for these reactions can be summarized in analogy with the matrix equation (7.2) for intracellular reactions:

$$\Delta \mathbf{Z} = \mathbf{0} \tag{7.66}$$

where Δ is a stoichiometric matrix. It is assumed that the metamorphosis reactions do not involve any change in the total mass, and the sum of all stoichiometric coefficients in each reaction is therefore equal to zero.

To describe the rate of the metamorphosis reactions a forward reaction rate u_i is introduced for the ith reaction, and the rates of all the metamorphosis reactions are collected in the rate vector **u**. Besides formation from other morphological forms, a given morphological form may also be synthesized from substrate through intracellular reactions (and different metabolic products may also be formed by different morphological forms). An intracellularly structured model may describe these reactions, but in order to reduce the model complexity one will normally use a simple unstructured model for description of the growth and product formation of each cell type, e.g. the Monod model describes the specific growth rate of the qth morphological form. When the specific growth rate has been specified for each morphological form, the specific growth rate of the total biomass is given as a weighted sum of the specific growth rates of the different morphological forms:

$$\mu = \sum_{i=1}^{Q} \mu_i Z_i \tag{7.67}$$

The rate of formation of each morphological form is determined both by the metamorphosis reactions and by the growth associated reactions for each form, and the mass balance for the qth morphological form can be derived in analogy with Eq. (7.14) (see Nielsen and Villadsen (1992) for details):

$$\frac{dZ_q}{dt} = \Delta_q^T \mathbf{u} + (\mu_q - \mu) Z_q \tag{7.68}$$

The first term accounts for the net formation of the qth morphological form by the metamorphosis reactions (the vector Δ_q specifies the stoichiometric coefficients for the qth morphological form in all the metamorphosis reactions). The second term accounts for growth of the qth morphological form and the last term accounts for dilution due to growth of the biomass (this is a consequence of the normalization of the concentrations of the morphological forms). The concept of morphologically structured models is illustrated in Example 7.8.

Example 7.8 A simple morphologically structured model describing plasmid instability

A potential obstacle to commercial application of recombinant bacteria and yeasts is plasmid instability. Sometimes a daughter cell that does not contain plasmids is formed upon cell division, and since the metabolic burden is higher for plasmid-containing cells, the plasmid-free cell will grow faster than the plasmid-containing cells. Even a small plasmid instability will therefore ultimately result in the appearance of a large fraction of nonprotein-producing cells. Plasmid stability can be improved by increasing the plasmid copy number or by designing the host-plasmid system in a way that ensures that plasmid-free cells cannot survive. Modeling of plasmid instability can be done using the concept of morphologically structured models. Thus, we assume that when plasmid containing cells Z_p are dividing a certain fraction δ of the cells are converted to plasmid free cells Z_h, whereas the remaining fraction of the cells maintains the plasmid. This can be described by the metamorphosis reaction:

$$(1-\delta)Z_p + \delta Z_h - Z_p = 0 \tag{1}$$

The stoichiometric coefficient δ (often called the *segregation parameter*) is equal to the probability of formation of a plasmid-free cell upon growth of plasmid containing cells. The stoichiometry in Eq. (1) is

Modeling of Growth Kinetics

illustrative since it shows that when one unit of recombinant cells divides, a fraction δ of plasmid-free cells is formed. However, since no new cell mass is formed by the metamorphosis reaction $\delta Z_p = \delta Z_h$ and in terms of stoichiometry (2) is simplified to

$$Z_h - Z_p = 0 \qquad (2)$$

However, the rate of the metamorphosis reaction (2) is different from that of reaction (1). Since the metamorphosis reaction (1) specifies growth of the plasmid containing cells the forward rate of this reaction is equal to $\mu_p Z_p$, whereas the rate of the metamorphosis reaction is equal to $\delta \mu_p Z_p$.

In order to apply the general mass balance for the two morphological forms, we first specify the stoichiometric matrix Δ:

$$\Delta = (-1 \quad 1) \qquad (3)$$

Now we set up the mass balance for the plasmid containing cells:

$$\frac{dZ_p}{dt} = -\delta \mu_p Z_p + (\mu_p - \mu) Z_p = [(1-\delta)\mu_p - \mu] Z_p \qquad (4)$$

and for the plasmid free cells:

$$\frac{dZ_h}{dt} = \delta \mu_p Z_p + (\mu_h - \mu) Z_h \qquad (5)$$

For each morphological form, there is a contribution from the metamorphosis reaction, a contribution from growth of the form, and finally a contribution accounting for dilution due to the expansion of the total biomass. The term for the dilution is analogous to the dilution term in the intracellularly structured models, whereas the term accounting for formation of intracellular components in intracellularly structured models is replaced by two terms in the morphologically structured models; i.e., one for exchange between forms and one for growth.

The specific growth rate for the total biomass is given by:

$$\mu = \mu_p Z_p + \mu_h Z_h \qquad (6)$$

Normally the specific growth rate for plasmid containing cells is lower than that of plasmid free cells, and the specific growth rate for the total biomass will therefore also be smaller than μ_h. Consequently both terms on the right hand side in the mass balance (5) are positive, and the fraction of plasmid free cells will continuously increase due to two factors: there is a constant formation of plasmid free cells from plasmid containing cells, and as $\mu_h > \mu_p$ the plasmid free cells will outgrow the plasmid containing cells.

With the formation of plasmid-free cells a culture with two different strains develops, namely the recombinant strain and the parental (or host) strain. Thus the culture is *mixed*, with two different strains of the same species. The metamorphosis reaction (2) may also be used to describe the spontaneous occurrence of mutants during cellular growth, but here the forward reaction rate is not necessarily similar to the specific growth rate of the original strain. Occurrence of spontaneous mutants is a phenomenon often observed for filamentous fungi, especially when they are grown in continuous cultures for long periods, and the resulting mutants are normally referred to as *colonial mutants*.

In case the growth of the individual morphological forms is to be described by an intracellularly structured model the complexity of morphologically structured models increases substantially. It is possible to derive a general mass balance for the concentration of the intracellular variables in the different morphological forms (Nielsen and Villadsen, 1994), but as these equations have limited practical use we will not elaborate on this topic. Instead we will focus on two different applications of morphologically structured models: quantitative description of oscillating yeast cultures and growth of filamentous microorganisms.

7.6.1 Oscillating Yeast Cultures

Whereas the division of unicellular bacteria is symmetric, with the formation of two almost identical cells, the cell division of yeast is asymmetric, with formation of a so-called mother and a so-called daughter cell (see Fig. 7.17). The daughter cell is converted to a mother cell within the time t_d. After a period of maturation (t_m) a new bud emerges on the mother cell, and after a further time period t_b the bud has obtained a critical size, resulting in cell division. At cell division a *bud scar* is formed on the cell envelope of the mother cell, and it is believed that the cell cannot form a new bud at this position. t_d, t_m, and t_b are functions of the cellular composition (and therefore also of the environmental conditions).

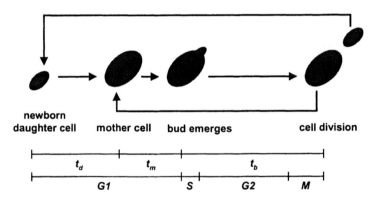

Figure 7.17 The cell cycle of budding yeast. The period of maturation (i.e., $t_d + t_m$ for a daughter cell and t_m for a mother cell) is called the *G1-phase*, and in this phase the cells prepare for budding (e.g., carbohydrate storage is built up). The budding itself (lasting t_b) consists of three phases. First the chromosomes are duplicated in the *S-phase*. Thereafter the cells prepare for cell division (the *G2-phase*), and finally they enter the *M-phase* (M for mitosis), where cell division occurs.

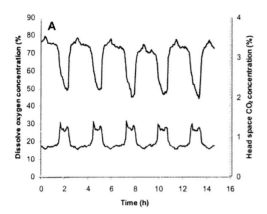
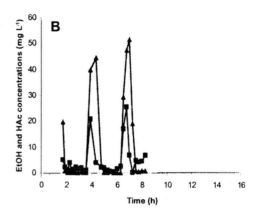

Figure 7.18 Spontaneous oscillation of *S. cerevisiae*.
A. Measurements of carbon dioxide in the exhaust gas and the dissolved oxygen concentration (DOT) in % of the equilibrium value.
B. Measurements of the concentrations of ethanol (■) and acetate (▲).
The data are taken from Frandsen (1993).

The cellular metabolism of the daughter cells (or *unbudded cells*) and the mother cells (or *budded cells*) is very different, and to give a correct description of the overall growth kinetics of yeast cultures it is therefore necessary to apply a morphologically structured model. In many cases one may, however, describe the average metabolism of the culture with an intracellularly structured model, i.e., assume that the population is homogeneous and still get a satisfactory description of many experimental data (see Example 7.3). These models can, however, not describe a fascinating phenomenon observed in a glucose-limited chemostat with *S. cerevisiae*: Spontaneous oscillations of many easily measured process variables such as the dissolved oxygen, ethanol and glucose concentrations; oxygen uptake rate; and carbon dioxide evolution rate, as well as intracellular variables (e.g., NADH and DNA). Figure 7.18 shows some typical results. The literature contains numerous experimental investigations of the oscillations [see in particular Strässle *et al.* (1989) and Münch (1992)]. These oscillations can be maintained for many generations, especially when one uses high-performance bioreactors, in which the environment is practically homogeneous throughout the reactor (Strässle *et al.* 1988; Duboc *et al.*, 1996; Duboc and von Stockar, 2000; Beuse *et al.*, 1999).

Spontaneous oscillations can only be maintained if the growth of the culture is *synchronous*, since after completion of one cycle the fraction of each morphological form in the cell mass must be the same as it was at $t - T_{osc}$, where T_{osc} is the length of the oscillation period. With the asymmetric cell division of *S. cerevisiae* it is not immediately clear how the synchrony can be maintained for many generations. If, however the population can be divided into a discrete number of subpopulations (e.g mother cells m and daughter cells d), and if the generation time of one subpopulation is an integer multiple of the generation time of the other then synchrony can be obtained. Thus, if the creation of a mother cell from a new born daughter cell takes exactly 2 ($t_b + t_m$) then the ratio of mother cells to daughter cells in the population will be constant and equal to $(1 + \sqrt{5})/2$ as shown by Hjortso and Nielsen (1994).

Note 7.7 Relation between T_{osc} and the dilution rate in continuous culture
The result of Hjortso and Nielsen (1994) is a special case of the following:
Let m be the population of mother cells, d the population of daughter cells. New mother- and daughter cells are born in each generation. Thus at $t = (n + 1) T_{osc}$:

$$m_{n+1} = m_n + d_n \quad (\text{or } m_{n+2} = m_{n+1} + d_{n+1}) \tag{1}$$

$$d_{n+1} = p \, m_n \tag{2}$$

if the mother cell has the probability p to develop a new daughter cell exactly at $t = (n+1) T_{osc}$. Eliminating d_{n+1} between (1) and (2) yields a second order difference equation:

$$m_{n+2} - m_{n+1} - p \, m_n = 0 \tag{3}$$

But the whole culture grows exponentially with time, and consequently after one cycle time T_{osc} in a continuous culture:

$$m_{n+1} = m_n \exp(\mu \, T_{osc}) \tag{4}$$

When (4) is inserted in the difference equation one obtains the following algebraic equation for $\rho = \exp(D \, T_{osc})$:

$$\rho^2 - \rho - p = 0 \quad \text{or} \quad \rho = \tfrac{1}{2} (1 + (1 + 4 p)^{1/2}) \tag{5}$$

from which a relation between D and T_{osc} is derived :

$$T_{osc} = \ln \rho / D \tag{6}$$

For $p = 1$ (i.e all mother cells start to make daughter cells immediately after they have lost a bud) the result is identical to what is found in the Hjortso and Nielsen (1994) model. With the extra parameter p Duboc and von Stockar (2000) could fit experimental data better. For $p = \tfrac{1}{2}$ the value of ρ decreases from 1.618 to 1.366 and at a given D a smaller oscillation time T_{osc} is predicted In both cases the oscillation time is, however predicted to decrease inversely proportional to D. This last feature is confirmed by experiments, but the experiments also seem to show that different values of p between 2 and ½ should be used at different D values (see figure 7 of Duboc and von Stockar (2000)). The value $p = 2$ is of course not allowed in the Duboc and von Stockar hypothesis, but the resulting relation $T_{osc} = \ln 2 / D$ corresponds to the cell cycle of prokaryotes where daughter cells start to develop into mother cells immediately after cell division has taken place. For the reasons outlined below stable oscillations are not likely to develop for these organisms.

Finally it should be mentioned that Beuse et al. (1999) explained synchrony in the culture by a hypothesis that is a mirror image of the Duboc-von Stockar hypothesis. Beuse *at al.* propose that some daughter cells may take more than 2 ($t_b + t_m$) to develop into a mother cell while all mother cells can give birth to daughter cells after $t_b + t_m$.

Both hypotheses do however suffer from the weakness that it is difficult to understand why a fixed proportion of the mother cells (or daughter cells) should choose to take "an extra day off" and that they should do so, generation after generation.

If there is not really a finite number of subpopulations with stringently kept rules for the generation time then the whole concept of synchrony falls apart. In a real culture of cells there must be a statistical distribution of e.g generation times for the cells, and if synchrony should prevail in the real culture there must be some mechanism by which e.g "slow" cells catch up with faster proliferating cells. In mathematical terms there must exist an *attractor*, and in the case of bioreactions the nature of the attractor must be biochemical. This is really where the physiological interest in oscillating cultures is to be found: Through the study of spontaneous oscillations in continuous cultures one may learn something about non-linear, feed forward cell processes that influence the cell cycle. Hjortso and Nielsen (1994) proposed that the critical cell size for budding or for cell division might oscillate. Experimentally it has been shown that the critical cell size depends on the ethanol concentration. At high ethanol concentration the critical mass for budding decreases while the critical mass for cell division increases (Martegani *et al.*, 1990). Consequently, the attractor may function because of an oscillating ethanol concentration, and it is known from experiments that this is actually so. At the start of the *S* phase (Fig. 7.17) there is a burst of catabolic activity: A peak in CO_2 production and a corresponding ethanol peak (more difficult to observe on the background of the ethanol already present in the medium) signals the beginning of the "gestation phase". Storage carbohydrates, accumulated by the mother cells in the t_m phase, are released to fuel the burst of catabolic activity. The daughter cells (in the t_d phase) may take up the excreted ethanol and use it to speed up their passage through the G_1 phase. This was the hypothesis of Strässle *et al.* (1988), of Martegani *et al.* (1990) and of Münch (1992) in their attempt to explain how the synchronization of the culture could happen.

There are, however many other groups who in the last ten years have contributed to the search for a biological explanation of spontaneous oscillations in yeast. In general oscillations have never been observed in anaerobic yeast cultures. Most studies are done in aerobic cultures with glucose as substrate, but there are also examples of oscillations in aerobic yeast cultures fed with ethanol (Murray *et al.*, 2001). In the group of Kuriyama in Japan a number of biological attractors have been suggested. It could be CO_2 that possibly influences transport through the cell membrane (but not ethanol in the exterior medium, since a pulse of ethanol added to the medium does not change the oscillatory behavior (Keulers *et al.*, 1996)), or oscillations could have a connection with the sulfur metabolism of yeast ((Sohn and Kuriyama, 2001). It is tempting to suggest that the possible inhibition of the respiratory system when ethanol is produced (but not by mere addition of ethanol to the medium) by overflow at the pyruvate branch point could constitute a feed back regulation that could lead to oscillations. As yet the correct mechanism (and there may be several contributing mechanisms) has not been found, but in view of the importance of the yeast system as host for recombinant protein production the subject certainly merits much more fundamental research.

Example 7.9 A simple morphologically structured model for oscillating yeast cultures
Cazzador *et al.* (1990) presented a computer model that predicts spontaneous oscillations of a budding yeast culture. A previously published computer model (Cazzador and Mariani, 1988) for cell-size distribution at transient growth is combined with a model for characterization of cells by their mass m and genealogical age a_g. Unlike chronological age, genealogical age a_g is a discrete variable and $a_g = 0, 1, \ldots$ represents the number of bud scars on the cell. The mass is discretized into a number of groups (or morphological forms), and in each group cells have an average mass characteristic for that particular group. The critical mass for budding is assumed to be constant, whereas the critical mass for cell division (m_{div}) is taken to be a monotonically increasing function of the glucose concentration s as shown in the empirical expression Eq. (1) which also includes an effect of a_g. k_1 and k_2 are positive constants, smaller than 1. In the model the specific growth rate was assumed to follow the same Monod expression for both unbudded and budded cells. The yield coefficient of biomass on glucose is, however, different for the two morphological forms, and it is thereby possible to describe oscillations in the glucose concentration when the budding index, the fraction of budded cells, varies. On the other hand, the glucose concentration affects the critical mass for cell division, and oscillations in the glucose concentrations result in oscillations in the budding index. The computer model has been used to simulate spontaneous oscillations at different dilution rates, and the simulations give a qualitative fit to experimental data. Despite the incorporation of many of the mechanisms described by the verbal model, the key formula Eq. (1) is, as mentioned above, completely empirical, but nothing better can at the present be suggested.

$$m_{div}(a_g, s) = \left[m_{div,min} + (m_{div,max} - m_{div,min}) \frac{s^3}{s^3 + K^3} \right] \left(1 + k_1 \frac{1 - k_2^{a_g}}{1 - k_2} \right) \quad (1)$$

From the more detailed computer model Cazzador (1991) derived a simple morphologically structured model [see also Cazzador and Mariani (1990)]. Despite variations of the intracellular composition within the groups of budded and unbudded cells considered in the detailed computer model, one can collect all the cells in each of these two groups of cells into two morphological forms in order to obtain a simple morphologically structured model that may describe spontaneous oscillations. The two forms are: unbudded cells (Z_u) and budded cells (Z_b). Thus there is no discrimination between daughter and unbudded mother cells, but this kind of simplification is acceptable in a simple model. The metamorphosis reactions are

$$-Z_b + Z_u = 0 \quad ; \quad u_b = k_b Z_b \quad (2)$$

$$-Z_u + Z_b = 0 \quad ; \quad u_u = k_u Z_u \quad (3)$$

The reaction shown in Eq. (2) describes cell division, whereas the reaction shown in Eq. (3) describes budding. Thus the rate of the first metamorphosis reaction (u_b) may be interpreted as the rate of cell division of budded cells, and the rate of the second metamorphosis reaction (u_u) may be interpreted as the rate of budding of unbudded cells. Each of the two morphological forms is also synthesized from the substrate, at a rate μ_i,

$$-\alpha_i s + Z_i = 0 \quad ; \quad \mu_i \text{ where } i = u, b \quad (4)$$

Cazzador (1991) performed a stability analysis of this simple morphologically structured model, and he found that a necessary condition for instability is either Eq. (5) or Eq. (6) (see also Problem 7.4). However, a stable limit cycle, i.e., presence of sustained oscillations, is obtained only with Eq. (4). Equation (4) specifies that if the specific growth rate is the same for the two morphological forms, the yield of cell mass from the substrate has to be larger for the budded cells than for the unbudded cells [notice that this constraint is used in the computer model of Cazzador et al. (1990) described above]. Equation (6) specifies that if the yields are the same for the two morphological forms, the unbudded cells must have a higher specific growth rate than the budded cells if instability is to be maintained. However, this constraint does not ensure sustained oscillations, which is probably a consequence of the simple model structure.

$$\alpha_u > \alpha_b \quad \text{if} \quad \mu_u = \mu_b \tag{5}$$

$$\mu_u > \mu_b \quad \text{if} \quad \alpha_u = \alpha_b \tag{6}$$

Based on the computer model of Cazzador et al. (1990), Cazzador (1991) could specify rates for the two metamorphosis reactions (2) and (3) such that the morphologically structured model could describe spontaneous oscillations. The rates of these reactions are functions of the specific growth rate and the critical mass for cell division [which again is a function of the glucose concentration, according to Eq. (1)]. With the condition of Eq. (5), Cazzador finds the rates of the metamorphosis reactions to be given (see Problem 7.4) by

$$k_b = \mu \frac{h-1}{h \ln(h)} \tag{7}$$

$$k_u = \mu \frac{1}{\ln(1/(h-1))} \tag{8}$$

where h is defined by Eq. (9):

$$h = \frac{m_{div}}{m_{bud}} \tag{9}$$

m_{bud} is the critical mass of budding, which is taken to be constant, and m_{div} is calculated by Eq. (1), where the genealogical age a_g is taken to be zero. The model predicts spontaneous oscillations, but the predicted oscillation period is too long compared with experimental data, probably because of the extremely simplified expressions for the rates of the metamorphosis reactions. In the Cazzador model these rates are functions only of the glucose concentration in the medium, and due to the empiricism of Eq. (1), which is used in Eqs. (7)-(9), the model does not reveal much of the physiology behind oscillations. The simple model is, however, valuable for analytical studies of the oscillations.

7.6.2 Growth of Filamentous Microorganisms

The mechanisms for growth of filamentous microorganisms are very different from those of unicellular microorganisms, since the cells are connected in so-called hyphal structures (see Fig. 7.19). All cells within these multicellular structures may contribute to the growth process, i.e., production of protoplasm, but extension of the hyphae occurs only at the tips. The number of tips in a mycelium is therefore a characteristic morphological variable. Even though the linear rate of tip extension has an upper limit, the total length of a mycelium may increase exponentially due to the formation of new tips along the hyphae. The frequency of formation of new tips is determined by the rate of production of protoplasm within the mycelium and the number of tips to which the material is distributed. The ratio between the size of the mycelium and the number of tips is therefore another characteristic morphological variable, which Caldwell and Trinci (1973) called the *hyphal growth unit*. They originally defined it as the total mycelium length divided by the number of tips (called the *hyphal growth unit length* l_{hgu}) but it may also be defined on the basis of total mycelium mass (called *hyphal growth unit mass*). At conditions that support rapid growth, a densely branched mycelium with a large hyphal diameter is observed, whereas a less branched mycelium with a small hyphal diameter is observed at poor growth conditions (Nielsen, 1992), where the mycelium will extend itself in the hope of reaching an environment where the growth conditions are better, and it therefore forms long threads with very few branch points.

In a hyphal element several cells behind the tip are involved in the tip extension process, since they supply the necessary cellular material for tip extension, e.g., cytoplasmic material and building blocks for wall synthesis. These cells are not separated by a septum (the wall between the individual cells), and they therefore share a common cytoplasm in which the nuclei of all the cells are found. The part of the hyphal element between the tip (or apex) and the first septum is called the *apical compartment*. The cells just behind the apical compartment have an intracellular composition very similar to that of the apical cells, and this part of the hyphal element is called the *subapical compartment*. Despite the presence of a septum between the apical and subapical cells, there may be an exchange of protoplasm since the septa are often perforated. When one moves further away from the tip, one finds cells containing large vacuoles. These cells do not participate directly in the tip extension process, but they are believed to be of importance in creating an intracellular pressure sufficient to ensure transport of protoplasm toward the tip section. This part of the hyphal element is referred to as the *hyphal compartment*.

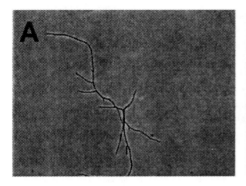
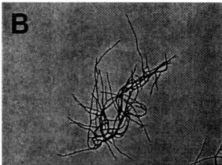

Figure 7.19 Some typical pictures of hyphal element of filamentous fungi.
A. Single hyphal element of *Aspergillus nidulans*
B. Agglomerate of several hyphal elements of *Aspergillus nidulans*

In filamentous fungi there is a substantial accumulation of small vesicles at the apex, and the vesicles are likely to play an important role in the tip extension. They are believed to contain wall subunits, lytic enzymes, and synthetic enzymes that are transported with the vesicles to specialized regions of the endomembrane system in the apical and subapical compartments (McIntyre et al., 2001). The vesicles, each carrying its load of enzymes and/or wall precursors, are transported by unknown mechanisms through the cytoplasm to the tip section of the apical cell, normally referred to as the *extension zone*. When a vesicle comes into contact with the cell membrane at the apex, it fuses with the membrane, and the vesicle content is released into the wall region. The excreted lytic enzymes attack the microfibrillar skeleton in the cell wall, resulting a plastification of the wall structure, which thereby becomes unable to withstand the inner pressure from the cytoplasm. The microfibrils therefore become stretched, and the surface area of the wall increases. In filamentous procaryotes (typically *Streptomyces* species) the wall material is soluble in the cytoplasm and is probably transported to the apex by molecular diffusion. It has been found that the wall section at the apex is more susceptible to compounds affecting wall synthesis and assembly than other sections. This indicates that the lytic enzymes in filamentous procaryotes are positioned in the wall section at the apex (Prosser and Tough, 1991).

When a new tip is formed, it initially grows and its growth corresponds to an increase in the size of the apical compartment. When the apical compartment has attained a certain size, a septum is formed behind the tip, and some of the old apical cell mass becomes new subapical cell mass. Under constant environmental conditions, the size of the apical compartment remains constant, and the net result of tip extension is therefore formation of subapical cells. The control of septum formation has been studied in filamentous fungi, and Fiddy and Trinci (1976) introduced the term *duplication cycle* to describe the events that lead to the net formation of a whole new apical compartment. For *Aspergillus nidulans* the duration of the duplication cycle has been found to be identical with the doubling time of the biomass.

Branching is the mechanism by which new apical compartments are formed, and it occurs at certain preferred branching points on a hyphal element. In filamentous fungi, it has been suggested that branching occurs at locations where for one reason or another there is an accumulation of vesicles, whereas branch formation in filamentous prokaryotes does not result from accumulation of material (Prosser and Tough, 1991). Since vesicles are synthesized both in the subapical and in the apical compartment in hyphae of filamentous fungi, it seems reasonable that at positions where the protoplasmic flow is reduced, e.g., at the position of a septum, there is an accumulation of vesicles. Branching may therefore be associated with septum formation, and for the filamentous fungus *Geotrichum candidum* more than 70% of the observed branch points in a subapical compartment are positioned close to the septum separating this compartment from the apical compartment (Trinci, 1984). For other species of filamentous fungi there is, however, a more equal distribution of the branch points throughout the subapical compartment. Branching is observed mainly in the subapical compartment, but in some filamentous species apical branching may occur.

Originally the growth mechanisms of filamentous microorganisms was studied using surface cultures, but here the morphology is completely different from that found in a submerged culture. Through the use of automated image analysis it has, however, become possible to analyze a large number of hyphal elements in submerged cultures, and hereby information on the hyphal morphology may be obtained (Cox *et al.*, 1998). Furthermore, through the use of flow-through cells that are positioned directly under a microscope equipped with an automated image analysis system it is possible to follow the outgrowth of single hyphal elements and hereby study the growth kinetics in great detail (Spohr *et al.*, 1998; Christiansen *et al.*, 1999).

Not many models specifically address the growth mechanisms of filamentous microorganisms (for a recent review see Krabben and Nielsen (1998)) since normally the focus is on the primary metabolism, where it is not necessary to consider the hyphal structure explicitly; i.e., many of the intracellularly structured models described in Sections 7.3-7.4 may also be used to describe the primary metabolism of filamentous microorganisms. To model the formation of secondary metabolites, which may be determined by the cellular differentiation, it is, however, often necessary to consider morphological structure. Furthermore, when a description of the morphology of the hyphal elements is the objective, one must of course include morphological structure in the model.

Megee *et al.* (1970) described the first morphologically structured model for filamentous fungi. The model was used to describe growth and production formation of *Aspergillus awamori*. Five separate morphological forms are considered in the model:

Z_A - Apical compartment in actively growing hyphae
Z_H - Subapical compartment in actively growing hyphae
Z_C – Conidiophore[4] developing hyphae
Z_B - Black spores
Z_M - Matured spores

With these five morphological forms it was possible to describe the complete life cycle of so-called *imperfect fungi* (fungi with no sexual reproduction). The model includes product formation as a

---

[4] Conidiophores are modified hyphae on which the asexual spores are formed.

Modeling of Growth Kinetics

result of the differentiation processes, a reasonable hypothesis for many secondary metabolites. The model describes several general observations concerning growth of *A. awamori*, and it is a nice example of how morphological structure can be used to describe a very complex system. A disadvantage of the model is the large number of parameters, but for simulation of submerged growth one may neglect spore formation and consider only actively growing hyphae, i.e., the morphological forms Z_A and Z_H. Thereby the original model is substantially simplified, as illustrated in Example 7.10.

Example 7.10 A simple morphologically structured model for growth of filamentous microorganisms
Based on the growth mechanisms described above Nielsen (1993) derived a simple morphologically structured model including the three morphological forms shown in Fig. 7.20:
- Apical cells (Z_A)
- Subapical cells (Z_S)
- Hyphal cells (Z_H)

The model is a progeny of the Megee *et al.* model. The verbal formulation of the model is:
Active growth, i.e., uptake of substrates and formation of biomass, occurs only in apical and subapical cells. When the tip extends, an apical cell is converted to a subapical cell, whereas a new apical cell is produced from subapical cell material when a branch point is formed in the subapical compartment. When the subapical cells become more and more vacuolated, they change into inactive hyphal cells.

The mathematical formulation is given in Eqs. (1) and (2). Three metamorphosis reactions described in matrix form in Eq. (1) are considered. They represent branching, tip growth, and differentiation, respectively. The kinetics of all three metamorphosis reactions is taken to be first order in the morphological form which disappears. Furthermore, formation of inactive hyphae is assumed to be inhibited by high substrate concentrations. The Monod model describes growth of both the apical and subapical cells, where s is the extracellular glucose concentration.

$$\begin{pmatrix} 1 & -1 & 0 \\ -1 & 1 & 0 \\ 0 & -1 & 1 \end{pmatrix} \begin{pmatrix} Z_A \\ Z_S \\ Z_H \end{pmatrix} = \begin{pmatrix} 0 \\ 0 \\ 0 \end{pmatrix} \quad ; \quad \begin{pmatrix} u_1 \\ u_2 \\ u_3 \end{pmatrix} = \begin{pmatrix} k_1 Z_S \\ k_2 Z_A \\ k_3 Z_S /(sK_3 + 1) \end{pmatrix} \quad (1)$$

$$\begin{pmatrix} -\alpha_A \\ -\alpha_S \\ 0 \end{pmatrix} s + \begin{pmatrix} 1 & 0 & 0 \\ 0 & 1 & 0 \\ 0 & 0 & 0 \end{pmatrix} \begin{pmatrix} Z_A \\ Z_S \\ Z_H \end{pmatrix} = \begin{pmatrix} 0 \\ 0 \\ 0 \end{pmatrix} \quad ; \quad \begin{pmatrix} \mu_A \\ \mu_S \\ \mu_H \end{pmatrix} = \begin{pmatrix} k_A s/(s+K_s) \\ k_S s/(s+K_s) \\ 0 \end{pmatrix} \quad (2)$$

Inserting (1) and (2) in (7.67), (7.68) one obtains the specific growth rate of the total biomass and the mass balances for the three morphological forms:

$$\mu = (k_A Z_A + k_S Z_S) \frac{s}{s + K_s} \quad (3)$$

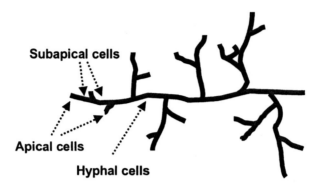

Figure 7.20 Structure of a densely branched hyphal element with indication of apical, subapical, and hyphal cells.

$$\frac{d}{dt}\begin{pmatrix} Z_A \\ Z_S \\ Z_H \end{pmatrix} = \begin{pmatrix} u_1 - u_2 + (\mu_A - \mu)Z_A \\ -u_1 + u_2 - u_3 + (\mu_S - \mu)Z_S \\ u_3 - \mu Z_H \end{pmatrix}$$

(4)

The rates of the metamorphosis reactions and the morphological variables have been shown to correlate with specific measures of the hyphae (Nielsen, 1993). Thus, the hyphal diameter is given by:

$$d_{hyphae} = a_1 \frac{\mu}{Z_A}$$

(5)

where a_a is a physiological constant determined by the water content and the density of the hyphae. Equation (5) is based on the assumption that only precursors synthesized in the apical compartment contribute to tip growth. For some species of filamentous microorganisms, precursor synthesized in the subapical compartment may also be transported to the apex and incorporated in the hyphal wall, but the contribution from the subapical compartment to the total precursor synthesis is assumed to be small and is neglected in the model. The hyphal growth unit length may also be derived from the model:

$$l_{hgu} = \frac{\mu}{a_2 u_1}$$

(6)

Modeling of Growth Kinetics

where a_2 is physiological constant determined by the mass of apical compartment per tip. Finally the tip extension rate q_{tip} can be derived from the hyphal growth unit length and the specific growth rate:

$$q_{tip} = \mu l_{hgu} \tag{7}$$

With the morphologically structured model the fraction of the morphological forms can be calculated and using Eqs. (5)-(7) it is then possible to calculate the development of directly measurable morphological variables.

The model was compared with experimental data for *G. candidum*, *Streptomyces hygroscopicus*, and *Penicillium chrysogenum*, and in Fig. 7.21 and 7.22 the results of the comparison with data for *G. candidum* are shown.

In Fig. 7.21, measurements of the hyphal diameter, the hyphal growth unit volume, and the hyphal growth unit length in a steady-state chemostat are shown as functions of the dilution rate D. The hyphal diameter increases with D, and since the hyphal growth unit volume is approximately constant, l_{hgu} decreases with the dilution rate. Thus, when the glucose concentration decreases, the hyphal elements become less branched and form long hyphae. In Fig. 7.22, measurements of the total hyphal length, the number of tips, and the hyphal growth unit length during outgrowth of a single spore on a solid medium are shown, together with model simulations. The total hyphal length is observed to increase exponentially, whereas there is a lag phase before the first branch point is formed, i.e., the number of tips increases from one to two. Due to the sudden formation of new tips the hyphal growth unit length oscillates until the number of tips becomes large and an approximately constant value for l_{hgu} is obtained. The modeling concept illustrated in this example has been applied for simulation of many other systems, e.g. for enzyme production by *Aspergillus oryzae* (Agger *et al.*, 1998). Recent reviews on modeling of filamentous fungi are given by Nielsen (1996) and Krabben and Nielsen (1998).

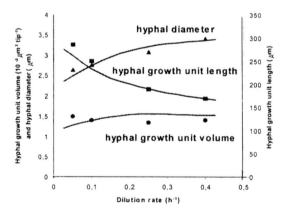

Figure 7.21 Morphology of *G. candidum* as a function of the specific growth rate in a submerged culture. Hyphal growth unit length, hyphal growth unit volume, and the hyphal diameter are shown as functions of the dilution rate in a chemostat. Lines are model simulations. The experimental data are taken from Robinson and Smith (1979) and model simulations are from Nielsen (1993).

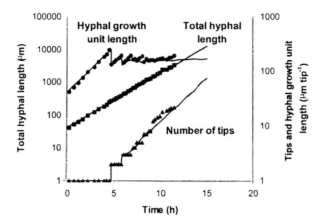

Figure 7.22 The morphology of *G. candidum* during the outgrowth of a single spore. Measurements of the total hyphal length (l_t) in μm, the number of tips, and the hyphal growth unit length (l_{hgu}) in μm per tip are compared with model simulations (lines). The data are taken from Trinci (1974) and model simulations are from Nielsen (1993).

PROBLEMS

Problem 7.1 Estimation of parameters in the Monod model

From measurements of the residual glucose concentration in a steady-state chemostat at various dilution rates, you can find the following results:

D (h^{-1})	s (mg L^{-1})
0.13	11
0.19	14
0.23	18
0.36	38
0.67	85
0.73	513

Calculate by linear regression the parameters in the Monod model.

You want to check the value of μ_{max} determined above and therefore increase the dilution rate in the chemostat to $D = 1.1$ h^{-1}. This results in a rapid increase in the glucose concentration, and after a while $s \gg K_s$. The result of the change in dilution rate is a decrease in the biomass concentration, and during the washout you measure the biomass concentration as a function of time, and obtain the following results:

Time (h)	x (g L^{-1})
0	5.1
0.5	4.5
1.0	3.7
2.0	2.8
3.0	2.1
4.0	1.4

Determine μ_{max} from this experiment. Discuss the applied method [see also Esener *et al.* (1981c)].

Problem 7.2 Inhibitory effect of lactic acid

Bibal *et al.* (1988, 1989) studied the inhibition of lactic acid on *Streptococcus cremoris*, and in this exercise we will analyze their data.

a. The influence of lactic acid on the growth of *S. cremoris* was examined by measuring the maximum specific growth rate during batch growth of the bacterium in media containing various concentrations of lactic acid (p). The results are summarized below:

p (g L^{-1})	μ (h^{-1})
0	0.90
12.0	0.68
39.0	0.52
55.0	0.13

According to Section 2.1.2.1 (see also Example 2.1), it is mainly the undissociated form of lactic acid that can pass the cellular membrane, and we will therefore assume that it is only the undissociated acid that has a toxic effect on the cells. Plot the relative specific growth rate, i.e., $\mu_{max}(p)/\mu_{max}(p=0)$, versus the concentration of the undissociated acid concentration (in mM). pH = 6.3 was used and pK_a for lactic acid is 3.88. Assume that the inhibition model given by Eq. (7.22) holds. Find the inhibition constant K_i. Plot the model together with the experiments.

b. From the results in a. you conclude that Eq. (7.22) is not well suited for description of the experimental data, since the inhibition by lactic acid seems to be stronger, especially at high values of undissociated lactic acid concentrations (p_u), i.e., there seems to be a certain maximum concentration of undissociated acid above which growth stops. Specify another simple one-parameter model for the influence of p_u on μ, and estimate the model parameter. What concentration of lactic acid (p) does this maximum value correspond to?

c. Plot the maximum specific growth rate as a function of the pH in a medium containing 1 g L^{-1} and 10 g L^{-1} of lactic acid (total concentration), using the model found in (b).

d. Measuring the yield coefficient on lactose in a steady-state chemostat at different concentrations of lactic acid, Bibal *et al.* (1988; 1989) found the data below

p (g L^{-1})	Y_{sx} (g DW g^{-1})
0	0.16
7.5	0.16
13.0	0.14
18.5	0.14
21.0	0.14
32.0	0.12
38.5	0.11
45.0	0.10
48.5	0.09

How can you explain the decrease in the yield coefficient with increasing lactic acid concentration?

e. Assume that the maintenance coefficient m_s is 0.05 h^{-1}. Calculate the true yield coefficient stoichiometric coefficient in eq. (7.26) for $p = 0$. Using the model derived in (b), calculate the maintenance coefficient as a function of p_u. Explain the results.

Problem 7.3 Modeling of the lac-operon in *E. coli*

We will now consider the model for the lac-operon described in Section 7.5.1.

a. The repressor has four binding sites for the inducer (lactose), but in the derivation of Eq. (7.52) only the repressor-inducer complex where all four sites are occupied is considered. We now consider binding at all four sites. Specify all the equilibria and the definitions of the association constants. The association constant for formation of X_rS_i is termed K_{1i}, and that for formation of $X_OX_rS_i$ is termed K_{4i}. Binding of the inducer to the repressor operator complex can be neglected (i.e., the equilibrium in eq. (7.44c) is not considered).

Assume that the affinity for the binding of the repressor to the operator is approximately the same whether no, one, two or three inducers are bound to the repressor, i.e., $K_{41} = K_{42} = K_{43} = K_{44}$. This assumption is reasonable since the repressor probably changes its conformation only when the last inducer is bound to it. With this assumption show that the fraction of repressor-free operators is given by

$$Q_1 = \frac{[X_O]}{[X_O]_t} = \frac{1 + K_1[S_{lac}]_t^4}{1 + K_1[S_{lac}]_t^4 + K_2[X_r]_t + K_S K_{11}[X_r S_{lac}][S_{lac}]_t(1 + K_{12}[S_{lac}]_t + K_{12}K_{13}[S_{lac}]_t^2)} \quad (1)$$

where $K_1 = K_{41} K_{42} K_{43} K_{44}$.

We now assume that the $K_{11} \approx K_{12} \approx K_{13} \approx K_{14}$, i.e., the association constant for the fourth inducer is much stronger than the corresponding constants for the first three sites. This assumption follows from our assumption above that the conformation of the repressor changes only when the fourth inducer is bound and when the conformation changes the repressor-inducer complex becomes very stable. What other assumptions are required for reducing Eq. (1) to Eq. (7.52)?

b. Lee and Bailey (1984d) also modeled the lac-operon, but they included binding of the repressor to a nonspecific binding site in the chromosome (X_d). Again we neglect binding of inducer to the repressor-operator complex, and the equilibria are therefore

$$X_r + nS_{lac} \overset{K_1}{\leftrightarrow} X_r nS_{lac} \qquad (2a)$$

$$X_o + X_r \overset{K_2}{\leftrightarrow} X_o X_r \qquad (2b)$$

$$X_o + X_r S_{lac} \overset{K_4}{\leftrightarrow} X_o X_r S_{lac} \qquad (2c)$$

$$X_d + X_r \overset{K_5}{\leftrightarrow} X_d X_r \qquad (2d)$$

$$X_d + X_r S_{lac} \overset{K_6}{\leftrightarrow} X_d X_r S_{lac} \qquad (2e)$$

By assuming that $[X_d]_t \approx [X_d]$, show that

$$Q_1 = \frac{[X_O]}{[X_O]_t} = \frac{1 + K_5[X_d]_t + K_1[S_{lac}]_t^n (1 + K_6[X_d]_t)}{1 + K_5[X_d]_t + K_1[S_{lac}]_t^n (1 + K_6[X_d]_t) + K_2[X_r]_t} \qquad (3)$$

c. Lee and Bailey (1984d) specified the parameters in Eq. (3) to be

$K_1 = 10^7 \text{ M}^{-1}$, $K_2 = 2 \cdot 10^{12} \text{ M}^{-1}$, $K_4 = 2 \cdot 10^9 \text{ M}^{-1}$
$K_5 = 10^3 \text{ M}^{-1}$, $K_6 = 1.5 \cdot 10^9 \text{ M}^{-1}$

Furthermore, they state that

$[X_d]_t = 4 \cdot 10^{-2}$ M   and $[X_r]_t = 2 \cdot 10^{-8}$ M

Plot the value of Q_1 using both eq. (3) and Eq. (7.52) as a function of the inducer concentration $[S_{lac}]_t$. Comment on the result.

The parameters given by Lee and Bailey are for IPTG (isopropyl-β-D-thiogalactosidase), a frequently used inducer in studies of the lac-operon. Assume that the parameters are the same for lactose as inducer and calculate the concentration of lactose (in mg L^{-1}) to give $Q_1 = 0.5$. Discuss why even this low concentration of lactose leads to induction of the lac-operon.

d. Show that for an antiinducer (neglect binding of the repressor to nonspecific sites)

$$Q_1 = \frac{[X_o]}{[X_o]_t} = \frac{1+K_1[S_{lac}]_t^m}{1+K_1[S]_t^n + K_1K_4[S]_t^n[X_r]_t} \tag{4}$$

Problem 7.4 Oscillating yeast

We now want to analyze the simple morphologically structured model for oscillating yeast described by Cazzador (1991). The metamorphosis reactions are given in Eqs. (2)-(3) of Example 7.9, and the growth of each morphological form is given by Eq. (4) of the example. The rate constants of the metamorphosis reactions (k_b and k_u) and the specific growth rates of the two morphological forms (μ_b and μ_u) are all assumed to be functions of the limiting substrate concentration only.

    a. For a chemostat with dilution rate D, write the steady-state balances for the two morphological forms and for the limiting substrate. The concentration of the limiting substrate in the feed is s^f, and the biomass concentration in the chemostat is x.

    b. By combining the steady-state balances for the two morphological forms, it is possible to derive an equation that relates D, μ_b, μ_u, k_b, k_u. Specify this relation. If it is assumed that for any s there is one and only one admissible dilution rate which satisfies this relation, show that this leads to the constraint

$$\frac{k_b}{\mu_b} + \frac{k_u}{\mu_b} > 1 \tag{1}$$

    c. To analyze the stability of a given steady state ($\tilde{Z}_b, \tilde{Z}_u, \tilde{s}, \tilde{x}$) [corresponding to the rates ($\tilde{u}_b, \tilde{u}_u, \tilde{\mu}_b, \tilde{\mu}_u, \tilde{D}$)], you have to linearize the model, i.e.,

$$\mathbf{J} \begin{pmatrix} Z_b \\ Z_u \\ s \end{pmatrix} = \mathbf{0} \tag{2}$$

where $\mathbf{J}$ is the Jacobi matrix. Show that the Jacobi matrix is given by

$$\mathbf{J} = \begin{pmatrix} -b_1 & k_u & r_b - \phi \\ k_b & -b_2 & r_u + \phi \\ -\mu_b \alpha_b x & -\tilde{\mu}_u \alpha_u x & -b_3 \end{pmatrix} \tag{3}$$

and specify b_1, b_2, b_3, r_b, r_u and ϕ as functions of the model parameters and variables in the steady state considered (see Example 7.9). What can be inferred about the signs of b_1, b_2, b_3, r_b and r_u? Show that the eigenvalues of the Jacobi matrix are given as zeros of the polynomial in Eq. (4)

$$F(\lambda) = \lambda^3 + a_1\lambda^2 + a_2\lambda + a_3 \tag{4}$$

and specify the coefficients in $F(\lambda)$ as functions of the model parameters and values of the variables (or partial derivatives of these) taken at the considered steady state.

d. Show by using the steady-state balances for the two morphological forms that

$$r_b - \phi = \tilde{\mu}'\tilde{Z}_b + \tilde{Z}'_b(b_1 + k_u) \tag{5}$$

$$r_b + \phi = \tilde{\mu}'\tilde{Z}_u - \tilde{Z}'_b(b_2 + k_b) \tag{6}$$

where the prime is used to indicate the derivative with respect to the substrate concentration. By combining eqs. (5) and (6), it is possible to derive an expression for $\tilde{\mu}'$

$$\tilde{\mu}' = r_b + r_u + \tilde{Z}_b(\mu_b - \mu_u) \tag{7}$$

Use eqs. (5)-(7) to eliminate ϕ from the expressions for the coefficients in the polynomial in eq. (4).

e. It can be shown that the requirements for stability (i.e., only eigenvalues with a negative real part) are

$$a_1 > 0 \; ; \; a_2 > 0 \; ; \; a_3 > 0 \; ; \; a_4 = a_1 a_2 - a_3 > 0 \tag{8}$$

If you are interested in mathematics, you may derive the four inequalities. There is a bifurcation, i.e., change from a stable system to a non-stable system, when the parameters a_1 to a_4 are all positive except one, which changes sign. If a_4 becomes negative, the bifurcation is dynamic (so-called Hopf bifurcation) and sustained oscillations are obtained. This corresponds to the situation where a pair of complex conjugate eigenvalues cross the imaginary axis while the third eigenvalue is real and negative. What does a_1, a_2, a_3 and $a_4 < 0$ imply concerning the parameters in the morphologically structured model when $\tilde{Z}'_b > 0$?

Analyze the following three cases:
- $\mu_b = \mu_u$ and $\alpha_b = \alpha_u$
- $\mu_b = \mu_u$ and $\alpha_b \neq \alpha_u$
- $\mu_b \neq \mu_u$ and $\alpha_b = \alpha_u$

For the last case a requirement for instability leads to $\tilde{D}' < 0$. Is this physiologically reasonable?
Final note: with second case of above, Cazzador derived analytical expressions for the rates of the metamorphosis reactions (see Problem 8.5) and showed that sustained oscillations can be simulated.

REFERENCES

Agger, T., Spohr, A. B., Carlsen, M. and Nielsen, J. (1998) Growth and product formation of *Aspergillus oryzae* during submerged cultivations: Verification of a morphologically structured model using fluorescent probes. *Biotechnol. Bioeng.* 57, 321-329

Bailey, J. E. and Ollis, D. F. (1986). *Biochemical Engineering Fundamentals*, 2d. ed., McGraw-Hill, New York.

Baltzis, B. C., Fredrickson, A. G. (1988). Limitation of growth by two complementary nutrients: Some elementary, but neglected considerations, *Biotechnol. Bioeng.* 31, 75-86.

Benthin, S., Nielsen, J., Villadsen, J. (1991). A simple and reliable method for the determination of cellular RNA content, *Biotechnol. Techniques* **5**, 39-42.

Beuse, M., Kopmann, A., Diekmann, H., and Thoma, M. (1999). Oxygen, pH value and carbon source induced changes in the mode of oscillation in synchronous continuous culture of *Saccharomyces cerevisiae*. *Biotechnol.Bioeng.* **63**,410-417

Bibal, B., Goma, G., Vayssier, Y., Pareilleux, A. (1988). Influence of pH, lactose and lactic acid on the growth of *Streptococcus cremoris*: a kinetic study, *Appl. Microbiol. Biotechnol.* **28**, 340-344.

Bibal, B., Kapp, C., Goma, G., Pareilleux, A. (1989). Continuous culture of *Streptococcus cremoris* on lactose using various medium conditions, *Appl. Microbiol. Biotechnol.* **32**, 155-159.

Caldwell, I. Y. and Trinci, A. P. J. (1973). The growth unit of the mould *Geotrichum candidum*, *Arch. Mikrobiol.* **88**:1-10.

Carlsen, M., Jocumsen, K. V., Emborg, C., Nielsen, J. (1997) Modelling the growth and Proteinase A production in continuous cultures of recombinant *Saccharomyces cerevisiae*. *Biotechnol. Bioeng.* **55**, 447-454

Cazzador, L. (1991). Analysis of oscillations in yeast continuous cultures by a new simplified model, *Bull. Math. Biol.* **5**:685-700.

Cazzador, L. and Mariani, L. (1988). A simulation program based on a structured population model for biotechnological yeast processes, *Appl. Microbiol. Biotechnol.* **29**:198-202.

Cazzador, L. and Mariani, L. (1990). A two compartment model for the analysis of spontaneous oscillations in *S. cerevisiae*, Abstract book (European Congress on Biotechnology, Copenhagen) **5**, 342.

Cazzador, L., Mariani, L., Martegani, E., and Alberghina, L. (1990). Structured segregated models and analysis of self-oscillating yeast continuous culture, *Bioproc. Eng.* **5**:175-180.

Christiansen, T., Spohr, A., Nielsen, J. (1999) On-line study of growth kinetics of single hyphae of *Aspergillus oryzae* in a flow-through cell. *Biotechnol. Bioeng.* **63**, 147-153

Cox, P. W., Paul, G. C., Thomas, C. R. (1998) Image analysis of the morphology of filamentous micro-organisms. *Microbiol.* **144**, 817-827

Dhurjati, P., Ramkrishna, D., Flickinger, M. C., Tsao, G. T. (1985). A cybernetic view of microbial growth: Modeling of cells as optimal strategists, *Biotechnol. Bioeng.* **27**, 1-9.

Domach, M. M., Leung, S. K., Cahn, R. E., Cocks, G. G., Shuler, M. L. (1984). Computer model for glucose-limited growth of a single cell of *Escherichia coli* B/r-A, *Biotechnol. Bioeng.* **26**, 203-216.

Duboc,P. and von Stockar, U. (2000) Modeling of oscillating cultivations of *Saccharomyces cerevisiae*: Identification of population structure and expansion kinetics based on on-line measurements.*Chem.Engr.Sci.* **55**,149-160

Egli, T. (1991). On multiple nutrient limited growth of microorganisms with special reference to dual limitation by carbon and nitrogen substrates, *Antonie van Leeuwenhoek* **60**, 225-234.

Esener, A. A., Roels, J. A., Kossen, N. W. F. (1981a). The influence of temperature on the maximum specific growth rate of *Klebsiella pneumoniae*, *Biotechnol. Bioeng.* **23**, 1401-1405.

Esener, A. A., Roels, J. A., Kossen, N. W. F. (1981b). Fed-batch culture: Modeling and applications in the study of microbial energies, *Biotechnol. Bioeng.* **27**, 1851-1871.

Esener, A. A., Roels, J. A., Kossen, N. W. F., Roozenburg, J. W. H. (1981c). Description of microbial growth behaviour during the wash-out phase; determination of the maximum specific growth rate, *Eur. J. Appl. Microbiol. Biotechnol.* **13**, 141-144.

Esener, A. A., Veerman, T., Roels, J. A., Kossen, N. W. F. (1982). Modeling of bacterial growth; Formulation and evaluation of a structured model, *Biotechnol. Bioeng.* **29**, 1749-1764.

Fiddy, C. and Trinci, A. P. J. (1976). Mitosis, septation, branching and the duplication cycle in *Aspergillus nidulans*, *J. Gen. Microbiol.* **97**:169-184.

Frandsen, S. (1993). Dynamics of *Saccharomyces cerevisiae* in continuous culture, Ph.D. thesis, Technical University of Denmark, Lyngby.

Fredrickson, A. G. (1976). Formulation of structured growth models, *Biotechnol. Bioeng.* **18**, 1481-1486.

Harder, A., Roels, J. A. (1982). Application of simple structured models in bioengineering, *Adv. Biochem. Eng.* **21**, 55-107.

Herbert, D. (1959). Some principles of continuous culture, *Recent Prog. Microbiol.* **7**, 381-396.

Herendeen, S. L., van Bogelen, R. A., Neidhardt, F. C. (1979). Levels of major proteins of *Escherichia coli* during growth at different temperatures, *J. Bacteriol.* **139**, 185-194.

Hjortso, M. A. and Nielsen, J. (1994). A conceptual model of autonomous oscillations in microbial cultures, *Chem. Eng. Sci.* **49**:1083-1095.

Ingraham, J. L., Maaløe, O., Neidhardt, F. C. (1983). *Growth of the Bacterial Cell*, Sinauer Associates, Inc., Sunderland.

Jöbses, I. M. L., Egberts, G. T. C., van Baalen, A., Roels, J. A. (1985). Mathematical modeling of growth and substrate conversion of *Zymomonas mobilis* at 30 and 35°C, *Biotechnol. Bioeng.* **27**, 984-995.

Keulers, M, Satroutdinov, A.D., Suszuki, T. and Kuriyama, H (1996). Synchronization affector of autonomous short period sustained oscillation of *Saccharomyces cerevisiae*. *Yeast* **12** 673-682

Kompala, D. S., Ramkrishna, D., Tsao, G. T. (1984). Cybernetic modeling of microbial growth on multiple substrates, *Biotechnol. Bioeng.* **26**, 1272-1281.

Kompala, D. S., Ramkrishna, D., Jansen, N. B., Tsao, G. T. (1986). Investigation of bacterial growth on mixed substrates: Experimental evaluation of cybernetic models, *Biotechnol. Bioeng.* **28**, 1044-1055.

Krabben, P., Nielsen, J. (1998) Modeling the mycelium morphology of *Penicillium* species in submerged cultures. *Adv. Biochem. Eng./Biotechnol.* **60**, 125-152

Lee, S. B., Bailey, J. E. (1984a). A mathematical model for λdv plasmid replication: Analysis of wild-type plasmid, *Plasmid* **11**, 151-165.

Lee, S. B., Bailey, J. E. (1984b). A mathematical model for λdv plasmid replication: Analysis of copy number mutants, *Plasmid* **11**, 166-177.

Lee, S. B., Bailey, J. E. (1984c). Analysis of growth rate effects on productivity of recombinant *Escherichia coli* populations using molecular mechanism models, *Biotechnol. Bioeng.* **26**, 66-73.

Lee, S. B., Bailey, J. E. (1984d). Genetically structured models for lac promoter-operator function in the *Escherichia coli* chromosome and in multicopy plasmids: lac operator function, *Biotechnol. Bioeng.* **26**, 1372-1382.

Lee, S. B. and Bailey, J. E. (1984e). Genetically structured models for lac promoter-operator function in the *Escherichia coli* chromosome and in multicopy plasmids: lac promoter function, *Biotechnol. Bioeng.* **26**, 1381-1389.

Martegani, E., Porro, D., Ranzi, B. M. and Alberghina, L. (1990). Involvement of a cell size control mechanism in the induction and maintenance of oscillations in continuous cultures of budding yeast, *Biotechnol. Bioneg.* **36**:453-459.

McIntyre, M., Müller, C. Dynesen, J., Nielsen, J. (2001) Metabolic engineering of the morphology of *Aspergillus*. *Adv. Biochem. Eng./Biotechnol.* **73**, 103-128

Megee, R. D., Kinishita, S., Fredrickson, A. G., and Tsuchiya, H. M. (1970). Differentiation and product formation in molds, *Biotechnol. Bioeng.* **12**:771-801.

Monod, J. (1942). *Recherches sur la croissance des cultures bacteriennes*, Hermann et Cie, Paris.

Monod, J., Wyman, J., Changeux, J.-P. (1963). Allosteric proteins and cellular control systems. *J. Mol. Biol.* **6**, 306-329

Murray, D.B., Engelen, F., Lloyd, D., and Kuriyama, H. (1999) Involvement of glutathione in the regulation of respiratory oscillation during a continuous culture of *Sacchasromyces cerevisiae*. *Microbiology* **145**, 2739-2745

Müller-Hill, B. (1996) *The lac Operon: A Short History of a Genetic Paradigm*. Walter de Gruyter & Co., Berlin

Münch, T. (1992). Zellzyklusdynamik von Saccharomyces cerevisiae *in Bioprozessen*, Ph.D. thesis, ETH, Zürich.

Nielsen, J. (1992). Modelling the growth of filamentous fungi, *Adv. Biochem. Eng. Biotechnol.* **46**: 187-223.

Nielsen, J. (1993). A simple morphologically structured model describing the growth of filamentous microorganisms, *Biotechnol. Bioeng.* **41**:715-727.

Nielsen, J. (1996) Modelling the morphology of filamentous microorganisms. *TIBTECH* **14**:438-443

Nielsen, J., Villadsen, J. (1992). Modeling of microbial kinetics, *Chem. Eng. Sci.* **47**, 4225-4270

Nielsen, J., Villadsen, J. (1994). Bioreaction Engineering Principles. Plenum Press, New York

Nielsen, J., Nikolajsen, K., Villadsen, J. (1991a). Structured modeling of a microbial system 1. A theoretical study of the lactic acid fermentation, *Biotechnol. Bioeng.* **38**, 1-10.

Nielsen, J., Nikolajsen, K., Villadsen, J. (1991b). Structured modeling of a microbial system 2. Verification of a structured lactic acid fermentation model, *Biotechnol. Bioeng.* **38**, 11-23.

Nielsen, J., Pedersen, A. G., Strudsholm, K., Villadsen, J. (1991c). Modeling fermentations with recombinant microorganisms: Formulation of a structured model, *Biotechnol. Bioeng.* **37**, 802-808.

Packer, H. L., Keshavarz-Moore, E., Lilly, M. D., and Thomas, C. R. (1992). Estimation of cell volume and biomass of *Penicillium chrysogenum* using image analysis, *Biotechnol. Bioeng.* **39**: 384-391.

Peretti, S. W. and Bailey, J. E. (1986). Mechanistically detailed model of cellular metabolism for glucose-limited growth of *Escherichia coli* B/r-A, *Biotechnol. Bioeng.* **28**, 1672-1689.

Peretti, S. W., Bailey, J. E. (1987). Simulations of host-plasmid interactions in *Escherichia coli*: Copy number, promoter strength, and ribosome binding site strength effects on metabolic activity and plasmid gene expression, *Biotechnol. Bioeng.* **29**, 316-328.

Pirt, S. J. (1965). The maintenance energy of bacteria in growing cultures, *Proc. Royal Soc. London Ser. B.* **163**, 224-231.

Pronk, J. T.; Steensma, H. Y., van Dijken, J. P. (1996). Pyruvate metabolism in *Saccharomyces cerevisiae*. *Yeast* **12**, 1607-1633.

Powell, E. O. (1967). The growth rate of microorganisms as a function of substrate concentration, in *3 Int. Symposium on Microbial Physiology and Continuous Culture*, E. O. Powell, ed., 23-33.

Prosser, J. I. and Tough, A. J. (1991). Growth mechanisms and growth kinetics of filamentous microorganisms, *Crit. Rev. Biotechnol.* **10**:253-274.

Ramkrishna, D. (1982). A cybernetic perspective of microbial growth, in *Foundations of Biochemical Engineering: Kinetics and Thermodynamics in Biological Systems*, American Chemical Society, 161-178.

Ramkrishna, D., Fredrickson, A. G., Tsuchiya, H. M. (1967). Dynamics of microbial propagation: Models considering inhibitors and variable cell composition, *Biotechnol. Bioeng.* **9**, 129-170.

Ramkrishna, D., Kompala, D. S., Tsao, G. T. (1984). Cybernetic modeling of microbial populations. Growth on mixed substrates, in *Frontiers in Chemical Reaction Engineering*, Vol. 1, Wiley Eastern Ltd., New Delhi, 241-261.

Ramkrishna, D., Kompala, D. S., Tsao, G. T. (1987). Are microbes optimal strategists? *Biotechnol. Prog.* **3**, 121-126.

Rieger, M., Kappeli, O., Fiechter, A. (1983). The role of limited respiration in the incomplete oxidation of glucose by Saccharomyces cerevisiae, *J. Gen. Microbiol.* **129**, 653-661.

Robinson, P. M. and Smith, J. M. (1979). Development of cells and hyphae of *Geotrichum candidum* in chemostat and batch culture, *Proc. Br. Mycol. Soc.* **72**:39-47.

Roels, J. A. (1983). *Energetics and Kinetics in Biotechnology*, Elsevier Biomedical Press, Amsterdam.

Roels, J. A., Kossen, N. W. F. (1978). On the modeling of microbial metabolism, *Prog. Ind. Microbiol.* **14**, 95-204.

Seo, J.-H., Bailey, J. E. (1985). Effects of recombinant plasmid content on growth properties and cloned gene product formation in *Escherichia coli*, *Biotechnol. Bioeng.* **27**, 1668-1674.

Shuler, M. L., Domach, M. M. (1982). Mathematical models of the growth of individual cells, in *Foundations of Biochemical Engineering: Kinetics and Thermodynamics in Biological Systems*, American Chemical Society Publications, 93-133.

Shuler, M. L., Leung, S. K., Dick, C. C. (1979). A mathematical model for the growth of a single bacterial cell, *Ann. N. Y. Acad. Sci.* **326**, 35-55.

Sohn, Ho-Yong, and Kuriyama, H.(2001) Ultradian metabolic oscillation of *Saccharomyces cerevisiae* during aerobic continuous culture: Hydrogen sulphide, a population synchronizer, is produced by sulphite reductase. *Yeast* **18**, 125-135

Sonnleitner, B. and Kappeli, O. (1986). Growth of *Saccharomyces cerevisiae* is controlled by its limited respiratory capacity: Formulation and verification of a hypothesis, *Biotechnol. Bioeng.* **28**: 927-937.

Spohr, A. B, Mikkelsen, C. D., Carlsen, M., Nielsen, J., Villadsen, J. (1998) On-line study of fungal morphology during submerged growth in a small flow-through cell. *Biotechnol. Bioeng.* **58**, 541-553

Strässle, C., Sonnleitner, B., and Fiechter, A. (1988). A predictive model for the spontaneous synchronization of *Saccharomyces cerevisiae* grown in continuous culture I. Concept, *J. Biotechnol.* **7**:299-318.

Strässle, C., Sonnleitner, B., and Fiechter, A. (1989). A predictive model for the spontaneous synchronization of *Saccharomyces cerevisiae* grown in continuous culture II. Experimental verification, *J. Biotechnol.* **9**:191-208.

Strudsholm, K., Nielsen, J., Emborg, C. (1992). Product formation during batch fermentation with recombinant E. coli containing a runaway plasmid, *Bioproc. Eng.* **8**, 173-181.

Sweere, A. P. J., Giesselbach, J., Barendse, R., de Krieger, R., Honderd. G., Luyben, K. Ch. A. M. (1988). Modeling the dynamic behaviour of Saccharomyces cerevisiae and its application in control experiments, *Appl. Microbiol. Biotechnol.* **28**, 116-127.

Trinci, A. P. J. (1974). A study of the kinetics of hyphal extension and branch initiation of fungal mycelia, *J. Gen. Microbiol.* **81**:225-236.

Trinci, A. P. J. (1984). "Regulation of hyphal branching and hyphal orientation". In *The Ecology and Physiology of the Fungal Mycelium*, D. H. Jennings and A. D. M. Rayner, eds., Cambridge University Press, Cambridge, UK.

Tsao, G. T., Hanson, T. P. (1975). Extended Monod equation for batch cultures with multiple exponential phases, *Biotechnol. Bioeng.* **17**, 1591-1598.

Turner, B. G., Ramkrishna, D. (1988). Revised enzyme synthesis rate expression in cybernetic models of bacterial growth. *Biotechnol. Bioeng.* **31**, 41-43.

Varner, J., Ramkrishna, D. (1999). Metabolic engineering form a cybernetic perspective: Aspartate family of amino acids. *Metabolic Eng.* **1**, 88-116

Williams, F. M. (1967). A model of cell growth dynamics, *J. Theoret. Biol.* **15**, 190-207.

8

Population Balance Equations

In Chapter 7, cell population balances are written in terms of a distribution of mass fractions of the total biomass. This allows a direct combination of intracellularly structured models and population models. However, the population balances based on mass fractions do not permit the incorporation into the model of specific events in the cell cycle, and the single-cell models of Section 7.5.2 can therefore not be used in connection with these population balances. Since there are numerous examples that show a direct influence of certain specific events in the cell cycle on the overall culture performance, e.g., the distribution of plasmids to daughter cells on cell division in recombinant cultures, we need to derive a population balance based on cell number to obtain a correct description of these processes.

In a population balance based on cell number, the basis is the individual cells. Thus, the cellular content of the intracellular components has the unit grams per cell, and we can therefore not use the composition vector **X** (unit: grams per gram dry weight). Instead the properties of the cell are described by the vector **y**, which may also contain information about the cell's age, size, etc. The distribution of cells in the population is given by $f(\mathbf{y},t)$, where $f(\mathbf{y},t)\,d\mathbf{y}$ represents the number of cells per unit volume within the property space **y** to $\mathbf{y} + d\mathbf{y}$ at time t. Thus the total number of cells per unit volume in the population is given by

$$n(t) = \int_{V_y} f(\mathbf{y},t)\,d\mathbf{y} \tag{8.1}$$

where V_y is the total property space. In general, n is determined from a mass balance for the limiting substrate, as illustrated in Note 8.1.

Note 8.1 Determination of the total number of cells from a substrate balance
For a distribution of cells with different substrate uptake kinetics, the volumetric rate of substrate consumption for a single limiting substrate is given by,

$$q_s(t) = -\int_{V_y} r_s(\mathbf{y},s) f(\mathbf{y},t)\,d\mathbf{y} \tag{1}$$

where $r_s(\mathbf{y},s)$ is the rate of substrate consumption per cell per time. The mass balance for the limiting substrate is therefore

$$\frac{ds}{dt} = D(s_f - s) - \int_{V_y} r_s(\mathbf{y},s) f(\mathbf{y},t)\, d\mathbf{y} \qquad (2)$$

If $r_s(\mathbf{y},s)$ is taken to be independent of the cellular state, the steady-state solution to the mass balance gives

$$n = \frac{D(s_f - s)}{r_s(s)} \qquad (3)$$

Thus if the substrate concentration is known, the total number of cells can be calculated. If the single-cell kinetics is described as a function of the limiting substrate concentration, s can be calculated from the parameters in the single-cell kinetic model (see Problem 8.1).

For a homogeneous system (or for a given homogeneous volume element), the dynamic balance for the distribution function is Eq. (8.2):

$$\frac{\partial f(\mathbf{y},t)}{\partial t} + \nabla_\mathbf{y}[\mathbf{r}(\mathbf{y},t) f(\mathbf{y},t)] = h(\mathbf{y},t) - D f(\mathbf{y},t) \qquad (8.2)$$

$\mathbf{r}(\mathbf{y},t)$ is the rate of change of properties, i.e., r_i is the rate along the ith property axis in the total property space V_y. $h(\mathbf{y},t)$ is the net rate of formation of cells with the property $\mathbf{y}$ due to cell division, and D is the dilution rate in the bioreactor. It is assumed that there are no cells in the liquid stream entering the bioreactor (or the considered volume element), i.e, $f_\text{in}(\mathbf{y},t) = 0$. The first term is the accumulation term. The second term accounts for the formation and removal of elements with the given properties due to cellular processes, e.g. growth etc. The first term on the right hand side accounts for net formation of elements/cells with the property $\mathbf{y}$, e.g. upon cell division there is a net formation of new cells. The last term on the right hand side accounts for washout of elements/cells from the bioreactor.

The population balance eq. (8.2) holds only for a homogeneous bioreactor; i.e., the distribution function is the same in each volume element in the bioreactor. This assumption is reasonable for laboratory-scale bioreactors whereas it is doubtful for large-scale bioreactors. In Note 8.2, the population balance is generalized to consider variations in the distribution function throughout the three-dimensional physical space.

Note 8.2 General form of the population balance
With a non-homogeneous physical space, the distribution function also becomes a function of position in the space (i.e., $f(\mathbf{z},\mathbf{y},t)$, where $\mathbf{z}$ is the physical state space. $f(\mathbf{z},\mathbf{y},t)\, d\mathbf{z}d\mathbf{y}$ is the number of cells within the physical space between $\mathbf{z}$ and $\mathbf{z} + d\mathbf{z}$ and within the property space between $\mathbf{y}$ and $\mathbf{y} + d\mathbf{y}$, and the total number of cells per unit volume in the population is therefore given by

$$n(t) = \int_{V_y} \int_{V_z} f(\mathbf{z},\mathbf{y},t)\, d\mathbf{z}d\mathbf{y} \qquad (1)$$

Population Balance Equations

The generalized form of the population balance is given by Eq. (2) where $\mathbf{v}(\mathbf{z},t)$ is the rate of liquid flow at position $\mathbf{z}$ in the physical space:

$$\frac{\partial f(\mathbf{z},\mathbf{y},t)}{\partial t} + \frac{1}{V}\frac{dV}{dt}f(\mathbf{z},\mathbf{y},t) + \frac{1}{V}\nabla_z[\mathbf{v}(\mathbf{z},t)f(\mathbf{z},\mathbf{y},t)] + \nabla_y[\mathbf{r}(\mathbf{z},\mathbf{y},t)f(\mathbf{z},\mathbf{y},t)] = h(\mathbf{z},\mathbf{y},t) \quad (2)$$

For a homogeneous system, the distribution function is the same throughout the physical space, i.e.,

$$f(\mathbf{z},\mathbf{y},t) = f_h(\mathbf{y},t) = \int_{V_z} f(\mathbf{z},\mathbf{y},t)\,d\mathbf{z} \quad (3)$$

and consequently

$$\frac{\partial f_h(\mathbf{y},t)}{\partial t} + \frac{1}{V}\frac{dV}{dt}f_h(\mathbf{y},t) + \frac{1}{V}f_h(\mathbf{y},t)\int_{V_z}\nabla_z\mathbf{v}(\mathbf{z},t)\,d\mathbf{z} + \nabla_y[\mathbf{r}(\mathbf{y},t)f_h(\mathbf{y},t)] = h(\mathbf{y},t) \quad (4)$$

Now, applying the divergence theorem of Gauss [see, e.g., Kreyszig (1988)]

$$\int_{V_z}\nabla_z\mathbf{v}(\mathbf{z},t)\,d\mathbf{z} = \int_s \mathbf{n}(\mathbf{z},t)\mathbf{v}(\mathbf{z},t)\,dS \quad (5)$$

where $\mathbf{n}(\mathbf{z},t)$ is the outward normal on the system's surface S. Normally the transport across the surface of the system is characterized by two flows, one ingoing flow v_{in} and one outgoing flow v_{out}. We therefore have

$$\int_{V_z}\nabla_z\mathbf{v}(\mathbf{z},t)\,d\mathbf{z} = v_{in}f_{in}(\mathbf{y},t) - v_{out}f_h(\mathbf{y},t) \quad (6)$$

and by inserting Eq. (6) in Eq. (4) we find

$$\frac{\partial f_h(\mathbf{y},t)}{\partial t} + \nabla_y[\mathbf{r}(\mathbf{y},t)f_h(\mathbf{y},t)] = h(\mathbf{y},t) + \frac{1}{V}\left[v_{in}f_{in}(\mathbf{y},t) - \left(v_{out} + \frac{dV}{dt}\right)f_h(\mathbf{y},t)\right] \quad (7)$$

For a chemostat and a batch reactor the volume is constant, i.e., $dV/dt=0$ and $v_{out}/V=D$ for the chemostat. For a fed-batch reactor, $v_{out}=0$ and $(1/V)\,dV/dt = D$. With no cells in the ingoing stream, i.e., $f_{in}(\mathbf{y},t)=0$, Eq. (7) therefore reduces to Eq. (8.2).

As mentioned above the formation of cells with property $\mathbf{y}$ is described by the function $h(\mathbf{y},t)$ in the population balance of Eq. (8.2). Thus, the population balance equations allows to describe discrete events occurring, e.g. at cell division through the function. The function $h(\mathbf{y},t)$ is often split into two terms:

$$h(\mathbf{y},t) = h^+(\mathbf{y},t) - h^-(\mathbf{y},t) \quad (8.3)$$

where $h^+(\mathbf{y},t)$ represents the rate of formation of cells with property $\mathbf{y}$, and similarly $h^-(\mathbf{y},t)$ represents the rate of disappearance of these cells as they divide. Cell division is normally a singular event, which occurs quite independently of what happens to the other cells in the population. Let $b(\mathbf{y},t)$ represent the division frequency (or *breakage frequency*), i.e., $b(\mathbf{y},t)\,dt$ is the probability that a cell with property $\mathbf{y}$ at time t divides in the interval t to $t + dt$. Thus

$$h^-(\mathbf{y},t) = b(\mathbf{y},t)f(\mathbf{y},t) \qquad (8.4)$$

In order to identify $h^+(\mathbf{y},t)$, we must consider the (average) number of cells arising from division of a cell with property $\mathbf{y}$. This is normally 2, independent of the cellular properties and the environmental conditions, i.e., two new cells are formed upon cell division.[1] Next we define the function $p(\mathbf{y},\mathbf{y}^*,t)$ to represent the probability of the formation of cells with properties $\mathbf{y}$ and $\mathbf{y}^* - \mathbf{y}$, respectively, upon division of a cell with property $\mathbf{y}^*$.[2] The rate of formation of cells with property $\mathbf{y}$ is then given by

$$h^+(\mathbf{y},t) = 2\int_{V_y} b(\mathbf{y}^*,t)p(\mathbf{y},\mathbf{y}^*,t)f(\mathbf{y}^*,t)\,d\mathbf{y}^* \qquad (8.5)$$

The function $p(\mathbf{y},\mathbf{y}^*,t)$ is called the *partitioning function*. It satisfies the constraints $p(\mathbf{y},\mathbf{y}^*,t) = 0$ whenever one of the elements y_i in the property vector $\mathbf{y}$ is larger than y_i^*, and it is scaled by

$$\int_{V_y} p(\mathbf{y},\mathbf{y}^*,t)\,d\mathbf{y} = 1 \qquad (8.6)$$

Combining Eqs. (8.3)-(8.5) gives

$$h(\mathbf{y},t) = 2\int_{V_y} b(\mathbf{y}^*,t)p(\mathbf{y},\mathbf{y}^*,t)f(\mathbf{y}^*,t)\,d\mathbf{y}^* - b(\mathbf{y},t)f(\mathbf{y},t) \qquad (8.7)$$

No direct influence of the environmental conditions is included in Eq. (8.7). However, both r and h are normally functions of the concentrations of substrate and metabolic products in the surrounding medium. Application of the population balance and an example of the breakage frequency and the partitioning function are illustrated in Example 8.1.

Example 8.1 Specification of the partitioning function and the breakage frequency

---

[1] For the meiosis of eucaryotes, four cells are formed in a cell cycle, but this special situation will not be considered here.
[2] This holds only when the cell properties are conserved upon cell division. There are many cell properties for which this is not the case, e.g., cell age and surface area. However, here the h function can often be described explicitly (e.g., as a Dirac delta function) as illustrated in Example 6.2.

Population Balance Equations

Kothari et al. (1972) applied a population model originally derived by Eakman et al. (1966) for description of the size distribution of the yeast *Schizosaccharomyces pombe* at steady state in a chemostat. Thus they use a one-dimensional distribution function $f(m, t)$, which at steady state is given by

$$\frac{d[f(m)r(m,s)]}{dm} = 2\int_{m}^{\infty} b(m^*,s)p(m^*,m)f(m^*)\,dm^* - b(m,s)f(m) - Df(m) \tag{1}$$

or

$$\frac{df(m)}{dm} = \frac{2}{r(m,s)}\int_{m}^{\infty} b(m^*,s)p(m^*,m)f(m^*)\,dm^* - \frac{1}{r(m,s)}\left[b(m,s) + \frac{dr(m,s)}{dm} + D\right]f(m) \tag{2}$$

where $r(m,s)$ is the growth rate for cells with mass m (grams per cell per hour). Both the growth rate and the breakage function are taken to be functions of the substrate concentration s in the surrounding medium. In the model it is assumed that the distribution of division masses around the mean division mass m_d is of a Gaussian type, and the breakage function is therefore given by

$$b(m,s) = \frac{2e^{-[(m-m_d)/\varepsilon]^2} r(m,s)}{\varepsilon\sqrt{\pi}\,erfc[(m-m_d)/\varepsilon]} \tag{3}$$

For the partitioning function, it is furthermore assumed that the distribution of daughter cell mass m is also of the Gaussian type, with a median of half the mass of the parent cell at division, m^*. Thus

$$p(m^*,m) = \frac{e^{-[(m-0.5m^*)/\xi]^2}}{\xi\sqrt{\pi}\,erf(m^*/2\xi)} \tag{4}$$

These definitions of the breakage and partitioning functions give the right trends. It is, of course, not biologically reasonable that $b(0,s) \neq 0$ and $p(m^*,m) \neq 0$, but this does not influence the conclusions drawn by Kothari et al. (see below).

By comparing model calculations with experimental data for the mass distribution (obtained using a Coulter counter), Kothari et al. estimated the parameters in the model, i.e., the average mass at division m_d and the standard deviations (ε and ξ) for the functions in Eqs. (3) and (4). Furthermore, they examined different models for $r(m)$ (s is constant at a certain dilution rate in the chemostat), and found that a model where $r(m)$ is constant and independent of m, i.e., $r(m)=k$ corresponding to zeroth-order growth kinetics for the cell mass, gave the best fit to the experimental data obtained at different dilution rates (see Fig. 8.1). With a model where $r(m)$ is first-order in m, i.e., $r(m)=km$ corresponding to exponential growth of the single-cell mass, the calculated distribution function could not be fitted to the measured profile. Thus the application of the population model based on number revealed that *the growth rate of individual cells is not proportional to their mass*. This does not contradict the dictum that the growth rate of a microbial culture is proportional to the *total* biomass concentration, since

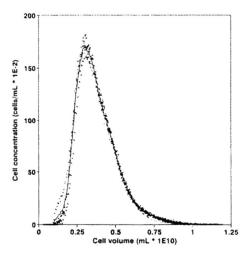

Figure 8.1 Cell size distribution for *S. pombe*. The data points are measurements obtained using a Coulter counter, and the line reflects model simulations. [Reprinted by permission from I. R. Kothari *et al.* (1972), Estimation of parameters in population models for *Schizosaccharomyces pombe* from chemostat data, *Biotechnol. Bioeng.* **14**, 915-938.]

$$q_x = \int_0^\infty kf(m)\,dm = kn = \frac{k}{\langle m \rangle} x \qquad (5)$$

where $\langle m \rangle$ is the average cell mass for the population (equal to x/n). Since $k/\langle m \rangle$ is equal to the specific growth rate for the population, Eq. (5) is equal to the standard expression for q_x. Equation (5) shows that when r is independent of m, the specific growth rate for the culture is a function of the average cell mass. This would not be the situation with exponential growth of the individual cells, since here we find

$$q_x = \int_0^\infty kmf(m)\,dm = k\langle m \rangle n = kx \qquad (6)$$

[for evaluation of the integral, see Eq. (8.10)].

Example 8.1 illustrates how the population balance can be used to examine the behavior of the single cells in a culture: By comparison of the calculated distribution function with an experimentally determined distribution function, different models for the behavior of the individual

Population Balance Equations

cells can be evaluated. In their analysis, Kothari et al. (1972) used the cell mass distribution function, which can be obtained using a Coulter counter or other size-distribution measurement system. Today it is also possible to obtain distribution functions for many other cellular properties by flow cytometry [see, e.g., Bailey and Ollis (1986) for a description]. With this technique one may measure the single-cell content of macromolecules like proteins, chromosomal DNA, carbohydrates, and plasmid DNA by applying specific fluorescent dyes that label the macromolecular pool of interest. Furthermore, by measuring the accumulation of an intracellular fluorescent product formed by the action of a certain enzyme, it is possible to quantify the cellular content of a specific enzyme. In addition to measurement of the cellular content by fluorescent techniques, it is also possible to measure the cell size by light scattering, and most modern flow cytometers are equipped with both light-scattering and fluorescent-measurement facilities whereby a two-dimensional distribution function for a population can be obtained (see Fig. 8.1).

With measurements of both cell size and cell composition it is possible to calculate the distribution function, based on mass fractions $\psi(X)$ from the two-dimensional distribution function $f(y,m)$ (y is the content of the measured component in grams per cell). Cells with the composition X (grams per gram dry weight) are found on the curve $y=Xm$ in the y-m plane, and the total concentration of cells with a composition X is therefore given by

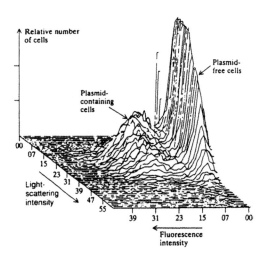

Figure 8.2 Two-dimensional distribution function for a population of recombinant *S. cerevisiae*. The fluorescence intensity is correlated to the content of plasmid DNA, and the light-scattering intensity is correlated to the cell size. It is observed that the plasmid-free and plasmid-containing cells are easily distinguished. [Reprinted by permission from J. E. Bailey and D. F. Ollis (1986), *Biochemical Engineering Fundamentals*, McGraw-Hill.]

$$\psi(X)x = \int_0^\infty mf(m, Xm)\, dm \qquad (8.8)$$

Thus, flow cytometry measurements may also be used to obtain the distribution function based on mass fraction.

The population balance in Eq. (8.2) must satisfy an initial condition describing the state of the population, i.e., $f(y,0)$ should be known. Generally, this is sufficient since the model normally satisfies consistency criteria, i.e., that the flux is zero at the boundaries of the property space. On some occasions, boundary conditions do enter the analysis, but this depends on how the problem is formulated. The solution to the dynamic balance Eq. (8.2) can be found by one of several weighted residual methods, as has been illustrated for a one-dimensional distribution function by Subramanian and Ramkrishna (1971) [see also Ramkrishna (1985) for a discussion of various solution methods for the dynamic population balance].

In many situations it is, however, sufficient to obtain qualities pertaining to the average composition and the standard deviation for the population. These can be calculated from the moments of the distribution functions, where the nth moment of a one-dimensional distribution function is defined by

$$M_n(t) = \int_{V_y} y^n f(y,t)\, dy \qquad (8.9)$$

The zeroth moment is equal to the total number of cells per unit volume [see Eq. (8.1)], and from the first moment the average cell composition can be calculated:

$$\langle y(t) \rangle = \frac{M_1(t)}{M_0(t)} = \frac{\int_{V_y} y f(y,t)\, dy}{n(t)} \qquad (8.10)$$

From the definition of the variance σ^2 of the distribution function,

$$\sigma(t)^2 = \int_{V_y} [y - \langle y(t) \rangle]^2 f(y,t)\, dy \qquad (8.11)$$

we find a relationship between the variance and the second moment of the distribution function:

$$\sigma(t)^2 = \frac{M_2(t)}{n(t)} - \langle y \rangle^2 = \frac{\int_{V_y} y^2 f(y,t)\, dy}{n(t)} - \langle y \rangle^2 \qquad (8.12)$$

The definitions of the moments can easily be extended to the case of a multidimensional distribution function. Thus, for a two-dimensional property space (see also Example 8.4),

Population Balance Equations

$$M_{ij}(t) = \int_{V_{y_1}} \int_{V_{y_2}} y_1^i y_2^j f(y_1, y_2, t)\, dy_1\, dy_2 \tag{8.13}$$

Besides its application to single-cell populations, the population balance of Eq. (8.2) may also be used for many other systems [see Ramkrishna (1979) for a review of population balances], and typical applications in connection with bioprocesses are

1. Single-cell populations (illustrated in Examples 8.1, 8.2, and 8.3).
2. Populations of hyphal elements in connection with cultures of filamentous microorganisms (Example 8.4).
3. Populations of pellets in cultures of filamentous microorganisms and immobilized cells.
4. Description of the bubble-size distribution in the liquid during aerated fermentation processes.

For processes where agglomeration is involved instead of breakage (or cell division), the function $h(\mathbf{y},t)$ is not given by Eq. (8.7) [see Ramkrishna (1985) for details]. In the following, we illustrate the application of the population balance of Eq. (8.2) for some microbial systems.

Example 8.2 Population balance for recombinant *E. coli*

Seo and Bailey (1985) examined the distribution of plasmid content in recombinant *E. coli* cultures by means of a population balance. They first considered the steady-state age distribution of the cells:

$$\frac{d[r(a)f(a)]}{da} = h(a) - Df(a) \tag{1}$$

With age being the considered cell property, we have (by definition) that $r(a) = 1$ and

$$h(a) = 0; \quad a < t_d \tag{2}$$

where t_d is the length of the cell cycle (assumed to be constant in the model). The balance of Eq. (1) therefore reduces to

$$\frac{df(a)}{da} = -Df(a) \tag{3}$$

At cell division it is assumed that there is a certain probability θ for the formation of a plasmid-free cell (the *segregation-parameter*), and a cell balance relating newborn cells to the dividing cells (the so-called renewal equation) therefore gives

$$f(0) = (2 - \theta) f(t_d) \tag{4}$$

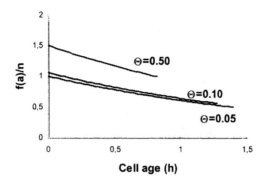

Figure 8.3 The age distribution for recombinant cells at $D = 0.5$ h^{-1} and for different values of the parameter θ.

The solution to the balance in Eq. (3), subject to Eqs. (4) and (8.1) is

$$f(a) = \frac{2-\theta}{1-\theta} Dn\, e^{-Da} \qquad (5)$$

Equation (5) gives the steady-state age distribution for plasmid-containing cells, and the distribution is observed to be a function of the specific growth rate and the segregation parameter θ (see Fig. 8.3). The doubling time t_D is also a function of the segregation parameter:

$$t_D = \frac{\ln(2-\theta)}{\mu} \qquad (6)$$

Thus, for a given specific growth rate μ of the culture it is seen that the cell cycle time decreases when the segregation parameter increases, i.e., the cells have to speed up their growth rate in order to compensate for the loss of a certain category of cells at cell division.

In order to find the distribution of plasmid content in the population of recombinant cells, it is necessary to specify the distribution of plasmids to the two daughter cells upon cell division and the rate of plasmid synthesis in the cells. First, the age distribution of cells with p_b plasmids at birth has a form similar to Eq. (5), i.e.,

$$f_b(a) = c_b n\, e^{-Da} \qquad (7)$$

where c_b is determined by the applied model for the distribution of plasmids to the daughter cells. Three models for c_b were examined by Seo and Bailey:

- **Model 1.** In this model it is assumed that plasmids are distributed to daughter cells randomly at cell division and that all cells contain N plasmids at cell division.
- **Model 2.** The assumptions underlying this model are similar to those for Model 1, but here it is

Population Balance Equations

furthermore assumed that $2M$ plasmids are evenly distributed to the daughter cells, whereas the rest are distributed at random.
- **Model 3.** In this model it is assumed that all the plasmids are distributed randomly and that K plasmids are synthesized for each cell cycle.

Obviously, Model 2 does not result in instability of the recombinant culture, since each daughter cell always receives M plasmids from the dividing cell. From the age distribution specified for each subpopulation of cells with p_b plasmids at birth, it is possible to find the distribution of plasmid content p in cells with p_b plasmids at birth:

$$f_b(p) = f_b(a) \left| \frac{da}{dp} \right| \tag{8}$$

By assuming zeroth-order kinetics for the plasmid replication within each cell, the relationship between the plasmid content and the cell age is given by

$$p = p_b + ca \tag{9}$$

where c is the plasmid replication rate constant. Combining Eqs. (7), (8), and (9), we find

$$f_b(p) = f_b(a)\frac{1}{c} = c_b \frac{1}{c} \exp\left(-D\frac{p-p_b}{c}\right) \tag{10}$$

Finally, the distribution function for all plasmid-containing cells is found from summation of all the subpopulations, i.e.,

$$f(p) = \sum_i f_i(p) \tag{11}$$

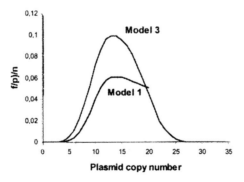

Figure 8.4 The calculated distribution of cells with varying plasmid copy number for Model 1 and Model 3.

With this model Seo and Bailey calculated the plasmid content distribution for the population of recombinant cells using Models 1 and 3. The results are shown in Fig. 8.4. For Model 1, there is a distinct maximum plasmid copy number of $N = 20$, whereas for Model 3 there is broad distribution of plasmid copy number in the population.

Using the plasmid copy number distribution found from the above model, Seo and Bailey calculated the productivity of the recombinant culture, i.e., the rate of formation of the recombinant product. For the individual cells the product formation kinetics was taken to be

$$r_e(p) = k_e p \left(1 - \frac{p}{p_{max}}\right) \tag{12}$$

which is an analogue of Eq. (7.64). The total productivity for the culture is

$$q_e = \int_0^\infty f(p) r_e(p)\, dp \tag{13}$$

and with the plasmid copy number distribution in Fig. 8.4, q_e was found to have a maximum when it is specified as a function of the average plasmid copy number (calculated as the first moment of the distribution function). Thus, for production of a recombinant protein there is an optimal average plasmid copy number for the population.

Finally, Seo and Bailey examined the instability of the recombinant population by using the dynamic age distribution for two subpopulations: (1) plasmid-containing cells and (2) plasmid-free cells. Depending on the parameters N and K in Models 1 and 3, the stability of the recombinant culture was examined. Obviously the stability increases for increasing values of both parameters.

The Seo and Bailey model represents a detailed analysis of recombinant cultures, and it has served as a useful guide for setting up simple models where homogeneity in the cell population is assumed.

Example 8.3 Age distribution model for *S. cerevisiae*
The asymmetric cell division of budding yeast (see Fig. 7.16) has been modeled by Hjortso and Bailey (1982). Their model is based on an age distribution. Cells having an age smaller than $a_1 = t_d$ are defined as daughter cells, and cells with an age larger than a_1 are called mother cells. Since cell division occurs only at $a = a_1 + a_2$ (where $a_2 = t_m + t_b$), a balance similar to that of Eq. (3) of Example 8.2 holds for the steady-state age distribution:

$$\frac{df(a)}{da} = -Df(a) \tag{1}$$

In their model Hjortso and Bailey used a normalized distribution function

$$\phi(a) = \frac{f(a)}{n} \tag{2}$$

Population Balance Equations

The zeroth moment of this distribution function is equal to 1, and the first moment is the average cell age for the population. The balance for the normalized distribution function is similar to Eq. (1):

$$\frac{d\phi(a)}{da} = -D\phi(a) \tag{3}$$

The cell balances relating to cell division, i.e., the renewal equations, are

$$\phi(0) = \phi(a_1 + a_2) \tag{4}$$

$$\phi(a_1^+) = \phi(a_1^-) + \phi(a_1 + a_2) \tag{5}$$

The solution to the balance in Eq. (3) with the boundary conditions of Eq. (4)-(5) and the normalization of the distribution function is given by

$$\phi(a) = \begin{cases} D(e^{Da_2} - 1)\, e^{-Da}; & 0 < a < a_1 \\ D\, e^{D(a_2 - a)}; & a_1 < a < a_1 + a_2 \end{cases} \tag{6}$$

where

$$a_1 = -\frac{1}{D}\ln(e^{Da_2} - 1) \tag{7}$$

From the age distribution function it is possible to calculate other distribution functions by using Eq. (8). This equation is a generalization of Eq. (8) in Example 8.2:

$$\phi(w) = \phi(a(w))\left|\frac{da(w)}{dw}\right| \tag{8}$$

$$\phi(a) = \begin{cases} \dfrac{D}{r(w)} \exp\left(-D\displaystyle\int_{w_0}^{w} \frac{dy}{r(y)}\right); & w_0 < w < w_1 \\ \dfrac{D\, e^{Dt_2}}{r(w)} \exp\left(-D\displaystyle\int_{w_1}^{w} \frac{dy}{r(y)}\right); & w_1 < w < w_2 \end{cases} \tag{9}$$

where $\phi(w)$ is a distribution function for another characteristic cellular variable w, e.g., cell mass. If w is synthesized at a rate $r(w)$ which is independent of cell age one finds that $\phi(w)$ is given by Eq. (9). w_0, w_1, and w_2 are values of w for cells with an age of respectively 0, a_1, and $a_1 + a_2$.

With w being the mass m of the individual cell, $r(w)$ is given by Eqs. (10) and (11) for, respectively, first- and zeroth-order growth kinetics for the single cell.

$$r(m) = km \tag{10}$$

$$r(m) = k \tag{11}$$

With these two models for the single-cell growth rate the steady-state distribution function $\phi(w)$ can be calculated, and the results are shown in Fig. 8.5. It is observed that the distribution function is completely different in the two models, and with measurements of the cell size distribution (e.g., by flow cytometry), it should be possible to distinguish between the two models.

With the distribution function in Eq. (9), one can calculate the number fraction of daughter and mother cells, respectively, (here termed Z^* in order to avoid confusion with the mass fraction Z used in Section 7.6):

$$Z_d^* = \int_0^{a_1} \Phi(a)da = 2 - e^{\mu a_2} \tag{12}$$

$$Z_b^* = \int_{a_1}^{a_1+a_2} \Phi(a)da = e^{\mu a_2} - 1 \tag{13}$$

The length t_b of the budding phase is often considered to be independent of the specific growth rate ($t_b \approx 1.8$ h), even though it decreases slightly with μ (see Fig. 8.6). If the same holds for the time for maturation, i.e., t_m approximately constant, then it is concluded that a_2 is independent of the specific growth rate. Thus from Eq. (13) the fraction of mother cells in the culture increases with the specific growth rate. This corresponds well with experimental data given by Lievense and Lim (1982), who found that the budding index, i.e., the fraction of budded cells, increases with D in a steady-state chemostat (Fig. 8.6).

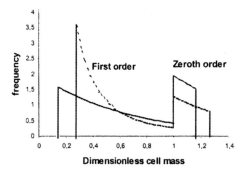

Figure 8.5 Cell mass distribution function with, respectively, first- and zeroth-order growth kinetics of the individual cells. The model parameters are: $D = 0.2$ h^{-1}; k (zeroth) = 0.13 and k (first) = 0.2. The cell mass has been rendered dimensionless by w_1, which is set to 1.

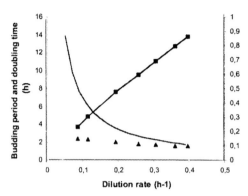

Figure 8.6 Measurements of the length of the budding period t_d (■) and the fraction of budded cells Z_b^* (▲) as functions of the dilution rate in a steady-state chemostat. Assuming that $a_2 \approx t_b$, the fraction of budded cells is calculated using Eq. (13) (shown as line). Also shown is $t_D = \ln 2/D$. The data are taken from Lievense and Lim (1982).

From Fig. 8.6, it is observed that the doubling time approaches the length of the budding period for high specific growth rates. Consequently, t_d decreases for increasing μ - showing that the size of the daughter cells resulting from the cell division increases with μ, and for high specific growth rates the cell division becomes almost symmetric (two cells of almost equal size are formed).

Due to the formation of a bud scar on the cell envelope of the mother cell upon cell division it is expected that mother cells would have a maximum age. Hjortso and Bailey (1982) introduced the concept of the genealogical age of mother cells and calculated the distribution of cells with varying numbers of bud scars. Thereby the effect of various hypotheses concerning the growth ability of mother cells with many bud scars could be examined. Hjortso and Bailey (1983) also carried out experiments involving transient situations by using the dynamic balance Eq. (14), which is derived from Eq. (8.2) with $h(a)=0$ for all $a \neq a_1 + a_2$ and $r(a) = 1$.

$$\frac{\partial \phi(a)}{\partial t} + \frac{\partial \phi(a)}{\partial a} = -D\phi(a) \qquad (14)$$

The dynamic balance was solved by the method of characteristics and again the influence of changing the single-cell kinetics from zeroth to first order was examined.

Hjortso and Bailey (1984a) extended their segregated population model to predict plasmid stability at steady-state growth. They assumed that the culture is under selection pressure, whereby only plasmid-containing cells can survive in the environment. The population balance of Eq. (3) still holds, but the cell balances relating to cell division in Eqs. (4) and (5) are modified to those of Eqs. (15) and (16) in order to account for plasmid loss. θ_m and θ_d (the segregation parameters) are the probabilities of formation of, respectively, a plasmid-free mother and a plasmid-free daughter cell at cell division.

$$\Phi(0) = (1-\theta_d)\Phi(a_1+a_2) \tag{15}$$

$$\Phi(a_1^+) = \Phi(a_1^-) + (1-\theta_m)\Phi(a_1+a_2) \tag{16}$$

With these boundary conditions, the steady-state solution of the population balance becomes

$$\phi(a) = \begin{cases} D\dfrac{e^{D(a_1+a_2)}-1+\theta_m}{1-\theta_d-\theta_m}e^{-Da}; & 0 < a < a_1 \\ D\dfrac{e^{D(a_1+a_2)}}{1-\theta_d-\theta_m}e^{-Da}; & a_1 < a < a_1+a_2 \end{cases} \tag{17}$$

where

$$a_1 = \frac{1}{D}\ln\left(\frac{1-\theta_d}{e^{Da_2}-1+\theta_m}\right) \tag{18}$$

From Eq. (18E6.3), it is observed that the length of the unbudded period decreases with increasing values of θ_m and θ_d. The reason is that when a large portion of the cells loses their plasmid at cell division (and thereby die due to selection pressure from the environment), the culture has to speed up its specific growth rate at the cellular level. This is observed as formation of larger daughter cells during the budding period, and the length of the unbudded period for the daughter cells is therefore shortened. The decrease in a_1 results in a narrower age distribution, with a larger fraction of the cells being mother cells.

In the model above, only the age distribution for the plasmid containing cells is considered. However, for prediction of productivity of a recombinant product one needs information on the plasmid copy number distribution, and Hjortso and Bailey therefore extended their model to describe the plasmid copy number distribution (by an approach similar to that illustrated in Example 8.2). Two models for plasmid replication during the cell cycle were examined: (1) Sufficient plasmids are synthesized during the cell cycle to ensure that the copy number is always N at cell division, and (2) N plasmids are synthesized during the cell cycle. The plasmid copy number distribution was calculated, and the distribution is quite different for the two models (one being bimodal and the other looking qualitatively like an exponential decay).

Finally, Hjortso and Bailey (1984b) extended their model for the recombinant yeast to dynamic conditions and examined a shift to a nonselective medium. In a chemostat, the fraction of the population that contains plasmids approaches zero asymptotically as growth proceeds. The decrease in the fraction of plasmid-containing cells depends on the value of a_1, which is a function of the specific growth rate. For any $a_1 > 0$, asymmetric division results in a faster decrease in the fraction of plasmid-containing cells than does binary fission. This shows that the asymmetric division of budding yeast results in a different behavior than that observed for recombinant bacteria.

Example 8.4 Population model for hyphal elements
In submerged cultures of filamentous fungi, there is a population of hyphal elements, which may be characterized by many different properties (e.g., their length and the number of tips), but the relative content

Population Balance Equations

of different cell types may also be important (see Section 7.6.2). We now assume that each hyphal element is characterized completely by its total length (l) and the number of actively growing tips (n); i.e., $f(l,n)\,dl\,dn$ is the number of hyphal elements with length l and n actively growing tips. The hyphal diameter is normally constant (at least for certain environmental conditions) and similarly holds for the hyphal density. The hyphal length is therefore proportional with the hyphal mass, and the total tip extension rate for the hyphae is therefore given by the $\mu(l,n,t)l \cdot n$ is in reality an integer, but it is here taken to be a real number. The length of the hyphal element increases due to growth with the specific rate $\mu(l,n,t)$, and new actively growing tips are formed due to branching with the frequency $\phi(l,n,t)$. It is now assumed that both the specific growth rate of the hyphal element and the branching frequency are independent of the hyphal element properties, i.e., they are not functions of l and n. Consequently, the rate of change of length is $\mu(t)l$ and the rate of change of actively growing tips is $\phi(t)$. The dynamic mass balance for the distribution function $f(m,n)$ is therefore given by

$$\frac{\partial f(l,n,t)}{\partial t} + \frac{\partial}{\partial l}\left[\mu(t)lf(l,n,t)\right] + \frac{\partial}{\partial n}\left[\phi(t)f(l,n,t)\right] = h(l,n,t) - Df(l,n,t) \tag{1}$$

The dynamic mass balance is illustrated in Fig. 8.7, where it is shown that for a given control volume, i.e. the distribution function, there are inputs and outputs due to tip extension and branching. Furthermore, there are inputs due to formation of new hyphal elements by fragmentation and spore germination. Finally there are hyphal elements leaving the control volume due to hyphal fragmentation and washout.

We now consider growth of a mycelium in an agitated tank where shear stress causes the hyphae to break up (hyphal fragmentation). With binary fission of hyphal elements, the net rate of formation of hyphal elements with property $\{l,n\}$ formed upon fragmentation is given by Eq. (8.7), i.e.,

$$h(l,n,t) = 2\int_{V_m}\int_{V_n} b(l^*,n^*,t)p(\{l^*,n^*\},\{l,n\})f(l^*,n^*,t)dn^*dl^* - b(l,n,t)f(l,n,t) \tag{2}$$

$b(l,n,t)$ is the breakage function, which describes the rate of fragmentation of hyphal elements with property $\{l,n\}$, and $p(\{l^*,n^*\},\{l,n\})$ is the partitioning function.

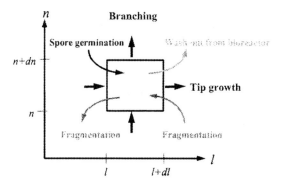

Figure 8.7 Illustration of the different elements in the population balance equation for individual hyphal elements in a filamentous fungal culture.

Within the population of hyphal elements there will be more or less fragmentation at different positions in the individual hyphal elements. This fragmentation occurs when the local shearing forces become larger than the tensile strength of the hyphal wall (van Suijdam and Metz, 1981). The tensile strength of the hyphal wall depends on the morphological state of the individual cell in the hyphal element, e.g., the tensile strength of the hyphal cells could be smaller than that of apical cells, but here it is assumed that it is constant and independent of the state of the hyphae. The breakage function and the partitioning function are therefore both taken to be independent of the morphological state of the individual cells.

In a culture with a very large number of hyphal elements, the average ratio of hyphal length to the number of actively growing tips in the two "daughter" fragments is identical, i.e., $\langle l \rangle / \langle n \rangle = \langle l^* - l \rangle / \langle n^* - n \rangle$, if fragmentation occurs with equal probability at any position on the hyphal elements. It is therefore assumed that $l/n = (l^* - l)/(n^* - n)$ (or $l/n = l^*/n^*$), whereby the partitioning function can be stated as a function of l^* and l only. Furthermore, since the tensile strength of the hyphal wall is assumed to be constant, there is an equal probability of fragmentation at any position in the hyphal element. The partitioning function is therefore given by

$$p(\{l^*,n^*\},\{l,n\}) = p^*(l^*,l) = \begin{cases} 1/l^* & \text{if } l^* > l \text{ and } l/n = l^*/n^* \\ 0 & \text{in all other cases} \end{cases} \quad (3)$$

The breakage function specifies the rate of fragmentation, and this is taken to be a linear function of the total length of the hyphal element, i.e.,

$$b(l,n,t) = \psi(t)l \quad (4)$$

$\psi(t)$ is the specific rate of fragmentation, which is here taken to be a function of the environmental conditions only. $\psi(t)$ is determined by the number of times the hyphal elements enter the zones where the local shearing forces are larger than the tensile strength of the hyphal element, e.g., the impeller zone, and it is therefore determined both by the circulation pattern and the shear force distribution in the bioreactor. However, in a simple model describing growth in a stirred tank reactor, one can assume that $\psi(t)$ is a function only of the energy input (which can be calculated from the stirring speed), i.e., it is constant for constant stirring speed (van Suijdam and Metz, 1981).

By inserting the partitioning function of Eq. (3) and the breakage function of Eq. (4) in Eq. (2) we get

$$h(l,n,t) = 2\psi(t) \int_l^\infty f(l^*,n^*,t)dl^* - \psi(t)lf(l,n,t) \quad (5)$$

In order to solve the dynamic balance of Eq. (1) for the distribution function, it is necessary to specify proper boundary conditions, and these are given in Eqs. (6)-(8). Fragmentation never results in the formation of hyphal elements with zero length or no actively growing tips, and the boundary conditions in Eq. (6) therefore hold. Since the breakage function is linearly dependent on the length of the hyphal element, no hyphal elements have a infinite length. Furthermore, due to the assumption $l^*/n^* = l/n$ there are also no hyphal elements with an infinite number of tips. Finally, $f_0(m,n)$ specifies the distribution function at time $t = 0$:

$$f(0,n,t) = 0 \; ; \quad f(m,0,t) = 0 \tag{6}$$

$$\lim_{l \to \infty} f(l,n,t) = 0 \; ; \quad \lim_{n \to \infty} f(l,n,t) = 0 \tag{7}$$

$$f(m,n,0) = f_0(m,n) \tag{8}$$

Solving the dynamic balance for the distribution function is complicated, since the property space is two dimensional and since an integration step is involved in the calculation of $h(l,n,t)$. For comparison of the population model with measurements obtained in a continuous bioreactor at steady state, it is, however, sufficient to find the steady-state solution of the population balance. For a population balance with a one-dimensional property space Singh and Ramkrishna (1977) used the method of weighted residuals to find the steady-state solution. This method could probably be extended to find the steady-state solution for the population balance derived here. Alternatively one may introduce discrete variables and then solve differential equations for each discrete variable (Krabben et al., 1997). However, an institutively simpler approach is to apply Monte-Carlo simulations where a large number of individual hyphal elements are simulated, and the elements may then together form the distribution function. Krabben et al. (1997) applied this approach to simulate the distribution function for different fragmentation kinetics, i.e. different partitioning functions and different breakage functions. It is practically impossible to obtain sufficient experimental data – even with fully automated image analysis, on the distribution function in two dimensions to validate different models. Krabben et al. (1997) therefore took a different approach. Based on the simulated distribution functions they generated contour plots for the function. These contour plots were then statistically used to evaluate experimentally obtained data for the hyphal element properties. Hereby it was possible to evaluate the different models, and based on this it was concluded that hyphal fragmentation only takes place for hyphal elements above a certain size, and then follows second order kinetics. It was, however, not possible to discriminate between two models for the partitioning function, i.e. whether there is largest probability for fragmentation at the centre of the hyphae or whether there is equal probability for fragmentation throughout the hyphae.

Even though the Monte Carlo simulations offer a simple method for simulation of even complex models, it is often sufficient to look at the average properties of the hyphal elements, as these may be directly compared with measured data. Using the balance for the distribution function of above Nielsen (1993) derived dynamic balances for these variables for the average hyphal length and the average number of tips. These balances are given by:

$$\frac{de}{dt} = (\psi - D)e \tag{9}$$

$$\frac{d\langle l \rangle}{dt} = (\mu - \psi)\langle l \rangle = \varphi \langle n \rangle - \psi \langle l \rangle \tag{10}$$

$$\frac{d\langle n \rangle}{dt} = \phi - \psi \langle n \rangle \tag{11}$$

where φ is the average tip extension rate for all the hyphal tips in the population. Using these balances it is possible to obtain information about the growth kinetics, i.e. the branching frequency and the tip extension rate from measurements of the average total hyphal length and the average number of tips. The total biomass concentration x is given by $x = \langle l \rangle e$, and combining the dynamic balances of Eqs. (9) and (10), the well-known mass balance for biomass is found:

$$\frac{dx}{dt} = (\mu - D)x \qquad (12)$$

With the technique of image analysis, it is possible to determine experimentally the average properties of the hyphal elements, and the simple model based on average properties may be valuable for extracting the growth kinetics of hyphal elements. The variance of the properties of the hyphal element is, however, normally quite large, i.e., the relative standard deviation of the hyphal length is often 50% or more of the average values. The estimation of the average properties should therefore be based on a large number of single estimations, i.e., many individual hyphal elements have to be measured. The variance of the distribution function can be calculated from Eq. (8.11), and if an explicit expression for the variance could be derived the model predicted variance could be compared with the experimentally determined variance. It is, however, not possible to derive simple expressions to be used in the calculation of the variances, and the model is therefore evaluated only by comparison with experimental data for the average properties. It should, however, be noted that the kinetics derived based on the above model is based on average properties, and it may therefore not necessarily specify something about the kinetics of individual hyphal elements. Here it is necessary to study the growth kinetics of individual hyphal elements in a growth chambers positioned under a microscope equipped with an image analysis system. Using such a system the growth kinetics of *Aspergillus oryzae* has been studied in great detail (Spohr et al., 1998; Christiansen et al., 1999).

Nielsen (1993) compared the population model derived above with experimental data for the total hyphal length and the number of tips obtained during fermentations with *P. chrysogenum*. The model was also used to examine the influence of the energy input on the morphology of *P. chrysogenum* by comparing model simulations with experimental data given by Metz et al. (1981) and van Suijdam and Metz (1981) [the data were obtained both during a batch fermentation and in a steady-state chemostat (D =0.05 hr^{-1})]. In Fig. 8.8, data for the total hyphal length are shown as a function of the energy input for the two cases. The data for the batch fermentation are the measurements obtained in the exponential growth phase, where μ =0.12 hr^{-1}. The energy input E (units of W L^{-1}) was calculated from the stirrer speed as specified by van Suijdam and Metz (1981). By assuming the linear relation in Eq. (13) between the specific rate of fragmentation and the energy input, the total hyphal length was calculated from the steady-state form of the balance for $\langle l \rangle$ in Eq. (10) and a hyphal diameter of 3.6 μm (Metz et al., 1981).

$$\psi = 2.1 \cdot 10^6 \cdot E + 50.0 \cdot 10^6 \qquad (13)$$

The model is observed to correspond well with the experimental data, except for the two measurements at very low energy input in the chemostat. Considering the data scatter, and the presence of only two measurements in this range, it is reasonable to conclude that the rate of fragmentation is linearly correlated to the energy input (note that the same correlation holds for the two sets of independent experimental data).

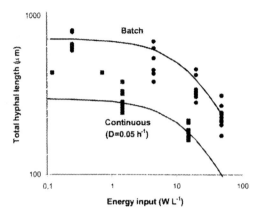

Figure 8.8 Effect of energy input (W L^{-1}) on the total hyphal in μm of *P. chrysogenum* in a submerged culture. The two data sets (Mertz, 1976) are from a series of batch fermentations (batch) and from a series of chemostat experiments (continuous, $D = 0.05$ h^{-1}). The lines are model simulations.

PROBLEMS

Problem 8.1 Derivation of single-cell mass distribution functions
Consider an organism for which division occurs at the cell mass M, birth at the cell mass $M/2$, and the single-cell mass growth rate follows first-order kinetics, i.e., $r(m) = km$.

a. Find the normalized, steady-state cell mass distribution $\phi(m)$ for this organism in a chemostat with dilution rate D.

b. Find the relation between D and k, and use it to eliminate k from the expression for $\phi(m)$.

c. Find the distribution function for zeroth-order kinetics (i.e., $r(m) = k$), and compare the distribution functions for, respectively, zeroth- and first-order kinetics.

d. Assuming a constant yield Y_{sx} of mass per individual cell from the substrate, write a steady-state substrate balance and simplify this to obtain an equation among D, k, and n (the total cell number).

e. Assume that the substrate dependence of k follows a Monod-type expression. Find the substrate and cell number concentrations as functions of the dilution rate.

Problem 8.2 Linear single-cell kinetics
Consider an organism for which division occurs at cell mass M, birth at cell mass $M/2$, and the single-cell mass growth rate follows $r(m) = k_1 m + k_2$.

a. Find the normalized, steady-state cell mass distribution $\phi(m)$ for this organism in a chemostat with dilution rate D.

b. Find the relation among D, k_1, and k_2.

c. Because there is more than one parameter in the rate expression for the single-cell kinetics, it is not possible to use the last result to simplify the expression for the distribution of states and write it solely as a function of the dilution rate in the chemostat. There are simply not enough equations to solve for k_1 and k_2 in terms of D. This problem can be resolved if one postulates kinetic expressions for both parameters in terms of the concentration of the limiting substrate. From these equations one can eliminate the substrate concentration and obtain one of the parameters in terms of the other. This result can then be used with the expression obtained in (a) to find all the parameters in terms of the dilution rate. Demonstrate this procedure for the model, above assuming that

$$k_1 = \frac{k_{1,\max} s}{s + K} \quad (1)$$

$$k_2 = \frac{k_{2,\max} s}{s + K} \quad (2)$$

where s is the concentration of the growth-limiting substrate and the parameters $k_{i,\max}$ and K must be determined from experiments. Express k_2 as a function of k_1 and substitute the result into the expression obtained in (b), solve for k_1 as a function of D, and eliminate k_2 and k_1 from the expression for $\phi(m)$.

Problem 8.3 Continuous plant cell cultures

For plant cell cultures it has been found experimentally that the cell concentration in the outlet stream from a chemostat is less than the concentration in the vessel. This is caused by the large size of plant cells, which occasion them to sediment out of the outlet stream and back into the vessel. Of course, this effect is more pronounced for large cells than for small cells, and the phenomenon can therefore be expected to affect the cell mass distribution.

We can model this by assuming that the cell mass distribution in the outlet stream can be calculated from the distribution in the vessel by the following expression:

$$\phi_{outlet}(m) = \phi_{vessel}(m) \frac{a}{m + a} \quad (1)$$

where a is a constant parameter.

Derive a population balance equation for this situation, and find the normalized steady-state cell mass distribution in the vessel, assuming that cells divide when they reach the cell mass M, cells are born with cell mass $M/2$, and the cell mass growth rate follows zeroth-order kinetics.

Problem 8.4 Cell death

Consider an organism which, when it attains the age a_d either dies (with a probability θ) or divides.

a. Write the cell balance over dividing cells and solve for the normalized, steady-state age

Population Balance Equations

distribution, $\phi(a)$, in a chemostat with dilution rate D.

b. Show that the doubling time, defined as the duration of the cell cycle a_d, does not equal the doubling time defined on the basis of the specified growth rate of the population.

Problem 8.5 Derivation of conversion rates in a yeast model

Cazzador (1991) evaluated the parameter k_u and k_b in his morphologically structured model (see Section 5.2 and Problem 5.3). In this evaluation he assumed
- The critical mass for budding is assumed to be constant and equal to 1, i.e., $m = 1$.
- The critical mass for cell division is assumed to be a known function of the substrate concentration $m_{div}(s)$ such that $1 < m_{div} < 2$.
- The size of the mother cells after division is assumed to be $m = 1$ and that of newborn daughters $m = m_{div} - 1$.
- The specific growth rate of the single cell is assumed to be μ_u in the first (unbudded) phase and μ_b in the second (budded) phase of the cell cycle.

a. Write the steady-state balances for the distribution function of cell mass for cells in the two phases. Specify the renewal equations.

b. Solve the balances for the distribution functions.

c. The transfer rates between the two morphological forms can be expressed as

$$k_u x Z_u = \mu_u f_u(1) \tag{1}$$

$$k_b x Z_b = (m_{div} - 1)\mu_b m_{div} f_b(m_{div}) \tag{2}$$

Use these equations to derive expressions for k_u and k_b as functions of m_{div} (which again is a function of the limiting substrate concentration) in the two cases

i. $\mu_u = \mu_b = \mu$
ii. $\mu_u \neq \mu_b$

d. Now let m_{div} and μ be given as

$$m_{div} = 1.2 + 0.5\frac{s^8}{s^8 + 10^8} \tag{3}$$

$$\mu = 0.3\frac{s}{s+4} \tag{4}$$

Plot k_u and k_b as functions of the limiting substrate concentration s. Simulate using a PC the dynamic equations for Z_u, Z_b, s (see Problem 7.4), and x in a chemostat operated with $D = 0.2$ h^{-1}, $s_f = 400$ g L^{-1}, $a_u = 10$, and $a_b = 2$. Show that a stable limit cycle is obtained.

REFERENCES

Bailey, J. E. and Ollis, D. F. (1986). *Biochemical Engineering Fundamentals,* 2nd ed., McGraw-Hill, New York.

Cazzador, L. (1991). Analysis of oscillations in yeast continuous cultures by a new simplified model, *Bull Math. Biol.* **5**, 685-700.

Christiansen, T., Spohr, A., Nielsen, J. (1999). On-line study of growth kinetics of single hyphae of *Aspergillus oryzae* in a flow-through cell. *Biotechnol. Bioeng.* **63**, 147-153

Hjortsø, M. A. and Bailey, J. E. (1982). Steady-state growth of budding yeast populations in well-mixed continuous-flow microbial reactors, *Math. Biosci.* **60**, 235-263.

Hjortsø, M. A. and Bailey, J. E. (1983). Transient responses of budding yeast populations, *Math. Biosci.* **63**, 121-148.

Hjortsø, M. A. and Bailey, J. E. (1984a). Plasmid stability in budding yeast populations: Steady state growth with selection pressure. *Biotechnol. Bioeng.* **26**, 528-536.

Hjortsø, M. A. and Bailey, J. E. (1984b). Plasmid stability in budding yeast populations: Dynamics following a shift to nonselective medium. *Biotechnol. Bioeng.* **26**, 814-819.

Kothari, I. R., Martin, G. C., Reilly, P. J., Martin, P. J., and Eakman, J. M. (1972). Estimation of parameters in population models for *Schizosaccharomyces pombe* from chemostat data. *Biotechnol. Bioeng.* **14**, 915-938.

Eakman, J. M., Fredrickson, A. C., and Tsuchiya, H. M. (1966). Statistics and dynamics of microbial cell populations. *Chem. Eng. Prog. Symp. Ser.* **62**, 37-49.

Krabben, P., Nielsen, J., Michelsen, M. L. (1997). Analysis of single hyphal growth and fragmentation in submerged cultures using a population model. *Chem. Eng. Sci.* **52**, 2641-2652

Kreyszig, E. (1988). *Advanced Engineering Mathematics,* 6th ed., John Wiley & Sons, New York.

Lievense, J. C. and Lim, H. C. (1982). The growth and dynamics of *Saccharomyces cerevisiae, Ann. Report Ferm. Proc.* **5**, 211 - 262.

Metz, B. (1976). *From Pulp to Pellet,* Ph.D. thesis. Technical University of Delft, Delft.

Metz, B., Bruijn, E. W., and van Suijdam, J. C. (1981). Method for quantitative representation of the morphology of molds. *Biotechnol. Bioeng.* **23**, 149 - 162.

Nielsen, J. (1993). A simple morphologically structured model describing the growth of filamentous microorganisms. *Biotechnol. Bioeng.* **41**, 715 - 727.

Ramkrishna, D. (1979). Statistical models for cell populations. *Adv. Biochem. Eng.* **11**, 1 - 48.

Ramkrishna, D. (1985). The status of population balances. *Rev. Chem. Eng.* **3**, 49 - 95.

Seo, J.-H. and Bailey, J. E. (1985). A segregated model for plasmid content on growth properties and cloned gene product formation in *Escherichia coli*. *Biotechnol. Bioeng.* **27**, 156 - 166.

Singh, P. N. and Ramkrishna, D. (1977). Solution of population balance equations by MWR. *Comp. Chem. Eng.* **1**, 23 - 31.

Spohr, A. B., Mikkelsen, C. D., Carlsen, M., Nielsen, J., Villadsen, J. (1998). On-line study of fungal morphology during submerged growth in a small flow-through cell. *Biotechnol. Bioeng.* **58**, 541-553

Subramanian, G. and Ramkrishna, D. (1971). On the solution of statistical models of cell populations. *Math. Biosci.* **10**, 1 - 23.

van Suijdam, J. C. and Metz, B. (1981). Influence of engineering variables upon the morphology of filamentous molds. *Biotechnol. Bioeng.* **23**, 111 - 148.

9

Design of Fermentation Processes

In all the previous chapters the bioreactor has been invisibly present, waiting in the wings for a cue to enter central stage. The discussion of rate measurements in chapters 3 and 5 and of kinetics in chapters 6 to 8 would not have been meaningful without the short introduction to basic mass balancing for stirred tank reactors in steady state (chapters 3 and 5) and extended to include also transients in Chapter 7. Now the bioreactor itself will be given full attention. The kinetics of Chapter 7 will be used as part of the description of the behavior of typical reactor configurations used in the laboratory and in industry. Steady state- and transient operation of bioreactors will be the main subject of the present chapter, and optimization problems, typical topics of texts on chemical reaction engineering, will be treated. In particular we shall solve the mass balances for the most popular reactor configurations and look for the maximum productivity $(q_p)_{max}$ rather than for the maximum specific productivity $(r_p)_{max}$. Also optimal start-up procedures for bioreactors are discussed, and it will be shown how stability of the process can be assured and how infection of the culture or spontaneous mutation of the producing strain will influence the outcome of the process.

The bioreactor is still claimed to be "ideal" following the definition given in Section 3.1. There is no shunt of substrate from inlet to outlet, no dead zones or clumps of undissolved solid substrate floating around. A drop of substrate is instantaneously distributed throughout the entire reactor volume, and the sparger provides an intimately mixed gas-liquid medium with no air bubbles sliding up along the reactor wall. Some laboratory reactors approach the ideal. Mixing time is on the order of 1-2 s, and the gas-liquid mass transfer rate is very high [see, e.g., Sonnleitner and Fiechter (1988)]. These reactors may be abundantly equipped with on-line measuring and control systems, and one is able to follow the effect of steep transients imposed on the microbial environment. These units, the true bioreactors, are used for scientific investigations, to learn more about the cell metabolism, and to study the cell as the ultimate biochemical reactor. Other experiments are carried out - often without involvement of an actual fermentation - in the equipment that is going to be used for industrial production. Here the interaction between mechanical devices such as agitators, draught tubes, static mixers with or without corrugated surfaces, and a fluid of given properties can be studied. The outcome is a series of time constants for mixing, for circulation, for gas-to-liquid transport and the like. As discussed in Chapter 11, it may be hoped that both the chemical interactions between the cell and its microenvironment, and the physical interactions between the cell or the cell culture and the macroenvironment will eventually be clarified in enough detail to allow a new bioprocess to be designed with only minimal scale-up problems.

Since the focus is now on the reactor rather than on the bioreaction we shall preferably use kinetic models that are quite simple. The Monod model and its progeny (Section 7.3.1) is used most of the time both when steady state and non-steady state operation of the bioreactor is discussed. In Section 7.3 it was emphasized that the Monod model is unsuited for description of cell behaviour during rapid transients, but it will nevertheless be used here as part of a transient bioreactor model. The reason is that we do not wish to mess up the analysis of phenomena associated with transient operation of bioreactors, such as the analysis of the stability of a given steady state by including at the same time the complexities embodied in the structured models of Chapter 7. The general principles of bioreactor performance are equally well explained when simple kinetic expressions are used. In a research situation the complexities of structured kinetics and of non-ideal bioreactors (Chapter 11) must of course eventually be considered.

Reactors used in the chemical industry are roughly divided into two groups: Stirred Tank Reactors and Plug Flow (or Tubular) Reactors. Reactors used in the bioindustry for fermentation of microorganisms can be divided into the same two groups, but the stirred tank reactor is far more popular. In the chemical and petrochemical industry it is the other way round. The reason is that the volumetric rate of a bioreaction is proportional to the biomass concentration and therefore the apparent reaction order in the substrate concentration is negative down to very low levels of the substrate. Conventional chemical reactions are typically of order greater than zero in the "limiting" reactant. Also aerobic fermentations cannot really be carried out in plug flow reactors due to the very large difference in time constants between the bioreaction and the passage of the gas through the reactor. This is of course similar to liquid phase chlorination of hydrocarbons in the chemical industry where some kind of stirred tank must be better than a plug flow reactor. A specific problem with the use of plug flow reactors in the bioindustry is that the feed must contain biomass—otherwise no reaction will take place unless biomass has been immobilized on particles that remain in the reactor. A continuous feed of non-sterile medium to the reactor would create all sorts of problems with infection and would not be chosen. Therefore plug flow reactors are only used if the substrate conversion needs to be almost complete as is the case in bioremediation of very toxic wastewater. Here the plug flow reactor is placed after a stirred tank reactor as discussed in Section 9.2 to do the final clean-up of the waste stream.

The term "stirred tank reactor" is applied to the liquid phase alone. In an aerobic process it is advantageous to contact the well mixed liquid phase with a gas phase that as far as possible passes the reactor in plug flow since this will give the highest rate of mass transfer of a valuable substrate from the gas phase and contribute to a high utilization of the gas phase or to lower cost of compression of the gas phase if air is used. Consequently, a bioreactor may operate as a stirred tank with respect to the biomass and liquid phase substrates such as glucose while it is desirable to have plug flow for gaseous substrates. The design of bioreactors with both liquid phase and gas phase substrates is an important topic to be discussed in Section 9.1.4

9.1 The Stirred Tank Bioreactor

As discussed in the introduction almost all bioreactors are stirred tanks operating at or close to atmospheric pressure. The reactors are manufactured out of stainless steel, and they are equipped with devices to obtain good mixing of the liquid phase and an efficient cooling of the medium.

Design of Fermentation Processes

Table 9.1. Advantages and disadvantages of different modes of operation of the stirred tank reactor.

Mode of operation	Advantages	Disadvantages
Batch	**Versatile**: Can be used for different reactions every day. **Safe**: Can be properly sterilized. Little risk of infection or strain mutation. Complete conversion of substrate is possible.	**High labor cost**: Skilled labor is required. **Much idle time**: Sterilization, growth of inoculum, cleaning after the fermentation. **Safety problems**: When filling, emptying and cleaning.

Continuous at steady state	**Works all the time**: low labor cost, good utilization of reactor.
	
	Often efficient: due to the autocatalytic nature of microbial reactions, the productivity can be high.
	
	When the production of the desired compound is catabolite repressed, continuous production or fed-batch is obligatory
	
	Automation may be very appealing.
	
	Constant product quality.
		Often disappointing: Promised continuous production for months fails due to:
		  a. Infection, e.g., a short interruption of the continuous feed sterilization.
		  b. Spontaneous mutation of the microorganism to a non-producing strain.
		
		Very inflexible: can rarely be used for other productions without substantial retrofitting.
		
		Downstream: The downstream process equipment must be designed for low volumetric rate, continuous operation-unless holding tanks are used.

Fed-batch	Combines the advantages of batch and continuous operation. Excellent for control and optimization of a given production criterion.	Some of the disadvantages of both batch and continuous operation-but advantages far outweigh the disadvantages, and fed-batch is used to produce both biomass (baker's yeast), industrial enzymes and antibiotics.

Substrates from the gas phase are transferred to the liquid phase by spargers or through high efficiency nozzles. Coalescence of gas bubbles is minimized by forcing the gas-liquid dispersion through static mixers or by redispersion in the vicinity of the mechanical mixers as further discussed in Chapter 10. The numerous patents describing the placement of mechanical mixers, of air lift reactors, and reactors with forced flow through static mixers or with rapid recirculation of liquid, injection of gas in the high-pressure recirculation loop and dispersion of the gas liquid mixture into the main reactor will not at all be discussed. All these reactors operate in *Batch Mode,* in *Continuous Mode,* or as *Semi-Batch*-called *Fed Batch* reactors in the bioindustry. The following discussion will center on these three general modes of operation. Table 9.1 summarizes the main reasons for using one or the other type of operation.

The entire mathematical treatment of tank reactors in the following paragraphs revolves around the first-order differential equation, Eq. (9.1), in the vector variable $\mathbf{c} = (\mathbf{s}, \mathbf{p}, x)$.

$$\frac{d(V\mathbf{c})}{dt} = V(\mathbf{q}^t + \mathbf{q}) + v_f \mathbf{c}_f - v_e \mathbf{c}_e \tag{9.1}$$

$\mathbf{c}$ contains the liquid-phase concentrations, subscript *f* indicates feed while *e* indicates effluent, *V* is the reactor volume, and *v* is the liquid flow to or from the reactor. $\mathbf{q}^t$ is the vector of volumetric mass transfer rates and $\mathbf{q}$ is the vector of volumetric reaction rates.

9.1.1. Batch Operation

With neither liquid nor gas flow into or out of the bioreactor, Eq. (9.1) simplifies to

$$\frac{d\mathbf{c}}{dt} = \mathbf{q}(\mathbf{c}) \tag{9.2}$$

with appropriate initial values for the variables. When all volumetric rates are proportional, one obtains the following mass balances for biomass x, substrate s, and product p:

$$\frac{dx}{dt} = \mu x; \quad x(t=0) = x_0 \tag{9.3}$$

$$\frac{ds}{dt} = -Y_{xs}\mu x; \quad s(t=0) = s_0 \tag{9.4}$$

$$\frac{dp}{dt} = Y_{xp}\mu x; \quad p(t=0) = p_0 \tag{9.5}$$

When the yield coefficients Y_{xs} and Y_{xp} are independent of *s*, *x*, and *p*, a key assumption in the unstructured models of Section 7.3, the three coupled first-order differential equations Eqs. (9.3) - (9.5), can be rearranged into one first-order differential equation in *x* and two algebraic equations from which *s* and *p* can be obtained directly once the differential equation has been solved.

For the simplest Monod kinetics in Eq. (7.16), the result is

$$s = s_0 - Y_{xs}(x - x_0) \tag{9.6a}$$

Design of Fermentation Processes

$$p = p_0 + Y_{xp}(x - x_0) \tag{9.7a}$$

$$\frac{dx}{dt} = \mu_{max}\frac{s}{s+K_s}x = \mu_{max}\frac{s_0 - Y_{xs}(x - x_0)}{s_0 - Y_{xs}(x - x_0) + K_s}x; \qquad x(t=0) = x_0 \tag{9.8a}$$

or, with dimensionless variables,

$$S = 1 - X + X_0 \tag{9.6b}$$

$$P = P_0 + X - X_0 \tag{9.7b}$$

$$\frac{dX}{d\theta} = \frac{1 - X + X_0}{1 - X + X_0 + a}X; \quad X(t=0) = X_0 = \frac{x_0}{Y_{xs}s_0} \tag{9.8b}$$

where

$$S = \frac{s}{s_0}; \quad X = \frac{x}{Y_{sx}s_0}; \quad P = \frac{p}{Y_{sp}s_0}; \quad \theta = \mu_{max}t; \quad a = \frac{K_s}{s_0} \tag{9.9}$$

At the end of the fermentation, $X = 1 + X_0, P = P_0 + 1$, and $S = 0$.

The differential equation, Eq. (9.8b), is of the separable type and can be integrated by a standard technique as further described in Note 9.1. The result is an expression in which θ is obtained as an explicit function of X:

$$\theta = \left(1 + \frac{a}{1+X_0}\right)\ln\left(\frac{X}{X_0}\right) - \frac{a}{1+X_0}\ln(1 + X_0 - X) \tag{9.10}$$

The last term in Eq. (9.10) increases from zero at $\theta = 0$ to infinity when $\theta \to \infty$. Neither Eq. (9.10) nor the identical expression

$$\exp(\theta) = \frac{X}{X_0}\left(\frac{X}{X_0(1 + X_0 - X)}\right)^{a/(1+X_0)} \tag{9.11}$$

looks particularly tractable by a graphical procedure in which the two kinetic parameters μ_{max} and K_s are to be retrieved from the time profile of a batch experiment. When data from actual fermentation experiments are inserted it is, however, often true that an accurate value for μ_{max} can be obtained while K_s — as expected from the discussion in Section 7.3.1.— is almost impossible to calculate based on the batch fermentation data.

Example 9.1 Kinetic data from a batch experiment
We now consider a batch fermentation where the initial cell density x_0 is 10 mg L^{-1}, s_0 = 10 g L^{-1}, Y_{sx} = 0.2 g g^{-1}, and K_s = 10 mg L^{-1}. Thus

$$X_0 = \frac{10}{0.2 \cdot 10^4} = 0.005 \qquad (1)$$

$$a = \frac{10}{10^4} = 10^{-3} \qquad (2)$$

$$\frac{a}{1+X_0} \approx 10^{-3} \qquad (3)$$

With these typical parameter values, the last factor in Eq. (9.11) is equal to 1.002 for $z = x/x_0 = 10$ and 1.004 for z = 50. Consequently, during the time it takes for the biomass concentration to increase to 50 times its initial value (or 25% of the total increase of biomass concentration), z increases exponentially with time and the time constant is equal to $1/\mu_{max}$. When both a and X_0 are small, the value of K_s has virtually no influence on the value of μ_{max}, which can be determined from a plot of ln(z) versus time.

K_s can in principle be determined from the final part of the experiment, when $X \approx X_0 + 1$. Here one can reasonably take z to be constant equal to 1 + 1/X_0 and $a = K_s / s_0$ can be found from the slope of

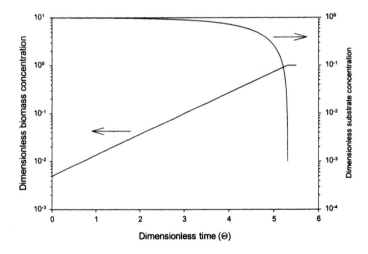

Figure 9.1. Profiles of dimensionless biomass and substrate concentration during a batch cultivation following Monod kinetics. (a = 0.001 and X_0 = 0.005).

Design of Fermentation Processes

$$\theta = \left\{ \frac{1}{1+X_0} \ln\left[\frac{X}{X_0(X_0+1-X)} \right] \right\} a + \ln\left(1 + \frac{1}{X_0}\right) \quad (4)$$

where X_0 is supposed to be known and μ_{max} is determined from the initial part of the fermentation experiment. Experimental scatter of the biomass concentration data or any small model error (there may be a tiny maintenance demand, or Y_{sx} may change slightly with s) will, however, make the determination of K_s very uncertain. It takes only $0.033/\mu_{max}$ (or less than 4 min when $\mu_{max} = 0.5 \, h^{-1}$) for S to decrease from $10a$ to $0.1a$, i.e., the transition from approximately zeroth-order kinetics to the observable end of the batch fermentation (see Fig. 9.1). With a maximum sampling frequency of 0.5 - 1 samples per minute, this allows only 2 - 4 measurements to be used in the determination of K_s. It is therefore concluded that it is practically impossible to determine K_s from a batch fermentation experiment.

Note 9.1 Analytical solution of biomass balance with unstructured growth kinetics
Let μ be of the form

$$\mu = \mu_{max} f_1(p) f_2(s) \quad (1)$$

and let the yield coefficients Y_{sx} and Y_{xp} be independent of s, as assumed in maintenance-free unstructured models. From the two algebraic expressions, Eqs. (9.6) and (9.7),

$$\frac{dx}{dt} = \mu_{max} f_1\left[p_0 + Y_{xp}(x-x_0)\right] f_2\left[s_0 - Y_{xs}(x-x_0)\right] x \quad (2)$$

With typical expressions like Eq. (7.21) to (7.23) for μ, the resulting differential equation is of the form

$$d(\mu_{max} t) = d\theta = \frac{P(z)}{zQ(z)} dz \quad (3)$$

where P and Q are polynomials of degree n_p and n_q, respectively, in the dimensionless biomass concentration $z = x/x_0 = X/X_0$. Thus, with product inhibition expressed in the manner of Eq. (7.22),

$$d(\mu_{max} t) = \frac{s + K_s}{sx}\left(1 + \frac{p}{K_i}\right) dx = \frac{X-1-X_0-a}{(X-1-X_0)X}\left(1 + \frac{Y_{sp}s_0}{K_i}(P_0 + X - X_0)\right) dX \quad (4)$$

where the dimensionless variables are defined in Eq. (9.9). With substrate inhibition according to Eq. (7.21),

$$d(\mu_{max} t) = \frac{s^2/K_i + s + K_s}{sx} dx = \frac{bS^2 + S + a}{Sx} dx = \frac{b(1-X+X_0)^2 + (1+X_0-X)+a}{(1+X_0-X)X} dX \quad (5)$$

where $b = s_0/K_i$ and the remaining parameters are defined in Eq. (9.9). Finally, with Eq. (7.23)

$$\mu = \mu_{max} \frac{s}{s+K_s}\left(1-\frac{p}{p_{max}}\right) \qquad (6)$$

one obtains

$$d(\mu_{max} t) = \frac{X-1-X_0-a}{(X-1-X_0)X} \frac{P_0+X_{max}-X_0}{X_{max}-X} dX \qquad (7)$$

where

$$X_{max} = \frac{p_{max}-p_0}{Y_{sp}s_0} + X_0 \qquad (8)$$

The standard technique for integration of a differential equation of the type of Eq. (3) is

1. Use synthetic division to separate the original expression into a number of terms of zero or higher degree in z (or X) and a final term of the form of Eq. (3) in which the numerator polynomial $P'(z)$ has a degree $n_p < n_q$.
2. The first terms can be integrated directly. The last term is decomposed into a number of partial fractions, one for each zero in $Q(z)$ [or $Q(X)$]. When a zero z_i is of degree n, it gives rise to n fractions of the form $(z-z_i)^{-n}, (z-z_i)^{-n+1}, \ldots, (z-z_i)^{-1}$. The final result is

$$\frac{P(z)}{xQ(z)} = a_1 z^m + a_2 z^{m-1} + \cdots + a_{m+1} + \frac{A_0}{z} + \frac{A_1}{z-z_1} + \cdots + \frac{A_{i1}}{(z-z_i)^n} + \cdots + \frac{A_{in}}{z-z_i} + \cdots \qquad (9)$$

3. The constants $A_0, A_1, \ldots, A_{i1}, A_{in}, \ldots$ are determined by identification of the coefficients to equal powers of z in $P'(z)$ and on the right-hand side of Eq. (9)
4. Each of the resulting terms is easily integrated.

As examples of the procedure, consider first Eq. (5) and next Eq. (7):

$$\begin{aligned}\frac{b(1+X_0-X)^2+(1+X_0-X)+a}{X(1+X_0-X)} &= b(1+X_0)\frac{1}{X}-b+\frac{1+X_0-X+a}{X(1+X_0-X)} \\ &= b(1+X_0)\frac{1}{X}-b+\frac{A_0}{X}+\frac{A_1}{1+X_0-X} \\ &= b(1+X_0)\frac{1}{X}-b+\frac{(A_1-A_0)X+A_0(1+X_0)}{X(1+X_0-X)}\end{aligned} \qquad (10)$$

or $A_1 - A_0 = -1$ and $A_0(1+X_0) = 1+X_0+a$. Hence the solution of Eq. (5) is

Design of Fermentation Processes

$$\theta = b(1+X_0)\ln\left(\frac{X}{X_0}\right) - b(X-X_0) + \left(1+\frac{a}{1+X_0}\right)\ln\left(\frac{X}{X_0}\right) - \frac{a}{1+X_0}\ln(1+X_0-X) \quad (11)$$

The last two terms of Eq. (11) is the solution Eq (9.10) for the simple Monod model with $b = 0$. For small t where $X \approx X_0$, one may use a linear approximation for $\ln(X/X_0)$, and the sum of the first two terms in Eq. (11) is

$$b(1+X_0)\left(\frac{X}{X_0}-1\right) - b(X-X_0) = b\left(\frac{X}{X_0}-1\right) \quad (12)$$

Since $b = s_0/K_i$ can be large when s_0 is large or K_i in Eq. (7.21) is small, an appreciable time may pass before $z = x/x_0$ has moved away from the vicinity of 1. Consequently, an experimentally observed lag phase in a batch fermentation could be modeled by a substrate inhibition term added to the simple Monod kinetics. Unless the time lag increases with increasing initial substrate concentration according to Eq. (11), the model has, however, no mechanistic foundation.

Decomposition of the right-hand side of Eq. (7) yields

$$(P_0 + X_{max} - X_0)\left(\frac{A_0}{X} + \frac{A_1}{X-1-X_0} + \frac{A_2}{X_{max}-X}\right)$$
$$= (P_0 + X_{max} - X_0)\frac{X-(1+X_0+a)}{X(X-1-X_0)(X_{max}-X)} \quad (13)$$

where

$$-A_0 - A_1 + A_2 = 0, \quad (X_{max}+1+X_0)A_0 + X_{max}A_1 - (1+X_0)A_2 = 1$$

and

$$-X_{max}(1+X_0)A_0 = -(1+X_0+a)$$

or,

$$A_0 = \frac{1+X_0+a}{X_{max}(1+X_0)}; \quad A_1 = \frac{a}{(1+X_0)(X_{max}-1-X_0)}; \quad A_2 = \frac{X_{max}-1-X_0-a}{X_{max}(X_{max}-1-X_0)} \quad (14)$$

and the solution of Eq. (7) is therefore

$$\theta = (P_0 + X_{max} - X_0) \left[\frac{1 + X_0 + a}{X_{max}(1 + X_0)} \ln\left(\frac{X}{X_0}\right) - \frac{a}{(1 + X_0)(X_{max} - 1 - X_0)} \ln(X_0 + 1 - X) \right.$$

$$\left. - \frac{X_{max} - 1 - X_0 - a}{X_{max}(X_{max} - 1 - X_0)} \ln\left(\frac{(X_{max} - X)}{(X_{max} - X_0)}\right) \right] \quad (15)$$

When p_{max} is large, X_{max} is much larger than $1 + X_0$, the largest value obtained by complete conversion of substrate to biomass. In this case the inhibition term is negligible for all t, and Eq. (15) degenerates to Eq. (9.10). If, however, $(p_{max} - p_0)/s_0 Y_{sp} < 1$, the fermentation stops when $X = X_{max}$ and the corresponding biomass concentration x can be found from Eq. (9.7a) with $p = p_{max}$.

The two examples illustrate that an analytical solution of the mass balances is possible for a number of μ expressions of the form of Eq. (1). The analytical solution has several appealing features; the influence of key parameter combinations is emphasized, and limiting solutions are often extractable. The quest for an analytical solution should, however, not be pursued too ardently, because standard computer programs easily find the solution to the mass balances just as accurately as insertion in the final expressions. Furthermore—as will become evident in most of the examples of this chapter—it is rarely possible to obtain an analytical solution at all.

When the yield coefficients Y_{xs} and Y_{xp} vary with μ due to maintenance demands, the mass balances in Eqs. (9.3) - (9.5) for the batch reactor are modified to

$$\frac{dx}{dt} = \mu x; \quad x(t=0) = x_0 \quad (9.12)$$

$$\frac{ds}{dt} = -(Y_{xs}^{true} \mu + m_s)x; \quad s(t=0) = s_0 \quad (9.13)$$

$$\frac{dp}{dt} = (Y_{xp}^{true} \mu + m_p)x; \quad p(t=0) = p_0 \quad (9.14)$$

where the maintenance kinetics of Eqs. (7.24) and (7.25) have been inserted for r_s and r_p. Quite often there is a simple relation between m_s and m_p (e.g., $m_s = m_p$ for homofermentative, anaerobic lactic acid fermentation, since a given amount of sugar used for cell maintenance is retrieved quantitatively as the product, lactic acid).

It is not possible as in Eqs. (9.6) - (9.8) to rewrite Eqs. (9.12) - (9.14) as one differential equation in x and two explicit expressions for s and p. Certain important results concerning the total yield of biomass that can be obtained from a given amount of substrate can, however, as illustrated in Example 9.2, be calculated approximately. One qualitative result of general validity is that

Design of Fermentation Processes

maintenance has little influence when the specific growth rate is high since the ratio between m_s and μ_{max} is the relevant parameter in the dimensionless substrate balance in Eq. (4) of Example 9.2. Consequently, it is in the continuous stirred tank reactor or in the fed batch reactor rather than in the batch reactor that one should be worried about maintenance losses (that is, if biomass and not the metabolite is the desired product).

Example 9.2 Effect of maintenance on distribution of substrate between biomass and product

Assume that $\mu = \mu_{max}$ throughout the fermentation. Whether this approximation is reasonable or not depends on the value of $a = K_s / s_0$. If a is less than 0.01, virtually all biomass is formed with a constant specific growth rate. With the typical values for kinetic parameters and operating conditions used in Example 9.1, the maximum possible increase of biomass concentration is $Y_{sx} s_0 = 2$ g L^{-1}. When 98% of the available substrate has been consumed, S is still 20 times as large as a ($s = 200$ mg L^{-1} and $K_s = 10$ mg L^{-1}), and the approximation $\mu = \mu_{max}$ is reasonable. Whether μ is constant or starts to decrease for smaller values of s has no visible influence on the total biomass yield, which can at most increase by a further 2%–and probably considerably less, since much of the remaining substrate goes to maintenance.

With substitution of Y_{sx}^{true} for Y_{sx} in the definition of the dimensionless variables given by Eq. (9.9) we have:

$$\frac{dX}{dt} = \mu_{max} X \tag{1}$$

$$\frac{dS}{dt} = -X_0 \mu_{max} \left(1 + \frac{m_s Y_{sx}^{true}}{\mu_{max}}\right) \frac{X}{X_0} \tag{2}$$

for which the solution is

$$X = X_0 \exp(\theta) = X_0 \exp(\mu_{max} t) \tag{3}$$

$$S = 1 - X_0 (1 + b) [\exp(\theta) - 1] \tag{4}$$

where

$$b = \frac{m_s Y_{sx}^{true}}{\mu_{max}} \quad \text{and} \quad \theta = \mu_{max} t \tag{5}$$

Similarly, for a metabolic product,

$$P = P_0 + \left(1 + \frac{m_p Y_{sx}^{true}}{Y_{sp}^{true} \mu_{max}}\right) X_0 [\exp(\theta) - 1] \tag{6}$$

Table 9.2. Changes in the distribution of a given amount of consumed substrate between biomass and product[a] for various values of m_s.

m_s (h^{-1})	$X_{final} - X_0$	$P_{final} - P_0$	θ_{final}
0	1	1	5.303
0.08	0.961	1.0096	5.264
0.16	0.926	1.0186	5.227
0.32	0.862	1.0344	5.156

[a] μ is assumed to be equal to μ_{max}, and $m_s = m_p$.

Using Eqs. (3) - (6), it is possible to relate the changes in the total biomass yield and product yield to the dimensionless maintenance parameter. Table 9.2 shows the calculation for some reasonable parameter values $Y_{sx}^{true} = 0.2$, $Y_{sp}^{true} = 0.8$, $\mu_{max} = 0.4 \text{ h}^{-1}$, and $m_s = m_p = 0, 0.08, 0.16$ and 0.32 h^{-1}, respectively. Other parameter values are taken from Example 9.1.

From the first two rows of the table it is easily checked that when $Y_{sx}^{true} + Y_{sp}^{true} = 1$, the sum of produced biomass and product is s_0 irrespective of the value of $m_s = m_p$. When the maintenance requirement increases, more substrate is directed toward product formation.

The higher degree of complexity of biochemically structured kinetic models as compared with unstructured models usually makes batch reactor mass balances based on structured kinetics intractable by analytical methods—but again it must be stressed that computer solutions are easy to find, and a study of the numerical solution for various operating conditions may certainly reveal characteristic features of the model. The lactic acid fermentation kinetics of Eqs. 3 to 5 in Example 7.6 will be used to illustrate how a structured kinetic model can be coupled to an ideal batch reactor model. The final model becomes

$$\frac{dx}{dt} = \mu x = [\gamma_{21} r_2 - (1 - \gamma_{32}) r_3] x$$
$$= \left[\gamma_{21} k_2 \frac{s}{s + K_2} - (1 - \gamma_{32}) k_3 \frac{s}{s + K_3}\right] \frac{s_N}{s_N + K_N} X_A x \quad (9.15a)$$

$$\frac{ds}{dt} = (-r_1 - \alpha_2 r_2) x = -\left(k_1 \frac{s}{s + K_1} + \alpha_2 k_2 \frac{s}{s + K_2} \frac{s_N}{s_N + K_N}\right) X_A x \quad (9.15b)$$

$$\frac{dp}{dt} = r_1 x = k_1 \frac{s}{s + K_1} X_A x \quad (9.15c)$$

$$\frac{ds_N}{dt} = -r_2 x = -k_2 \frac{s}{s + K_2} \frac{s_N}{s_N + K_N} X_A x \quad (9.15d)$$

Design of Fermentation Processes

$$\frac{dX_A}{dt} = -\frac{dX_G}{dt} = \gamma_{21} r_2 - r_3 - \mu X_A$$

$$= \left\{ \left(\gamma_{21} k_2 \frac{s}{s+K_2} - k_3 \frac{s}{s+K_3} \right) \right.$$

$$\left. - \left[\gamma_{21} k_2 \frac{s}{s+K_2} - (1-\gamma_{32}) k_3 \frac{s}{s+K_3} \right] X_A \right\} \frac{s_N}{s_N + K_N} X_A \quad (9.15e)$$

Analytical solution of Eq. (9.15) is impossible, and computer simulations are in general necessary when comparing the predictions of the model and experimental results. Since the new feature introduced in the structured model is the division of cell mass into an active and a nonactive component we shall, however, focus on this aspect. Hence, s and s_N are both assumed to be much larger than the saturation constants. X_A has the value X_{A0} at $t = 0$. All other variables have initial values as previously defined. The simplified form of the key equation for X_A is

$$\frac{dX_A}{dt} = \{(\gamma_{21}k_2 - k_3) - [\gamma_{21}k_2 - (1-\gamma_{32})k_3]X_A\}X_A$$

$$= (c_1 - c_2 X_A)X_A \; ; \quad X_A(t=0) = X_{A0}$$

⇓

$$\frac{dX_A}{X_A(c_1 - c_2 X_A)} = \frac{1}{c_1}\left(\frac{1}{X_A} + \frac{c_2}{c_1 - c_2 X_A}\right)dX_A = dt \; ; \quad X_A(t=0) = X_{A0} \quad (9.16)$$

⇓

$$c_1 t = \ln\left[\frac{X_A(c_1 - c_2 X_{A0})}{X_{A0}(c_1 - c_2 X_A)}\right]$$

The explicit solution for X_A is

$$\frac{X_A}{X_{A0}} = \frac{\alpha \exp(c_1 t)}{\alpha - 1 + \exp(c_1 t)} \quad (9.17a)$$

where $\alpha = c_1 / c_2 X_{A0}$. Clearly the final value of X_A is

$$X_{A,\text{final}} = \frac{c_1}{c_2} = \frac{\gamma_{21}k_2 - k_3}{\gamma_{21}k_2 - (1-\gamma_{32})k_3} \quad (9.17b)$$

The biomass concentration profile is given by

$$\ln\left(\frac{x}{x_0}\right) = \frac{c_2 X_{A0}}{c_1} \int_0^t \frac{\alpha \exp(c_1 t)}{\alpha - 1 + \exp(c_1 t)} d(c_1 t) = \ln\left[\frac{\alpha - 1 + \exp(c_1 t)}{\alpha}\right]$$
$$\Downarrow$$
$$\frac{x}{x_0} = \frac{\alpha - 1 + \exp(c_1 t)}{\alpha} \tag{9.18}$$

Finally, the solution of the product mass balance is

$$\frac{p - p_0}{x_0} = \frac{k_1 [\exp(c_1 t) - 1]}{c_2 \alpha} \tag{9.19}$$

The calculations on a greatly simplified structured model demonstrate that these models can give an insight into the fermentation process far beyond what is offered by the unstructured models:
- Exponential growth of both biomass and product is eventually obtained. The time constant is determined by the net rate of formation of active component in the cell, i.e., the constant c_1 in the substrate-independent model.
- A time lag is introduced in a biologically reasonable way. Cells with a low initial activity—e.g., taken from a stirred tank reactor with a very small dilution rate—have a low initial value of X_A and a large time lag, as seen from the horizontal shift between the growth curves in Fig. 9.2.

9.1.2 The Continuous Stirred Tank Reactor

The mass balance for a continuous, ideally mixed, constant-volume tank reactor is

$$\frac{d\mathbf{c}}{dt} = \mathbf{q}^t + \mathbf{q}(\mathbf{c}) + \frac{v}{V}(\mathbf{c}_f - \mathbf{c}) \tag{9.20}$$

The influence of mass transfer on the operation of the steady state stirred tank reactor is considered in section 9.1.4, while the dynamics of stirred tank continuous reactors is deferred to Section 9.3. Thus, in the present section the accumulation term and q^t are both zero.

Monod kinetics will again be used to illustrate some general features of the steady-state stirred tank reactor model. The feed to the stirred tank reactor is sterile and contains no product. In accordance with the major assumption of the ideal tank reactor, the concentration c_i of any substrate or metabolic product is constant at every point of the reactor. In most cases the effluent concentration is also c_i, but this is not necessarily correct, e.g., if the reactor system contains some kind of separation equipment as will be discussed in section 9.1.3.

Design of Fermentation Processes

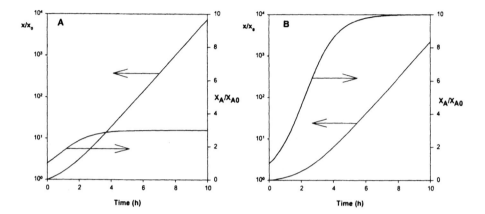

Figure 9.2. Structured model for lactic acid fermentation in a batch reactor. For two different values of a (Eq 9.17 a) the growth of biomass x/x_0 and of the active compartment X_A are shown as functions of time. **A** : $a = 3$, **B** : $a = 10$. c_l (Eq 9.16) = 1.

With Monod kinetics and sterile, product-free feed, the steady-state balances already introduced in Chapter 3 are:

$$q_x = \mu x = \mu_{max} \frac{s}{s + K_s} x = Dx \qquad (9.21)$$

$$q_s = -Y_{xs}\mu x = -D(s_f - s) \qquad (9.22)$$

$$q_p = Y_{xp}\mu x = Dp \qquad (9.23)$$

or

$$D = \mu = \mu_{max} \frac{s}{s + K_s}; \qquad x = Y_{sx}(s_f - s); \qquad p = Y_{sp}(s_f - s) \qquad (9.24)$$

For a given value of D, the concentrations of substrate, biomass, and metabolic product are calculated from Eq. (9.24). Obviously, any one of the concentrations could also be used as the key variable from which values of the remaining variables could be calculated.

Introduction of dimensionless variables S and X

$$S = \frac{s}{s_f} \,; \quad X = \frac{x}{Y_{sx}s_f} \tag{9.25}$$

yields the following expressions for D and for the overall mass balance:

$$D = \frac{\mu_{max} S}{S+a}\,; \quad X = 1-S \tag{9.26}$$

where $a = K_s / s_f$ is a parameter quite similar to that defined in Eq. (9.9) for batch fermentation. Now the natural scale factor is s_f rather than s_0. The largest possible value of S is 1, and consequently, the largest possible dilution rate is

$$D_{max} = \frac{\mu_{max}}{1+a} \tag{9.27}$$

and *washout* occurs when D exceeds D_{max}.

These basic features of steady-state operation of a continuous stirred tank reactor were already introduced in Section 3.1 and discussed in Example 7.1. For any given function $\mu = k\, f_1(p) f_2(s)$, the maximum dilution rate D_{max} can be found analogously to Eq. (9.27), which holds for maintenance-free Monod kinetics only. Typically D_{max} is found for $S = 1$, but the maximum value of D can also be found for an S value in the open interval $]0;1[$. Thus for the substrate inhibition kinetics given by Eq. (7.21),

$$\mu = \frac{\mu_{max} s}{s^2/K_i + s + K_s} = \frac{\mu_{max} S}{bS^2 + S + a} \tag{9.28}$$

where
$$a = \frac{K_s}{s_f}\,; \quad b = \frac{s_f}{K_i}$$

the function $\mu(S)$ has an extremum:

$$\mu_{extr} = \frac{\mu_{max}}{2\sqrt{ab}+1} = \frac{\mu_{max}}{2\sqrt{K_s/K_i}+1} \quad \text{at} \quad S_{extr} = \sqrt{\frac{a}{b}} = \frac{\sqrt{K_i K_s}}{s_f} \tag{9.29}$$

Since μ is an increasing function of S for $0 < S < S_{extr}$, the maximum dilution rate is obtained from one of the two expressions in Eq. (9.30):

Design of Fermentation Processes

$$D_{max} = \begin{cases} \mu_{extr} & \text{for } a/b \leq 1 \\ \mu_{max}/(b+1+a) & \text{for } a/b \geq 1 \end{cases} \qquad (9.30)$$

The value of μ_{extr} is independent of s_f, while for any given values of the kinetic parameters K_i and K_s it is seen that S_{extr} is inversely proportional to s_f (whereas, of course, s_{extr} is independent of s_f).

When μ is a monotonously increasing function of S, operation of the stirred tank reactor at a D value close to D_{max} leads to very low productivity. The effluent substrate concentration is close to s_f and, consequently, there is neither much biomass nor much product in the effluent stream. For nonmonotonic kinetics such as Eq. (9.28), the highest specific growth rate and the highest productivity may both be found at D values where S is appreciably smaller than 1; cf. the result in Eq. (9.29) and the corresponding result for maximum productivity calculated in Example 9.13.

The *productivity* of a given species i in the stirred tank reactor is defined as

$$P_i = D c_i = q_i \qquad (9.31)$$

The highest productivity is in general *not* found where the specific growth rate is at its maximum value. With simple Monod kinetics and sterile feed,

$$P_x = Dx = \frac{\mu_{max} S}{S+a} Y_{sx} s_f (1-S) \qquad (9.32)$$

or

$$\frac{P_x}{\mu_{max} Y_{sx} s_f} = \frac{S(1-S)}{S+a} \qquad (9.33)$$

for which the maximum value is

$$\left(\frac{P_x}{\mu_{max} Y_{sx} s_f}\right)_{max} = \left(\sqrt{1+a} - \sqrt{a}\right)^2 \quad \text{for} \quad S_{opt} = -a + \sqrt{a^2 + a} \qquad (9.34)$$

The corresponding D value is

$$D_{opt} = \mu_{max} \frac{-a + \sqrt{a^2 + a}}{\sqrt{a^2 + a}} = \mu_{max}\left(1 - \frac{1}{\sqrt{1+1/a}}\right) \qquad (9.35)$$

The results in Eqs. (9.33) - (9.35) can be expressed verbally as follows:

- It is desired to treat a feed stream v of substrate with concentration s_f in a stirred tank reactor of given volume V. The kinetics follows a Monod expression with $K_s = a s_f$. Now the maximum amount (in kilograms per hour) of substrate that can he converted is

$$-V q_{s,\max} = V D (s_f - s_{opt}) = s_f D V (1 - S_{opt})$$

  where $s = S_{opt}$ is calculated from Eq. (9.34) and $D = D_{opt}$ from Eq. (9.35).

- If a given feed stream value v is larger than $D_{opt} V$, the best policy is to divert some of the feed stream—a "shunt" around the bioreactor—or to use cell recirculation as discussed in section 9.1.3.

By the same reasoning, the highest cell productivity (in kilograms per hour) for the given volume V is calculated from Eq. (9.34). The same formula can be used to calculate $(P_p)_{\max}$ since $P_p = Y_{xp} \cdot P_x$.

Needless to say, the term "optimal" used in Eqs. (9.34) and (9.35) would be highly misleading in situations where the cost of substrate has to be weighted against the value of the biomass or product. Environmental constraints might make it necessary to go to lower effluent values of the substrate concentration than that given by Eq. (9.34), at the cost of increasing V, to have enough capacity to process the given feed stream. The above mathematical treatment is used solely to illustrate how one of many optimization problems associated with bioreactor operation can be solved.

Example 9.3 Optimal productivity in steady- state stirred tank bioreactors.
In figure 9.3 the problem of finding the maximum steady state productivity in the stirred tank has been solved graphically for several typical unstructured kinetic models. Substrate inhibition, product inhibition or maintenance added to a basic kinetic rate model always lowers the maximum productivity, which is proportional to q_x. It is easily seen that the area of the hatched rectangle is $(1-S) Y_{sx} s_f q_x^{-1} = \mu^{-1} = D^{-1}$. With the graphical representation of the problem it is also clear that a given value of s (or D) corresponds to one and only one value of q_x, while a given value of q_x can be obtained for two different values of s (or D), one with low x and high D and the other with high x and low D.

  Hatched area : Holding time to achieve maximum productivity.
  A : 3.37 h ; B : 7.17 h ; C : 4.69 h ; D : 4.29 h.

Substrate inhibition shifts S_{opt} towards lower values because of the negative influence of S on μ at high S values. Product inhibition, and to a smaller extent maintenance demands shifts S_{opt} towards higher S values, in the first case to avoid that too much product is formed and in the last case because the relative amount of substrate that goes to biomass decreases for small D values (i.e. small S).

Design of Fermentation Processes

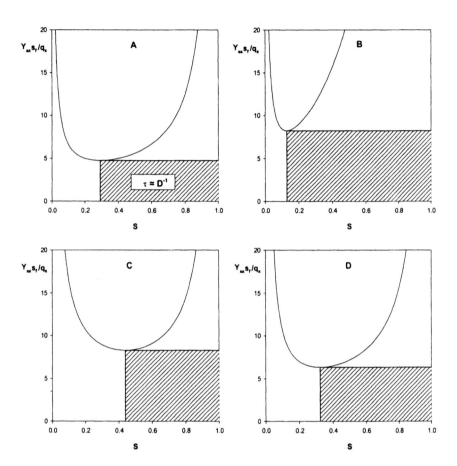

Figure 9.3. Productivity of a steady state stirred tank reactor.
  A. Monod kinetics : $\mu_{max} = 0.5$ h^{-1}, $K_s = 0.4$ g L^{-1}. $s_f = 2$ g L^{-1}, $Y_{sx} = 0.5$ g L^{-1}.
  B. Substrate inhibition : As in A, and $K_i = 0.25$ g L^{-1}, i.e. $b = 8$ in Eq (9.28).
  C. Product inhibition (Eq (7.23)): As in A, and $s_f Y_{sp}/p_{max} = 2/3$.
  D. Maintenance (Eq (9.13),(9.14)): As in A, and $b = 0.2$ in Eq (5 E9.2).

When maintenance kinetics is coupled to a Monod expression for $\mu = D$:

$$q_s = -D(s_f - s) = -(Y_{xs}^{true} D + m_s) x \quad \text{or} \quad X = x/(s_f Y_{sx}) = \frac{S(1-S)}{S + b(S+a)} \quad (1)$$

where the new parameter

$$b = m_s / (\mu_{max} Y_{xs}^{true}). \quad (2)$$

When maintenance is added to any kinetics $\mu(S)$ the biomass concentration decreases to zero for $D \to 0$, whereas $X = 1 - S$ for maintenance free kinetics.

For the kinetics of figure (9.3.D) the largest value of X that can be obtained is:

$$X_{max} = \frac{\left(\sqrt{1+a^*} - \sqrt{a^*}\right)^2}{b+1} \quad \text{for} \quad S = -a^* + \sqrt{a^{*2} + a^*} \quad \text{where } a^* = ab/(1+b) \quad (3)$$

The value of q_x or of $P_x = q_x / (\mu_{max} s_f Y_{xs}^{true})$ can always be found as the product of D and x (or X):

Monod:

$$P_x = \frac{S(1-S)}{S+a} \quad (4)$$

Where the explicit solutions for $(P_x)_{max}$ and for S_{opt} are shown in eq (9.34).

Substrate inhibition:

$$P_x = \frac{S(1-S)}{S + a + bS^2} \quad \text{where } b = s_f K_i^{-1}.$$

$$S_{opt} = -c + (c^2 + c)^{1/2} \quad \text{and} \quad c = a/(1+b) \quad (5)$$

Product inhibition:

$$P_x = \frac{S(1-S)(1 - d(1-S))}{S + a} \quad \text{where } d = Y_{sp} s_f / p_{max} \quad (6)$$

Maintenance:

$$P_x = \frac{S^2(1-S)}{(S+a)(S+b(S+a))} \quad \text{(figure 9.3 D)} \quad (7)$$

Design of Fermentation Processes

In (4) and (5) relatively simple algebraic expressions are obtained for S_{opt}. This is not so for (6) and (7), but algebraic solution of these problems is hardly worth the effort since an automatic calculation of q_x^{-1} as in figure 9.3 gives the answer for any given values of the parameters if only an expression for q_x as a function of S can be set up.

The productivity problems solved here for unstructured kinetics may also have limited value in practice. Thus, in aerobic fermentation of yeast q_x increases practically proportional to D for any reasonable value of s_f since K_s is in the mg L^{-1} range. At $D = D_{crit}$ the stoichiometry changes abruptly with the onset of ethanol production. To obtain maximum productivity of biomass and of a growth associated protein product it is desirable to work at D only slighter lower than D_{crit}. This mode of operation leads to severe control problems as discussed by Andersen et al. (1997).

9.1.3 Biomass Recirculation

In the preceding discussion it has been shown that for given fermentation kinetics (with or without maintenance) and for a given feed composition it is possible for a steady state continuous reactor to calculate the complete effluent composition based on a single measured quantity. This may be either the dilution rate or the concentration of substrate, biomass, or product in the effluent. If in Eq. (9.20) an extra degree of freedom is introduced by relaxation of the assumption that the biomass concentration x_e in the effluent is equal to the biomass concentration x in the reactor, then a new set of design problems arises. For all other reactants and products the assumption of ideal homogeneity of the stirred tank reactor is maintained, but the value of D that corresponds to given feed and effluent concentrations will now depend also on x. When $x > x_e$, the reactor is able to process more feed than in the basic situation with $x = x_e$ since the rate of biomass formation is proportional to x. Enrichment of the reactor medium relative to the effluent stream can be achieved by means of a cell centrifuge installed after the reactor and *recirculation of cells* to the inlet of the reactor through an exterior loop. The same effect can also be obtained by means of a filter installed inside the reactor—typical on-line probes for removal of cell free medium to analysis are extreme cases of complete cell-medium separation. The cells may undergo a partial sedimentation in the reactor—here, cells immobilized on granules of inactive carrier material such as sand particles or flocs of microorganisms are typical examples. Finally, a more-or-less loose wall growth of cells leads to enrichment of cells in the reactor, although cells growing on the walls or on granules may exhibit different kinetic behaviour than do freely suspended cells.

With maintenance-free Monod kinetics, the modification of Eq. (9.21) to Eq. (9.23) reads

$$D x_e = \mu x \tag{9.36}$$

$$D(s_f - s) = Y_{xs}\mu x \tag{9.37}$$

$$Dp = Y_{xp}\mu x \tag{9.38}$$

In figure 9.4 A define $f = x_e/x < 1$. Now from (9.36):

$$Df = \mu \tag{9.39}$$

while the substrate and product balances are

$$D(s_f - s) = Y_{xs} Df \frac{x_e}{f} \quad \text{or} \quad x_e = Y_{sx}(s_f - s) = Y_{sx}(s_f - s_e) \tag{9.40}$$

$$Dp = Y_{xp} Df \frac{x_e}{f} \quad \text{or} \quad p = Y_{sp}(s_f - s) \tag{9.41}$$

Equations (9.40) and (9.41) show that the overall mass balances for x_e and $p = p_e$ are identical with those of Eq. (9.24). The cell enrichment model differs from the basic model in one respect only: For a given value of D it is possible to obtain a specified effluent composition (s_e, x_e, p_e) by a suitable choice of f in Eq. (9.39).

We shall now relate f to the operating variables of a cell recirculation design. In Fig. 9.4 A a hydrocyclone separates the effluent from the reactor into a product stream v and a recirculation stream Rv, where R is the recirculation factor. The cell concentration is x_R in the recirculation stream, and $x_R = x\beta$, with $\beta > 1$. Neither product concentration nor substrate concentration are affected by the hydrocyclone.

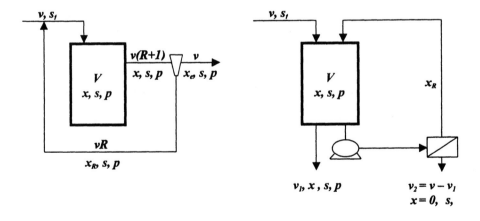

A **B**
Figure 9.4 Biomass recirculation using a hydrocyclone (A) and an ultrafilter (B).
In (B) the permeate $v_2 = v - v_1$ is cell free. The bleed stream v_1 is taken from the reactor.

Design of Fermentation Processes

A mass balance for cells at the hydrocyclone gives

$$v(1+R)x = vx_e + Rvx_R \tag{9.42}$$

or

$$1 + R = f + R\beta \quad \text{or} \quad f = 1 - R(\beta - 1) \tag{9.43}$$

The fraction of cells recirculated to the reactor is

$$\alpha = \frac{Rvx_R}{v(1+R)x} = \frac{R\beta}{1+R} \tag{9.44}$$

and the cell concentration in the reactor is

$$x = \frac{x_e}{1 - R(\beta - 1)} \tag{9.45}$$

Next consider the ultrafilter version in figure 9.4 B:

A mass balance for biomass is

$$V\mu(s)x = v_1 x \quad \text{or} \quad \mu = v_1/V = (v - v_2)/V = D(1 - \Omega) \tag{9.46}$$

where Ω is the filter separation factor defined as v_2/v and $D = v/V$ as usual.
It is seen that unit B can work at a dilution rate $(1-\Omega)^{-1}$ higher than the ordinary continuous stirred reactor for which $D(\Omega=0) = \mu$. Since μ is chosen to be the same with or without recycling of biomass, the biomass productivity is increased by a factor $(1-\Omega)^{-1}$. The biomass appears in a usually small stream $v_1 = (1-\Omega)v$, and with a much higher concentration $x = x(\Omega=0)(1-\Omega)^{-1}$ than without recycling. The product is in a cell-free stream v_2 and ready for down-stream processing. Also here the productivity has increased by a factor $(1-\Omega)^{-1}$ if the product from v_1 can be retrieved. If the stream v_1 is regarded as waste the productivity of P is increased by a factor $\Omega(1-\Omega)^{-1}$.

We can shortly describe the functioning of the cell recirculation set up as follows:
μ and consequently the effluent concentrations of substrate and product, which are both functions of μ alone are chosen by the designer of the reactor at values which are optimal by some criterion – it may be that a low effluent s-value is needed, or the yield of product may be small if s is too high. To obtain a satisfactory volumetric production rate recirculation by either scheme A or scheme B is chosen. The two design parameters (f in A, and Ω in B) are chosen such that the value of D (chosen to satisfy a given production in the reactor) matches the desired value of μ. In both schemes the higher productivity is due to the higher cell density in the reactor. This result is at least in theory independent of whether there is a loss of substrate due to maintenance.

Certainly more substrate is "lost" due to maintenance, but at the same time a much larger feed stream has been treated in the system. Relatively speaking the same percentage of substrate is "lost" (the maintenance loss depends only on μ). The stress on the cells caused by rapid pumping of medium through the recirculation loop and the filter may, however, give an independent contribution to the maintenance demand of the cell. In that case the recirculation solution may be at a disadvantage. Otherwise the only negative aspect of cell recirculation is that the pumping costs may outweigh the savings in capital investment for the reactor. This is treated in Example 9.6.

Example 9.4. Design of cell recirculating system
It is desired to process 1 m³ h⁻¹ of a feed with s_f = 4 g L⁻¹ of the growth-limiting substrate. Assume Monod kinetics with K_s = 1 g L⁻¹ and μ_{max} = 1 h⁻¹. Y_{sx} = 0.5 g/gram of substrate. The reactor volume is 500 L. What is the maximum cell productivity with and without cell recirculation?

a. $x = x_e$: D = 1 / 0.5 = 2 h⁻¹, and since $a = K_s/s_f$ = 0.25 and μ_{max} = 1 h⁻¹, the productivity is zero because $D > \mu_{max}/(a+1) = 0.8$ h⁻¹.

b. $x > x_e$: To avoid washout, Df must be smaller than 0.8 h⁻¹ and, consequently,

$$f = 1 - R(\beta - 1) < \frac{0.8}{2} = 0.4 \tag{1}$$

The maximum productivity is, of course, obtained when $S = 0$, i.e., when $x_e = s_f \cdot Y_{sx} = 2$ g L⁻¹ and $(P_x)_{max} = Dx_e = 2 \cdot 2 = 4$ g L⁻¹ or 2 kg h⁻¹ in a 500-L reactor. When $S \to 0$, the specific growth rate tends to zero, and from Eq. (9.39) with a given $D = 2$ h⁻¹ one obtains $f \to 0$, which means that the reactor cell concentration $x \to \infty$. This is not a feasible solution.

To illustrate the remaining calculations, choose $S = 0.309$, for which $\mu = 0.553$ h⁻¹ and, consequently,

$$f = 1 - R(\beta - 1) = \frac{\mu}{D} = 0.276 \tag{2}$$

The biomass concentration in the effluent is

$$x_e = 4 \cdot 0.5 (1 - 0.309) = 1.382 \text{ g L}^{-1} \tag{3}$$

and

$$P_x = Dx_e = 2.764 \text{ g L}^{-1}\text{h}^{-1} \tag{4}$$

or 1.38 kg h⁻¹. The cell concentration in the reactor is

Design of Fermentation Processes

$$x = \frac{x_e}{f} = 5.01 \text{ g L}^{-1} \tag{5}$$

and the fraction of cells recirculated depends on the effectiveness of the cell separator. Thus for $R = 2$ one obtains $\beta = 1.36$ and $\alpha = 90.8\%$. The cell concentration in the recycle loop is $x_R = \beta x = 6.81 \text{ g L}^{-1}$. For a less efficient operation of the centrifuge, $R = 3$, one obtains $\beta = 1.24$, $\alpha = 93.1\%$, and $x_R = 6.21 \text{ g L}^{-1}$.

Example 9.5 Design of a recirculation system—with maintenance requirement

We shall add on to Example 9.4 to include the influence of maintenance on the solution of the optimal productivity problem. Take $m_s = m_p = 0.1 \text{ h}^{-1}$ and $Y_{sp}^{\text{true}} = Y_{sx}^{\text{true}} = 0.5$, i.e., complete conversion of substrate to either product or biomass. All other kinetic and operational parameters are as in Example 9.4. Formulas for calculating X_{max} and the corresponding S value are derived in Example 9.3.

$$X_{max} = \frac{1}{b+1}\left(\sqrt{1+a^*} - \sqrt{a^*}\right)^2 = 0.7660 \tag{1}$$

$$S_{max, x} = -a^* + \sqrt{a^{*2} + a^*} = 0.09785 \tag{2}$$

Since D is given, the maximum productivity is obtained when $X = X_{max}$, i.e.,

$$(P_x)_{max} = DX_{max} s_f Y_{sx}^{\text{true}} = 2 \cdot 0.7660 \cdot 4 \cdot 0.5 = 3.06 \text{ g L}^{-1} \text{ h}^{-1} \tag{3}$$

which is obtained for $\mu = 0.2813 \text{ h}^{-1}$, $f = 0.1406$, and $x = 0.7660 \cdot 0.5 \cdot 4/0.1406 = 10.90 \text{ g L}^{-1}$. For $R = 2$, the value of β is calculated from Eq. (9.43). $\beta = 1.43$ and $x_R = 15.58 \text{ g L}^{-1}$. $\alpha = 95.3\%$ of the cells are recycled, and the recycle stream is much thicker than for the corresponding maintenance-free example.

Example 9.6 Design of an integrated lactic acid production unit.

4.5 wt. % lactose in a waste stream from a large cheese factory is to be converted to lactic acid in a stirred tank continuous reactor. The feed stream is 15 m³ h⁻¹ to which enzymatically hydrolyzed whey protein is added as nitrogen source. A high yield of lactic acid on lactose is not a requirement *per se*, but the product is going to be used as the monomer for production of polylactate, and to avoid prohibitory down-stream costs it is necessary to convert almost all the sugar, and at the same time to work with a high enough D value to avoid formation of byproducts (the "mixed acids" of Example 5.6). The choice of a continuous process is obligatory. Otherwise the value of the product (88 wt.% polymer grade lactic acid) will not match the cost of raw materials, labor and energy.

The feed is sterile, and in a conventional stirred tank, continuous reactor the effluent product concentration is p_o. The reactor volume is 10 m³.

In an ultrafilter cell recirculation system the separation factor Ω is chosen to 0.9, and v_1 and v_2 are 1.5

and 13.5 m³ h⁻¹, respectively.

A conservatively designed ultrafilter supports a flux $F = 10$ L m⁻² h⁻¹, and consequently the required filter area is 1350 m².

From an ultrafilter brochure one obtains the information that the necessary volumetric flow to the filter in order to sustain a flux $F = 10$ L m⁻² h⁻¹ through the filter is 0.74 m³ (m² filter)⁻¹. Hence a recirculation flow of $0.74 \cdot 1350 = 1000$ m³ h⁻¹ is needed.

Assume that a pressure drop of 4 bar across the filter is needed to support the flux, and that the pump has a mechanical efficiency of 65 %. Then the energy input is $4 \cdot 10^5 (0.65 \cdot 3600)^{-1} = 171$ kW.

The reactor capacity has been increased by a factor 10 by cell recirculation, but at the cost of a considerable energy input. A filter must be installed anyhow in order to separate the product stream from the biomass and remaining proteins, but a conventional drum filter working on only 15 m³ h⁻¹ is bound to be cheaper than an ultrafilter treating 1350 m³ h⁻¹.

A 190 000 t lactic acid per year plant (Dow –Cargill) has recently been commissioned in Omaha, Nebraska. Economic considerations of the kind illustrated in this example are typical for the design of large, integrated production plants.

9.1.4 The Stirred Tank with a Substrate Extracted From a Gas Phase

When one of the substrates is admitted with a gas phase that is sparged to the reactor there are, as already indicated in Chapter 3 two possible candidates for the rate limiting process, the gas-to-liquid transport and the bioreaction in the liquid phase.
If the transport process is rate limiting then:

$$k_l a (s_l^* - s_l) - Y_{xs_l} \mu(s_l) x = D (s_l - s_{lf}) \tag{9.47}$$

where s_l is the liquid phase concentration of the reactant from the gas phase and $\mu(s_l)$ is the specific growth rate, expressed as a function of the limiting substrate concentration s_l. As usual for steady state continuous reactors $D = \mu(s_l)$.
Both s_l and s_{lf} can safely be set equal to zero compared with s_l^*, and (9.47) degenerates to

$$q_x = D x = Y_{s_1 x} s_l^* k_l a \tag{9.48}$$

If on the other hand a substrate S_2 in the liquid phase is rate limiting then (for a Monod rate expression):

$$x = Y_{s_2 x}(s_{2f} - s_2) \text{ and } q_x = D x \tag{9.49}$$

The obvious choice is to work at conditions where q_x calculated from either (9.48) or (9.49) is the same.

Design of Fermentation Processes

This can be used to calculate the optimal value of the operating variable $k_l a$ for given D and s_{2f}.

$$(k_l a)_{opt} = D \frac{Y_{s_2 x}}{Y_{s_1 x}} \frac{s_{2f}}{s_1^*} \left(1 - \frac{K_2 D/s_{2f}}{\mu_{max} - D}\right) \tag{9.50}$$

Example 9.7 Optimal design of a single cell production.
Let S_1 be methane which is fed to the stirred tank reactor as natural gas (90% methane) mixed with oxygen in volumetric ratio 1 to 1.31, and $P = 1$ atm total pressure.
$s_1^* = 4.67$ mg L^{-1} and $Y_{s_1 x} = 0.8$ g g^{-1}.
The second substrate is ammonia. $\mu_{max} = 0.35$ h^{-1}, $K_2 = 50$ mg L^{-1}.
The biomass composition is $CH_{1.8}O_{0.5}N_{0.2}$.

Let the dilution rate be 0.2 h^{-1}, and s_{2f} 200 mg L^{-1}.

From (9.50) one obtains

$$(k_l a)_{opt} = 0.2 \frac{24.6/(0.2 \cdot 17)}{0.8} \frac{200}{4.67}\left(1 - \frac{50 \cdot 0.2/200}{0.35 - 0.20}\right) = 51.6 \text{ h}^{-1}$$

$$x = \frac{0.8 \cdot 4.67 \cdot 51.6}{0.2} = 965 \text{ mg L}^{-1} \tag{1}$$

It would be useless to increase $k_l a$ above 51.6 h^{-1} since the liquid phase reaction would then become rate limiting.

If the biomass concentration calculated in (1) is considered to be too low to obtain a reasonable volumetric production rate it becomes interesting to redesign the process for e.g. $x = 10$ g L^{-1} From (9.48) the required mass transfer coefficient is calculated to be $k_l a = 51.6 \cdot (10 / 0.965) = 535$ h^{-1}. But at the same time the ammonia feed concentration must be increased; otherwise the higher rate of methane consumption cannot be sustained.

Solution of (9.50) for s_{2f} when $k_l a$ is 535 h^{-1} yields $s_{2f} = 1449$ mg L^{-1}.

The conclusion is clearly that if one feed stream is "improved" then the other feed stream must also be "improved".

An overall maximum productivity is obtained when the volumetric rate of the bioreaction is maximized according to Eq (9.34-9.35) and the matching $k_l a$ value is calculated.

With the given data for the liquid phase reaction and with $s_{2f} = 1449$ mg NH$_3$ L^{-1} one obtains

$$D_{opt} = 0.2861 \text{ h}^{-1}, \quad x = Y_{s_2 x}(s_{2f} - s_2) = 8865 \text{ mg L}^{-1}, \quad q_x = D_{opt} x = 2.54 \text{ g L}^{-1}\text{h}^{-1}. \tag{2}$$

Finally from Eq (9.48):

$$k_l a = 2540 / (0.8 \cdot 4.67) = 679 \text{ h}^{-1} \qquad (3)$$

A mass transfer coefficient of 679 h^{-1} can be obtained, also in an industrial reactor designed for good mass transfer.

Consequently an over all maximum productivity improvement of 27 % compared with the case $D = 0.2$ h^{-1}, $x = 10$ g L^{-1} can be obtained by simultaneous optimization of the bioreaction and the mass transfer process.

When the liquid phase concentration of one of the reactants is constrained by a mass transfer balance one should be wary of using one of the popular rate expressions in which the specific growth rate is given as the product of two Monod expressions, one for each of the substrates.

Let S_2 be the liquid phase substrate and S_1 the substrate transferred from the gas phase.

The yield coefficients are Y_{s1x} and Y_{s2x} respectively. The mass transfer coefficient has a known value $k_l a$.

The amount of S_1 transferred from the gas phase is $q_1' = k_l a (s_1^* - s_1)$ while the rate of the bioreaction is $q_x = q_1' Y_{s1x}$.

Consequently q_x can be calculated in two ways:

$$q_x = Y_{s1x} k_l a (s_1^* - s_1) = \mu_{max} Y_{s2x} \frac{s_2}{s_2 + K_2} \frac{s_1}{s_1 + K_1} (s_{2f} - s_2) \qquad (9.51)$$

Eq (9.51) looks like an equation from which s_2 can be found when the value of s_1 is given. But this cannot be true for all values of s_1 since the right hand side of the equation is zero for $s_1 = 0$. There is clearly a lower value for s_1 below which no solutions for s_2 can be found. The right hand side of (9.51) has a maximum value determined from Eq(9.34), and consequently we must look for the lowest value of s_1 for which the left hand side of (9.51) is smaller than or equal to this maximum value. If s_1^* is increased (e.g by increasing the total pressure) then the lowest permissible s_1 value will also increase. If μ_{max} is increased then $(s_1)_{min}$ also decreases towards zero, and the process is controlled by mass transfer alone. The two values obtained for s_2 when s_1 is above its minimum value is no surprise. The same value for q_x can be obtained both for a small D and a large x, and for a large D and a small x.

Design of Fermentation Processes

9.1.5 Fed-Batch Operation

As indicated in table 9.1 the operation of bioreactors in semi batch mode is very popular in the fermentation industry. The reactor is started as a batch, and a suitably large biomass concentration is obtained by consumption of the initial substrate. The rate of product formation is typically low or even zero during the batch cultivation. At a certain time, usually when the substrate level has decreased to a very low level, a feed of (usually very concentrated) substrate is initiated. At the same time an *inducer* may be added to switch on the metabolic pathways that lead to the desired product. During the whole *fed batch* period no product is withdrawn from the reactor, and the medium volume keeps increasing. At the end of the fed batch period a certain portion of the reactor volume may be withdrawn and sent to down-stream processing. New substrate is added, and after a batch period a second fed batch period is started. This is *repeated fed batch* operation. It will work satisfactorily if the remainder of biomass used to grow up a new culture has not been weakened, and perhaps producing the desired product with a low yield.

When a time-varying feed stream $v(t)$ containing one or more of the substrates at a constant concentration c_f is admitted to a stirred tank reactor without withdrawal of a corresponding effluent stream, the mass balances become

$$\frac{d(\mathbf{c}V)}{dt} = V\frac{d\mathbf{c}}{dt} + \mathbf{c}\frac{dV}{dt} = V\frac{d\mathbf{c}}{dt} + \mathbf{c}v(t) = V\mathbf{q}(\mathbf{c}) + \mathbf{c}_f v(t) \tag{9.52}$$

The mass balances for the fed-batch operational mode therefore become

$$\frac{d\mathbf{c}}{dt} = \mathbf{q}(\mathbf{c}) + \frac{v(t)}{V(t)}(\mathbf{c}_f - \mathbf{c}) \tag{9.53}$$

There is a striking similarity between Eq. (9.53) and the mass balances for a stirred tank reactor. In Eq. (9.53), the dilution rate $D = v(t)/V(t)$ is, of course, a function of time, and the fed-batch model is in principle a purely transient model, although most results can safely be derived assuming that the growth is balanced.

One may visualize the fed-batch as a control problem: Subject to certain constraints it is possible to choose the control function $v(t)$ such that a given goal is reached. This goal may be defined at the end of the fermentation process where $V(t)$ has reached a specified value. This end-point control problem is a classical problem of control theory: e.g., to choose s_f and the initial values s_0 and x_0 that characterize the state when the fed-batch process $v(t)$ is initiated so that a given state (x,s) is reached in the shortest possible time. The chemical engineering literature abounds with solutions of this kind of problem. Typical references are Menawat *et al.* (1987) and Palanki *et al.* (1993). We shall, however, choose to study two simpler problems where the control action is applied with the purpose of achieving certain metabolic conditions for the cell culture at every instant during the fermentation. The concept of an instantaneous control action is illustrated in sufficient generality with only one growth-limiting substrate and the biomass as the state vector. To simplify the discussion maintenance free kinetics will be used.

The two most obvious feed policies are

- Choose $v(t)$ so that $s = s_0$ throughout the fermentation.
- Choose $v(t)$ so that $q_x = q_x^0$ throughout the fermentation.

The two policies correspond to fermentation at constant specific growth rate and at constant volumetric rate of biomass production, respectively. Both policies have obvious practical applications. When s is kept at a level below that at which part of the added substrate is converted to undesired products, a large amount of biomass (together with an associated protein which may be the real product) is produced at a reasonably high rate and a high final biomass concentration can be obtained. Neither the continuous stirred tank reactor (one is afraid that the subtle qualities of the yeast that gives the optimal leavening properties of the product will be lost in a long fermentation run) nor the batch reactor (diversion of glucose to ethanol, which inhibits the growth and represents a considerable loss of substrate) are suitable for baker's yeast production, but fed-batch operation is universally applied. The constant volumetric rate policy is important if removal of the heat of reaction is a problem or if the capability to supply another substrate, e.g., oxygen is exceeded when $q_x > q_x^0$.

Calculation of $v(t)$ corresponding to constant $s = s_0$ (which means that μ is constant and equal to μ_0) is quite simple. From Eq. (9.52), and with $x_f = 0$:

$$\frac{d(xV)}{dt} = \mu_0 xV \Rightarrow xV = x_0 V_0 \exp(\mu_0 t) \tag{9.54}$$

while the substrate balance taken from (9.53) reads:

$$\frac{ds}{dt} = 0 = -Y_{xs}\mu_0 x + \frac{v(t)}{V(t)}(s_f - s) \tag{9.55}$$

or

$$v(t) = \frac{Y_{xs}\mu_0}{s_f - s_0} xV = \frac{Y_{xs}\mu_0}{s_f - s_0} x_0 V_0 \exp(\mu_0 t) \tag{9.56}$$

$V(t)$ is found by integration of (9.56) from $t = 0$, and $x(t)$ by inserting $V(t)$ in (9.54).

$$\frac{V}{V_0} = 1 - bx_0 + bx_0 \exp(\mu_0 t) \tag{9.57}$$

where $b = Y_{xs}/(s_f - s_0)$, and

Design of Fermentation Processes

$$\frac{x}{x_0} = \frac{\exp(\mu_0 t)}{1 - bx_0 + bx_0 \exp(\mu_0 t)} \tag{9.58}$$

Equations (9.56) - (9.58) provide the complete explicit solution to the constant specific growth rate problem. v is seen to increase exponentially with time. The biomass concentration x is a monotonically increasing function of time with an upper limit $1/b = Y_{sx}(s_f - s_0)$ for $V \to \infty$. The value of x for a specified V/V_0 is calculated from (9.58) using a value of t obtained by solution of Eq. (9.57).

The constant q_x fed batch fermentation can also be designed quite easily. One would of course not voluntarily abandon the constant μ policy that gives the maximum productivity, consistent with the constraint that no byproducts should be formed, unless forced to do so. But with the increasing biomass concentration x (Eq (9.58)) a point may be reached at $t = t^*$ when e.g. the rate of oxygen transfer or the rate of heat removal from the reactor can no longer match the increasing q_x. From that point on we must work with $q_x = q_0 = q(x^*)$.

In the constant μ period the value of $s = s_0 = s(\mu_0)$ is usually many orders of magnitude smaller than s_f. In the continued fermentation with constant q_x from V^* to V_{final} the substrate concentration in the reactor decreases even further since μx is constant and there would be no purpose of continuing beyond t^* if x did not increase.

If we assume that the reactor volume keeps increasing exponentially also after t^* *then* the transient mass balance for the biomass from $t_1 = t - t^*$ when the constant μ period ends is:

$$\frac{dx}{dt_1} = \mu x - \frac{v(t)}{V(t)} x = q_0 - kx \tag{9.59}$$

The exponential increase of V is given by:

$$V = V^* \exp(k t_1) \tag{9.60}$$

The value of the parameter k will be determined shortly.

The substrate balance reads:

$$\frac{ds}{dt_1} = -Y_{sx}\mu x + \frac{v(t)}{V(t)}(s_f - s) = -Y_{sx}q_0 + k(s_f - s) \tag{9.61}$$

Integration of a weighted sum of the substrate and biomass balances yields

$$x + Y_{sx} s = (x^* + Y_{sx} s_0 - Y_{sx} s_f) \exp(-k t_1) + Y_{sx} s_f \tag{9.62}$$

Since s and s_0 are negligible compared to s_f (9.62) can be simplified to

$$x \approx (x^* - Y_{sx} s_f) \exp(-k t_1) + Y_{sx} s_f \qquad (9.63)$$

Integration of (9.59) yields another expression for x:

$$x = (x^* - q_0/k) \exp(-k t_1) + q_0/k \qquad (9.64)$$

The two expressions for x become identical if

$$q_0/k = Y_{sx} s_f \quad \text{or} \quad k = q_0/(Y_{sx} s_f) . \qquad (9.65)$$

The design of the constant q_x policy is therefore quite explicit. k is chosen according to (9.65) where q_0 as well as $Y_{sx} s_f$ are known. Hence the time t_1 to reach V_{final} is calculated from (9.60) and the corresponding x value from (9.64). The approximation in (9.63) is without any consequence for the result.

The constant q_x period can be shown to end with the same biomass concentration as would have been obtained if the constant μ policy could have been maintained until V_{final} was reached, but the processing time $t^* + (t_1)_{final}$ is longer, and hence the productivity is somewhat smaller.

Example 9.8 Fed batch fermentation to produce baker's yeast.
The biomass grows aerobically on glucose (S) with NH_3 as nitrogen source and

$$\mu = \frac{0.4 s}{s + 150 \,(\text{mg L}^{-1})} \qquad (1)$$

For $\mu \leq 0.25 \text{ h}^{-1}$ ($s \leq 250$ mg L^{-1}) the growth is purely respiratory and $Y_{xo} = 0.6836$ mol O_2 (C-mole biomass)$^{-1}$.

It is desired to design an optimal fed batch process starting at the end of a preliminary batch period in which the biomass concentration has increased to $x_0 = 1$ g L^{-1} and the glucose concentration has decreased to $s_0 = 250$ mg L^{-1}. The feed concentration during the fed batch operation is 100 g glucose L^{-1}. At $t = 0$ the reactor volume is V_0, and the fed batch process stops when $V = 4 V_0$. The temperature is 30° C and the oxygen is fed as air with 20.96% O_2.

Obviously the constant μ policy will select $\mu = \mu_0 = 0.25$ h^{-1}, the largest value of the specific growth rate for which no byproducts are formed.

Y_{sx} is calculated from a redox balance:

$$(1 - 1.05 \, Y_{sx}) = Y_{so} = 0.6836 \, Y_{sx} \quad \text{or} \quad Y_{sx} = 0.5768.$$

From Eq (9.56-9.58) with $b = Y_{sx}/(s_f - s_0) = (24.6/30)/(100 \cdot 0.5768) = 0.02114$ L g^{-1}

Design of Fermentation Processes

one obtains the following:

$$v(t) = 0.25 \cdot 0.02114 \cdot 1 \cdot V_0 \, exp(0.25\, t) \quad (2)$$

$$V(t) = V_0 (1 - b + b\, exp(0.25\, t)) \quad (3)$$

$$x(t) = exp(0.25t)\, V_0 / V \quad \text{for} \quad x_0 = 1 \text{ g L}^{-1} \quad (4)$$

At the time t_{final} when $V = V_{final} = 4\, V_0$ one obtains from (3) that $(1 - b + b\, exp(0.25\, t)) = 4$ and t_{final} is calculated to 19.84 h.

From (4) the corresponding x-value is determined to 35.7 g L^{-1}. $\quad (5)$

Assume that the largest attainable value of $k_l a$ is 650 h^{-1} and that the oxygen tension in the medium needs to be 10 % of the saturation value.

$$(q')_{max} = 650(1.16 \cdot 10^{-3} \cdot 0.2096) \cdot 0.9 = 0.1422 \text{ mol O}_2 \text{ h}^{-1} \text{L}^{-1}.$$

This oxygen uptake can support a volumetric biomass growth rate

$$(q_x)_{max} = (0.6836)^{-1}\, 24.6\, (q')_{max} = 5.116 \text{ g L}^{-1} \text{h}^{-1} \text{ corresponding to } x = x^\bullet = 20.46 \text{ g L}^{-1}.$$

Solving (4) for $x = 20.46 \text{ g L}^{-1}$ yields $t^\bullet = 14.26$ h, and from (3) $V = V^\bullet = 1.73\, V_o$.

From (9.65) $k = 5.116/(0.4730 \cdot 100) = 0.1082 \text{ h}^{-1}$.

Thus from $t = t^\bullet$ ($t_l = 0$) to t_{final} the reactor volume increases as $V = V^\bullet \, exp\,(0.1082\, t)$, and for $V = 4V_0$ (i.e. $V/V^\bullet = 2.31$) one obtains $t_l = 7.73$ h and $t_{final} = 22.0$ h.

x_{final} is calculated from (9.64) to 35.62 g L^{-1}, and apart from the permissible approximation in (9.62) this is the same as the concentration reached at the end of a constant—μ fed batch fermentation. The increase in production time from 19.8 to 22 h is not large.

The optimal design of a fed batch fermentation that gives the maximum productivity and yet satisfies the constraint imposed by a limited oxygen transfer should follow the lines illustrated in this example. One should, however, not be misled to believe that this is the over all best production policy. A continuous steady state fermentation has a far greater productivity.

Let the reactor volume be $4\, V_0$ since this volume must be available at the end of the fed batch process. If a continuous production of biomass with $x = 35.62 \text{ g L}^{-1}$ is to be maintained in the reactor then s_f should be $35.62/ Y_{sx} = 61.75 \text{ g L}^{-1}$ when the miniscule effluent glucose concentration is neglected. If $D = 0.25 \text{ h}^{-1}$, the highest D value for which no ethanol is produced, then $q_x = 0.25 \cdot 35.62 = 8.905 \text{ g L}^{-1} \text{h}^{-1}$ which cannot be supported by the available mass transfer coefficient. To obtain $q_x = q_{x\, max} = 5.116 \text{ g L}^{-1} \text{h}^{-1}$ the dilution rate must be lower, namely $D = 0.1436 \text{ h}^{-1}$. But still, a much higher volume of glucose can be processed to give $x = 35.62 \text{ g L}^{-1}$: $v = 4\, V_0 \cdot 0.1436 \text{ L h}^{-1}$ or in 22 hours $= t_{final}$ a total volume of $12.64\, V_0$ compared to only $4\, V_0$ by the optimal fed batch process.

9.2 The Plug Flow Reactor

The basic model for the tubular reactor is the so-called plug flow reactor model in which no concentration (or temperature) gradients in the radial coordinate are admitted. Hence there is only one spatial dimension, the distance z along the reactor axis. A mass balance for reaction component i in a volume element Adz where A is the reactor's cross-sectional area yields

$$\frac{\partial c_i}{\partial t} = -v_z \frac{\partial c_i}{\partial z} + q_i(c) \qquad (9.66)$$

where v_z is the linear velocity in the z direction.

The transient mass balance for the plug flow tubular reactor is a hyperbolic partial differential equation which can be solved using the method of characteristics, as described in Aris and Amundson (1973). $c_i(z,t=0)$ as well as $c_i(z=0,t)$ must be known in order to solve Eq. (9.66). If $c_i(z=0,t)$ is constant in time, a steady-state profile $c_i(z)$ will gradually develop. An inlet disturbance, e.g., a substrate pulse, travels along the reactor axis as a pulse of diminishing amplitude if c_i is being consumed. With these few comments on the transient equation, Eq. (9.66), we shall devote the remainder of this section to the steady-state solution.

If, as assumed in the plug flow model v_z is constant in the cross-section, the steady-state model becomes

$$\frac{d\mathbf{c}}{dt'} = \mathbf{q}(\mathbf{c}) \qquad (9.67)$$

where

$$\mathbf{c}(t'=0) = \mathbf{c}^0 \quad \text{and} \quad t' = z/v_z \qquad (9.68)$$

Equation (9.67) is mathematically identical to Eq. (9.2) with t', the time it takes a liquid element to travel a distance z along the reactor axis, replacing t. Physically speaking, the two reactor operations are, of course, quite different. The batch reactor is inoculated at time $t = 0$, and the condition in the ideally mixed tank changes as a function of time. The plug flow reactor is studied at steady state, it operates in a continuous mode, and no mixing even between neighboring fluid elements is admitted.

A continuous injection of cells at $z = 0$ is impracticable, and inoculation of the continuous plug flow reactor is done by placing it downstream from a continuous stirred tank. By this arrangement an efficient reactor system for conversion of substrate to a very low residual concentration is obtained. Recirculation of cells from the reactor outlet to $z = 0$ is another way of operating a continuous plug flow reactor.

The integrated form of Eq. (9.67) can be taken from the corresponding expressions for the batch

Design of Fermentation Processes

reactor in Section 9.1.1. Thus, for maintenance-free Monod kinetics, Eq. (9.10) is used to calculate the effluent concentration from the plug flow reactor.

$$\theta = \mu_{max} \cdot \frac{V}{v} = \left(1 + \frac{a}{1 + X_0}\right) \ln\left(\frac{X}{X_0}\right) - \frac{a}{1 + X_0} \ln(1 + X_0 - X) \qquad (9.69)$$

V is the total reactor volume and v the volumetric flow rate. The ratio V/v is termed the residence time (τ in chemical engineering literature).

As explained in connection with Eq. (9.10), the scaling factor used in a and X_0 is the substrate concentration s_0 at the start of the batch. Here s_0 is the inlet substrate concentration to the plug flow reactor. When, as often happens in practice, the plug flow reactor is installed downstream from a stirred tank reactor, it is more practical to use the inlet substrate concentration s_f of the sterile feed to the stirred tank as scaling factor. Thus, for maintenance-free kinetics

$$x_0 = Y_{sx}(s_f - s_0); \quad X_0 = \frac{x_0}{Y_{sx} s_0} = \frac{s_f}{s_0} - 1 \qquad (9.70)$$

and

$$a = \frac{K_s}{s_0}; \quad \frac{a}{1 + X_0} = \frac{K_s}{s_f} = a_f \qquad (9.71)$$

Furthermore,

$$1 + X_0 - X = \frac{s_f}{s_0} - \frac{x_0 + Y_{sx}(s_0 - s)}{Y_{sx} s_0} = \frac{s_f}{s_0} - \frac{s_f - s_0 + s_0 - s}{s_0} = \frac{s}{s_0} \qquad (9.72)$$

Consequently, Eq. (9.69) can be written in the more convenient form

$$\mu_{max} \tau = (1 + a_f) \ln\left(\frac{x}{x_0}\right) - a_f \ln\left(\frac{s}{s_0}\right) \qquad (9.73)$$

for the case of an unbroken chain of reactors. If side streams are admitted, the simplifications of Eq. (9.73) cannot be applied, but Eq. (9.69) still holds.

Example 9.9.   A chemostat followed by a plug flow reactor
It is desired to reduce the concentration of substrate in a sterile stream with $s_f = 60$ to 3 g m^{-3}. This is to be done in a combination of two reactors, a chemostat followed by a plug flow reactor. The kinetics is

$$q_x = \frac{4}{3}\frac{s}{4+s}x \text{ g m}^{-3}\text{h}^{-1} \qquad (1)$$

i.e., the Monod constant is 4 g m^{-3} while μ_{max} = 4/3 h^{-1}. Furthermore,

$$q_s = -10(\text{g g}^{-1})\cdot q_x \qquad (2)$$

and

$$v = 2.5 \text{ m}^3\text{h}^{-1} \qquad (3)$$

With two bioreactors in series, it is possible to minimize the total reactor volume necessary to reduce s from 60 to 3 g m^{-3}. The chemostat should be run at an S value for which q_x is maximum. i.e.,

$$S = -a_f + \sqrt{a_f^2 + a_f} = -\frac{4}{60} + \sqrt{\left(\frac{4}{60}\right)^2 + \frac{4}{60}} = 0.2 \qquad (4)$$

or

$$s_0 = 0.2\cdot 60 = 12 \text{ g m}^{-3}; \quad x_0 = 0.1\cdot(60-12) = 4.8 \text{ g m}^{-3} \qquad (5)$$

and

$$D = \frac{4}{3}\cdot\frac{4}{4+12} = 1 \text{ h}^{-1} \qquad (6)$$

i.e., V_I = 2.5 m^3. With this degree of preconversion, the plug flow reactor takes over at the point where q_x starts to decrease with decreasing s i.e. at the minimum on the curves in figure 9.3. It is known from any textbook on reaction engineering that a plug flow reactor is the best reaction vessel whenever the rate of conversion is a monotonically increasing function of the reactant concentration.

From Eq. (9.73),

$$\frac{4}{3}\tau = \left(1 + \frac{1}{15}\right)\ln\left(\frac{0.1\cdot(60-3)}{4.8}\right) - \frac{1}{15}\ln\left(\frac{3}{12}\right) \Rightarrow \tau = 0.2068 \text{ h} \qquad (7)$$

i.e., V_2 = 0.206 · 2.5 = 0.517 m^3.

No other chemostat-plus-plug flow reactor combination could give a smaller total reactor volume than $V_1 + V_2$ = 3.017 m^3 if the substrate content of a feed stream 2.5 m^3 h^{-1} is to be reduced from 60 to 3 g m^3

We shall now assume that another stream v_I = 0.5 m^3 h^{-1} of s = 30 g m^3 and x = 0 is introduced after the chemostat. The combined streams are to be treated in the plug flow reactor to give an effluent of s = 3 g m^{-3}. The chemostat is still operated so that the effluent is $(x, s) = (4.8, 12)$ g m^3, i.e., at its maximum production rate. Conditions at the inlet to the plug flow reactor are

Design of Fermentation Processes

$$x_0 = 4.8 \cdot 2.5/3 = 4 \text{ g m}^{-3} \quad ; \quad s_0 = \frac{12 \cdot 2.5 + 30 \cdot 0.5}{3} = 15 \text{ g m}^{-3} \tag{8}$$

And the exit conditions are $s = 3$ g m^{-3}, $x = x_0 + 0.1(15 - 3) = 5.2$ g m^{-3}. Thus

$$\frac{4}{3}\tau = \left[1 + \frac{4/15}{1 + 4/(0.1 \cdot 15)}\right] \ln\left(\frac{5.2}{4}\right) - \frac{4}{55}\ln\left(1 + \frac{40}{15} - \frac{52}{15}\right) = \frac{59}{55}\ln\left(\frac{5.2}{4}\right) - \frac{4}{55}\ln\left(\frac{1}{5}\right) = 0.3985 \tag{9}$$

i.e., $V_2 = 3/4 \cdot 0.3985 \cdot 3 = 0.897$ m^3 or $V_1 + V_2 = 3.397$ m^3. With an extra stream, the minimum of $V_1 + V_2$ is not quite at the point where the chemostat operation is optimized. The cell-free side stream requires a few more cells to be delivered from the chemostat to operate the plug flow reactor better. The true optimum is obtained for $(x, s) = (4.87, 11.3)$ g m^3 from the chemostat and $(V_1 + V_2)_{\min} = 3.3939$ m^3.

A tubular reactor can operate on a sterile feed if, once inoculated, a portion of the effluent stream is returned to the inlet. The general situation is shown in Fig. 9.5. The feed to the reactor contains both cells and substrate, and a portion $v_R = vR$ of the net stream is returned to the inlet and mixed with the feed stream. A mass balance at the point where the recycle stream joins the feed stream gives:

$$vs_f + Rvs = (1 + R)vs_1 \tag{9.74}$$

$$vx_f + Rvx = (1 + R)vx_1 \tag{9.75}$$

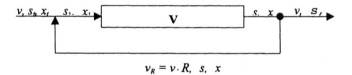

Figure 9.5 Schematic of a plug flow reactor with external recycle.

From Eqs. (9.74) and (9.75), the inlet concentrations to the plug flow reactor are obtained as functions of s_f, x_f, s, and x. The true flow through the reactor is $v(1 + R)$, and Eq. (9.69) can be applied directly to calculate the reactor volume. In the important case $x_f = 0$, one obtains from Eq. (9.73)

$$\mu_{\max} \frac{V}{v(1+R)} = (1 + a_f) \ln\left(\frac{x}{x_1}\right) - a_f \ln\left(\frac{s}{s_1}\right) \tag{9.76}$$

where $a_f = K_s / s_f$ as before.

Many industrial loop reactors closely resemble the recycle reactor shown schematically in Fig. 9.5. At some point of the loop, substrate is injected (e.g., industrial wastewater and oxygen), and at some point, the biomass containing effluent with low substrate content is removed. Apart from a head space where a gaseous reactant or product is separated from the liquid, the whole unit works as a plug flow reactor, quite often with substantial residence time in the loop (50-100 m tube length). The recycle ratio $R = v_R / v$ is often very high.

When R approaches infinity, the reactor is nothing more than a stirred tank continuous reactor ($D = v / V$). If $R = 0$, the reactor does not work when the feed is sterile. Since s is often desired to be $<< s_f$ and a stirred tank has a poor performance (low μ) for small s, it is intuitively clear that there must be an optimal recycle rate $R < \infty$ at least for some values of s / s_f. Thus, for a given value of s (or x) we shall determine the value of R for which $\tau = V / v$ is a minimum, i.e., where a given reactor volume V is capable of treating the highest possible feed stream v. For $x_f = 0$, one obtains from Eqs. (9.67) and (9.75)

$$\tau = (R+1) \int_{Rx/(R+1)}^{x} \frac{dx'}{q_x(x')} = (R+1) I(R) \tag{9.77}$$

Since x is a given quantity, the integral and hence τ are functions of R only. The minimum value of τ can be found by application of Leibnitz's rule:

$$I(p) = \int_{a(p)}^{b(p)} f(x,p) \, dx$$
$$\Downarrow \tag{9.78}$$
$$\frac{dI}{dp} = \frac{db}{dp} f[b(p),p] - \frac{da}{dp} f[a(p),p] + \int_{a(p)}^{b(p)} \frac{\partial f(x,p)}{\partial p} dx$$

In Eq. (9.78), the integrand f and both boundaries of the integral are assumed to be functions of the parameter p. $f[a(p), p]$ and $f[b(p), p]$ are the values of the integrand when the lower and upper boundary respectively, are inserted for the integration variable x. Applied to Eq. (9.77), where only the lower boundary is a function of R, we get

$$\frac{d\tau}{dR} = \int_{x_1}^{x} \frac{dx'}{q_x(x')} + (R+1)\left[-\frac{x}{(R+1)^2} \frac{1}{q_x(x_1)} \right] \tag{9.79}$$

or for $\dfrac{d\tau}{dR} = 0$:

$$\int_{x_1}^{x} \frac{dx'}{q_x(x')} = \frac{x}{R+1} \frac{1}{q_x(x_1)} = \frac{x - x_1}{q_x(x_1)} \tag{9.80}$$

The last expression of Eq. (9.80) uses the fact that

Design of Fermentation Processes

$$x - x_1 = x - \frac{Rx}{R+1} = \frac{x}{R+1}, \text{ or } S_1 = \frac{s_1}{s_f} = \frac{RS+1}{R+1} \quad (9.81)$$

Equation (9.80) allows a beautiful geometrical interpretation of the solution to the optimal recycle problem, an interpretation that is quite independent of the form of $q_x(x)$ (see Fig.9.6). The optimal choice of s_1 is that for which the area under the $1/q_x$ curve from s to s_1 is equal to the area of the rectangle with sides $1/q_x(s_1)$ and (s_1-s). It is immediately clear that an optimal solution to the recycle reactor design cannot be found when s is greater than the value of s where $q_x(s)$ attains its maximum value. For simple Monod kinetics, $R_{opt} \to \infty$ when $S = 1 - X \to -a + \sqrt{a^2 + a}$, the value of s / s_f that has been seen to give the highest cell-production rate in a stirred tank using sterile feed. For all larger values of s / s_f, a stirred tank works better than any type of recirculation reactor. Since the above value of S is always below 0.5 (for $a \to \infty$), one should, of course, always choose a continuous stirred tank when less than 50% conversion of substrate is desired.

The graphic solution of the reactor optimization problem works for any functional relation $\mu(s)$, but an algebraic solution may be more convenient if available. Thus for simple Monod kinetics and $x_f = 0$, one obtains the following expression for the integral in Eq. (9.80):

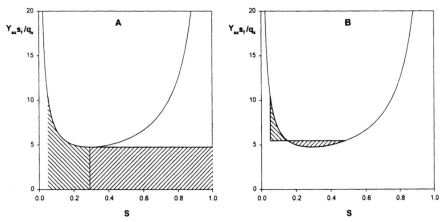

Figure 9.6 A. Plug flow reactor placed after a stirred tank reactor to give the highest productivity in the total reactor system. B. Selection of the best value of $S_1 = s_1 / s_f$ (= 0.4828) to give the optimal recycle ratio R (= 1.195) in a plug flow reactor with recycle.
In both A and B the kinetic data are taken from Figure 9.3 A (Monod kinetics). The desired value of S is 0.05.

$$\mu_{\max} \int_{x_1}^{x} \frac{dx'}{q_x(x')} = (1+a_f)\ln\left[\frac{x}{Rx/(1+R)}\right] - a_f \ln\left[\frac{s}{(s_f+Rs)/(1+R)}\right]$$

$$= (1+a_f)\ln\left(\frac{R+1}{R}\right) - a_f\left\{\ln\left[\frac{(R+1)s}{Rs}\right] - \ln\left[\frac{s_f}{Rs}+1\right]\right\}$$

$$= \ln\left(\frac{R+1}{R}\right) + a_f \ln\left(\frac{s_f+Rs}{Rs}\right) \tag{9.82}$$

$$= \ln\left(\frac{R+1}{R}\right) + a_f \ln\left(\frac{1+RS}{R}\right)$$

and the right-hand side of Eq. (9.80) is

$$\mu_{\max} \frac{x-x_1}{q_x(x_1)} = \frac{1}{R} + \frac{a_f(1+R)}{R(1+Rs/s_f)} \tag{9.83}$$

For given values of s, a_f, and s_f, one may determine R_{opt} by solution of Eq. (9.80) with Eqs. (9.82) and (9.83) inserted.

Example 9.10. Recycle reactor design
Determine the optimal recycle ratio for a tubular reactor that reduces a feed stream substrate concentration from 60 to 3 g m^{-3}. The kinetics are given in Example 9.9. The optimal recycle ratio is determined by solution of

$$\ln\left(\frac{R+1}{R}\right) + \frac{1}{15}\ln\left(\frac{20+R}{R}\right) = \frac{1}{R} + \frac{R+1}{R}\frac{1/15}{1+R/20} \tag{1}$$

Any kind of reasonable iteration method gives the solution $R = 2.50$ in a few iterations. From Eq. (9.77), with Eq. (9.82) inserted for the integral,

$$\mu_{\max}\tau = (R+1)\left[\ln\left(\frac{R+1}{R}\right) + \frac{1}{15}\ln\left(\frac{s_f+Rs}{Rs}\right)\right] \tag{2}$$

or

$$\tau = 0.75 \cdot 3.5\left[\ln(1.4) + \frac{1}{15}\ln(9)\right] = 1.26 \text{ h} \tag{3}$$

This result is obviously better than what would have been obtained in a chemostat:

$$\tau = \frac{1}{D} = \left(\frac{3+4}{\frac{4}{3}\cdot 3}\right) = 1.75 \text{ h} \tag{4}$$

Design of Fermentation Processes

which is excessively expensive when an almost quantitative conversion of substrate is desired.

The solution that gives minimum total residence time is, however, still the chemostat-plug flow reactor combination discussed in Example 9.9.

$$\tau = 1 + 0.2068 = 1.207 \tag{5}$$

The choice between two reactors in series (τ = 1.21 h) or one reactor (τ = 1.26 hr) with a recirculation pump is difficult. The first design is somewhat more complicated, while the pumping cost may be significant in the second design (R = 2.5).

In the present treatment of the tubular reactor, only the simplest form of the mass balance has been used. Much more complicated forms are found in the literature. Thus, if a gaseous substrate is fed in cross-flow (e.g., from a sparger mounted in the reactor axis), a transport term $k_l a(c_i^* - c_i)$, where c_i^* is the interfacial saturation concentration of the substrate, appears on the right-hand side of Eq. (9.54). Another term, the so-called axial dispersion term

$$D_{eff} \frac{d^2 c_i}{dz^2} \tag{9.84}$$

is also added on the right-hand side of Eq. (9.67). A theoretical foundation for this term can be deduced for unaerated fermentations. With the generally low reaction rates of bioreactions, the fluid has to move very slowly through the reactor and the velocity profile is probably parabolic $\left[v_z = v_0 (1 - r^2 / R^2) \right]$. Here radial concentration gradients exist, and through a perturbation solution of the underlying parabolic partial differential equation that describes the steady-state profile in the z and r directions, the influence of the radial transport is converted to a perturbation term like Eq. (9.84) in a one-dimensional differential equation [see, e.g., Villadsen and Michelsen (1978)]. Here D_{eff} is related to the molecular diffusivity of c_i in the liquid phase. In the design of industrial plug flow or loop reactors the overall mixing effect of gas bubbles, baffles, static mixers, and the like can be described in terms of the axial dispersion coefficient D_{eff}. Correlations for the effect of scale-up on reactor performance are often based on the concept of axial dispersion as discussed in Chapter 11. One may, on the other hand, have some reservations concerning the widespread use of axial diffusion terms in scientific studies of tubular (bio-)reactor performance. The simple model of Eq. (9.67) brings forward the main aspect of the subject without obfuscation by the numerical and conceptual difficulties tied in with the conversion of Eq. (9.67) from a first-order initial-value problem to a second-order boundary-value problem. The dispersion model contains more parameters and may give a better fit to a given set of laboratory data, but the theoretical basis of the model is weak, and lack of detail in the transient description of the microbial kinetics easily invalidates any attempt to give a physical interpretation of the model parameters.

9.3 Dynamic Analysis of Continuous Stirred Tank Bioreactors

In Section 9.1 and 9.2 the design of fermentation processes using a stirred tank or a plug flow reactor has been discussed. The design was primarily for steady state operation, although of course both the batch reactor and the fed batch reactor operate in a transient mode.

Although time is a variable in the batch and in the fed batch design problem the physiological state of the culture hardly changes during the entire exponential growth phase of a batch fermentation, and the lag-phase can be adequately analyzed using the simple compartment models discussed in Chapter 7 (Example 7.6). In the fed batch cultivation one would as explained in Section 9.1.5 choose to work with a constant μ (and consequently at constant physiological state) unless insufficient mass- or heat transfer makes it necessary to switch to a constant q_x operation. s decreases in the constant q_x period, but the physiological state of the cells is not likely to change-certainly not for an aerobic yeast fermentation, and not either in most other cases.

A dynamic analysis of the plug flow reactor is beyond the scope of this book, and the results of such an analysis would have much less value than the results of a similar analysis of a typical plug flow reactor for catalytic gas phase reactions in the chemical industry. For a bioprocess the dynamic analysis would be overlaid with many unpredictable phenomena due to the complexity of the bioreaction.

Fermentation in a stirred tank, continuous reactor is, however, very sensitive to disturbances in the environment. Continuous stirred tank reactors can safely be predicted to have an increasing appeal in future bioprocessing of cheap raw materials to make low value products. It is consequently important, especially for the process control of these reactors to analyze the dynamics of stirred tank continuous reactors. Since continuous stirred tanks are also the preferred choice for high quality scientific studies of cell physiology it is, as already indicated in Chapter 3, very desirable to include transients in the experimental study. To extract physiological information from the transients one must be able to model the reactor response to different transients.

This analysis is the subject of the present section. First the response of the reactor to sudden changes in the environment will be studied. Some fundamental aspects of reactor stability will be the final result. Thereafter the response of the culture in a continuous stirred tank to infection and to changes in the morphology of the culture will be described. The dynamics associated with changes in the culture composition are of great importance to the fermentation industry.

9.3.1 Dynamic Response of the Reactor for Simple, Unstructured Kinetic Models

A dynamic response is typically observed after a step change of the dilution rate D, after a change in the feed substrate concentration s_f, or after a pulse of substrate has been added to the reactor.

Design of Fermentation Processes

Each of these responses can be studied in some generality if the kinetic model is very simple. First of all it will be assumed, that the rates are proportional, a key assumption in the introductory discussion of growth kinetics in Section 7.3.1. Thus a substrate demand for maintenance is not considered at first, and the stoichiometry of the black box model is taken to be independent of the environment. These assumptions are unrealistic, and after having discussed the concepts of dynamic reactor modelling more realistic models will be considered.

When the feed contains neither biomass nor product, and when the yield coefficients are constants the following dynamic mass balances are obtained:

$$\frac{dx}{dt} = \mu x - Dx$$
$$\frac{ds}{dt} = -Y_{xs}\mu x + D(s_f - s) \qquad (9.85\ a,\ b,\ c)$$
$$\frac{dp}{dt} = Y_{xp}\mu x - Dp$$

The specific growth rate is assumed to depend on both s and p, and at $t = 0$, $(x, s, p) = (x_0, s_0, p_0)$. The mass balances can be linearly combined to give:

$$\frac{d(x + Y_{sx}s)}{dt} = -D(x + Y_{sx}s) + DY_{sx}s_f$$
$$\frac{d(x - Y_{px}p)}{dt} = -D(x - Y_{px}p) \qquad (9.86\ a,b)$$

with the solution

$$x + Y_{sx}s = A\exp(-Dt) + Y_{sx}s_f$$
$$x - Y_{px}p = B\exp(-Dt) \qquad (9.87\ a,\ b)$$

The two arbitrary constants A and B are found from the initial conditions that apply at $t = 0-$:

a D is changed from D_0 to D at $t = 0+$; s_f is unchanged:

$$x_0 + Y_{sx}s_0 = A + Y_{sx}s_f$$

and since at the steady state before the change of D the left hand side is equal to $Y_{sx}s_f$ then the arbitrary constant A must be zero. Next B is determined from (9.87 b):

$$x_0 - Y_{px} p_0 = x_0 - Y_{px} Y_{sp} (s_f - s_0) = x_0 - Y_{sx} (s_f - s_0) = 0 = B.$$

b Feed substrate concentration changed from s_f^0 to s_f; D is unchanged:

$$x_0 + Y_{sx} s_0 = A + Y_{sx} s_f^0 + Y_{sx} (s_f - s_f^0) \quad \text{or} \quad A = - Y_{sx} (s_f - s_f^0).$$

$$x_0 - Y_{px} p_0 = x_0 - Y_{sx} (s_f^0 - s_0) = 0 = B.$$

c A pulse of substrate added at $t = 0+$. D and s_f are unchanged.

$$x_0 + Y_{sx} s_0 + \Delta s = A + Y_{sx} s_f, \quad \text{or} \quad A = \Delta s.$$

$$x_0 - Y_{px} p_0 = x_0 - Y_{sx} (s_f - s_0) = 0 = B.$$

In all three situations B is zero, and during the whole transient $p = Y_{xp} x$. The substrate- and biomass balances are also uncoupled. For a step change in D the relation between s and x is an algebraic equation $s = s_f - Y_{xs} x$. When s_f is changed (case **b**) or when a pulse of substrate is added (case **c**) then

$$x + Y_{sx} s = Y_{sx} s_f - (s_f - s_f^0) Y_{sx} \exp(-D t)$$

$$x + Y_{sx} s = \Delta s \exp(-D t)$$

(9.88 a, b)

For $s_f > s_f^0$ the sum of x and $Y_{sx} s$ is always below its final value $Y_{sx} s_f$. In case **c** the sum is always positive.

Since D is unchanged during the transient then at the end of the transient when s and p have reached their final values s_∞ and p_∞:

$$D = \mu (s_0, p_0, s_f^0) = \mu (s_\infty, p_\infty, s_f) \tag{9.89}$$

If μ does not depend on p then s must return to its initial value, and $s_\infty = s_0$. Then from (9.88 a):

$$x_\infty = Y_{sx} s_f - Y_{sx} s_\infty = Y_{sx} (s_f - s_0) = x_0 + Y_{sx} (s_f - s_f^0) \tag{9.90}$$

When μ depends on p then s_∞ is not equal to s_0 since s_f will appear in μ when p is eliminated using a mass balance relation between s and p. The final value of x is obtained from

$$x_\infty = Y_{px} p_\infty, \text{ where } p_\infty = Y_{sp} (s_f - s_\infty) \text{ and } s_\infty \text{ is determined by solution of (9.89). (9.90)}$$

In the pulse experiment both x and s return to their initial values following (9.88 b) *provided that the initial steady state is stable,* and that the added pulse is not too large. This aspect will be

Design of Fermentation Processes

discussed in Section 9.3.2.

Example 9.11 Calculation of final values of s, p and x after a change in s_f.
Consider the situation of figure 9.3 C with $s_f^0 = 2$ g L^{-1} and $s_f = 4$ g L^{-1}. $Y_{sp}=Y_{sx}= \frac{1}{2}$.

For s_f^0 the parameter a = 0.2 and for s_f the parameter is 0.1.
Likewise the parameter $s_f Y_{sp}/p_{max}$ changes from = ⅔ to 1 ⅓.

For D = 0.2 h^{-1} the algebraic equation (1) is solved for $s_f = 2$ and 4 g L^{-1}, respectively:

$$D = \frac{\mu_{max} s}{s + K_s}\left(1 - \frac{Y_{sp}(s_f - s)}{p_{max}}\right) \tag{1}$$

For $s_f = 2$ g L^{-1} : S = 0.4 and s = 0.8 g L^{-1}. x = 0.5 (2 – 0.8) = 0.6 g L^{-1} = p.

For $s_f = 4$ g L^{-1} : S = 0.6 and s = 2.4 g L^{-1}. x = p = 0.8 g L^{-1}.

Due to the product inhibition the steady state value of s is much higher at $s_f = 4$ g L^{-1} than at 2 g L^{-1}. The relation (9.90) is not valid anymore.

After this general discussion of the relationship between x, s and p during the transient, the time-profile will be calculated for a few examples.

a First consider a change in D. The algebraic relations between s and x and between p and x are inserted in the biomass balance (9.85 a) and this is solved by separation of variables from $x(t=0) = x_0$.

$$\frac{dx}{dt} = \mu(s(x), p(x))x - Dx \quad ; \quad x(t=0) = x_0$$

$$dt = \frac{dx}{(\mu(x) - D)x} \tag{9.92}$$

For Monod kinetics:

$$\mu = \frac{\mu_{max}(s_f - Y_{xs}x)}{K_s + (s_f - Y_{xs}x)} \tag{9.93}$$

(9.93) is inserted in (9.92) which is integrated to

$$t = \frac{1}{D_{max} - D}\left(\ln z - \left(1 - \frac{D_{max} - D}{\mu_{max} - D}\right)\ln\left(\frac{z - z_\infty}{1 - z_\infty}\right)\right) \quad (9.94)$$

$z = x/x_0$, and D_{max} is the wash-out dilution rate $\mu_{max}(a+1)^{-1}$ in Eq (9.27).

The parameter z_∞ is given by:

$$z_\infty = \frac{D_{max} - D}{D_{max} - D_0}\frac{\mu_{max} - D_0}{\mu_{max} - D} \quad (9.95)$$

When $n = D_{max}/D > 1$ then z_∞ is the final value of x/x_0 at the end of the transient. If $n < 1$ the final value of x is of course zero since the culture is washed out. In that case z_∞ is to be regarded as just another parameter in the expression for the transient.

For the special case of $n = 1$ Eq (9.96) is used instead of Eq (9.95) to calculate the transient.

$$t = \frac{Y_{sx}s_f}{x_0}\frac{1+a}{D_{max}a}\left(\frac{1}{z} - 1\right) + \frac{1}{D_{max}a}\ln z \quad (9.96)$$

The final value of x is also zero for $n = 1$.

Example 9.12 Change of D for maintenance-free Monod kinetics
Consider Monod kinetics with the parameters used in Example 9.4:

$$\mu_{max} = 1\,h^{-1} \quad ; \quad Y_{sx} = 0.5 \quad ; \quad K_s = 1\,g\,L^{-1} \quad (1)$$

and the operating parameters $s_f = 4\,g\,L^{-1}$, $D = 0.4\,h^{-1}$ before $t = 0$, and 0.6, 0.8, or 4/3 h^{-1} after $t = 0$.

First, the steady state before $t = 0$ is calculated:

$$s_0 = \frac{K_s D}{\mu_{max} - D} = \frac{1 \cdot 0.4}{1 - 0.4} = \tfrac{2}{3}\,g\,L^{-1} \quad ; \quad x_0 = 0.5\left(4 - \tfrac{2}{3}\right) = \tfrac{5}{3}\,g\,L^{-1} \quad (2)$$

$$\frac{x_0}{Y_{sx}s_f} = \frac{\tfrac{5}{3}}{0.5 \cdot 4} = \tfrac{5}{6} \quad (3)$$

Next, we consider the transient where $D = 0.6\,h^{-1}$ after $t = 0$. Since

Design of Fermentation Processes

$$n = \frac{D_{max}}{D} = \frac{0.8}{0.6} = \tfrac{4}{3} > 1 \tag{4}$$

the transient ends in a new steady state where

$$z_\infty = \frac{x_\infty}{x_0} = \frac{0.8-0.6}{0.8-0.4} \cdot \frac{1-0.4}{1-0.6} = \tfrac{3}{4} \tag{5}$$

and

$$t = 5\left\{\ln\left(\frac{x}{x_0}\right) - \tfrac{1}{2}\ln\left[4\left(\frac{x}{x_0} - \tfrac{3}{4}\right)\right]\right\} \tag{6}$$

If instead of $D = 0.6$ h^{-1}, we choose $D = 0.8$ h^{-1} after $t = 0$, then $n = 1$ and

$$t = \tfrac{6}{5} \cdot \tfrac{25}{4}\left(\frac{x_0}{x} - 1\right) + 5\ln\left(\frac{x}{x_0}\right) = 7.5\left(\frac{x_0}{x} - 1\right) + 5\ln\left(\frac{x}{x_0}\right) \tag{7}$$

Finally, take $D = 4/3$ h^{-1}, for which $n = 3/5$:

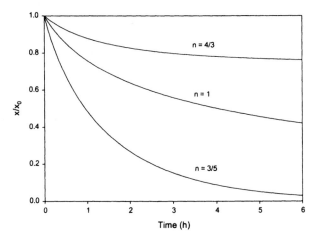

Figure 9.7. Simulation of transient experiments in a stirred tank reactor for Monod kinetics. At time zero, D is changed from 0.4 h^{-1} to either 0.6, 0.8 or 4/3 h^{-1}.
$\mu_{max} = 1$ h^{-1}, $a = K_s/s_f = 0.25$; $X_0 = x_0/(Y_{sx}s_f) = 5/6$.

$$t = -\tfrac{15}{8}\left\{\ln\left(\frac{x}{x_0}\right) + \tfrac{3}{5}\ln\left[\tfrac{5}{7}\left(\tfrac{12}{5} - \frac{x}{x_0}\right)\right]\right\} \tag{8}$$

The three transients illustrating the approach to a new nonzero steady state ($n = 4/3$), for the washout dilution rate $D = D_{max}$, and for D in excess of D_{max} ($n = 3/5$) are shown in Fig. 9.7. The three curves approach the new steady state at very different rates, the smallest rate being that corresponding to $n = 1$. Although the kinetics used here are too simple to imitate a real transient, the observed difference in the approach to a new steady state after a change in D is found also with other kinetics expressions. Whereas an approach to within 90% of the new steady state is reached after about 4-6 hours for $D = 0.6$ h^{-1} and $D = 4/3$ h^{-1}, 56 hours is required for $D = D_{max} = 0.8$ h^{-1}.

A comparison of the relative magnitude of the two terms in Eq. (9.94) results in an accurate method for determination of μ_{max} from transient wash-out experiments. Thus for $n < 1$ the first term eventually dominates when $x/x_0 \to 0$.

$$t \approx \frac{1}{D_{max} - D}\ln\left(\frac{x}{x_0}\right) + c \quad \text{for} \quad n < 1 \quad \text{or} \quad \ln\left(\frac{x}{x_0}\right) = (D_{max} - D)t + c^* \tag{9.97}$$

The slope of the straight line $\ln(x/x_0)$ vs. t is the difference between D_{max} and the D value chosen in the wash-out experiment. When using (9.97) to find D_{max} one must keep in mind that the basis for the method is an unstructured kinetic model. Results such as those shown in Figure 7.14 are obtained if the initial state of the biomass is far from that of a fully active biomass.

b Change of the feed concentration from s_f^0 to s_f.

Although x and s can be separated using (9.88 a) the resulting differential equation cannot be solved analytically, and as is almost always the case the dynamic mass balance must be solved by numerical integration.

Design of Fermentation Processes

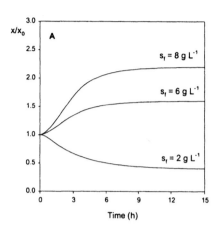
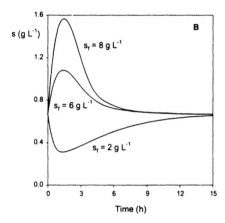

Figure 9.8. Transient in a constant-D, stirred tank continuous reactor following a change in substrate feed concentration from $s_f^0 = 4$ g L^{-1}. Kinetic parameters as in Example 9.12, and $D = 0.4$ h^{-1}. A: $z = x / x_0$ for $s_f = 8$, 6 and 4 g L^{-1}. B: substrate concentration, g L^{-1}.

Figure 9.8 A and B shows the transients of $z = x / x_0$ and of s for the same set of kinetic parameters as were used in figure 9.7. $D = 0.4$ h^{-1} and $s_f^0 = 4$ g L^{-1}. It is noted that s returns to its initial value ($s_0 = ⅔$ g L^{-1}) as is the case whenever μ is a function of s alone, and D is kept constant during the experiment. In the beginning there is an overshoot in s when $s_f > s_f^0$. There is not enough biomass available to cope with the larger influx of substrate, but eventually x increases to its final value $x_\infty = x_0 + Y_{sx}(s_f - s_f^0)$, i.e
$z \to (s_f - s_0) / (s_f^0 - s_0)$.

The curves on figure 9.8 were simulated using Matlab to solve

$$\frac{dz}{dt} = \mu_{max} \frac{-fX_0 z + (f-1)\exp(-Dt) + 1}{a - fX_0 z + (f-1)\exp(-Dt) + 1} z - Dz \qquad (9.98)$$

with $z = 1$ for $t = 0$; $f = s_f^0 / s_f$ and $X_0 = x_0 / (Y_{sx} s_f^0)$. $D = 0.4$ h^{-1}.

c Transients caused by pulse addition of substrate.

When a pulse of substrate is added to a continuous stirred tank reactor s immediately jumps to $s_0 + \Delta s$. Extra biomass is formed, but according to Eq (9.88 b) the sum $x + Y_{sx} s$ decreases exponentially towards $Y_{sx} s_f$ with a time constant D^{-1}. Finally x and s return to their original steady state values x_0 and s_0, unless the original steady state is unstable or the pulse Δs is too large.

A subject of great interest in bioreactions as well as in any other physical system governed by non linear dynamics is the analysis of the stability of a given steady state to perturbations in the state variables, here x, s and p. In our discussion perturbations in s are of specific interest since addition of biomass will always lead to a return to the steady state (s is consumed faster, the rate of biomass production decreases while the loss of biomass to the effluent is higher than at steady state until x has returned to x_0). Likewise an addition of a pulse of product will typically lead to a lower growth rate since the product inhibits growth, less product is formed and finally the original steady state is restored. This is not always so for addition of a substrate pulse. If μ decreases with increasing s as is the case for substrate inhibited kinetics (Eq 7.21) beyond the maximum in $\mu(s)$ then an addition of s leads to a lower growth rate, a further increase of s and finally to wash out of the biomass.

This verbal description of stability of a steady state to pulses in the state variables can be translated to a rigorous mathematical analysis by standard methods from mathematical physics.

9.3.2. Stability Analysis of a Steady State Solution

To illustrate the concepts of this analysis the stability of a steady state solution to Eq (9.85 a, b) will be discussed. Including a product balance and substrate consumption for maintenance can be done without further complications, but this case is left as a problem (Problem 9.6).

Thus, if we define the *state vector* $\mathbf{c}$ as (x,s) and a vector function $\mathbf{F}$ of the vector variable $\mathbf{c}$ as the right hand sides of Eq (9.85 a, b): $F_1 = \mu x - D x$; $F_2 = -Y_{xs} \mu x + D(s_f - s)$ then

$$\mathbf{F} \approx \mathbf{F}_0 + \left(\frac{\partial \mathbf{F}}{\partial \mathbf{c}}\right)_{c_0} d\mathbf{c} \tag{9.100}$$

$\left(\dfrac{\partial \mathbf{F}}{\partial \mathbf{c}}\right)_{c_0}$ is defined as $\quad \mathbf{J} = \begin{pmatrix} \dfrac{\partial F_1}{\partial x} & \dfrac{\partial F_1}{\partial s} \\ \dfrac{\partial F_2}{\partial x} & \dfrac{\partial F_2}{\partial s} \end{pmatrix}_{x_0, s_0}$ (9.101)

$\mathbf{F}_0 = \mathbf{F}(x_0, s_0) = \mathbf{0}$, since in the steady state both F_1 and F_2 are zero.

Consequently we obtain the following *linear* differential equations in the *deviation variable* $\mathbf{y} = \mathbf{c} - \mathbf{c}_0$:

Design of Fermentation Processes

$$\frac{dy}{dt} = J \cdot y \qquad (9.102)$$

This was of course just what was done in Example 6.5 when the two non-linear rates r_1 and r_2 were linearized around a given state (x^0, s^0).

The solution of the two linear differential equations with constant coefficients is a weighted sum of exponentials. If both exponentials have negative arguments then the perturbation will die out, and the steady state will be exponentially approached.

The *eigenvalues* λ_i of the *Jacobian Matrix* **J** determine the exponentials of the solution.

Matrix **J** of dimension (N × N) has N eigenvalues. These are determined as the N zeros of the polynomial obtained by calculating the determinant of the matrix **J** $- \lambda$ **I**.

For our case of N = 2 the determinant is a polynomial of degree 2 in λ.

$$Det \begin{pmatrix} \mu - D - \lambda & \mu_s x \\ -Y_{xs}\mu & -Y_{xs}\mu_s x - D - \lambda \end{pmatrix}_{x_0, s_0} \qquad (9.103)$$

μ_s is defined as $\left(\frac{\partial \mu}{\partial s}\right)_{ss}$ and the subscript ss stands for the steady state that is investigated.

In the steady state $\mu = D$ (while μ must be different from D during the transient since s is different from s_0). With this simplification the following equation is obtained for λ:

$$\lambda(Y_{xs}\mu_s x + D + \lambda) + Y_{xs} x D \mu_s = 0 \qquad (9.104)$$

$$\lambda = \begin{cases} -D \\ -Y_{xs} x \mu_s \end{cases}$$

There is always one negative eigenvalue $= -D$, and if μ_s is positive for the steady state the other eigenvalue is also negative.

For Monod kinetics μ increases monotonically with s, and μ_s is positive for any steady state. For substrate inhibition kinetics, Eq (9.28), all steady states to the left of the maximum of $\mu(s)$ at $s = (K_i K_s)^{1/2}$ have $\mu_s > 0$, while $\mu_s < 0$ to the right of the maximum

The stability of the steady state is determined by the following

- If all eigenvalues have a negative real part then the steady state is *asymptotically stable*.

This means that the state is stable, at least for infinitely small perturbations of the state variables.
- If just one eigenvalue has a positive real part then the steady state is unstable.
- If the imaginary part b_i of all eigenvalues $\lambda_i = a_i \pm i\, b_i$ is zero then the deviation variable y increases exponentially away from y_0 ($a_i > 0$) or decreases exponentially to $\mathbf{0}$ ($a_i < 0$). If any $b_i \neq 0$ then the movement away from y_0 or towards $\mathbf{0}$ is oscillatory.

By these rules all steady states are asymptotically stable for Monod kinetics while steady states with $s > (K_i K_s)^{1/2}$ are unstable for the substrate inhibition kinetics. Oscillations will not occur, neither for Monod kinetics nor for substrate inhibition kinetics.

The asymptotic stability analysis gives no clue as to the final goal of the path away from an asymptotically unstable steady state. It does not either tell what happens if an asymptotically stable steady state is submitted to a perturbation of finite magnitude.

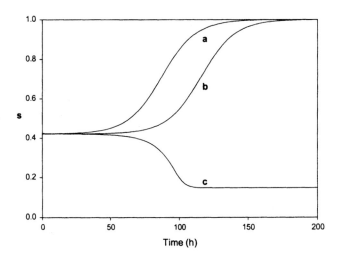

Figure 9.9. Time profiles of the response of a continuous stirred tank reactor which is disturbed by applying a pulse ΔS to the unstable steady state. Substrate inhibition kinetics with $\mu = 0.5\, S / (S^2 + S + 0.0625)$; $S = s / s_f = 0.42162$ at the steady state $D = 0.3185$ h^{-1}.
$\Delta S = 0.00338$ (a), 0.00038 (b), and -0.00062 (c).

Design of Fermentation Processes

Figure 9.9 gives the time profiles of $S = s/s_f$ after perturbation of an unstable steady state with substrate inhibition kinetics. S tends to either $S = 1$ (for a positive perturbation of S) or (for a negative perturbation of S) to the asymptotically stable steady state which accompanies the unstable steady state for any D value below μ_{extr} of Eq (9.29).

Those two stable steady states are the *attractors* for the solution of the dynamic mass balances. In Figure 9.11 a much more general representation of the dynamics is given in terms of a phase diagram for the same example as discussed in Figure 9.9. Here the track of (S,X) is followed from any starting point (S_0,X_0), and the diagram clearly shows which initial conditions will lead to a transient that ends up in each of the two attractors. The diagram also shows that a large positive S-perturbation of the asymptotically stable steady state will bring the transient "over the hill" and force it towards the stable (but trivial) steady state at $S = 1$.

Maintenance substrate demands coupled to more complex kinetics is easily included in the analysis if numerical solution of the dynamic mass balances is acceptable. Only in a few exceptional cases will a more complicated rate expression $\mu(x,p)$ admit to an analytical treatment. We shall briefly discuss the solution of the mass balances when $q_s = -(Y_{xs}^{true} \mu(s) + m_s)x$ as in Eqs (9.12)-(9.13). In this case, and in complete analogy with the derivation of (9.104) one obtains:

$$2\lambda = -(Y_{xs}\mu_s x + D) \pm \sqrt{(Y_{xs}\mu_s x + D)^2 - 4\mu_s x m_s} \qquad (9.105)$$

The solution of Eq. (9.105) is

$$\lambda = \begin{cases} -D - \delta \\ -Y_{xs}\mu_s x + \delta \end{cases} \quad \text{for} \quad \mu_s > 0 \quad \text{at} \quad (x_0, s_0) \qquad (9.106a)$$

$$\lambda = \begin{cases} -D + \delta \\ -Y_{xs}\mu_s x - \delta \end{cases} \quad \text{for} \quad \mu_s < 0 \quad \text{at} \quad (x_0, s_0) \qquad (9.106b)$$

where the value of the positive constant δ can be calculated for a given maintenance constant m_s, and it is assumed that

$$(Y_{xs}\mu_s x - D)^2 > 4\mu_s x m_s \quad \text{when} \quad \mu_s > 0 \qquad (9.107)$$

With reference to the previous example of substrate-inhibition kinetics it appears that maintenance destabilizes the nontrivial steady states when $\mu_s > 0$, while an unstable steady state would be somewhat stabilized. We shall give a little more detail for the very simplest case where μ is given by the Monod expression. For a given $S = s/s_f$, $D = \mu_{max} S/(S+a)$, and $X = x/(Y_{sx}s_f)$ given by Eq. (1) of Example 9.3:

$$\mu_s = \frac{1}{s_f}\frac{d\mu}{dS} = \frac{1}{s_f}\frac{a}{S(S+a)}D \qquad (9.108)$$

and Eq. (9.105) simplifies to

$$2\lambda = D\left\{-\left[X\frac{a}{S(S+a)}+1\right] \pm \sqrt{\left[X\frac{a}{S(S+a)}-1\right]^2 - \frac{4abX}{S^2}}\right\} \quad (9.109)$$

For all steady-state values (S, D), a real eigenvalue is always negative. There may be complex eigenvalues, but their real part is always negative. Consequently, all the steady states are stable, but oscillations may occur in the transient if for a given steady state

$$\left[X\frac{a}{S(S+a)}-1\right]^2 < \frac{4abX}{S^2} \quad (9.110)$$

For $a = 0.2$ and $b = 0.1$, the inequality of Eq. (9.110) is satisfied when $0.154 < S < 0.388$, i.e., for a steady state with an X value somewhat to the right of the maximum of the X-versus-S curve. A perturbation of any of these steady states results in a damped oscillation back toward the steady state. The largest amplitude of the oscillations is expected to be in the vicinity of the point where $\Phi = \mathrm{Im}(\lambda)/\mathrm{Re}(\lambda)$ is at a maximum. This occurs for $S \approx 0.42$ where $\Phi = 0.42$. Figure 9.10 shows the transient when the steady state (S_0, X_0) = (0.26, 0.62876) is perturbed by $\Delta S_0 = 0.24$ at $t = 0$. The solution is found by numerical integration of

$$\begin{aligned}
\frac{dX}{d\theta} &= \left(\frac{S}{S+a} - \frac{D}{\mu_{max}}\right)X \\
\frac{dS}{d\theta} &= -X\left(\frac{S}{S+a} + b\right) + \frac{D}{\mu_{max}}(1-S)
\end{aligned} \quad (9.111)$$

where $a = 0.2$, $b = 0.1$, $D/\mu_{max} = S_0/(S_0 + a) = 0.26/0.46 = 0.56522$. $S(\theta = 0) = 0.50$, and $x(\theta = 0) = 0.62876$. Only one overshoot of the steady state is noticed on the scale of the figure.

Design of Fermentation Processes

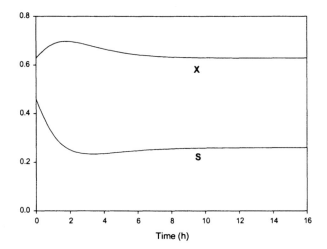

Figure 9.10 Perturbation of the steady state $(S_0, X_0) = (0.26, 0.62876)$ with a substrate pulse $\Delta S = 0.24$. Monod kinetics with maintenance ($b = 0.1$). $a = K_s/s_f = 0.2$.

Example 9.13 Startup of a chemostat

When the steady state is unstable for some S values in $]0, 1[$ the startup of a chemostat from a batch may require some thought. At the end of the batch fermentation, the substrate and biomass concentrations may be far from their steady-state values, and although the desired steady state may be stable, it is not certain that it will be reached after the switchover to continuous operation of the bioreactor.

Consider the substrate inhibition kinetics of Eq. (9.28) with the same values of the kinetic parameters $\mu_{max} = 0.5$ h^{-1}, $a = 1/16$, $b = 1$ as have been used before (Fig. 9.9). It is desired to operate the steady-state chemostat at the $S = s/s_f$ value for which maximum cell productivity is obtained:

$$q_x = P_x = \mu x \quad \text{or} \quad \frac{P_x}{\mu_{max} Y_{sx} s_f} = \frac{S(1-S)}{bS^2 + S + a} \tag{1}$$

The maximum value of P_x is obtained as in Example 9.3, Eq. (5).

$$S = -c + \sqrt{c^2 + c} \tag{2}$$

where

$$c = \frac{a}{1+b} = \tfrac{1}{32} \; ; \; S = 0.14826 \tag{3}$$

At this S value, the dilution rate for steady-state operation is

$$D = \frac{0.5S}{S^2 + S + \tfrac{1}{16}} = 0.3185 \, \text{h}^{-1} \tag{4}$$

Starting the batch operation with $X = x_0/(Y_{sx}s_f) = X_0$ and $S = s/s_f = S_0$, we shall after a certain batch time t_b switch to continuous operation with $D = 0.3185 \, \text{h}^{-1}$. For simplicity, let $S_0 = 1$, i.e., we start the batch with a substrate concentration equal to the feed concentration to be used in the ensuing continuous operation.

The transient chemostat operation is calculated by numerical solution of

$$\frac{dX}{d\theta} = \left(\frac{S}{bS^2 + S + a} - \frac{D}{\mu_{max}} \right) X \tag{5}$$

$$\frac{dS}{d\theta} = -\frac{S}{bS^2 + S + a} X + \frac{D}{\mu_{max}}(1-S) \tag{6}$$

where

$$\theta = \mu_{max} t = 0.5 t \; ; \; \frac{D}{\mu_{max}} = \frac{0.3185}{0.5} = 0.63702 \tag{7}$$

and $(X, S) = [X(t_b), S(t_b)]$ at the start ($t = 0$) of the continuous operation. Recalling the instability of steady states with $S > \sqrt{a/b}$, discussed in connection with Eq. (9.104), we may expect trouble if the startup of the chemostat is initiated when $S > (1/16)^{1/2} = 0.25$, i.e., at too short a batch time t_b, calculated from the explicit expression of Eq. 11 in note 8.1 where $X_0 = 0.05$ and $S = 0.14826$ at $t = t_b$ and $X(t_b)$ is found from Eq. (9.6b):

$$X(t_b) = 0.05 + (1 - 0.14826) = 0.90174 \tag{8}$$

$$t_b = 10.726 \, \text{h} \tag{9}$$

The "success" of the start up procedure can be judged with the help of figure 9.11.

Design of Fermentation Processes

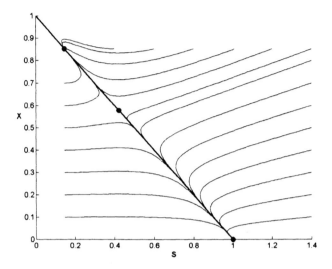

Figure 9.11. Phase diagram showing $X = x/(s_f Y_{sx})$ vs. $S = s/s_f$ for the substrate inhibited kinetics of Figure 9.9. The straight line shows the progress of a batch reaction started with an infinitely small inoculum. On the line are marked the two steady state solutions to the mass balances for $D = 0.3185$ h^{-1} and the trivial steady state solution at $S = 1$. The curves are trajectories followed after a switch to continuous cultivation at a given set of initial values for S and X. The trajectories always bend away from the saddle-point at the unstable steady state at $S = 0.42162$.

Starting the batch with larger inoculum (X_0 finite) moves the straight line to the right, and the asymptotically stable steady state is also reached when the batch is stopped at a larger value of S. A sufficiently large pulse ΔS (> 1) will also destabilize the asymptotically stable steady state.

In (8), (9) the initial S for the chemostat operation is equal to the desired steady-state value, and X is 0.05 higher than the corresponding steady-state value. The extra biomass is washed out and the steady state is reached with no difficulties.

When $X_0 = 0.05$ and $S(t_b) = 0.5325$ [or $X(t_b) = 0.5175$], batch operation has stopped at $t_b = 9$ h.

Figure 9.11 shows that the transient starting at this initial point ends at $S = 1, X = 0$.

The stability of a steady state continuous reactor, and in general the dynamics of the stirred tank reactor, can be studied numerically for any type of growth kinetics. In this way real problems can be investigated. In Section 9.3 we have chosen to highlight the fundamental aspects of reactor dynamics, and in the final section 9.3.3 the simultaneous presence of several competing (or mutually supporting) microbial species in the culture will introduce new facets of the subject without complicating the issue by using a "real" kinetic expression. Obviously the previous discussion is severely hampered by the knowledge of real kinetics that has become available in recent years. But some progress can still be made without introducing excessive complications.

The major objection that can be raised against the models treated hitherto is that the real culture does not respond instantaneously to a change in substrate concentration as has been tacitly assumed. The compartment models of Section 7.4.1, e.g. the lactic acid kinetics of example 7.6 do include a factor X_A which responds with a much larger time constant than s when the dilution rate or the feed concentration is changed. Just as was done for the batch reactor model in Section 9.1.1 (Eqs 9.15 to 9.19) the dynamics of the stirred tank can be studied if the correct rate expressions are used in Eq (9.85 a,b,c). A series of model studies have recently been made in which the concepts of the simple structured model of example 7.6 have been further developed. The main improvement is that the cellular processes are grouped into catabolic and anabolic processes which are taken to be uncoupled in a transient experiment. Hence two time constants, one for the catabolic processes (τ_{catab}) and one for the anabolic processes (τ_{anabol}) are introduced. Both time constants are much larger than the time constant for mixing new feed into the medium. In the model of Duboc et al.(1998) it was hypothesized that a culture that has operated in steady state at a low D value has a "hidden catabolic capability" which springs into action as soon as more substrate becomes available due to an up-shift of D, or when a substrate pulse is added. Experiments confirm the existence of this rapid- access catabolic capacity, since the rate of CO_2 production immediately increases when D is up-shifted from a low value. The jump in r_{catab} is furthermore independent of D in an anaerobic yeast fermentation, but depends only on D_0. If an anthropomorphic explanation is permitted one could say that the first duty of the metabolism of an organism that suddenly finds itself in a more pleasant environment is to start ATP production (the jump in r_{catab}), next to use the ATP to build up more catabolic machinery, and finally to use the reconstituted catabolic machinery to increase cellular growth. The experimental evidence supports that this happens. τ_{catab} may be around 1 hour while τ_{anabol} is more than 2 hours.

In a recent study by Melchiorsen et al. (2001) the same concept is used. Here the dynamics of the pyruvate metabolism is investigated in shift-up or shift-down experiments. A culture of *Lactococcus lactis* that has grown at steady state with a small D value immediately stops producing mixed acids when D is changed to a higher value. Obviously the PFL enzyme is inhibited by the higher glucose flux through the EMP pathway. But also, and with a time constant of about 1 hour, the catabolic machinery increases (as witnessed by an increasing lactic acid production). More slowly the specific growth rate catches up. The same phenomenon is observed when the D value is decreased.

There is no doubt that the subject of bioreactor dynamics will continue to supply many interesting research topics. The experimental techniques are now so strong that one is able to discriminate between models and thereby introduce a new level of investigation of microbial physiology. At the same time phenomena that at the present seem mysterious, such as the oscillations of a continuous aerobic yeast culture (Section 7.6.1) may find an acceptable explanation. New model studies, particularly of dynamic models, will reveal if assumptions that we make concerning the cell physiology (such as "the respiratory capacity of the yeast cell decreases when it starts to make ethanol by the overflow mechanism") can be supported by model simulation studies and finally confirmed by experiments.

9.3.3 Dynamics of the Continuous Stirred Tank for a Mixed Microbial Population

When more than one microbial species grows on the same substrate or if one species preys on another species, a rich variety of interesting dynamic problems emerges, and the solution of these problems is often of great practical importance for the operation of fermentation processes. In the following we shall examine some general features of the models for mixed populations. The concepts are adequately illustrated with a population of two microbial species x_1 and x_2, and since product formation is a simple function of the concentrations of the two species, we shall not consider the product mass balance. Consequently, the mass balances for maintenance-free kinetics are

$$\frac{dx_1}{dt} = \mu_1 x_1 - q_{12} + q_{21} - Dx_1 \tag{9.112a}$$

$$\frac{dx_2}{dt} = \mu_2 x_2 - q_{21} + q_{12} - Dx_2 \tag{9.112b}$$

$$\frac{ds}{dt} = -(Y_{x_1 s}\mu_1 x_1 + Y_{x_2 s}\mu_2 x_2) + D(s_f - s) \tag{9.112c}$$

Here both species grow on the same limiting substrate s. An interconversion between the two species—the metamorphosis reactions of Section 7.6—is included. q_{12} is the volumetric rate of conversion of species 1 to species 2, while q_{21} is the volumetric rate of conversion of species 2 to species 1.

A number of widely different situations can be modeled by a suitable interpretation of the interspecies reaction rates, and a few examples will be considered. But first the simplest case of $q_{12} = q_{21} = 0$ will be treated. This case is extremely important since it describes what happens after an *infection* of the continuous stirred tank reactor. Dimensionless mass balances for the case of simple infection and Monod kinetics are

$$\frac{dX_1}{d\theta} = \frac{\mu_{max,1}}{D} \frac{SX_1}{S+a_1} - X_1 \tag{9.113a}$$

$$\frac{dX_2}{d\theta} = \frac{\mu_{max,2}}{D} \frac{SX_2}{S+a_2} - X_2 \tag{9.113b}$$

$$\frac{dS}{d\theta} = -\left(\frac{\mu_{max,1}}{D} \frac{SX_1}{S+a_1} + \frac{\mu_{max,2}}{D} \frac{SX_2}{S+a_2}\right) + (1-S) \tag{9.113c}$$

where $\theta = Dt$, since D^{-1} is the common scale factor in the three equations. S and (X_1, X_2) are defined as in Eq. (9.25). Local asymptotic stability of a given steady state (S_0, X_{10}, X_{20}) is examined by the same method used in Eq. (9.102). The analysis turns out to be remarkably simple since the only nontrivial steady state with both species coexisting is that for which

$$\frac{\mu_{max,1}S_0}{S_0+a_1} = \frac{\mu_{max,2}S_0}{S_0+a_2} = D$$

$$\Downarrow \qquad (9.114)$$

$$S_0 = \frac{\mu_{max,1}a_2 - \mu_{max,2}a_1}{\mu_{max,2} - \mu_{max,1}} \quad ; \quad D = \frac{\mu_{max,1}a_2 - \mu_{max,2}a_1}{a_2 - a_1}$$

$X_{10} + X_{20} = 1 - S_0$, whereas the distribution between X_{10} and X_{20} is unknown. The Jacobian of the system consisting of Eqs. (9.113 a,b,c) is

$$\mathbf{J} = \begin{pmatrix} \frac{\beta_1 S_0}{S_0+a_1} - 1 & 0 & \frac{a_1\beta_1 X_{10}}{(S_0+a_1)^2} \\ 0 & \frac{\beta_2 S_0}{S_0+a_2} - 1 & \frac{a_2\beta_2 X_{20}}{(S_0+a_2)^2} \\ \frac{-\beta_1 S_0}{S_0+a_1} & \frac{-\beta_2 S_0}{S_0+a_2} & -\left(\frac{a_1\beta_1 X_{10}}{(S_0+a_1)^2} + \frac{a_2\beta_2 X_{20}}{(S_0+a_2)^2} + 1\right) \end{pmatrix} \qquad (9.115)$$

where $\beta_1 = \mu_{max,1}/D$ and $\beta_2 = \mu_{max,2}/D$. From Eq. (9.114), one obtains

$$\beta_1 \frac{S_0}{S_0+a_1} = \beta_2 \frac{S_0}{S_0+a_2} = 1 \qquad (9.116)$$

Consequently, the eigenvalues of Eq. (9.115) are zeros of

$$F(\lambda) = \lambda \left\{ \lambda^2 + \left[\left(\frac{a_1 X_{10}}{\beta_1} + \frac{a_2 X_{20}}{\beta_2}\right)\frac{1}{S_0^2} + 1 \right]\lambda + \left(\frac{a_1 X_{10}}{\beta_1} + \frac{a_2 X_{20}}{\beta_2}\right)\frac{1}{S_0^2} \right\} \qquad (9.117)$$

or

$$\lambda = \begin{cases} 0 \\ -1 \\ -\left(\frac{a_1 X_{10}}{\beta_1} + \frac{a_2 X_{20}}{\beta_2}\right)\frac{1}{S_0^2} \end{cases} \qquad (9.118)$$

Design of Fermentation Processes

The eigenvalue -1 is related to the time it takes before an added mass pulse of either S, X_1, or X_2 has been washed out of the continuous stirred tank reactor. This is seen by adding the three equations in 9.113 and integrating as in Eq. (9.88b), i.e.,

$$S + X_1 + X_2 = (\Delta S + \Delta X_1 + \Delta X_2)\exp(-Dt) + 1 \qquad (9.119)$$

Here ΔS, ΔX_1, and ΔX_2 are the excess concentrations of substrate and of the two biomass components at the start of the transient at $t = 0^+$. One of the remaining eigenvalues associated with the steady state with both X_1 and X_2 is clearly negative, but the eigenvalue of zero tells us that the steady state is only *conditionally stable*, as discussed in Example 9.14.

Example 9.14 Competing microbial species
Let $s_f = 100$ mg L^{-1} and

$$\mu_1 = \frac{0.4s}{s+10}; \quad \mu_2 = \frac{0.5s}{s+20} \qquad (s \text{ in mg L}^{-1}; \mu_1, \mu_2 \text{ in h}^{-1}) \qquad (1)$$

The steady state where x_1 and x_2 coexist is determined by Eq. (9.114):

$$S_0 = \frac{0.4 \cdot 0.2 - 0.5 \cdot 0.1}{0.5 - 0.4} = 0.3; \quad X_{10} + X_{20} = 0.7; \quad D = 0.3 \text{ h}^{-1} \qquad (2)$$

$\mu_1(S)$ and $\mu_2(S)$ are shown in Fig. 9.12.

Figures 9.13 A,B show the transients resulting after a perturbation of the steady state $S_0 = 0.3$. In A a pulse of substrate $\Delta S = 0.2$ added to a chemostat with $X_{10} = 0.1$ and $X_{20} = 0.6$ causes an initial increase in both X_1 and X_2, which return to their initial steady-state values after the pulse has been washed out. S undershoots the steady-state value 0.3 exactly one time. In B, the effect of a pulse of X_1, $\Delta X_1 = 0.2$ (i.e., $X_1 = 0.3$ and $X_2 = 0.6$) at $t = 0^+$ is examined. After S has returned to 0.3, the biomass composition has changed to $X_1 = 0.25$, $X_2 = 0.45$. Whereas the pulse addition of substrate gives an equal advantage to the two competing species and leaves the final biomass composition unchanged, the addition of a pulse of one of the species selectivity favors the rate of formation of this species, and one ends up with a biomass richer in the favored species. This is an implication of the zero eigenvalue of the Jacobian: only S is fixed at the steady state, while the partition of the remaining mass $1 - S$ is arbitrary.

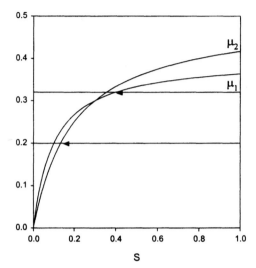

Figure 9.12. Competition between two microbial species X_1 and X_2 that coexist at $\mu = 0.3$ h^{-1}. The specific growth rates μ_1 and μ_2 are given in Eq (1). Arrows indicate what happens after an infection of a monoculture at $D = 0.32$ h^{-1} and 0.2 h^{-1} respectively.

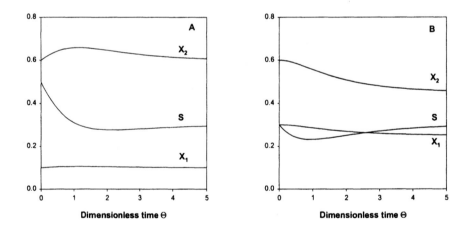

Figure 9.13. Disturbance of the steady state, Figure 9.12, at $\mu = 0.3$ h^{-1} with (**A**): $\Delta S = 0.2$ and (**B**): $\Delta X_1 = 0.2$

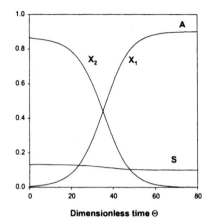
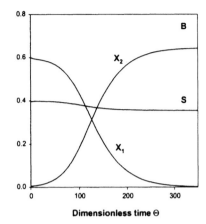

Figure 9.14. Washout of the slowest-growing species. In **A**: $D < 0.3$ h^{-1}, and X_2 is washed out. In **B**: $D > 0.3$ h^{-1}, and X_1 is washed out. Infection: $\Delta X_i = 0.005$ in both **A** and **B**. $D = 0.20$ h^{-1} in **A** and 0.32 h^{-1} in **B**.

Figures 9.14 A and B show what happens after infection of the reactor with the growth-favored microbial species. At $D < 0.3$, the specific growth rate of x_1 is higher than that of x_2. A stable steady-state cultivation of x_2 can be maintained only when x_1 is absent, in which case Eq. (9.113a) is automatically satisfied. Even a minute infection of the chemostat by x_1 leads to washout of x_2, and S decreases from 2/15 to the lower steady-state value 1/10 associated with growth of x_1 alone at the specified $D = 0.2$ h^{-1}. For $D > 0.3$ h^{-1}, infection of a monoculture of x_1 will correspondingly lead to a takeover by x_2, as illustrated in Fig. 9.13B for $D = 0.32$ h^{-1}.

Reversion of a producing strain to a nonproducing wild type strain of the same microorganism is a well-known cause of disappointment in the industrial application of microorganisms. The reversion may occur by mutation, or a valuable plasmid may be lost from a recombinant microorganism. Kirpekar et al. (1985) give an example of a Cephamycin-C producing strain which reverts to a nonproducing strain, and their analysis of experimental data from a chemostat culture to find a kinetic model for the metamorphosis reaction is discussed in Example 9.15. Another case study is discussed in Problem 9.8.

For growth kinetics of the monotonic Monod type and a dilution rate where species x_2 is growth-favored, even a small net rate of conversion of x_1 to x_2 by metamorphosis reactions will lead to washout of x_1, even when the culture was initially free of x_2. Conversely, a metamorphosis reaction in the other direction may help to stabilize the growth-handicapped species in the continuous culture.

The most frequent outcome of competition or parasitism in a stirred tank continuous reactor is that one of the microbial species is washed out except at a certain value of the dilution rate. The transient of a perturbation usually follows a straight (exponentially damped) path toward the final steady state, but oscillations can occur, and if the model has suitably complex growth kinetics the oscillations may even become chaotic. Example 9.16 illustrates a case in which undamped oscillations around the steady state with coexistence of both species is the outcome of a perturbation of this particular steady state. Problem 9.9 is offered as a further case study.

Example 9.15 Reversion of a desired mutant to the wild type

Let x_1 be the mutant (or plasmid-containing microorganism) that decays to the wild-type variant x_2 of the microorganism by a metamorphosis reaction. Both x_1 and x_2 grow on the same substrate, but with different specific growth rates.

The mass balances for the continuous stirred tank reactor are:

$$\frac{dX_1}{d\theta} = \frac{\mu_1}{D}X_1 - \frac{q_{12}}{D} - X_1 \tag{1}$$

$$\frac{dX_2}{d\theta} = \frac{\mu_2}{D}X_2 - \frac{q_{12}}{D} - X_2 \tag{2}$$

$$\frac{dS}{d\theta} = -\frac{1}{D}(\mu_1 X_1 + \mu_2 X_2) + 1 - S \tag{3}$$

$\Theta = Dt$, $X_1 = x_1 / (s_f Y_{sx})$, $X_2 = x_2 / (s_f Y_{sx})$, where for simplicity the yield coefficient Y_{sx_1} has been set equal to Y_{sx_2}. q_{12} is the rate of the irreversible metamorphosis reaction by which x_1 is converted to x_2. Addition of the three equations and integration shows that $S + X_1 + X_2 = 1$ at the end of any transient.

Following Kirpekar et al. (1985), we shall assume that the growth kinetics is studied at conditions of high substrate concentration where $\mu_1 = \mu_{max,1}$ and $\mu_2 = \mu_{max,2}$. In this way we can focus on the influence of the metamorphosis reaction. Thus

$$\mu_1 / D = \beta_1 \ ; \ \mu_2 / D = \beta_2 \ \text{and} \ \beta_2 / \beta_1 = c > 1$$

The metamorphosis reaction is usually very slow compared to the reactor dynamics (time constant $t(q_{12}) \gg 1/D$), and it is not a bad assumption that the total biomass concentration $x_1 + x_2$ in the effluent is constant in a reactor operated with constant D and s_f. What happens is that $f = x_2 / (x_1 + x_2)$ slowly increases toward 1 because the growth kinetics and the metamorphosis reaction favor x_2.
Consequently, by addition of Eqs. (1) and (2),

$$\frac{d(X_1 + X_2)}{d\theta} = 0 = \beta_1 X_1 + \beta_2 X_2 - (X_1 + X_2) \tag{4}$$

or

$$\beta_1(1-f) + \beta_2 f = 1 \rightarrow \beta_2 = \frac{c}{1 - f + fc} \tag{5}$$

Design of Fermentation Processes

Furthermore

$$\frac{df}{d\theta} = \frac{d[X_2/(X_1+X_2)]}{d\theta} = \frac{1}{X_1+X_2}\frac{dX_2}{d\theta} + 0 = \frac{f}{X_2}\frac{dX_2}{d\theta} \qquad (6)$$

or

$$\frac{df}{d\theta} = \frac{f}{X_2}\left(\frac{cX_2}{1-f+fc} + \frac{q_{12}}{D} - X_2\right) = \frac{fc}{1-f+fc} + \frac{fq_{12}}{X_2 D} - f$$

$$= \frac{fc}{1-f+fc} + \frac{q_{12}}{X_1 D}f\frac{1-f}{f} - f = \frac{f(1-f)(c-1)}{1-f+fc} + \frac{q_{12}(1-f)}{X_1 D} \qquad (7)$$

The most reasonable metamorphosis kinetics is $q_{12} = kX_1$, i.e., the rate by which species x_1 is converted to species x_2 is proportional to the concentration of the reactant x_1. With this kinetics,

$$\frac{df}{d\theta} = \frac{f(1-f)(c-1)}{1-f+fc} + \frac{k}{D}(1-f) \qquad (8)$$

For $k = 0$ and integration by separation of variables,

$$\theta = \frac{1}{c-1}\left[\ln\left(\frac{f}{f_0}\right) - c\ln\left(\frac{1-f}{1-f_0}\right)\right] \qquad (9)$$

where f_0 is the fraction of x_2 in the biomass for $\theta = 0$ and $c = \mu_{max,2}/\mu_{max,1}$. f approaches 1 when $\theta \to \infty$, and the transient is independent of the dilution rate.

For $k \neq 0$ and $k' = k/D$,

$$\theta = -\frac{c}{c-1+ck'}\ln\left(\frac{1-f}{1-f_0}\right) + \frac{1}{(c-1+ck')(1+k')}\ln\left[\frac{(c-1)(1+k')f+k'}{(c-1)(1+k')f_0+k'}\right] \qquad (10)$$

Again, f approaches 1 as $\theta \to \infty$, but now the transient depends on the value of D. Kirpekar et al. (1985) simulate the progress of f for experimental runs with various values of D. For the given $c(= 1.25)$, they can fit the value of k to the experiments.
An acceptable fit of the f versus θ transient for the three D values 0.025, 0.036, and 0.045 h^{-1} is obtained with $k = 0.0035$ h^{-1} (note that $k/D << 1$).

They also investigate various other models for the metamorphosis reaction:

Model 1: $\quad q_{12} = k\mu_{max,1}X_1 \quad$ (k dimensionless) $\qquad (11)$

Model 2: $\quad q_{12} = kfX_1 = k\dfrac{X_1 X_2}{X_1+X_2} \quad$ (k in h^{-1}) $\qquad (12)$

Model 3: $$q_{12} = kf\mu_{max,1}X_1 = k\mu_{max,1}\frac{X_1 X_2}{X_1 + X_2} \tag{13}$$

In all three cases, analytical integration is possible. The authors believe that the kinetics of Eq. (13) is the most reasonable, but their data does not allow them to discriminate between the metamorphosis reaction models of Eqs. (12) and (13).

Example 9.16 Competition between a microbial prey and a predator
In the typical prey—predator situation, the prey grows on the substrate fed to the reactor while the predator or parasite grows on the prey organism and thereby diminishes its net rate of growth. Equations (1) - (3) form the simplest possible chemostat model for prey—predator interaction without substrate limitation for growth of the prey $x_1 (\mu_1 = \mu_{max,1})$, while the growth rate of the predator is proportional to both the concentration of prey and the concentration of the predator x_2:

$$\frac{dX_1}{d\theta} = (\beta_1 - 1)X_1 - \beta_2 x_{10} X_1 X_2 \tag{1}$$

$$\frac{dX_2}{d\theta} = \beta_2 Y_{sx} s_f X_1 X_2 - X_2 \tag{2}$$

$$\frac{dS}{d\theta} = -\beta_1 X_1 + 1 - S \tag{3}$$

The dimensionless variables are defined as follows:

$$\theta = Dt; \quad X_1 = \frac{x_1}{s_f Y_{sx}}; \quad X_2 = \frac{x_2}{x_{10} Y_{x_1 x_2}}; \quad S = \frac{s}{s_f} \tag{4}$$

$Y_{x_1 x_2}$ is the yield of predator (kilogram of predator per kilogram of prey consumed) and x_{10} is the initial concentration of the prey organism. β_1 and β_2 are dimensionless rate constants, $\beta_1 = \mu_{max,1} / D$ and $\beta_2 = k_{12} / D$. Comparison with the previous example shows that the present model is a variant of Eqs. 1 to 3 in Example 9.15 with $\mu_2 = 0$, a yield coefficient of the metamorphosis reaction different from 1, and kinetics of the metamorphosis reaction given by Eq. (12E9.15) or (13E9.15). In the present example, $X_1 + X_2$ is not assumed to be constant.

There are 3 possible steady states:

- x_1 and x_2 coexist, i.e., both X_1 and X_2 have nonzero steady-state values.
- x_1 is present in nonzero steady-state concentration, but x_2 is washed out.
- both x_1 and x_2 are washed out (i.e., $S = 1$ and $X_1 = X_2 = 0$).

Each of these steady states will be considered and its stability will be analyzed based on an analysis of the eigenvalues of the Jacobian:

Design of Fermentation Processes

$$\mathbf{J} = \begin{pmatrix} \beta_1 - 1 - \beta_2 x_{10} X_2 & -\beta_2 x_{10} X_1 & 0 \\ \beta_2 Y_{sx} s_f X_2 & \beta_2 Y_{sx} s_f X_1 - 1 & 0 \\ -\beta_1 & 0 & -1 \end{pmatrix} \quad (5)$$

The steady state at which X_1 and X_2 coexist is

$$X_{10} = \frac{1}{s_f Y_{sx} \beta_2} \; ; \; X_{20} = \frac{\beta_1 - 1}{\beta_2 x_{10}} \; ; \; S_0 = 1 - \frac{\beta_1}{s_f Y_{sx} \beta_2} \quad (6)$$

where

$$\beta_1 > 1 \quad (X_2 > 0) \quad \text{and} \quad \beta_1 < s_f Y_{sx} \beta_2 \quad (S > 0)$$

When Eq. (6) is inserted into Eq. (5), one obtains the eigenvalues as solution of

$$(\lambda + 1)\left[-\lambda(-\lambda) + \beta_1 - 1\right] = 0 \quad (7)$$

The eigenvalue $\lambda = -1$ is associated with the washout of the initial mass pulse by which the steady state is disturbed. Since $\beta_1 > 1$ for the steady state with x_1 and $x_2 > 0$, there are two purely imaginary eigenvalues. Consequently, once perturbed the state vector (X_1, X_2, S) never returns to the steady state (X_{10}, X_{20}, S) but performs undamped oscillations around the steady state. For very small perturbations [which must include a perturbation of either X_1 or X_2 since a perturbation of S alone leave the right-hand side of Eq. (1) and Eq. (2) at their initial values (equal to 0)], the cycle time T of the oscillations is $2\pi / \sqrt{\beta_1 - 1}$, and in a phase-plane plot of X_2 versus X_1 the point (X_1, X_2) moves along a circle with radius equal to the original disturbance, e.g., ΔX_1.

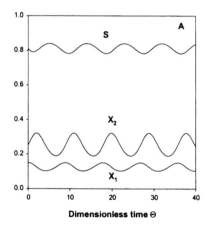
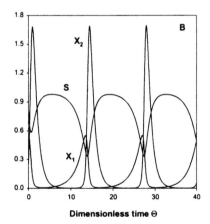

Figure 9.15. Coexistence of prey (X_1) and predator (X_2). **A**: $\Delta X_1 = 0.025$, **B**: $\Delta X_1 = 0.425$. The oscillations are around the steady state with: $\beta_1 = 1.5$, $\beta_2 = 0.1$ L g^{-1}, $x_{10} = 20$ g L^{-1}, $Y_{sx} s_f = 80$ g L^{-1} (see Eq 6).

The shape of the oscillation depends on the size of the perturbation and on the parameter values in Eqs. (1) - (3). Figure 9.15A shows very regular oscillations for a small initial disturbance $\Delta X_I = 0.025$ of the steady state $X_{10} = 0.125$, $X_{20} = 0.250$, $S_0 = 0.8125$. T is 8.97 (dimensionless), which is close to the limiting value $2\pi / \sqrt{\beta_1 - 1} = 2\pi\sqrt{2} = 8.886$. For $\Delta X_I = 0.425$ (Fig. 9.15B), the oscillations are severely distorted with sharp peaks of X_1 and X_2. The cycle time has increased to 13.87.

Figure 9.16 is similar to the phase plot Figure 9.11, but instead of the *saddle point* in Figure 9.11 we now have *limit cycles*. Figure 9.16 collects the essential information concerning the undamped oscillations that result after perturbing the above steady state with different ΔX_I. For the present, simple kinetics the shape of the phase-plane plots can be calculated analytically. The substitution $y_1 = \ln(X_1)$, and $y_2 = \ln(X_2)$ is introduced in the cell balances:

$$\frac{dy_1}{d\theta} = \beta_1 - 1 - \beta_2 x_{10} e^{y_2} \tag{8}$$

$$\frac{dy_2}{d\theta} = \beta_2 Y_{sx} s_f e^{y_1} - 1 \tag{9}$$

The two equations are differentiated once more with respect to θ and subtracted. After substitution of e^{y_1} and e^{y_2} from Eq. (8) to Eq. (9) in the resulting equation, one obtains

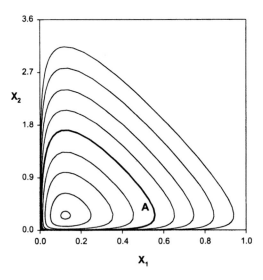

Figure 9.16. Coexistence of prey and predator. Phase plane diagram of X_1 and X_2 for different values of the initial disturbance ΔX_I. The extrema for the limit cycle marked A are shown in the table. Limit cycle

Design of Fermentation Processes

A corresponds to Figure 9.15 B

$$\frac{d^2 y_1}{d\theta^2} - \frac{d^2 y_2}{d\theta^2} + \frac{dy_1}{d\theta} + (\beta_1 - 1)\frac{dy_2}{d\theta} = 0 \tag{10}$$

which on integration from $\theta = 0$ and reintroduction of X_1 and X_2 yields

$$\beta_1 - \beta_2 x_{10} X_2 - \beta_2 Y_{sx} s_f X_1 + \ln(X_1) + (\beta_1 - 1)\ln(X_2) = K \tag{11}$$

where the value of K is determined from the initial condition at $t = 0^+$. If for $t = 0^+$, $X_1 = X_{10} + \Delta X_1$ and $X_2 = X_{20}$, one obtains

$$(\beta_1 - 1)\ln(X_2) = \beta_2 x_{10} X_2 + K - \left[\beta_1 - \beta_2 Y_{sx} s_f X_1 + \ln(X_1)\right] = \beta_2 x_{10} X_2 + \alpha \tag{12}$$

For a given value of ΔX_1, one may calculate the value of α.

The straight line $\beta_2 x_{10} X_2 + \alpha$ intersects $(\beta_1 -1)\ln(X_2)$ twice if α is smaller (more negative) than the value obtained from the intersection of the tangent of $(\beta_1 -1)\ln(X_2)$ with slope $\beta_2\, x_{10}$ and the ordinate axis. The slope of $(\beta_1 -1)\ln(X_2)$ is $\beta_2 x_{10}$ for $X_2 = (\beta_1 -1)\,/\,(\beta_2\, x_{10}) = X_{20}$. At this value of X_2, one obtains the following equation for X_1:

$$\beta_1 - \beta_2 Y_{sx} s_f X_1 + \ln(X_1) = \beta_1 - \beta_2 Y_{sx} s_f (X_{10} + \Delta X_1) + \ln(X_{10} + \Delta X_1) \tag{13}$$

One solution of Eq. (13) is obviously $X_1 = X_{10} + \Delta X_1$. The other solution is obtained by numerical solution of

$$F(X_1) = \ln(X_1) - \beta_2 Y_{sx} s_f X_1 - \ln(X_{10} + \Delta X_1) + \beta_2 Y_{sx} s_f (X_{10} + \Delta X_1) = 0 \tag{14}$$

A standard Newton iteration

$$X_{1,i+1} = X_{1,i} - \left(\frac{F}{dF/dX_1}\right)_{x_{1,i}} = X_{1,i} - \frac{X_{1,i} F(X_{1,i})}{1 - \beta_2 Y_{sx} s_f X_{1,i}} \tag{15}$$

will find the other zero of Eq. (14) in a few steps. Consequently, for a given ΔX_1 the phase-plane plot of X_2 versus X_1 extends in the X_1 direction between a value of X_1 found by solution of Eq. (15) and $X_1 = X_{10} + \Delta X_1$. The maximum extension in the X_1 direction is found for $X_2 = X_{20}$. ΔX_1 is large and positive, the solution of Eq. (15) is close to zero, and if $\Delta X_1 \approx -X_{10}$, the solution of Eq. (15) is large and positive.

The extension of the phase plane plot in the X_2 direction is found by an analogous procedure. X_1 must be equal to X_{10} and the two solutions of

$$(\beta_1 - 1)\ln(X_2) - \beta_2 x_{10} X_2 + \ln\left(\frac{X_{10}}{X_{10} + \Delta X_1}\right) + \beta_2 Y_{sx} s_f \Delta X_1 + \beta_2 x_{10} X_{20} - (\beta_1 - 1)\ln(X_{20}) = 0 \quad (16)$$

span the range of the phase plane plot in the X_2 direction.

In summary: For $\Delta X_1 = 0.4250$ [$X_1(t = 0^+) = 0.55$]—the situation depicted in Fig. 9.15B—the phase plane curve is delimited by the values listed in the following table:

X_1	0.55	0.00715	0.125	0.125
X_2	0.25	0.25	1.68642	0.001999

For any X_1 value between 0.55 and 0.00715, the corresponding two values of X_2 can be found by iterative solution of Eq. (11) at the given $\Delta X_1 = 0.4250$. The closed curve A in Fig. 9.16 is the result of these calculations. For large ΔX_1, the point (X_1, X_2) spends most of the very large cycle time T in the close vicinity of $(0, 0)$, while $S \approx 1$. An obvious question is why the transient is not caught by the steady-state solution $X_1 = X_2 = 0$. The answer is quite simply that this steady state is not stable when $\beta_1 > 1$, as is easily seen by insertion in Eq. (5). The eigenvalues are zeros of

$$(\lambda + 1)^2 (\beta_1 - 1 - \lambda) = 0 \quad (17)$$

and one eigenvalue is real and positive for $\beta_1 > 1$.

The third possible steady state with $X_1 > 0$ and $X_2 = 0$ is not stable either, since one eigenvalue is positive as long as $\beta_1 > 1$. As discussed further in Problem 9.9, a steady state with $0 < X_1 < 1$ and $X_2 = 0$ can also be obtained for $\beta_1 > 1$ if the growth kinetics for X_1 is changed to include substrate limitation $\mu_1 = \mu_{max,1} f(S)$. Now the steady state is determined by

$$\beta_1 f(S_0) = 1; \quad X_{20} = 0; \quad X_{10} = 1 - S_0 \quad (18)$$

for which the eigenvalues are solutions of

$$(\beta_2 Y_{sx} s_f X_{10} - 1 - \lambda)\left\{\lambda^2 + \left[\beta_1\left(\frac{\partial f}{\partial S}\right)_{S_0} \cdot X_{10} + 1\right]\lambda + \beta_1\left(\frac{\partial f}{\partial S}\right)_{S_0} \cdot X_{10}\right\} \quad (19)$$

i.e.,

$$\lambda = \begin{cases} \beta_2 Y_{sx} s_f X_{10} - 1 \\ -1 \\ -\beta_1\left(\frac{\partial f}{\partial S}\right)_{S_0} \cdot X_{10} \end{cases} \quad (20)$$

Design of Fermentation Processes

Thus, the steady state is stable if $(\partial f/\partial S) > 0$ at (X_{10}, S_0) and $\beta_2 Y_{sx} s_f X_{10} < 1$. With Monod kinetics, the steady state is stable if

$$\frac{\beta_2 Y_{sx} s_f (\beta_1 - 1 - a)}{\beta_1 - 1} < 1 \quad \text{and} \quad \beta_1 > 1 + a \tag{21}$$

This last result is a generalization of a previous result, Eq. (9.104). The last two eigenvalues of Eq. (20) are the same as the eigenvalues of Eq. (9.104) (but in dimensionless form), while the first eigenvalue of Eq. (20) accounts for the interaction between prey and predator.

An important piece of model work for heterogenous microbial cultures is to construct a map of the regions where one, several, or no microorganisms remain in the reactor. Typical operating conditions such as s_f and D can be used in a two-dimensional *bifurcation* diagram as illustrated in Problem 9.9. The result is of considerable value to the fermentation industry, since it indicates where the process can be operated without having a permanent infection of the reactor by a phage or a parasitic microorganism.

PROBLEMS

Problem 9.1 Chemostat operation

Assume the following kinetics for a fermentation process:

$$q_x = \frac{4}{3} \text{h}^{-1} \left(\frac{s}{4+s} \right) x; \quad q_s = -10 q_x \tag{1}$$

where s and x are in grams per cubic meter. The feed is sterile with $s_f = 60$ g m^{-3}.

a. Calculate the maximum feed flow v_{max} that can be handled by a chemostat of volume $V = 1$ m^3. For $V = 1$ m^3, calculate the maximum productivity of cells $P_{x,max}$.
b. An additional chemostat of volume V_1 is made available, and with the combined reactor volume $V + 1$ m^3 it is demanded that a feed stream $v = 2.5$ m^3 h^{-1} is treated such that the highest possible cell productivity (grams of cells produced per hour per total reactor volume) is obtained.

    1. Find the volume V_1 of the second chemostat if the two chemostats are installed in parallel.
    2. Calculate (by tabulation) the volume V of the second chemostat if the two chemostats are installed in series. Should the second reactor be upstream or downstream of the 1 m^3 reactor? (note that the calculation is explicit if you start with a set of substrate concentrations s_2 after the second reactor)

c. Now return to the case of a single chemostat with volume 1 m^3. For $t < 0$, the reactor is operating at $v = 1$ m^3 h^{-1}, and with $s_f = 60$ g m^{-3}. At $t = 0$, v is changed to (1) 1.5 m^3 h^{-1}, and (2) 0.75 m^3 h^{-1}. For cases (1) and (2), calculate and plot x / x_0 as a function of time. x_0 is the effluent cell concentration for $t < 0$.
d. Again, consider the 1- m^3 reactor operating at steady state and with $v = 1$ m^3 h^{-1}. Due to a failure of the feed sterilization system, the feed contains traces of a foreign microorganism with $\mu_{max} = 8/3$ h^{-1}

—otherwise the kinetics is given by Eq. (1). After 10 min, the concentration of the foreign microorganism is 10^{-2} g m^{-3}. At this point the steam sterilization of the feed is resumed, and thereafter the feed is sterile. Make a numerical simulation of x / x_0 and s / s_0 as functions of time.

e. You do of course observe that $\mu_2 = 2 \mu_1$ through the whole transient. Determine s and x_2 when x_1 has been washed out. Compared with the rapid changes in x_1 and x_2 the sum $x_1 + x_2$ does not change dramatically during the transient. A good approximation to the transients may conceivably be obtained by assuming that $x_1 + x_2$ stays constant. Consequently, introduce $f = x_2 /(x_1 + x_2)$ as in example 9.15 and follow the development in this example to show that the solution to the approximated infection problem is given by Eq 9 of the example.

Make a simulation of $f(t)$ for $0 < \theta < 20$. On the same plot show the difference $\Delta(\theta) = f(\text{approx}) - f(\text{exact})$. You will find that Δ is remarkably small for all θ.

Problem 9.2 Lactic acid production

As an employee in the R&D division of an agroindustrial company, you are entrusted with the solution of the following problem: The effluent from one of our cheese factories contains 45 kg of lactose per cubic meter. There is 4000 m^3 per annum of effluent (production time 7200 h), and the effluent is an environmental burden. Can this effluent be used to produce lactic acid? The market for lactic acid is weak and minimum investment cost of the reactor per kilogram produced of lactic acid is desirable.

Your preliminary research confirms that the lactose substrate is sterile and contains no lactic acid. From a literature search you come upon the following rate expressions:

$$q_x = \mu_{max}\left(1 - \frac{p}{p_{max}}\right)\frac{sx}{K_s + s}; \quad \mu_{max} = 0.5 \text{ h}^{-1};$$
$$p_{max} = 50 \text{ kg m}^{-3}; \quad K_s = 0.3 \text{ kg m}^{-3}$$
(1)

$$q_s = -Y_{xs}q_x; \quad Y_{sx} = \frac{1}{Y_{xs}} = 0.11 \text{ kg DW per kilogram of lactose}$$
(2)

$$q_p = bq_x + cx; \quad b = Y_{xp} = 8 \text{ kg of lactic acid per kilogram dry weight};$$
$$c = 0.1 \text{ kg of lactic acid per kilogram dry weight per hour}$$
(3)

On the basis of these pieces of information you decide to go ahead with the design of a chemostat in which the productivity P_p in kilograms of lactic acid per cubic meter per hour is maximized.

a. Your first inspiration is to neglect the complications of the kinetics, i.e., to assume that $p << p_{max}$ and $cx << bq_x$. With these assumptions, calculate the reactor volume for which maximum P_p is obtained. Also calculate the effluent concentrations of lactic acid, biomass, and lactose from the chemostat, and determine the maximum productivity $P_{p\,max}$ of lactic acid. Comment on the validity of the two assumptions.

b. With the slight suspicion that the assumption $p << p_{max}$ may not be realistic, you repeat the calculation of the effluent concentrations and of P_p for the reactor volume determined in (a), but now using the correct kinetic expression of Eq. (1). The assumption $cx << bq_x$ is still assumed to hold. You are allowed to be relieved that you did not build the reactor based on your approximate design.

c. With the approximate model of (b), you decide to calculate the reactor size which gives maximum productivity P_p. Analytical optimization is difficult, but tabulation is easy: For $s/s_f = 0.9, 0.8, ..., 0.1$, calculate D, x, p, and $P_p = Dp$. For the three entries closest to the maximum in P_p, find the optimal S by quadratic interpolation. Calculate V(opt). Does the assumption $cx \ll bq_x$ appear to be reasonable?

d. You decide to base your design on a reactor of volume $V = 2.40$ m³. The effluent from the reactor must, however, not contain more than 3 kg of lactose per cubic meter. Hence a centrifuge is installed at the outlet and 0.2 m³ of solution plus cells per hour is returned to the inlet. For $s = s_e = 3$ kg m⁻³, calculate the corresponding values of x_p and p using the whole of Eq. (1), but $cx \ll bq_x$. Determine the separation factor β for the centrifuge and the cell concentration x in the reactor.

e. The rather high value of x calculated in (d) reawakens your suspicion that the assumption $cx \ll bq_x$ may be invalid. Therefore you decide to redo the calculations of part (d), but now with the full model of Eqs. (1) - (3). You are in for several surprises:

1. It may be difficult to find a solution to (d).
2. This should prompt you to redo the calculations in (c), but using the full model.
3. Doing so may give you more solutions than you desire for some S and disappointingly few for other S values.

Do these calculations give you any reason to criticize the kinetic model in Eqs. (1) - (3)? What could be wrong with the model?

Problem 9.3 Lactic acid batch fermentation

On a certain yeast extract-casein peptone medium it has been reported that the growth-associated ATP consumption for lactic acid fermentation is

$$r_{ATP} = 31 \left(\frac{\text{mmoles of ATP}}{\text{gram dry weight}} \right) \mu(h^{-1}) + 17 \left(\frac{\text{mmoles of ATP}}{\text{gram dry weight} \cdot \text{hour}} \right) \quad (1)$$

The process is anaerobic, and lactic acid is the only metabolite formed. You are required to check the validity of this expression, and for this purpose you set up a fermentation experiment. The initial medium volume is $V_0 = 750$ mL, and the glucose concentration is 13 1/3 g L⁻¹. The medium is inoculated with $x_0 = 10$ mg of cells per liter, and exponential growth with $\mu = \mu_{max} = 0.6$ h⁻¹ starts immediately after inoculation. To keep pH constant at 6.80, the lactic acid produced by conversion of glucose has to be continuously titrated. A 1-M NaOH solution is used, and since the medium volume is consequently a function of time it becomes a little difficult to check the validity of the kinetics [i.e., Eq. (1) and the assumption $\mu = 0.6$ h⁻¹] by comparison of experimentally determined concentrations of biomass x (g L⁻¹) and lactic acid p (g L⁻¹), and the simulated results.

a. Assuming that Eq. (1) is valid and that $\mu = 0.6$ h⁻¹, derive an expression for $x(t)$ that can be used for comparison with the experimentally determined biomass concentration time profile. Also derive an expression for $V(t)$. Hint: It is easy to find xV, and $v(t)$ can be found from Eq. (1).
b. Derive an expression for $p(t)$, assuming that $p(t=0) = 0$.
c. Determine the time t_{end} at which all sugar is depleted, assuming that the glucose is quantitatively converted to lactic acid. What are x, p, and V when $T = T_{end}$? What is the relation between the apparent specific growth rate μ_{obs} and μ for large t_{end}?

Problem 9.4 Heat transfer limitation in SCP production.

a. In an SCP production 2.5 N m^3 CH$_4$ (N= 273 K, 1atm) is used to produce 1 kg of biomass X = CH$_{1.8}$ O$_{0.5}$ N$_{0.2}$ by continuous fermentation in a stirred tank. Determine Y_{so} and thereafter the total stoichiometry of the reaction, assuming that biomass, CO$_2$ and H$_2$O are the only products and that NH$_3$ is used as nitrogen source.
    Calculate the heat of reaction Q in kJ(kg biomass)$^{-1}$.

b. The bioreactor is cylindrical with diameter d and height h. $h = 3 d$. The inner cylinder area A (m^2), but not the bottom of the reactor is covered with heat exchange surface (see Figure 11.4). The effective temperature driving force is $\Delta T = 30°C$, and the heat transfer coefficient h_T is 300 kJ (m^2 h K)$^{-1}$. The heat transferred is consequently $Q' = h_T \Delta T A$ kJ h^{-1}. Show that –due to heat transfer limitation- the maximum productivity of biomass is

$$(q_x)_{max} = 0.60 \ V^{-1/3} \ \text{kg m}^{-3} \text{h}^{-1}, \text{ where } V \text{ is the reactor volume.} \quad (1)$$

For a sterile feed, and $x = 20$ g L^{-1} determine the maximum possible dilution rate, D_{max} for $V = 1$ L and for $V = 50$ m^3.

c. Laboratory experiments are conducted in a small reactor where heat transfer is no problem. The liquid feed is sterile and contains 5 g L^{-1} NH$_3$ and a sufficient supply of minerals. The gas feed is 1 N m^3 per m^3 reactor volume and per min. The composition is 25 vol% CH$_4$ and 75 vol% O$_2$. With this much O$_2$ in the feed the reaction is always limited by CH$_4$. At steady state the following data are obtained:

D (h^{-1})	CH$_4$ (vol% in exit gas)	Biomass in effluent : x g L^{-1}.
0.05	20.53	36.17
0.1	14.16	36.17
0.2	13.18	19.20
0.3	9.626	15.20
0.5	25.00	0

Assume that no water is transferred from gas to liquid phase or *vice versa*.
Show that $-q_{CH_4}$ in mol CH$_4$ consumed (L reactor min)$^{-1}$ and the gas-phase concentrations in the second column of the table are related by:

$$\frac{\text{vol \%}}{100} = \frac{0.25 \cdot 0.044642 + q_{CH_4}}{0.044642 + (1 + Y_{so})q_{CH_4} - Y_{sc}q_{CH_4}} \quad (2)$$

Add a column of $-q_{CH_4}$ to the table.

For $D = 0.2$ and 0.3 h^{-1} show that the yield coefficient Y_{sx} calculated in (a) is also found from the data of the table. Calculate the effluent NH$_3$ concentration. No NH$_3$ is stripped to the gas phase.
Finally explain the data for $D = 0.05$ and 0.1 h^{-1}. What happens at $D = 0.5$ h^{-1}?

Design of Fermentation Processes

Problem 9.5 Substrate inhibition kinetics

Substrate inhibition is often expressed as shown in Eq. (1)

$$q_x = \frac{\mu_{max} S}{S+a} \frac{b}{b+S} x = \mu x \qquad (1)$$

$S = s / s_{ref}$, and s_{ref} is a reference concentration, e.g., the feed-substrate concentration for a stirred tank continuous reactor. Let the cell-growth kinetics of Eq. (1) be combined with the substrate-consumption kinetics of Eq. (2), which includes some maintenance expressed through the constant c:

$$-q_s = Y_{xs} \mu_{max} x \left(\frac{S}{a+S} \frac{b}{b+S} + c \right) \qquad (2)$$

Introduce

$$X = \frac{x}{Y_{sx} s_{ref}}; \quad U = \frac{\mu}{\mu_{max}} \qquad (3)$$

a. For a chemostat operation ($s_{ref} = s_f$ and $x_f = 0$), calculate expressions for X and UX as functions of S, a, b, and c.
b. For $c = 0$, calculate the optimum effluent concentration S_{opt} for highest cell productivity P_x. What is S_{opt} for $a = 0.2$ and $b = 2$? What are the corresponding values of U and X?
c. The chemostat is started up by a fed batch procedure. The medium volume is V_0, with substrate and biomass concentrations s_0 and x_0 at $t = 0$. For $t > 0$, feed with substrate concentration s_f is supplied at the volumetric rate $v(t)$. Write the transient mass balances for RX and RS, where X and S are defined above ($s_{ref} = s_f$), and $R = V(t) / V_0$. Furthermore,

$$U = \frac{v(t)}{V_0 \mu_{max}}; \quad \theta = \mu_{max} t \qquad (4)$$

are other dimensionless variables to be used.
d. Show that, independent of the value of c, one obtains the maximum cell mass in the reactor at any instant of time if $S(t)$ is kept constant at $\sqrt{ab}$ (or $S = 1$ for $ab > 1$).
e. Show that, for $S = \sqrt{ab}$;

$$RX = X_0 \exp\left[\left(1 + \sqrt{\frac{a}{b}}\right)^{-2} \theta \right] \qquad (5)$$

Finally, determine $U(t)$ and $R(t)$.
f. Discuss a suitable start-up procedure if a given steady state [i.e., that of (a)] is to be obtained in the shortest possible time for a given value of x_0 and a given s_f.

Problem 9.6 Stability analysis with product inhibition, maintenance and cell recirculation.

If, in the dynamic mass balances for the stirred tank continuous reactor the separation factor Ω of the cell recirculation system with an ultrafilter (Eq 9.46) is introduced together with maintenance according to Eq (9.13)-(9.14) the following balances are obtained:

$$\frac{dx}{dt} = \mu x - (1-\Omega)Dx = 0 \quad (1)$$

$$\frac{ds}{dt} = D(s_f - s) - (Y_{xs}^{true}\mu + m_s)x \quad (2)$$

$$\frac{dp}{dt} = -Dp + (Y_{xp}^{true}\mu + m_p)x \quad (3)$$

Introduce in (1) to (3) dimensionless variables $S = s/s_f$, $X = x(1-\Omega)/(Y_{xs}^{true}s_f)$ and $P = p/(Y_{xp}^{true}s_f)$. Also let $\beta_s = Y_{xs}^{true} \cdot m_s$ and $\beta_p = Y_{xp}^{true} m_p$.

Consider a steady state solution $C_0 = (X_0, S_0, P_0)$ and introduce the deviation variable $y = C - C_0$ which is the solution of the linear differential equation (9.102).

Show that:
$$\text{Det}(J - \lambda I) = -\lambda \Delta_1 + (\mu + \beta_s)\Delta_2/(1-\Omega) + (\mu + \beta_p)\Delta_3/(1-\Omega) \quad (4)$$

$$\Delta_1 = (D + \lambda)^2 - (D + \lambda)(\mu_p - \mu_s)X/(1-\Omega)$$

$$\Delta_2 = -(D + \lambda)\mu_s X \quad (5)$$

$$\Delta_3 = (D + \lambda)\mu_p X$$

As in Section 9.3.2 μ_p and μ_s are the partial derivatives of μ with respect to p and s.

Continue the simplification to obtain the following equation for λ:

$$(D + \lambda)(-\lambda^2 + \lambda(-D + a) + D(1-\Omega)a + b) = 0 \quad (6)$$

In equation (6) $a = (\mu_p - \mu_s)X/(1-\Omega)$ and $b = (\beta_p\mu_p - \beta_s\mu_s)X/(1-\Omega)$.

Make an analysis of the stability of a steady state C_0 with different expressions $\mu(S,P)$.
What is the general effect of a product inhibition? The effect of cell recirculation?
You should make several simulation studies as part of your answer to this problem.
Note that the range of application of the text in Section 9.3.2 has been greatly expanded by means of this problem.

Problem 9.7 Production of SCP

Certain microorganisms can grow aerobically on methanol as the sole source of carbon and energy, even at 55 °C. These microorganisms are very well suited for production of single cell protein (SCP) in a hot climate such as that in Al Jubail (Saudi Arabia) where the cooling-water temperature is rarely below 30 °C. Two large methanol plants are located in the industrial complex at Al Jubail, and these plants can easily

Design of Fermentation Processes

supply cheap methanol for an SCP production of 50,000 tons year^{-1} (16 2/3 tons h^{-1} based on 3000 h yr^{-1} on stream).

Laboratory tests show that the biomass composition $CH_{1.8}O_{0.5}N_{0.2}$ and biomass yield on methanol $Y_{sx} = 9/16$ kg kg^{-1} are approximately independent of the growth rate. NH_3 is the nitrogen source, and only CO_2 and H_2O are produced besides the biomass.

One literature source states the following cell growth kinetics at 55 °C:

$$q_x = \frac{\mu_{max} s}{s + K_s} \frac{s_1}{s_1 + K_1} x \text{ (kg of cells m}^{-3} \text{ h}^{-1}\text{)} \quad (1)$$

s is the methanol concentration (kg m^{-3}), and s_l is the oxygen concentration (µM). $K_s = 0.0832$ kg m^{-3}, and $K_l = 2$ µM $= 2 \times 10^{-3}$ moles of O_2/ m^3. $\mu_{max} = 0.9$ h^{-1}.

a. Calculate the maximum steady-state productivity of cells in a stirred tank reactor fed with $s_f = 50$ kg m^{-3} sterile methanol. It is assumed that the oxygen concentration $s_l \gg K_l$. Calculate the corresponding minimum reactor volume for production of 16 2/3 tons of biomass h^{-1}. What is the O_2 requirement to sustain the production rate calculate above? Calculate the minimum value of $k_l a$ necessary to transfer the required O_2 to the liquid from air (21% oxygen). The saturation concentration s_l^* of oxygen in the liquid is 910 µM at 55 °C. Is it possible to transfer the required quantity of oxygen to the liquid from air using commercially available gas dispersers (spargers, agitators, etc.)?

b. Assume that $k_l a = 0.2$ s^{-1} is the highest mass transfer coefficient which can be obtained in the stirred tank reactor. With this value of $k_l a$ and the full expression (1) for the rate of cell production calculate corresponding values of q_x, s and V for different values of s_l. The s value calculated in question (a) seems to have a special significance. Why? Can $s_l = 0.2$ µM be used? Is there something wrong with the model (1)—or perhaps with some of the assumptions? See Section 9.1.4 for an explanation.

c. A much better design than the monstrous stirred tank of (b) can be devised: Pure oxygen is used rather than air, the reactor volume necessary to produce 16 2/3 tons of biomass h^{-1} is broken up into a number of smaller reactors, and the design of the individual reactor is improved. Assume that each reactor has a medium volume of 30 m^3. The reactor is designed as a tube formed into a loop equipped with a number of highly effective static mixers through which the liquid is pumped with a circulation time of 30 s. Pure oxygen enters near the top of the loop and follows the liquid through the loop while being constantly redispersed into small bubbles by the mixers. At the very top of the loop (just over the gas inlet), the spent gas leaves the reactor to enter the head space. In this apparatus, one can achieve $k_l a = 0.45$ s^{-1} as an average for the total circulated liquid volume when the liquid circulation time is 30 s, corresponding to a superficial liquid velocity of 1 m s^{-1}. The liquid-phase methanol concentration s is 1 kg m^{-3}, and $s_f = 50$ kg m^{-3}. s_l is taken to be 20 µM at all points in the loop (is this assumption critical?). Calculate the biomass productivity in each 30 m^3 reactor when (for economic reasons) 95% of the oxygen is to be taken up by the bioreaction during a single passage through the loop. How many reactors must be constructed to reach the productivity of 16 2/3 tons h^{-1}?

d. How would you modify the reactor design if the feed is methane and oxygen rather than methanol and oxygen. The inlet feed is pure oxygen and >99 % methane. The reaction stoichiometry is that of problem 9.4, and data for saturation concentrations of CH_4 is taken either from Example 9.7 or from tables.

Problem 9.8 Providing an experimental foundation for the design of a fermentation process.

Vara *et al.* (2002) made an experimental study of Teicoplanin production by *Actinoplanes teichomyceticus* in order to obtain design data for an industrial production of the antibiotic in continuous culture.
  a. You are required to make a short review of their experimental plan and to compare it with the experimental plan used in Example 6.1. The kinetic parameters used to construct table 6.1 were in fact extracted from the results of the above reference, thus illustrating the resemblance between mechanistically based kinetic expressions for enzymatic reactions and empirical rate expressions for cellular reactions. Which experimental plan is likely to lead to the best values for the kinetic parameters?
  b. Figure 6 of the reference shows simulations of the concentrations of substrate (glucose = s), biomass (x) and product (teicoplanin = p) as functions of dilution rate in a continuous cultivation. You are required to check the simulations using the kinetic model (including maintenance). Why does $x(D)$ decrease for small D while $p(D)$ continues to increase?
  c. Table 1 of the reference shows results of continuous experiments in a stirred tank with recirculation according to Figure 9.4 B. Find the connection between the parameter c in the reference and Ω used in Eq (9.46). Why are the data in the last column of the table almost independent of D? Confirm the conclusion of the paper that continuous operation with cell recirculation is the best mode of operation, and that an increase of productivity by a factor 3 compared to straight continuous operation is achieved.
  d. The microorganism gradually looses the ability to produce teicoplanin. Compare the deactivation model (10)-(11) used by the authors with that used in Example 9.15, and confirm that the model simulates the data in figure 8 of the reference with a high accuracy. Why could a simpler model taken from Example 9.15 not be used?

Problem 9.9 Prey-predator interaction

Consider the prey predator model of Eqs. 1 to 3 of Example 9.16, but with a general substrate limitation for growth of x_1, i.e., $\mu_1 / D = \beta_1 f(S)$ instead of $\mu_1 / D = \beta_1$.
  a. Let $\partial f / \partial S = f_s$, and show that the steady state

$$X_1 = 1/\beta_2 Y_{sx} s_f \ ; \quad X_2 = (\beta_1 f - 1)/\beta_2 x_{10} \ ; \quad S = 1 - \beta_1 f \, X_1 \qquad (1)$$

is stable if the zeros of

$$F(\lambda) = -(\lambda + 1)\left[\lambda^2 + \beta_1 f_s X_1 \lambda + (\beta_1 f_s X_1 + 1)(\beta_1 f - 1)\right] \qquad (2)$$

have negative real part.

  b. Can undamped oscillations occur?
  c. Take $f = S / (S + a)$, and show that (1) all eigenvalue of Eq. (2) have negative real part, and (2) x_1 and x_2 coexist for

$$\beta_1 > \frac{a + 1 - X_1}{1 - X_1} \qquad (3)$$

Design of Fermentation Processes

where

$$X_1 = \frac{1}{(Y_{sx}s_f\beta_2)} \quad (4)$$

d. On a diagram with $Y_{sx}s_f$ on the abcissa and β_1 on the ordinate, find the region in which the solution $X_2 = 0, X_1 > 0, S > 0$ is the only stable steady state.
e. On the same diagram, calculate the curve β_1 versus $Y_{sx}s_f$, which separates the region with exponentially damped oscillations from the region with pure exponential decay of a perturbation toward the steady state $(X_1, X_2, S) > 0$. Make the numerical calculations with $\beta_2 = 0.02$ L g^{-1} and $a = 0.2$.
f. Describe in your own words how you would plan the production of a valuable microorganism x_1 which can be attacked by a parasite or a phage. Can you choose any value of the dilution rate? of feed stream concentration s_f?
g. Illustrate with a few examples, using suitable values for the *true* kinetic parameters ($\mu_{max,1}$, K_s, etc.) and operating variables (s_f, D, etc.). This last part of the exercise is intended to help in the back-translation from the dimensionless variables and parameters, which are very helpful in the theoretical development but may be difficult to relate to an actual physical situation.

Answers to (d) and (e) are shown graphically in the figure below. The horizontal line below which washout is the only stable solution is given by $\beta_1 = 1 + a = 1.2$. Since in the figure s_f appears both in the abcissa and the ordinate, there may be some confusion if the figure is compared with Fig. 8 in Tsuchiya et al. (1972), who treated the same model. For $s_f \rightarrow 0$, the parameter $a \rightarrow \infty$ and the horizontal line in the figure will bend upward and approach the ordinate axis asymptotically, as in the corresponding figure in the original paper, where $1/D$ is used as ordinate.

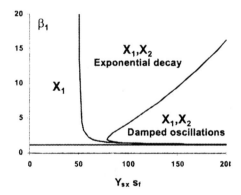

Problem 9.10 Production of a protein that is degraded by the action of proteases.

A valuable protein is produced by a growth associated process using a microorganism for which the volumetric growth rate is given by

$$q_x = \frac{0.2\,s}{s+1} x \text{ g biomass (L medium h)}^{-1} \quad (1)$$

In (1) the concentration of the growth limiting substrate is s g L^{-1}. x is the biomass concentration in g L^{-1}. The yield of biomass on the substrate is $Y_{sx} = 0.5$ g g^{-1}.

Production of the protein is severely catabolite repressed and the volumetric production rate is

$$q_p = \frac{1}{1+10s} q_x \quad \text{g protein (L medium h)}^{-1} \quad (2)$$

Production of the protein is to take place in a continuous, steady state stirred tank, and the glucose feed concentration is $s_f = 10$ g L^{-1}.

    a. Determine the largest value of D for which the steady state process will work. Thereafter determine the yield coefficient Y_{sp}. For which value of D (or s) is Y_{sp} at its maximum? Is that the same value for which q_p is at its maximum? Why is the continuous stirred tank process better than a batch reactor process?

    b. Make a sketch of q_p as a function of s for $0 < s < 10$ g L^{-1}. Calculate the largest productivity $(q_p)_{max}$ which can be obtained and determine the corresponding value of D.

    c. Equation (2) is an idealization of the real situation. The microorganism also produces proteases that to a considerable degree destroy the desired protein. The correct expression for q_p is:

$$q_p = \frac{1}{1+10s} q_x - kx \quad (3)$$

Determine the maximum value of k above which no net production of the protein will be obtained for any operating conditions. Derive an approximate relation between k and the maximum protein productivity. You should not attempt to obtain an algebraic solution, but use a graphical representation in which q_p given by (2) is plotted together with kx and make your approximate solution based on this graph.

    d. It is contemplated to install a cell separator at the exit of the reactor and recirculate cells to the inlet.
Discuss whether an increase of productivity (given by (3)) can be obtained in this way. Is there any difference between the two cases (A): The protein product is secreted 100 % to the medium and (B): The protein product remains quantitatively inside the cells? What is the potential increase in maximum productivity if a recirculation ratio $R = 0.5$ and a cell separation factor $\beta = 1.5$ (Eq (9.43)) is used?

Problem 9.11 Adhesion of cells can lead to serious underdesign of an industrial reactor.

In a continuous stirred tank reactor operating at a given dilution rate D and with sterile feed the concentration of biomass in the medium (the suspended culture) is x g L^{-1}. Some biomass adheres to the inner surface of the reactor. The concentration of adhering biomass (converted from a more "natural" unit g (m^2 reactor surface)$^{-1}$) is x_1 g L^{-1}. Suspended and adhering biomass grow with the same yield coefficient Y_{sx} and the same specific growth rate μ. Suspended and adhering biomass are exchanged by first order rate processes : $k_1 x$ (adhesion) and $k_2 x_1$ (attrition of biofilm).

    a. Write mass balances for the suspended and for the adhering biomass. Add a substrate balance to

Design of Fermentation Processes

the model. Derive the following relation between the steady state value of D and the corresponding value of μ:

$$D = \frac{k_1 + k_2 - \mu}{k_2 - \mu}\mu \tag{1}$$

Show that D is always larger than μ for realistic values of k_1 and k_2.

b. Let $\mu = 0.5 s / (s + 2)$ h^{-1}, $s_f = 10$ g L^{-1}, $Y_{sx} = 0.5$ g g^{-1}, $k_1 = 2$ h^{-1} and $k_2 = 1.5$ h^{-1}. The dilution rate is $D = 0.2$ h^{-1}.
Calculate s, x, and x_l and compare with the corresponding values for $k_l = 0$.
Assume that the above data were obtained with a small laboratory reactor of volume $V = 1$ L, height $h = 1.3$ diameter d. The inner surface of the reactor is 550 cm^2. For a film density $\rho = 1$ g cm^{-3} calculate the film thickness Δ.
(Answer: 123 μ m which is likely to go undetected in the experiments)
Make a simulation of x g L^{-1} as a function of D, both in the case of biomass adhesion and for $k_l = 0$ (no adhesion).

c. Consider a 50 m^3 industrial reactor for the same process. $h = 3 d$. The internal surface, including baffles, heat exchangers etc. is 100 m^2.
What would the film thickness be if k_l and k_2 have the same values as in (b)? In reality a film of this thickness would never be stable. It is more likely that the film thickness Δ is also 123 μ m in the industrial reactor. The adhesion mechanism is probably the same in both scales ($k_l = 2$ h^{-1}), but k_2 is much larger for the industrial reactor.
Determine the "effective" large scale value of k_2 and the relation between x and D. You will conclude that a serious underdesign of the large scale operation would result if the laboratory data were used uncritically as the design basis.

The data was taken from a real process, the production of lipases by a microorganism that hydrolyzes linseed oil in an aqueous emulsion of the vegetable oil. Some oil sticks to the reactor surface and part of the culture adheres to this film.

Problem 9.12 Plug flow reactor with recycle

The recycle reactor of Fig. 9.5 also functions for a sterile feed because part of the effluent from the plug flow reactor is recycled to the inlet. In Fig. 9.4, the operation of a chemostat is improved because part of the effluent is returned to the inlet with a higher cell concentration $x_R = \beta x$ than that which exists in the reactor. It is appealing to combine the two procedures by the installation of a cell separator (e.g., a cyclone) at the exit from the plug flow reactor and return cell-enriched medium to the inlet.
With reference to Fig. 9.5, let the exit concentration from the plug flow reactor be x_r while the cell concentration is $x_R = \beta x_r$ in the recycle stream $v_R = vR$. As in Fig. 8.4, it is assumed that the substrate concentration is s in all streams leading to or from the separator. $x_f = 0$, i.e., the feed stream is sterile.

a. Derive algebraic expressions for x_r, x_l, and s_l in terms of s_f, s, R, and β for maintenance-free Monod kinetics.

b. Show that Eq. (9.76) can be used to calculate τ if x_r is used in place of x and a' is used in place of $a_f = K_s / s_f$:

$$a' = \frac{1 - R(\beta - 1)}{1 - R(\beta - 1) s / s_f} a_f \tag{1}$$

c. Calculate (on a computer) a table of $\tau(R)$ for fixed values of s_f, s, K_s, μ_{max}, and β. Test the program using $s_f = 60$ g m^{-3}, $s = 0.6$ g m^{-3}, $K_s = 4$ g m^{-3}, $\mu_{max} = 4/3$ h^{-1}, and $\beta = 1.02$. Look sharply at the output around $R = 1.684$.

d. For the above example, what is R for $\beta = 1$? What is the global minimum of τ, and for what R value is that obtained? Make sketches of $\tau(R)$ for $\beta = 1.02$ and for $\beta = 1.05$—other parameter values are as in (c).

e. Can anything interesting be expected for $\beta < 1$? This mode of operation is perhaps useful if the cyclone is the first unit in a downstream operation to produce a cell concentrate. After all, the recycle reactor does function if only some cells are returned to the inlet. Explore the possibilities for various $\beta < 1$ and various s using fixed values of s_f, K_2, and μ_{max} as above.

f. Prove by differentiation of Eq. (9.76)—with the modifications in (b)—that an extremum of τ for a fixed value of β is found for an R value determined by solution of

$$(1 + a') \ln\left(\frac{R+1}{R\beta}\right) - a' \ln\left(\frac{s}{s_1}\right) + \frac{(R+1)(\beta - 1)(s_f - s)}{\left[s_f - sR(\beta - 1)\right]\left[1 - R(\beta - 1)\right]} a'$$
$$\times \left[\ln\left(\frac{R+1}{R\beta}\right) - \ln\left(\frac{s}{s_1}\right)\right] - \left(\frac{1}{R} + \frac{s_f}{s_1 R} a'\right) = 0 = F(R) \quad (2)$$

where a' is given by Eq. (1) and

$$s_1 = \frac{s_f + Rs}{R + 1} \quad (3)$$

Repeat the calculations of $\tau(R)$ for $\beta = 1.02$, but now include a tabulation of $F(R)$ in Eq. (2) to verify that $d\tau/dR = 0$ for $R \approx 1.684$ and the parameter values of (c). Convince yourself that Eq. (2) simplifies to Eqs. (9.82) – (9.83) for $\beta = 1$.

g. For $\beta = 1$, it was easy to derive Eq. (9.79) using the Leibnitz formula [Eq. (9.78)]. This procedure can also be used for $\beta \neq 1$. If you think that mathematics is fun, try to derive Eq. (2) following the procedure of Eqs. (9.78) - (9.83). The algebra is tough, and you must be careful to include all relevant terms of Eq. (9.78)—otherwise a lot of effort is wasted. Do not mistrust the result in Eq. (2)!

REFERENCES

Andersen, M.Y., Pedersen, N., Brabrand, H., Hallager, L., and Jørgensen, S.B. (1997).
Regulation of continuous yeast fermentation near the critical dilution rate using a productostat. *J. Biotechnol.* **54**, 1-14.

Aris, R. and Amundson, N. R. (1973). *First-Order Partial Differential Equations with Applications*, Prentice-Hall, Englewood Cliffs, NJ.

Duboc, P., von Stockar, U., and Villadsen, J. (1998). Simple generic model for dynamic experiments with *Saccharomyces cerevisiae* in continuous culture: Decoupling between anabolism and catabolism. *Biotechnol. Bioeng.* **60**, 180-189.

Kirpekar, A. C., Kirwan, D. J., and Stieber, R. W. (1985). "Modeling the stability of Cephamycin C producing N. *lactamdurans* during continuous culture," *Biotech. Prog.*, **1**, 231 - 236.

Levenspiel, O. (1999). *Chemical Reaction Engineering*, 3. ed., John Wiley & Sons, New York.

Melchiorsen, C.R., Jensen, N.B.S., Christensen, B., Jokumsen, K.V., and Villadsen, J. (2001). Dynamics of pyruvate metabolism in *Lactococcus lactis*. *Biotechnol.Bioeng.* **74**, 271-279.

Menawat, A., Muthurasan, R., and Coughanowr, D. R. (1987). "Singular control strategy for a fed-batch bioreactor: numerical approach," *AIChE J.* **33**, 776 - 783.

Nielsen, J. (1992). "On-line monitoring of microbial processes by flow injection analysis," *Proc. Con. Qual.* **2**, 371 - 384.

Palanki, S., Kravaris, C., and Wang, H. Y. (1993). "Synthesis of state feedback laws for end-point optimization in batch processes", *Chem. Eng. Sci.* **48**, 135 - 152.

Sonnleitner, B. and Fiechter, A. (1988). "High performance bioreactors: A new generation," *Anal. Chim. Acta* **213**, 199 - 205.

Tsuchiya, H. M., Drake, J. F., Jost, J. L., and Fredrickson, A. G. (1972). "Predator prey interactions of *Dictyosilium discoideum* and *Escherichia coli* in continuous culture," *J. Bacteriol.* **110**, 1147 - 1153.

Vara, A.G., Hochkoepple, A., Nielsen, J., and Villadsen, J. (2002). Production of Teicoplanin by *Actinoplanes teichomyceticus* in continuous culture. *Biotechnol.Bioeng.* **77**, 589-598.

Villadsen, J. and Michelsen, M. L. (1978). *Solution of Differential Equation Models by Polynomial Approximation*, Prentice-Hall, Englewood Cliffs, NJ.

10

Mass Transfer

The first requirement for any chemical reaction to take place is that the reactants are present at the site of reaction. In multiphase systems, the maximum rates of the transport processes are often lower than the maximum reaction rate, which in turn means that the overall reaction rate will be limited by the transport processes. In bioreactions, the transport of nutrients to the cell surface and the removal of metabolites from the cell surface to the bulk of the medium are rate processes with time constants not much smaller than those of the cellular reactions. Therefore, *mass transfer* must be included in an analysis of bioreactions alongside of stoichiometry and cellular kinetics.

In this book we have already referred to mass transfer in many places. Thus mass transfer from a gas phase to a liquid phase was included at the start of Chapter 3 in order to set up mass balances for the continuous steady state tank reactor and to calculate rates of bioreactions. In the analysis of the ideal bioreactor in Chapter 9 mass transfer was also included in order to provide a reasonable general design framework.

In the present chapter we shall review the physical foundation of mass transfer and discuss both experimental techniques and general methods for calculation of the rate of mass transfer. We shall refer to the vast body of empirical knowledge on mass transfer that is an essential part of the curriculum for chemical engineering students, and in order to make translation from chemical textbooks easier the standard nomenclature of these texts will be used as far as possible. The resulting change of nomenclature e.g. concentrations (c_s instead of s and p_A instead of π_A for partial pressure) should not represent any great problem for the reader.

Mass transfer takes place by two basic processes; convection and diffusion. A full treatment of mass transfer therefore in principle requires a fully known flow field. However, a simplified treatment in which the overall mass transfer is schematically divided into different transfer steps is normally used with good results. An overview of important mass transfer steps in a fermentation process is given in Fig. 10.1, which shows the individual steps involved in oxygen transport from a gas bubble to the reaction site inside the individual cells (Bailey and Ollis, 1986). The steps are:

1. Diffusion of oxygen from the bulk gas to the gas liquid interface.
2. Transport across the gas liquid interface.

3. Diffusion of oxygen through a relatively stagnant liquid region adjacent to the gas bubble i.e. from the gas-liquid interface to the well-mixed bulk liquid.
4. Transport of oxygen through the well-mixed liquid to a relatively unmixed liquid region surrounding the cells.
5. Diffusion through the stagnant region surrounding the cells.
6. Transport from the liquid to the pellet cell aggregate etc.
7. Diffusive transport of oxygen into the pellet etc.
8. Transport across the cell envelope (see Section 2.1.2).
9. Transport from the cell envelope to the intracellular reaction site e.g. the mitochondria.

For most processes one or more of these steps are in a pseudo-steady state. The transport through the well-mixed liquid is normally very rapid in laboratory-scale bioreactors because of the reasonable assumption of homogeneity in the medium (see Chapter 11). Furthermore Steps 5, 6, and 7 are relevant only for processes in which pellets or cell aggregates appear. Intracellular transport resistance is normally also neglected because of the small size of most cells. The most important mass transfer phenomena are thus of two kinds: (1) Gas liquid mass transfer (Section 10.1) and (2) Molecular diffusion of medium components into pellets or cell aggregates (Section 10.2). Both of these mass transfer phenomena are discussed at great length in chemical engineering textbooks (see e.g. Bird *et al.*, 2001 or Cussler, 1997), which should be consulted for an in-depth analysis of the subject. Here we will treat the subject only to enable the reader to combine the kinetic models of the previous chapters with simple models for the mass transport phenomena relevant to bioreactions.

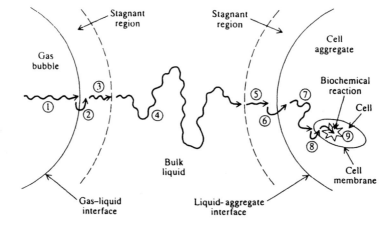

Figure 10.1. Overview of steps in the overall mass transfer of oxygen from a gas bubble to the reaction site inside the individual cells. (Reprinted by permission from J. E. Bailey and D. F. Ollis (1986), Biochemical Engineering Fundamentals, McGraw-Hill).

10.1 Gas-Liquid Mass Transfer

In aerobic processes, oxygen is a key substrate and because of its low solubility in aqueous solutions a continuous transfer of oxygen from the gas phase to the liquid phase is decisive for maintaining the oxidative metabolism of the cells (see Example 10.1). A few minutes without aeration of the medium has for example a serious impact on the ability of a culture of the mold *Penicillium chrysogenum* to produce the desired penicillin, whereas facultatively aerobic organisms, such as the yeast *Saccharomyces cerevisiae* or the bacterium *Escherichia coli*, will drastically change their product formation when deprived of oxygen.

However, oxygen gas-liquid mass transfer, is not the only important phase transfer process in fermentation processes. Carbon dioxide is formed during respiration, and also in most fermentative processes. A too high concentration of carbon dioxide may act inhibitory on the microorganism, and a continuous removal of the carbon dioxide formed is therefore needed (Jones and Greenfield, 1982). Methane and other light hydrocarbons can be used for the production of single-cell protein. In these processes, both oxygen and the sparingly soluble hydrocarbon must be transferred to the liquid phase at a rate sufficient to meet the requirements of the cells. Methane may also be a product gas in anaerobic wastewater treatment processes. In this case the necessary gas-liquid transfer involves removing the methane and carbon dioxide from the liquid phase.

Example 10.1 Oxygen requirements of a rapidly respiring yeast culture
To illustrate the requirements for a high gas-liquid mass transfer of oxygen, we consider the experimental data for *S. cerevisiae* analyzed in Example 3.5. For dilution rates below 0.25 h^{-1}, where the metabolism is purely respiratory, we have

$$Y_{so} = 0.425 \text{ moles O}_2 \text{ (C-mole glucose)}^{-1} \tag{1}$$

For a dilution rate of 0.2 h^{-1} there is virtually no glucose in the outlet from the chemostat (see Fig. 1), and the volumetric glucose uptake rate is therefore calculated to be

$$-q_s = 0.2 \text{ h}^{-1} \, 28 \text{ g L}^{-1} = 5.6 \text{ g L}^{-1} \text{ h}^{-1} = 0.1867 \text{ C-moles L}^{-1} \text{ h}^{-1} \tag{2}$$

Thus the volumetric oxygen uptake rate is

$$q' = -q_o = 0.425 \; 0.1867 \text{ C-moles L}^{-1} \text{ h}^{-1} = 79.3 \text{ mmoles O}_2 \text{ L}^{-1} \text{ h}^{-1} \tag{3}$$

If the dissolved oxygen concentration is at its maximum (approximately 0.26 mmoles O$_2$ L^{-1} when sparging with air, see Table 10.1) oxygen will be depleted within 12 s if the supply of oxygen is stopped. This illustrates the requirement for a continuous transfer of oxygen to the liquid medium. In Example 10.2, we will quantify the mass transfer necessary to keep the dissolved oxygen concentration constant.

Gas-liquid mass transfer is normally modeled by the two-film theory (see Fig. 10.2), which was introduced by Whitman (1923). The flux J_A of compound A through each of the two films is

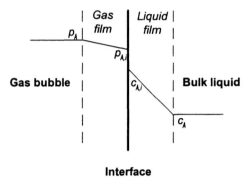

Figure 10.2. Concentration profiles in gas and liquid films for the transfer of the gaseous compound A into the liquid phase. The composition of the gas bulk and of the bulk liquid are assumed to be constant.

described as the product of the concentration difference across the film layer, i.e., a linear driving force, and a mass transfer coefficient, k. Thus the flux across the gas film is given by

$$J_{A,g} = k_g (p_A - p_{A,i}) \tag{10.1}$$

where p_A is the partial pressure of compound A in the gas bubble. Index i refers to the concentration at the gas-liquid interface. Similarly, for the flux across the liquid film

$$J_{A,l} = k_l (c_{A,i} - c_A) \tag{10.2}$$

In the dilute aqueous solutions normally used as fermentation media, the concentrations on each side of the gas-liquid interface can be related to each other by *Henry's law*:

$$p_{A,i} = H_A c_{A,i} \tag{10.3}$$

where H_A is Henry's constant for compound A (unit: atm L mole^{-1}). Table 10.1 lists the values of Henry's constant for a few components.

Table 10.1. Henry's constant for some gases in water at 25 °C.

Compound A	H_A (atm L mole^{-1})
Ammonia[a]	17.1 10^{-3}
Methane	745.5
Ethane	545.1
Carbon dioxide	29.7
Oxygen	790.6
Hydrogen	1270.0

[a] At a concentration of 0.2 g L^{-1} in the aqueous phase.

Mass Transfer

Since the interfacial concentrations are not directly measurable, we specify the overall flux of the considered component from the gas bubble to the liquid phase as an overall mass transfer coefficient multiplied by the driving force in the liquid phase, i.e.,

$$J_A = K_l(c_A^* - c_A) \tag{10.4}$$

where c_A^* is the saturation concentration in the bulk liquid corresponding to the bulk gas phase:

$$c_A^* = \frac{p_A}{H_A} \tag{10.5}$$

At steady state, $J_{A,g} = J_{A,l} = J_A$ and by inserting Eqs. (10.3) and (10.5) in Eq. (10.1), we find

$$\frac{1}{K_l} = \frac{1}{H_A k_g} + \frac{1}{k_l} \tag{10.6}$$

k_g is typically, considerably larger than k_l and for gases with large values of H_A such as oxygen and carbon dioxide (which have a small to moderate solubility in water) the gas-phase resistance is therefore negligible. Thus the overall mass transfer coefficient K_l is approximately equal to the mass transfer coefficient in the liquid film, i.e., k_l. Normally k_l is used for quantification of the mass transfer despite the fact that in practice only K_l can be measured.

To find the mass transfer rate of compound A per unit of reactor volume, i.e., the volumetric mass transfer rate q_A^t, we multiply the overall flux J_A by the gas-liquid *interfacial area* per unit liquid volume a (unit: m²/m³ = m⁻¹). Thus

$$q_A^t = J_A a = k_l a (c_A^* - c_A) \tag{10.7}$$

The product of the liquid mass transfer coefficient k_l and the specific interfacial area a is called the *volumetric mass transfer coefficient* or most often $k_l a$. Due to the difficulties in the determination of k_l and a individually, their product is normally used to specify the gas-liquid mass transfer. From Eq. (10.7), the volumetric mass transfer rate can be calculated if $k_l a$ and the driving force $(c_A^* - c_A)$ are known. The influence of the operating conditions on the value of $k_l a$ is discussed in the following sections.

In a well-mixed tank, c_A, has the same value at any position in the tank, whereas the value of c_A^* depends on the gas-phase concentration. Due to consumption or production, the inlet and outlet mole fraction of A will be different. A suitable approximation for the average driving force is the so-called logarithmic mean driving force, in which the known saturation concentrations at the inlet and exit from the tank are used in place of the true variable c_A^*.

$$(c_A^* - c_A) = \frac{(c_{A,inlet}^* - c_A) - (c_{A,outlet}^* - c_A)}{\ln(c_{A,inlet}^* - c_A) - \ln(c_{A,outlet}^* - c_A)} \qquad (10.8)$$

Example 10.2. Requirements for $k_l a$ in a laboratory bioreactor
We now return to the experimental data analyzed in Example 3.5. We want to calculate the minimum value of $k_l a$ needed to maintain the dissolved oxygen concentration in the medium above a desired value. Since the dissolved oxygen concentration in aqueous solutions is very low, the oxygen consumption approximately equals the mass transfer of oxygen at steady state, i.e.,

$$-q_O = q_O^t = k_l a (c_O^* - c_O) \qquad (1)$$

or

$$k_l a = \frac{q_O^t}{(c_O^* - c_O)} \qquad (2)$$

If the dissolved oxygen concentration is to be kept at 60% of the saturation value obtained by sparging with air containing 20.95% oxygen, we find

$$(c_O^* - c_O) = c_O^*(1 - 0.6) \qquad (3)$$

and if the bioreactor is operated at 1 atm, we find from Table 10.1

$$c_O^* = \frac{0.2095}{790.6} = 0.265 \cdot 10^{-3} \text{ moles L}^{-1} \qquad (4)$$

Inserting this value in Eq. (3) and using the calculated value of $-q_o$ in Example 10.1, we find by using Eq. (2)

$$k_l a = 748 \text{ h}^{-1} = 0.208 \text{ s}^{-1} \qquad (5)$$

This $k_l a$ value can normally be obtained in a well-stirred laboratory bioreactor. For larger stirred tank bioreactors, it may be difficult to obtain a $k_l a$ value above 500 h^{-1} and oxygen limitation may become a problem.

10.1.1. Models for k_l

The idea of expressing the mass transfer across surfaces by an overall mass transfer coefficient multiplied by a concentration difference is clearly a simplification of a complicated physical reality. It is for many reasons desirable to find a relation between the mass transfer coefficient and known physical variables such as e.g. the diffusivity of the solute. The first model of this kind was due to Nernst (1904), who assumed that mass transfer occurred by diffusion through a stagnant film.

Mass Transfer

Whitman (1923) took this a step further in his two-film theory, in which it was assumed that mass transfer from a gas bubble could be described by molecular diffusion through the two stagnant films. From *Fick's first law* with a constant concentration gradient through the film, we get for the mass transfer through the liquid film:

$$k_l = \frac{D_A}{\delta_f} \qquad (10.9)$$

where D_A is the diffusion coefficient for component A and δ_f is the thickness of the liquid film. The film theory gives a simple relation between the diffusivity and the mass transfer coefficient. However, it is not possible to calculate k_l from D_A, since the film thickness is normally not known. In fact, there is, strictly speaking, no stagnant film surrounding the bubbles, although the liquid velocity relative to the surface is low very close to the bubble. Furthermore, the assumption of a constant concentration gradient may be wrong, e.g. if the diffusing species A is consumed by a chemical reaction within the film. Despite these shortcomings, the film model is probably the most widely used model to illustrate the concept of mass transfer.

Not all models for mass transfer make use of a stagnant film. In the so-called surface renewal theory (Danckwerts, 1970), discrete liquid elements close to the gas-liquid interface are thought to be interchanged with a well-mixed bulk liquid. Each element stays close to the surface for a certain time, during which it is considered to be stagnant. The transfer of compound A from the gas phase to the element is determined by the exposure time of the liquid element by solving the unsteady-state form of Fick's law of diffusion (sometimes called Fick's second law of diffusion). The mean residence time at the surface, τ_{ren} is introduced in this model, and it can be shown that k_l is related to D_A through:

$$k_l = \sqrt{D_A/\tau_{ren}} \qquad (10.10)$$

It is seen from Eq. (10.10) that a slightly different relation between k_l and D_A than predicted from the film theory is obtained. The dependence of k_l on D_A is experimentally found to be between that predicted by the film theory and that predicted by the surface-renewal theory. Again, a calculation of k_l is not possible from the model, since it is difficult to find a value for the mean residence time at the surface.

A more rigorous approach is to connect the mass transfer coefficient to a solution of the flow field close to a surface. This can be done using boundary-layer theory (see e.g Cussler, 1997). The mass transfer coefficient, averaged over a length L (m), for transfer of a compound present in the solid phase of a sharp-edged plate, is given by:

$$\frac{k_l L}{D_A} = 0.646 \left(\frac{L u \rho}{\eta} \right) \left(\frac{\eta}{\rho D_A} \right)^{\frac{1}{3}} \qquad (10.11)$$

where u is the bulk flow velocity (m s^{-1}).

This results in the following relation between k_l and the diffusion coefficient:

$$k_l \propto D_A^{2/3} \tag{10.12}$$

where the proportionality factor is a function of the physical properties of the liquid and the liquid flow rate.

None of these theoretically based models are of much use for calculating the value of k_l in a real system, e.g., an agitated and sparged liquid. They do, however, tell us something about how the physical properties of the liquid influence the liquid mass transfer coefficient. Furthermore, the relative values of k_l for various compounds may be evaluated even if the diffusivities of the compounds are unknown (see also Note 10.3).

10.1.2. Interfacial Area and Bubble Behavior

In Eq. (10.7), the specific interfacial area a is based on the liquid volume, V_l, i.e.

$$a = \frac{A}{V_l} \tag{10.13}$$

where A is the total interfacial area in the gas liquid dispersion. This definition of the specific interfacial area is the most convenient when the volumetric rate of the mass transfer process is to be used together with mass balances for dissolved components, e.g., the dissolved oxygen. However, in empirical correlations for the volumetric mass transfer coefficient (see Section 10.1.4), one often uses the specific interfacial area based on the total volume of the gas-liquid dispersion, i.e.,

$$a_d = \frac{A}{V_d} = \frac{A}{V_l + V_g} \tag{10.14}$$

The two definitions of the specific interfacial area are related by

$$a_d = (1-\varepsilon)a \tag{10.15}$$

where

$$\varepsilon = \frac{V_g}{V_d} \tag{10.16}$$

Mass Transfer

is the *gas holdup* in the dispersion.

The specific interfacial area is a function of the bubble size distribution in the gas liquid dispersion, and a is obtained from

$$a = \frac{6\varepsilon}{(1-\varepsilon)d_{mean}} \quad (10.17)$$

where d_{mean} is the average bubble diameter, which may be calculated as the first moment of the bubble size distribution function. A surface-averaged diameter, the so-called *mean Sauter diameter*, is, however, often used:

$$d_{Sauter} = \frac{\sum n_i d_{b,i}^3}{\sum n_i d_{b,i}^2} \quad (10.18)$$

In a bioreactor, three main processes interact to determine the bubble size distribution (Fig. 10.3):
1. *Bubble formation*, determined by the breakup into discrete bubbles of the gas stream as it is sparged into the liquid phase.
2. *Bubble breakup*, determined by the competition between the stabilizing effect of the surface tension and the destabilizing effect of inertial forces.
3. *Bubble coalescence*, i.e., fusion of bubbles, determined by the properties of the gas-liquid interface.

A bubble is formed at the orifices of the sparger when the *buoyancy force* on the bubble exceeds the surface tension acting on the periphery of the orifice. Thus the initial bubble diameter $d_{b,i}$ is determined from a force balance:

$$\pi d_o \sigma = \frac{\pi}{6} d_{b,i}^3 g(\rho_l - \rho_g) \quad (10.19)$$

or

$$d_{b,i} = \left[\frac{6\sigma d_o}{g(\rho_l - \rho_g)}\right]^{\frac{1}{3}} \quad (10.20)$$

where σ is the surface tension of the liquid phase (unit: N m^{-1}), d_o is the orifice diameter, and g is the acceleration of gravity. Equation (10.20) holds only up to a certain gas flow rate over which the diameter of the bubbles starts to increase with the gas flow rate. For much higher gas flow rates, swarms of bubbles and finally an almost continuous jet flow will be formed. For viscous media, liquid viscosity rather than bubble surface tension provides the predominant resistance to bubble formation. For such systems an empirical correlation for the initial bubble diameter can be used (see Example 10.3).

Figure 10.3 Factors influencing the dynamic bubble-size distribution in a bioreactor.

When the bubbles have been formed at the orifices of the sparger, they circulate in the gas liquid dispersion until finally leaving the dispersion for the *head space*. The gas liquid dispersion is normally vigorously agitated, resulting in the formation of a *turbulent flow field*, in which there is a maximum bubble size $d_{b,max}$ determined by the balance of opposing forces:

1. Shear forces acting on the bubble tend to distort the bubble into an unstable shape so that the bubble disintegrates into smaller bubbles.
2. The surface tension force acting on the bubble tends to stabilize the spherical shape of the bubble.
3. In the dispersed phase, there is viscous resistance to deformation of the bubble.

In gas liquid dispersions, the viscous resistance on the gas side is negligible compared to the surface tension contribution, and at equilibrium we therefore have:

$$\tau_s = k_1 \frac{\sigma}{d_{b,max}} \tag{10.21}$$

where τ_s is the *shear stress*, i.e. force per unit area acting parallel to the surface (N m^{-2}), and k_1 is a dimensionless constant. If $d_b > d_{b,max}$ the shear force acting on the bubble is larger than the surface tension forces, and the result is bubble breakup. In order to calculate $d_{b,max}$ we need to determine a value for the shear stress. According to the statistical theory of turbulence, the dynamic shear stress acting on bubbles with diameter d_b is given by Eq. (10.22) for a turbulent flow field (see Note 10.1):

$$\tau_s = k_2 \rho_l^{\frac{1}{3}} \left(\frac{P_g}{V_l}\right)^{\frac{2}{3}} d_b^{\frac{2}{3}} \tag{10.22}$$

P_g/V_l is the power dissipation per unit volume (W m^{-3}), and k_2 is a dimensionless constant. In a low viscosity medium (like water), the dynamic shear stress given by Eq. (10.22) is much larger than the viscous shear stress. In such a case, a combination of Eqs. (10.21) and (10.22) will give the following expression for the maximum stable bubble diameter

Mass Transfer

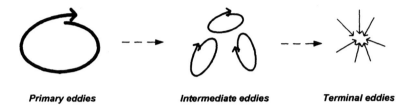

Primary eddies   Intermediate eddies   Terminal eddies

Figure 10.4. Energy transfer from primary eddies to terminal eddies.

$$d_{b,\max} = k \frac{\sigma^{0.6}}{\left(P_g/V_l\right)^{0.4} \rho_l^{0.2}} \tag{10.23}$$

Thus the maximum stable bubble diameter is reduced (giving a higher specific interfacial area) if the power input is increased. Based on theoretical predictions in the turbulent regime, Lehrer (1971) states $k = 1.93$ (with all parameters in SI units), whereas several researchers specify k as a function of the gas holdup ε. For pure water Lee and Mevrick (1970) suggest

$$k = 4.25\varepsilon^{\frac{1}{2}} \tag{10.24}$$

In the case of highly viscous media, also the viscous resistance needs to be taken into account and Eq. (10.23) is therefore less accurate.

Note 10.1. Calculation of maximum stable bubble diameter using the statistical theory of turbulence

A turbulent flow field is normally described by the statistical theory of turbulence, where the flow field is regarded as a distribution of superimposed *eddies* or velocity fluctuations characterized by their direction and magnitude. According to this theory, large *primary eddies* emerge due to the impeller action. The scale of these primary eddies is on the order of magnitude of the impeller diameter. These primary eddies are unstable and disintegrate into smaller eddies (called *intermediate eddies*), which are again unstable and therefore disintegrate further into still smaller eddies (called *terminal eddies*). The terminal eddies have completely lost their unidirectional nature and are therefore isotropic. Thus kinetic energy flows through the cascade of eddies until ultimately the energy is dissipated as heat (see Fig. 10.4). The size of the terminal eddies is (see, e.g., Moo-Young and Blanch, 1981)

$$l_{\min} = \frac{\eta^{\frac{3}{4}}}{\rho_l^{\frac{1}{2}}} \left(\frac{P_g}{V_l}\right)^{-\frac{1}{4}} \tag{1}$$

where η is the viscosity of the liquid.

Any given location will be passed by eddies of widely different velocities, and it is appropriate to introduce time-averaged values of the eddy velocity. For eddies of scale much smaller than the primary eddies but larger than the terminal eddies (of scale l_{min}), the time-averaged velocity of the eddies is given by

$$u(L) = c_2^{\frac{1}{2}} \left(\frac{P_g}{V_l}\right)^{\frac{1}{3}} \left(\frac{L}{\rho_l}\right)^{\frac{1}{3}} \qquad (2)$$

where u(L) is the velocity for a length scale L.

The shear stress acting on a bubble is largely determined by the velocity of eddies of about the same size as the bubble diameter. For τ_s we therefore use

$$\tau_s = \rho_l u(L = d_b)^2 \qquad (3)$$

Inserting Eq. (2) in Eq. (3), we obtain Eq. (10.22) with $k_2 = c_2$.

Description of energy input to a process by the theory of isotropic turbulence is valuable for understanding bubble behavior in a bioreactor. The theory may also be used to predict the influence of energy input on the fragmentation of hyphal elements and mycelial pellets. Local isotropic turbulence is, however, an idealization not always realized in practice. Furthermore, in the derivation of Eq. (2), the energy dissipated per unit liquid volume is used regardless of the means by which that energy is delivered (mechanical agitation or injection of compressed gas). This is an idealization, but an acceptable approximation for many systems.

Coalescence of bubbles can be considered as a three-step process (Moo-Young and Blanch, 1981).
1. Bubbles occasionally come into contact with each other within the liquid phase. This contact is characterized by a flattening of the contact surfaces, leaving a thin liquid film separating them.
2. The thickness of the separating liquid continuously decreases to a thickness of approximately 10^{-6} cm (Tse et al., 1998).
3. Finally, the film ruptures, which completes the coalescence process.

The entire process occurs in the milliseconds range, and the last step is practically instantaneous. What determines if coalescence will occur, is therefore if the time constant of the second step is smaller than the contact time of the bubbles. The rate of the second step is controlled by the properties of the liquid film. In a multicomponent liquid phase, interaction between molecules of different species leads to an enhanced concentration in the film layer of one or more of the soluted species. This results in an increased repulsion between two bubbles and therefore in a reduced coalescence. Especially with surface-active compounds, the enrichment in the film layer is considerable, even for very small concentrations in the bulk liquid. The influence of inorganic ions on the coalescence is illustrated in Fig. 10.5 where the mean Sauter diameter is shown as a function of the ionic strength of the electrolyte solution. Alcohols and other organic compounds have a similar influence on the coalescence. Here small alcohols are less efficient as coalescence reducers than larger alcohols, e.g., methanol has less influence on the mean bubble diameter than octanol at

Mass Transfer

the same concentration (Keitel and Onken, 1982). Surface active compounds also strongly influence the coalescence. Some (e.g., foam-stabilizing compounds like proteins) reduce the coalescence, whereas others (e.g., antifoam agents like fatty acids) increase the coalescence. Normally, surface-active materials reduce the mass transfer coefficient for the liquid film (i.e., k_l) and the overall effect of surface-active materials on the volumetric mass transfer coefficient k_la is therefore quite complex.

However, for foam-stabilizing compounds the increase in the specific interfacial area a (due to a smaller average bubble diameter) is normally larger than the decrease in k_l. In most fermentation media, the tendency for coalescence is smaller than for pure water. Water is therefore often called a *coalescing medium*, whereas many fermentation broths are *non-coalescing*.

It is quite clear that the mechanism of coalescence is not yet fully understood (Craig et al., 1993). Whereas the presence of certain ions in an aqueous medium reduces the coalescence, other ions seem to have no influence at all.

The combined effect of bubble breakup and coalescence on the average bubble diameter is determined by the relative rate of the two processes. If coalescence is very slow compared to bubble breakup, the average bubble diameter is determined by the breakup process, i.e. by Eq. (10.23). However, if the bubbles formed at the orifice are smaller than the maximum stable bubble diameter (i.e., $d_o < d_{b,max}$) the average bubble diameter is determined by the bubble formation process, i.e., Eq. (10.20). On the other hand, if coalescence occurs rapidly, bubbles formed at the orifices coalesce and grow larger until they exceed the maximum stable bubble size, after which bubble breakup occurs. Since bubble breakup depends on the local velocities of the eddies, there are local coalescence-breakup equilibria, resulting in a variation of bubble size throughout the bioreactor.

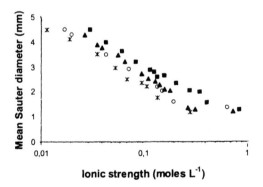

Figure 10.5 Mean Sauter bubble diameter as a function of ionic strength for aqueous solutions of the salts $Al_2(SO_4)_3$ (■), Na_2SO_4 (▲), NaCl (o) and NaOH (*). The data are from an air-water system in a bubble column (Keitel and Onken (1982)).

Assuming that the average bubble diameter is given by Eq. (10.23), we may obtain a correlation for the specific interfacial area by using Eqs. (10.15) and (10.17):

$$a_d = \frac{6\varepsilon}{k} \frac{\rho_l^{0.2}}{\sigma^{0.6}} \left(\frac{P_g}{V_l}\right)^{0.4} \quad (10.25)$$

In this correlation the gas holdup, ε, appears [not necessarily as a proportionality factor, since k may be a function of ε, according to Eq. (10.24)]. The gas holdup depends on the operating conditions, e.g., the dissipated energy and the gas flow rate, and normally an empirical correlation is applied for the gas holdup resulting in a correlation such as Eq. (10.26) for the specific interfacial area (Moo-Young and Blanch, 1981).

$$a_d = k u_s^\alpha \left(\frac{P_g}{V_l}\right)^\beta \quad (10.26)$$

u_s is the *superficial gas velocity* (the gas flow rate divided by the cross-sectional area of the tank. Unit: m s^{-1}). The parameters for this correlation are listed in Table 10.2 for both a coalescing and a noncoalescing medium. Theoretically, β should be equal to 0.4, but for a noncoalescing medium the dissipated energy influences the gas holdup, and β therefore becomes larger.

Example 10.3. Bubble size and specific interfacial area in an agitated vessel
We now consider aeration of a small pilot-plant bioreactor (total volume 41 L) by mechanical agitation (Pedersen *et al.*). Some of the data for the tank are summarized in Table 10.3.

Table 10.2 Parameter values for the correlation in Eq. (10.26).

	Coalescing	Noncoalescing
k	55	15
α	0.5	0.3
β	0.4	0.7

Table 10.3 Data for a sparged, mechanically mixed pilot plant bioreactor.

Symbol	Parameter	Value
d_o	Orifice diameter[a]	10^{-3} m
d_T	Tank diameter	0.267 m
V_l	Liquid volume	25 L
v_g	Gas flow rate	25 L min^{-1}
P_g	Power input	75 W

[a]The sparger is equipped with 10 orifices.

We first consider a system with water and air at 25 °C, i.e.,

ρ_l = 997 kg m^{-3}
ρ_g = 1.285 kg m^{-3}

$\sigma = 71.97 \cdot 10^{-3}$ N m^{-1}
$\eta = 1.00 \cdot 10^{-3}$ kg m^{-1} s^{-1}

First we want to calculate the initial bubble diameter using Eq. (10.20):

$$d_{b,i} = \left[\frac{6 \cdot 71.97 \cdot 10^{-3} \cdot 10^{-3}}{9.82 \cdot (997 - 1.285)}\right] = 3.53 \cdot 10^{-3} \text{ m} \tag{1}$$

From visual inspection of the system, however, we observe that a jet stream is formed at the orifices. We therefore search in the literature for a correlation for the initial bubble diameter, which may be more suitable for the high gas flow rate applied in the system. Bhavaraju et al. (1978) states that the correlation in Eq. (2) holds for gas flow rates up to $2 \cdot 10^{-4}$ m^3 s^{-1} (which is close to the $4.2 \cdot 10^{-4}$ m^3 s^{-1} used in the present system). Re_o is the Reynolds number for the gas stream at the orifice [given by Eq. (3)] and Fr_o is the *Froude number* for the gas stream at the orifice [given by Eq. (4)]:

$$d_{b,i} = 3.23 d_o Re_o^{-0.1} Fr_o^{0.21} \tag{2}$$

$$Re_o = \frac{4\rho_l \upsilon_g}{\pi \eta d_o} \tag{3}$$

$$Fr_o = \frac{\upsilon_g^2}{d_o^5 g} \tag{4}$$

With the operational values specified in Table 10.3, we find $Re_o = 5.3 \cdot 10^4$ and $Fr_o = 1.8 \cdot 10^5$ (the total gas flow is equally distributed to the ten orifices in the sparger), and therefore

$$d_{b,i} = (3.23 \cdot 10^{-3}) \cdot (5.3 \cdot 10^4)^{-0.1} (1.8 \cdot 10^5)^{0.21} = 13.8 \cdot 10^{-3} \text{ m} \tag{5}$$

This is a larger initial bubble diameter than found by using Eq. (10.20), and it corresponds better with the bubble size observed in the bioreactor when there is no agitation. Note that the correlation in Eq. (2) is insensitive to even large variations in the orifice diameter, and a change of the hole size of the sparger therefore has little effect on the initial bubble diameter. The two completely different values obtained tell us that correlations (both empirical and theoretically derived) should always be used with some caution, i.e., one should always check the range of validity for the correlation.

With the specified power input, we calculate the maximum stable bubble diameter, using Eq. (10.23). First we take k to be 1.93 m.

$$d_{b,i} = 1.93 \cdot \frac{(71.97 \cdot 10^{-3})^{0.6}}{(75/25 \cdot 10^{-3})^{0.4} (997)^{0.2}} = 4.07 \cdot 10^{-3} \text{ m} \tag{6}$$

Next we assume that the gas holdup is 0.1 (as is reasonable for the examined system) and use Eq. (10.24) to find $k = 1.34$. Thus from Eq. (10.23) we now find

$$d_{b,max} = 2.83 \cdot 10^{-3} \text{ m} \tag{7}$$

Again we find some deviation among the various correlations. The order of magnitude is, however, the same, and this is often sufficient for design purposes because of the large uncertainties of all the calculations. If the initial bubble diameter is 13.8 mm as found in Eq. (5), the breakup processes result in a rapid disintegration of the bubbles formed at the orifice, and we therefore do not observe the large bubbles formed right at the orifice when the dispersion is agitated.

10.1.3. Empirical Correlations for $k_l a$

A large number of different empirical correlations for the volumetric mass transfer coefficient $k_l a$ have been presented in the literature (for a review see e.g. Moo-Young and Blanch, 1981). Most of these correlations can be written in the form

$$k_l a_d = k u_s^\alpha \left(\frac{P_g}{V_l}\right)^\beta \tag{10.27}$$

which has a great similarity to Eq. (10.26) for the specific interfacial area. The parameters tend to depend on the considered system, i.e. the bioreactor design. Thus for different stirrers (see Chapter 11) and different tank geometry the parameter values may change significantly, and a certain set of parameters can be safely used only when studying a system, which resembles that from which the parameters were originally derived. Some parameter values reported in the literature for stirred tanks are listed in Table 10.4.

Normally, the correlation in Eq. (10.27) holds independently of whether mixing is performed mechanically in stirred tanks or pneumatically in bubble columns. Thus for the same power input per liquid volume, the magnitude of $k_l a$ is approximately the same in a stirred-tank reactor and in a bubble column. It is, however, possible to obtain much higher power input in stirred-tank reactors

Table 10.4 Parameter values for the empirical correlation given by Eq. (10.27)

Medium	k	α	β	Agitator	Reference
Coalescing	0.025	0.5	0.4	Six-bladed Rushton turbine	MooYoung and Blanch (1981)
	0.00495	0.4	0.593	Six-bladed Rushton turbine	Linek et al. (1987)
	0.01	0.4	0.475	Various agitators	Moo-Young and Blanch (1981)
	0.026	0.5	0.4	Not specified	van't Riet (1979)
Non-coalescing	0.0018	0.3	0.7	Six-bladed Rushton turbine	Moo-Young and Blanch (1981)
	0.02	0.4	0.475	Various agitators	Moo-Young and Blanch (1981)
	0.002	0.2	0.7	Not specified	van't Riet (1979)

Parameter values are specified in SI units, i.e., the power input is in W m^{-3} and the superficial gas flow rate is in m s^{-1}.

than in bubble columns, and stirred tanks are therefore traditionally used in aerobic fermentation processes where there is a high oxygen demand, e.g., antibiotic fermentations. New bioreactor designs based on cleverly designed static mixers or gas injection nozzles can, however, outperform the stirred tanks. A very high $k_l a$ value (up to 0.5 s^{-1}) can be obtained, but the corresponding power input is usually also very high.

When the range of process variables for which the correlation in Eq. (10.27) holds is studied in more detail, it is observed that the mass transfer coefficient $k_l a$ for a noncoalescing medium is greater by about a factor of 2 than that for a coalescing medium under the same operating conditions. These overall correlations are, however, very rough simplifications since they are made to fit data obtained in many different bioreactors. For a specific agitator system, e.g., a six-bladed Rushton turbine (see Section 11.2), the situation is more complex. Here it is found that the influence of the power input is larger in the noncoalescing medium, whereas the influence of the superficial gas velocity is smaller compared with a coalescing medium (e.g., pure water) (see Table 10.4).

In the derivation of Eq. (10.23), it was assumed that the dynamic shear stress caused by eddies was much larger than the viscous stress. This will not be true for highly viscous media, and the very simple correlations of the type given by Eq. (10. 27) will therefore no longer be valid. In general, the $k_l a$ value is found to decrease with increasing liquid viscosity, but the effect is small until $\eta > 50 \cdot 10^{-3}$ kg m^{-1} s^{-1}. For an in-depth review of mass transfer in highly viscous media see e.g. Schügerl (1981).

Example 10.4 Derivation of empirical correlations for $k_l a$ in a laboratory bioreactor

Pedersen (1992) examined the gas-liquid mass transfer in a stirred laboratory bioreactor. The volumetric mass transfer coefficient $k_l a$ was determined by the sulphite method (see Note 10.3), and the influence of aeration rate and stirring speed on $k_l a$ was examined. Data for the bioreactor and the operating conditions are summarized in Table 10.5.

The results of the measurements investigation using the sulfite method are shown as double logarithmic plots in Figs. 10.6 and 10.7. The volumetric mass transfer coefficient increases with increasing gas aeration rate v_g and with increasing stirring speed N, but there is an upper limit to the $k_l a$ value for the considered system. From each of the two series we find the correlations

Table 10.5. Data for a standard laboratory bioreactor and the range of operating conditions examined[a].

Variable	Value	Meaning
V_t	15 L	Tank volume
V_l	10 L	Liquid volume
d_t	0.20 m	Tank diameter
d_i	0.07 m	Stirrer diameter
N	4-25 s^{-1}	Stirring speed
v_g	2.2-25·10^{-5} m^3 s^{-1}	Gas flow rate

[a]The bioreactor was equipped with two Rushton turbines. For details of the design see Pedersen (1992).

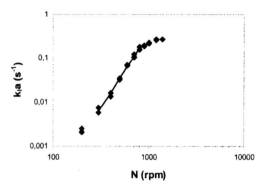

Figure 10.6 Double logarithmic plot of the influence of the stirring speed N (s^{-1}) on the volumetric mass transfer coefficient $k_l a$ (s^{-1}). The aeration rate is $v_g = 10^{-4}$ m^3 s^{-1}. The line is the regression line for the correlation in Eq. (1).

$$k_l a = 4.5 \cdot 10^{-5} N^{3.146} \quad (1)$$

and

$$k_l a = 27.0 v_g^{0.523} \quad (2)$$

From the correlation for the influence of the aeration rate, we find

$$k_l a = 27.0 \left(\frac{\pi d_t^2}{4}\right)^{0.523} \left(\frac{4 v_g}{\pi d_t^2}\right)^{0.523} = 4.42 u_s^{0.523} \quad (3)$$

and

$$k_l a = 4.42 u_s^{0.523} \left(\frac{N}{16.7}\right)^{3.146} = 6.3 \cdot 10^{-4} u_s^{0.523} N^{3.146} \quad (4)$$

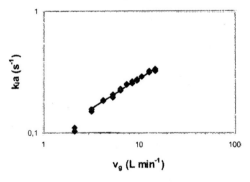

Figure 10.7. Double logarithmic plot of the influence of the aeration rate v_g on the volumetric mass transfer coefficient $k_l a$. The stirring speed is 16.7 s^{-1}. The line is the regression line for the correlation in Eq. (2).

Mass Transfer

Similarly, we find from the correlation for the influence of the stirring speed

$$k_l a = 4.5 \cdot 10^{-5} \left(\frac{u_s \pi d_t^2}{4 \cdot 10^{-4}} \right)^{0.523} \cdot N^{3.146} = 9.1 \cdot 10^{-4} u_s^{0.523} N^{3.146} \tag{5}$$

Thus the value of the numerical constant found for each of the two sets of experiments is slightly different. The correlation in Eq. (1) holds only for $N < 15$ s^{-1}, whereas the correlation in Eq. (2) is based on $N = 16.7$ s^{-1}. From Fig. 10.6 it is observed that for $N = 16.7$ s^{-1} the measured $k_l a$ value is lower than predicted by the correlation in Eq. (1), and this may explain the lower value for the constant in Eq. (4) compared with Eq. (5). Thus the correlation in Eq. (5) is probably the best, and to test the correlation another series of experiments was performed, with varying v_g at $N = 8.33$ s^{-1}. The results of this comparison are shown in Fig. 10.8.

If we compare the correlation derived in this example with Eq. (10.27), we see that the structure is the same since the power input is correlated to the stirring speed. For the examined bioreactor it was found that the power input (measured as the power drawn by the motor) is correlated with the stirring speed to the power 3, i.e.

$$P \propto N^3 \tag{6}$$

This indicates that the influence of the power input is stronger in the present system than reported in the literature (see Table 10.4). The influence of the superficial gas flow rate is also larger than reported in the literature (see value of α for noncoalescing medium in Table 10.4). The correlation derived here is based on measurement of $k_l a$ using the sulphite method, and since the sulphite concentration must be quite high (around 0.5 M) to obtain accurate measurements of the rate of sulphite consumption, the medium is strongly noncoalescent. This may explain the deviation between the correlation of the present example and similar correlations based on other measurement methods.

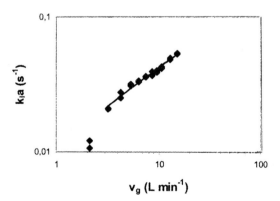

Figure 10.8. Double logarithmic plot of the influence of the aeration rate v_g (L min^{-1}) on the volumetric mass transfer coefficient $k_l a$ (s^{-1}). The stirring speed is 8.33 s^{-1}. The line is the regression line for the correlation in Eq. (5).

Linek *et al.* (1987) also applied the sulphite method to determine $k_l a$ in a stirred-tank reactor and found the correlation

$$k_l a = 0.00135 u_s^{0.4} \left(\frac{P_g}{V_l} \right)^{0.946} \tag{7}$$

where the influence of power input is much higher than indicated in table 10.4. Thus, application of the sulphite method may result in a correlation that is not valid for normal fermentation media (even when these are noncoalescing). The sulphite method is discussed further in Note 10.3.

In the experiments on which the present example is based, the power input was not measured directly, but calculated from the measured stirring speed and the correlation in Eq. (6). This correlation is, however, not generally valid, and it should be used only for preliminary calculations. Since determination of the power input requires measurement of the torque on the impeller shaft inside the bioreactor (see Section 10.3.1), it is convenient to use the stirring speed rather than the power input in empirical correlations for $k_l a$.

10.1.4. Mass Transfer Correlations Based on Dimensionless Groups

There is a tradition in the chemical engineering literature to express correlations for various transport coefficients in terms of dimensionless groups named after prominent members of the engineering community. Unfortunately, these dimensionless groups have come to work as filters, which tend to separate the treatment of these phenomena by chemical engineers from that done by their colleagues active in the fields of biology or chemistry. (One could even argue that the dimensionless groups to some extent separate the chemical engineers from their colleagues.)

Despite these drawbacks, there are several advantages to be gained from using dimensionless groups. Most importantly, these groups show *in what way physical variables interact* in their effect on the response variable. This is of great help in designing experiments for deriving empirical correlations, and also for obtaining a qualitative understanding of how transport phenomena may be influenced by operating conditions. A practical advantage is, furthermore, that dimensionless equations can be used equally well with barrels, gallons, or liters, with a minimum risk of unit conversion errors. Equations of the type given in Eq. (10.27), on the other hand, carry dimensions, which are sometimes not easily realized or sufficiently clearly stated.

Dimensionless groups may be derived by two, principally different routes. If known, the governing equation of the phenomenon of interest can relatively easily be transformed into a dimensionless form. In the thus transformed equation dimensionless groups will appear as coefficients. Well-known examples are e.g. the *Reynolds number* and *Froude number,* which appear in the dimensionless form of the Navier-Stokes equation (see section 11.3.6).

Mass Transfer 443

Table 10.6. Some important dimensionless groups for mass transfer correlations[a]

Definition	Name	Significance
$Sh = \left[\dfrac{k_l d}{D_A}\right]$	Sherwood number	mass transfer velocity relative to diffusion velocity
$Sc = \left[\dfrac{\eta}{D_A \rho_l}\right]$	Schmidt number	momentum diffusivity relative to mass diffusivity
$Re = \left[\dfrac{u d \rho_l}{\eta}\right]$	Reynolds number	inertial forces relative to viscous forces
$Gr = \left[\dfrac{d^3 g \rho_l (\rho_l - \rho_g)}{\eta^2}\right]$	Grashof number	bouyancy forces relative to viscous forces
$Pe = \left[\dfrac{du}{D_A}\right]$	Peclet number	flow velocity relative to diffusion velocity

[a] d is a length scale characteristic of the system which is studied, i.e. bubble, cell or cell aggregate. D_A is the diffusion coefficient for the considered species in the continuous phase; u is the linear velocity of bubble, cell, etc. relative to the continuous phase.

Also in the absence of a known governing equation it is, however, possible to derive dimensionless variables upon which to base empirical correlations. This can be done by *dimensional analysis*, i.e. by checking that correlations are dimensionally correct (see Example 10.5). The basis for this analysis is the Buckingham π-theorem, which is explained in detail in many chemical engineering textbooks (see e.g. Geankoplis, 1993).

Some of the more well-known dimensionless groups used in mass transfer correlations are summarized in Table 10.6.

Example 10. 5. Deriving dimensionless groups by dimensional analysis
The procedure for arriving at correlations based on dimensionless groups can be illustrated by the following example. Suppose that we need a correlation for the mass transfer coefficient, k_l, for a compound A, present in a bubble which moves rapidly with respect to its surrounding medium. To make a dimensional analysis, all relevant variables must be known, and these must, furthermore, be independent of each other. In the current example the following variables are regarded as important; diffusivity of A, D_A, bubble diameter, d_b, relative velocity of bubble, u_b, viscosity of liquid, η, and the density of the liquid, ρ_l. Our hypothesis is that we can express the mass transfer coefficient as a function of these variables, i.e.

$$k_l \propto D_A^{\alpha_1} u_b^{\beta} d_b^{\gamma} \eta^{\delta} \rho_l^{\varepsilon} \qquad (1)$$

where the α_1, β, γ, δ, and ε are unknown exponents to be determined.

Eq. 1 must be dimensionally correct. k_l has dimensions length (L) time (t)$^{-1}$, d_b has dimension L, u_b has dimensions L t^{-1}, D_A has dimensions L^2 t^{-1}, η has dimensions mass (M) L^{-1} t^{-1}, and ρ_l has dimensions M L^{-3}. Thus, we get 3 linear constraints on the exponents, which can be written

$$\begin{bmatrix} 1 \\ -1 \\ 0 \end{bmatrix} = \begin{bmatrix} 2 & 1 & 1 & -1 & -3 \\ -1 & -1 & 0 & -1 & 0 \\ 0 & 0 & 0 & 1 & 1 \end{bmatrix} \begin{bmatrix} \alpha_1 \\ \beta \\ \gamma \\ \delta \\ \varepsilon \end{bmatrix} \quad (2)$$

This leaves 3 degrees of freedom for the six variables, which implies that the correlation can be reformulated in terms of 3 dimensionless groups. (These groups were denoted "π_s" in the original formulation by Buckingham, thus the name of the theorem.) By choosing to express the other exponents in terms of α and β, we get

$$\gamma = \beta - 1 \quad (3)$$

$$\delta = 1 - \alpha_1 - \beta \quad (4)$$

$$\varepsilon = -(1 - \alpha_1 - \beta) \quad (5)$$

This is dimensionally consistent. However, it does not result in the most often used dimensionless groups in Table 10.6. To arrive at the more familiar dimensionless groups we set

$$\alpha_1 = 1 - \alpha \quad (6)$$

This gives:

$$\delta = \alpha - \beta \quad (7)$$

$$\varepsilon = \beta - \alpha \quad (8)$$

and thus

$$k_l \propto D_A^{1-\alpha} u_b^\beta d_b^{\beta-1} \eta^{\alpha-\beta} \rho_l^{\beta-\alpha} \quad (9)$$

which can be rearranged into:

$$\left[\frac{k_l d_b}{D_A}\right] \propto \left[\frac{\eta}{D_A \rho_l}\right]^\alpha \left[\frac{u_b d_b \rho_l}{\eta}\right]^\beta \quad (10)$$

or

$$Sh \propto Sc^\alpha Re^\beta \quad (12)$$

where d_b is used as the characteristic length and u_b as the characteristic velocity in the dimensionless groups. In the current example a high velocity of the bubble relative to its surrounding liquid was assumed.

Mass Transfer

Reported correlations for the Sherwood number are often of the kind

$$Sh = 2 + const \cdot Sc^\alpha Re^\beta \tag{10.28}$$

If the exponents α and β have the same value, the Schmidt and Reynolds number can be multiplied together forming the *Peclet number*, *Pe*. In cases where free convection dominates, the Grashof number appears instead of the Reynolds number (see Table 10.7). It can be seen from Eq 10.28 that as the Reynolds number approaches 0, i.e. for a bubble which is stagnant relative to the surrounding liquid, the Sherwood number will approach 2. This is in accordance with the analytical steady-state solution of the diffusion equation for a sphere in stagnant medium (see Note 10.2). For high values of *Re* the first term may be neglected compared to the second term. A key question is furthermore whether the bubble behaves as a rigid surface or not (see Table 10.7). For a rigid bubble, with an immobile interface, the boundary conditions at the surface will state a zero relative velocity between the liquid and the bubble surface. For a mobile interface, however, this boundary condition does not apply, and the mass transfer characteristics will be different (for a further discussion see e.g. Blanch and Clark, 1997).

Note 10.2. Derivation of the Sherwood number for a sphere in stagnant medium
For the case of steady-state mass transfer from a spherical particle with radius, $R_p = d_b/2$, in a stagnant medium, it is possible to derive an analytical value for the Sherwood number. The diffusion equation for species A at steady state is given by

$$\nabla^2 c_A = 0 \tag{1}$$

Due to the symmetry of the problem, it is convenient to use spherical coordinates, and with the radial distance denoted ξ. Eq. 1 can be expressed

$$\frac{1}{\xi^2}\frac{\partial}{\partial \xi}\left(\xi^2 \frac{\partial c_A}{\partial \xi}\right) = 0 \tag{2}$$

with the boundary conditions

$$c_A = c_{A,s} \text{ for } \xi = R_p \tag{3}$$

and

$$c_A \to 0 \text{ for } \xi \to \infty \tag{4}$$

Integrating Eq. 2 twice gives

$$c_A = c_{A,s} \frac{R_p}{\xi} \tag{5}$$

(Note: This is valid only for $\xi \geq R_p$)

The mass flux of A expressed by Fick's law at $\xi = R_p$ and that expressed using the mass transfer coefficient must be equal, *i.e.*

$$k_l(c_{A,s} - 0) = -D_A \frac{\partial c_A}{\partial \xi} \tag{6}$$

With the derivative of Eq. 5 inserted we get

$$k_l c_{A,s} = D c_{A,s} \frac{R_p}{R_p^2} \tag{7}$$

or

$$\frac{k_l d_b}{D_A} = 2 \tag{8}$$

This result, i.e. $Sh = 2$, is probably one of the best remembered pieces of knowledge by chemical engineering students. However, it must be kept in mind that it applies only for steady-state mass transfer from a single sphere into a stagnant medium.

Table 10.7. Some reported correlations for the Sherwood number

Correlation	Conditions	Remarks	Source
$Sh = 2 + 0.55 Re^{\frac{1}{2}} Sc^{\frac{1}{3}}$	$2 < Re < 1300$	Immobile gas-liquid interface	Froessling, 1938
$Sh = 0.82 Re^{\frac{1}{2}} Sc^{\frac{1}{3}}$	$200 < Re < 4000$	Immobile gas-liquid interface	Rowe *et al.*, 1965
$Sh = \left(4 + 1.21 Pe^{\frac{2}{3}}\right)^{\frac{1}{2}}$	$Re < 1$ $Pe < 10^4$	Immobile gas-liquid interface	Brian and Hales, 1969
$Sh = 0.42 Gr^{\frac{1}{3}} Sc^{\frac{1}{2}}$	Free convection	Mobile interface, large bubbles	Calderbank and Moo-Young, 1964
$Sh = 0.65 Pe^{\frac{1}{2}}$	$Re < 1$	Mobile gas-liquid interface	Blanch and Clark, 1997
$Sh = 0.65\left(1 + \frac{Re}{2}\right)^{\frac{1}{2}} Pe^{\frac{1}{2}}$	$1 < Re < 10$	Mobile gas-liquid interface	Blanch and Clark, 1997
$Sh = 1.13 Pe^{\frac{1}{2}}$	$Re \gg 1$	Mobile gas-liquid interface	Blanch and Clark, 1997

Example 10.6 Mass transfer to a single cell
The small dimension of the cell ensures that the lower limit of 2 for the Sherwood number can be applied. For transport of oxygen from the bulk liquid to a spherical cell of diameter 2 μm we find:

$$\frac{k_l d_{cell}}{D_{O2}} = 2 \text{ or } k_l = 2\frac{D_{O2}}{d_{cell}} \tag{1}$$

This gives

$$k_l = \frac{2 \cdot 2.1 \cdot 10^{-9} \text{ m}^2 \text{s}^{-1}}{2 \cdot 10^{-6} \text{ m}} = 2.1 \cdot 10^{-3} \text{ m s}^{-1} \tag{2}$$

The specific interfacial area per unit biomass is given by

$$a_{cell} = \frac{6}{d_{cell}(1-w)\rho_{cell}} \tag{3}$$

where w is the fractional water content. With $w = 0.7$ and a density of the cell of 10^6 g m^{-3}, the volumetric mass transfer coefficient for the cell is

$$\begin{aligned} k_l a_{cell} &= \frac{6k_l}{d_{cell}(1-w)\rho_{cell}} \\ &= \frac{6 \cdot 2.1 \cdot 10^{-3} \text{ m} \cdot \text{s}^{-1}}{2 \cdot 10^{-6} \text{ m}(1-0.7)10^6 \text{ g DW} \cdot \text{m}^{-3}} = 0.021 \text{ m}^3 \text{ g}^{-1} \text{ s}^{-1} \end{aligned} \tag{4}$$

With a bulk phase oxygen concentration equal to 60% of the saturation value, the maxium specific oxygen transport rate with air sparged through the reactor can be calculated assuming a concentration of oxygen at the cell surface very close to 0. This gives:

$$k_l a_{cell}(c_{O,bulk} - c_{O,surface}) \approx k_l a_{cell} c_{O,bulk} \tag{5}$$

and

$$k_l a_{cell} c_{O,bulk} = 0.021 \cdot 0.6 \cdot 0.2095 \cdot 1.22 \cdot 10^{-3} = 3.23 \cdot 10^{-3} \text{ mol O}_2 \text{ (g DW)}^{-1} \text{ s}^{-1} \tag{6}$$

Thus, mass transfer to the cells is very rapid compared with the oxygen requirement inside the cell (see Example 10.1), and it is therefore not necessary to consider this process in a model. Even if the mass transfer coefficient should be much lower than that used in Eq. (1), the maximum mass transfer rate from the bulk liquid to the cell is still very rapid compared to the cellular oxygen requirements.

10.1.5. Gas-Liquid Oxygen Transfer

Due to the importance of aerobic fermentation processes, gas-liquid mass transfer is almost synonymous with oxygen transfer from gas bubbles to the liquid medium. Since the volumetric rate of mass transfer is proportional to the difference between the equilibrium (saturation) concentration c_o^* and the actual value of the dissolved oxygen concentration in the medium c_o, it is necessary to know the saturation concentration. The solubility of oxygen, i.e. c_o^*, decreases with increasing temperature (see Table 10.8), and it decreases with the concentration of various solutes present in the medium (see Table 10.9). The solubility is also related to the partial pressure of oxygen in the gas phase. When Henry's law [Eq. (10.3)] can be applied, a change in the partial pressure will lead to a proportional change in the saturation concentration.

Table 10.8. Solubility of oxygen in pure water at an oxygen pressure of 1 atm.

Temperature (°C)	Solubility of O_2 (mmoles L^{-1})
0	2.18
10	1.70
15	1.54
20	1.38
25	1.26
30	1.16
40	1.09

Table 10.9. Solubility of oxygen at 25 °C and 1 atm O_2 in various aqueous solutions.

Component	Concentration (g L^{-1})	Solubility of O_2 (mmoles L^{-1})
Glucose	20	1.233
	50	1.194
	90	1.133
	180	0.990
Citric acid	25	1.242
	100	1.137
	200	0.983
Gluconic acid	25	1.210
	100	1.121
	200	0.991
Corn steep liquor	10	1.205
	50	1.189
	100	1.154
Yeast extract	5	1.255
	10	1.228
Xanthan	1	1.250
	5	1.251
Pullulan	1	1.266
	10	1.241
	20	1.240

The data are taken from Popovic *et al.* (1979).

Mass Transfer

According to the conclusion of Example 10.2, a very high k_la value is required to maintain the dissolved oxygen concentration at 60% of the saturation value in a rapidly respiring culture of *S. cerevisiae*. This high k_la value for oxygen can be obtained in a laboratory bioreactor (and perhaps in small-scale pilot-plant bioreactors), but will be difficult to obtain in a large-scale bioreactor. However, a substantially lower dissolved oxygen concentration than 60% is normally acceptable in industrial applications of *S. cerevisiae*. The simplest possible way to describe oxygen uptake kinetics is by using a Monod expression. The saturation constant, K_o, is in the range 1-10 μM, which corresponds to approximately 0.4-4.0% of the saturation concentration (at 25 °C and 1 atm of air). As long as the dissolved oxygen concentration is maintained well above this value, the oxygen consumption rate will therefore be zero order with respect to the dissolved oxygen concentration, i.e. oxygen limitation does not occur.

Note 10.3. Methods for determination of the volumetric mass transfer coefficient for oxygen

There are several different methods for measuring the volumetric mass transfer coefficient k_la for oxygen. Some of these methods can also be applied for other components, but others are specific for oxygen. Here a few of the best-known methods are described:

The direct method

Most bioreactors used for aerobic fermentation processes are equipped with exhaust gas analysis for oxygen and probes for measuring the dissolved oxygen concentration. From the measurement of oxygen content in the inlet and exhaust gases together with measurement of the gas flow rate it is possible at steady-state conditions, as already discussed in Chapter 3.1, to determine the volumetric oxygen transfer rate, q_O^t, which at steady-state, is equal to the volumetric oxygen uptake rate, $-q_O$. We have from Eq. (3.10)

$$q_O^t = \frac{1}{V_l}\left(\frac{p_O^{in} v_g^{in}}{RT^{in}} - \frac{p_O^{out} v_g^{out}}{RT^{out}}\right) \quad (1)$$

where R is the gas constant (= 8.314 J (mol K)$^{-1}$ = 0.08206 L atm (mol K)$^{-1}$), T is the temperature (K), p_0 the partial pressure of oxygen, v_g the gas flow rate.

Most gasanalyzers will give the result in terms of mole fraction of oxygen in the gas. When normal air is used, it is sufficient to analyze only the outlet gas composition. However, for large bioreactors it is important to take the pressure difference over the reactor into account. If also at the same time the dissolved oxygen concentration, c_o, is measured in the medium, it is possible to calculate the volumetric mass transfer coefficient from:

$$k_l a = \frac{q_O^t}{\left(c_O^* - c_O\right)} \quad (2)$$

Eq. 2 assumes that the saturation concentration, c_O^*, is the same throughout the reactor, *i.e.* that the pressure difference is small and that the decrease in partial pressure of oxygen in the gas phase is small. If that is not the case, a better option is to use Eq. 10.8 for the concentration driving force. This is a simple method, and has the major advantage that is can be applied during a real fermentation. However, accurate measurements of the oxygen content as well as of the gas flow rates are necessary and the solubility of oxygen in the medium must be known. Furthermore, the dissolved oxygen tension should not change during the measurement, since the assumption of a steady-state (or pseudosteady-state) is a strict requirement. See also problem 3.1 for experimental errors due to evaporation of e.g. water from the medium ($v_g^{in} \neq v_g^{out}$).

The dynamic method

The dynamic method is also based on measurement of the dissolved oxygen concentration in the medium. However, it does not require measurement of the gas composition and is therefore cheaper to establish. The dynamic mass balance for the dissolved oxygen concentration is:

$$\frac{dc_O}{dt} = k_l a(c_O^* - c_O) + q_O \qquad (3)$$

If the gas supply to the bioreactor is turned off, q_O^t is zero and the first term on the right hand side of the equation immediately drops to zero. The dissolved oxygen concentration decreases at a rate equal to the oxygen consumption rate, $-q_o$, which can therefore be determined from measurement of the dissolved oxygen concentration $c_O(t)$. If q_o is independent of c_0 the dissolved oxygen concentration decreases as a linear function of time. When the gas supply is turned on again, the dissolved oxygen concentration increases back to the initial level, and by using the estimated (average) value for q_0, the value of $k_l a$ can be determined from the measured profile of dissolved oxygen.
This method is simple, and it can be applied during a real fermentation. It is, however, restricted to situations in which q_0 can be determined correctly when the gas supply is turned off. Most available probes for measuring the dissolved oxygen concentration do, unfortunately, have a response time close to the characteristic time for the mass transfer process, and the measured concentrations are therefore influenced by the dynamics of the measurement device, e.g., an oxygen electrode.

The dynamic method can also be applied at conditions where there is no reaction, i.e. $q_O = 0$. This is interesting when studying the influence of operating parameters, e.g., the stirring speed and the gas flow rate, on the volumetric mass transfer coefficient in model media. After a step change in the concentration in the inlet gas, there are dynamic changes both in the dissolved oxygen concentration and in the oxygen concentration of the exhaust gas. Because of the slow dynamics of oxygen electrodes, it is preferable to apply the exhaust gas measurements for determination of the mass transfer coefficient. (Here a mass balance for oxygen in the gas phase should be used.) However, before this approach is applied, it is important to make a careful check of the dynamics of the gas analyzer.

The sulphite method

It is also possible to determine the oxygen transfer rate by using a model oxygen consuming chemical reaction. The traditional sulphite method is based on the oxidation of sulphite to sulphate by oxygen (see Eq. 4).

$$SO_4^{2-} - SO_3^{2-} - \frac{1}{2}O_2 = 0 \qquad (4)$$

This reaction is catalyzed by a number of metal ions, which may occur as impurities. However, one normally adds a known amount of copper (e.g. 10^{-3} M Cu^{2+}) or cobalt salt to the medium in order to make

Mass Transfer

the reaction almost instantaneous. In such a case, the rate of sulphite consumption is determined solely by the rate at which oxygen is transferred from the gas phase. Since, from Eq. (4)

$$q_{SO_3^{2-}} = 2q_o \tag{5}$$

we get

$$k_l a = -\frac{q_o}{c_o^* - c_o} = -\frac{q_{SO_3}}{2(c_o^* - c_o)} \tag{6}$$

The dissolved oxygen concentration, c_o, is virtually zero due to the very rapid reaction with sulphite. The reaction rate is determined from measuring the concentration of sulphite in samples taken at a set of time values after the start of the experiment. The sulphite concentration in the samples can be determined by adding an excess amount of iodine, and thereafter back-titrate with thiosulphate, using starch as an indicator. The reactions are given by Eqs. (7) and (8).

$$SO_4^{2-} + 2H^+ + 2I^- - SO_3^{2-} - I_2 - H_2O = 0 \tag{7}$$

$$S_4O_6^{2-} + 2I^- - I_2 - 2S_2O_3^{2-} = 0 \tag{8}$$

To apply the method, it is necessary to know the saturation concentration of oxygen in the strong sulphite solution. Since (for obvious reasons) this is not known, one may use the solubility of oxygen in sulphate solutions, which is 1.02 mM at 25 °C for 0.25 M SO_4^{2-} (Linek and Vacek, 1981). The sulphite method was often used in earlier days, since it is relatively easy to implement. However, it has a number of drawbacks. The sulphite oxidation may enhance the oxygen absorption, since the rapid chemical reaction may occur not only in the bulk liquid, but also in the liquid film. The assumption of a linear concentration profile in the film is therefore questionable. Another complication is the significant coalescence-reducing effects of sulphite on the bulk liquid, which results in a higher specific interfacial area. Both the enhancement of mass transfer due to reaction in the film, and the coalescence-reducing effect of sulphite, may lead to an overestimation of $k_l a$ compared to what is found in a normal fermentation media. This is a significant drawback of the method. A further practical disadvantage is that the sulphite method cannot be applied at all during a real fermentation, since the microorganisms would most likely be killed.

The hydrogen peroxide method

This method due to Hickman (1988) is, like the sulphite method, a chemical method. However, it has several advantages in comparison to the sulphite method. In the hydrogen peroxide method, the transfer of oxygen from the liquid to the gas phase is measured. Oxygen is generated by the enzyme-catalyzed decomposition of hydrogen peroxide (Eq. 9).

$$2H_2O_2 \xrightarrow{catalase} 2H_2O + O_2 \tag{9}$$

The reaction is first order with respect to both H_2O_2 and the enzyme, catalase. A constant volumetric flow of air passes the reactor, and after addition of a known amount of catalase, a continuous feed of H_2O_2 is applied to the reactor. Initially, hydrogen peroxide will accumulate in the reactor medium, but a steady-state will soon be established. At steady state the rate of decomposition of hydrogen peroxide, $-q_{H2O2}$, equals half the rate of oxygen transfer from the liquid phase to the gas phase:

$$k_l a(c_o^* - c_o) = \frac{q_{H_2O_2}}{2} \tag{10}$$

The volumetric decomposition rate of hydrogen peroxide is calculated from the volumetric addition rate and the concentration of hydrogen peroxide in the added liquid according to

$$q_{H_2O_2} = -\frac{v_{H_2O_2}^f c_{H_2O_2}^f}{V_l} \tag{11}$$

The interfacial concentration of oxygen is calculated from the gas phase partial pressure of oxygen. (Note that both terms in Eq. 10 are negative, since oxygen is generated in the liquid and thus DOT= $c_o / c_o^* >1$). An assumption concerning the mixing of the gas phase is necessary to determine the interfacial concentration of oxygen. Hickman used the assumption of complete backmixing of both the liquid and the gas phase, which gives

$$c_o^* = \frac{p_o^{exit}}{H_o} \quad \text{and} \quad k_l a = \frac{v_{H_2O_2}^f c_{H_2O_2}^f}{2V_l c_o^* (DOT - 1)} \tag{12}$$

The hydrogen peroxide method is easy to implement. All that is needed to estimate the $k_l a$ value is to measure the addition rate of hydrogen peroxide and the oxygen concentration (or DOT) in the liquid phase. An independent (but much less accurate) measurement of q_o^t is obtained from the difference in oxygen content between v_g^{in} and v_g^{out} as in the "direct method". In comparison to the sulphite method, a significant advantage of the hydrogen peroxide method is that the $k_l a$ value is not enhanced by the catalase concentration over a rather wide range (Hickman, 1988). The risk of enhancing the mass transfer by reaction in the liquid film is smaller, and the effect on coalescing properties by the catalase is apparently also smaller. The method has been applied with good results for $k_l a$ measurements in large industrial reactors also with viscous, non-Newtonian media (Pedersen, 1997).

Tracer methods

^{85}Kr is a volatile isotope emitting beta and gamma radiation. By injecting the isotope into the medium and then measuring the radioactivity in the exhaust gas, it is possible to determine the volumetric mass transfer coefficient for Kr (see Problem 10.3). Since it can be assumed that constant ratios exist between $k_l a$ values for different gases at the same conditions, the estimated $k_l a$ value for Kr can be used to calculate $k_l a$ for oxygen. Pedersen et al. (1994), used the ratio

$$\frac{(k_l a)_{Kr}}{(k_l a)_O} = 0.82 \tag{13}$$

Approximately the same value is found if one uses the ratio between the molecular diffusion coefficients for the two species (see Eq. 10.33). This method is easy to implement, and it is well suited for measurement both in model media and under real fermentation conditions [see, e.g., Pedersen et al. (1994)]. The main drawback is the radioactivity, which obviously limits the application in an industrial environment. However, the isotope is very volatile and the radiation has normally returned to the background level a few minutes after the injection. Kr is of course completely inert in connection with fermentations.

10.1.6. Gas-Liquid Mass Transfer of Components Other than Oxygen

The transport of oxygen is the most important gas-liquid mass transfer problem, and bioreactors to be used for aerobic processes are often designed to ensure a sufficiently high $k_l a$ value for oxygen. The mass transfer of other components is, however, also important in many processes. Perhaps the most obvious of these components is carbon dioxide, which is formed in many decarboxylation reactions, both in the TCA-cycle and in several fermentative pathways (see Chapter 2). The metabolically produced carbon dioxide will diffuse out of the cell, and will then be transferred from the liquid phase to the gas phase. Carbon dioxide is, however, not only a product of (catabolic) decarboxylation reactions, but it is also a substrate in several (anabolic) carboxylation reactions. Carboxylation reactions occur in the synthesis of amino acids and nucleic acids, and also in gluconeogenesis, during growth on substrates with a low number of carbon atoms (i.e. <6). With its dual function, it is not surprising that carbon dioxide may act either stimulatory or inhibitory to growth depending on its concentration. It has been reported that growth of. *E. coli* was stimulated when carbon dioxide was added to the inlet gas up to 5% (Lacoursiere *et al.*, 1986), but inhibitory at higher concentrations. Similarly, increasing the carbon dioxide to more than 3% in the inlet gas during cultivation of the filamentous fungi *Aspergillus niger* caused a decreased biomass yield (McIntyre and McNeill, 1997). The mechanism of inhibition is not clear, but may involve direct inhibition at specific enzymatic sites or interaction with the lipids in the cell membrane (Jones and Greenfield, 1982).

In contrast to the case for oxygen, the mass transfer of carbon dioxide is influenced by medium pH. The reason for this is that carbon dioxide participates in several liquid-phase reactions, and carbon dioxide may exist in the liquid phase in any of four forms: CO_2, H_2CO_3, HCO_3^- and CO_3^{2-}.

The total concentration of all forms of CO_2 is thus

$$[CO_2]_{tot} = [CO_2]_{aq} + [H_2CO_3] + [HCO_3^-] + [CO_3^{2-}] \approx [CO_2]_{aq} + [HCO_3^-] + [CO_3^{2-}] \tag{10.29}$$

Gaseous carbon dioxide is dissolved in water, where it undergoes a very slow reaction to form carbonic acid, which rapidly reacts to form a mixture of bicarbonate and carbonate. Combining the first two reactions into one, the following two equilibrium reactions are obtained:

$$CO_{2,aq} + H_2O \leftrightarrow H^+ + HCO_3^-; \qquad K_1 = \frac{[H^+][HCO_3^-]}{[CO_2]_{aq}} \tag{10.30}$$

$$HCO_3^- \leftrightarrow H^+ + CO_3^{2-}; \qquad K_2 = \frac{[CO_3^{2-}][H^+]}{[HCO_3^-]} \tag{10.31}$$

where the equilibrium constants are $K_1 = 10^{-6.3}$ M and $K_2 = 10^{-10.25}$ M. From these equilibria the total carbon dioxide concentration in the medium is found to be

$$[CO_2]_{tot} = [CO_2]_{aq}\left(1 + \frac{K_1}{[H^+]} + \frac{K_1 K_2}{[H^+]^2}\right) \quad (10.32)$$

$[CO_2]_{tot}$ is a strong function of the medium pH (see Fig. 10.9). For pH < 5, nearly all carbon dioxide is present as dissolved CO_2, while bicarbonate dominates when 7 < pH < 9 and carbonate for pH > 11. Thus, in neutral to alkaline media there is a coupling between chemical reaction and mass transfer.

To quantify the gas liquid mass transfer of other components one may use Eq. (10.33), which relates the volumetric mass transfer coefficients for two different components to the corresponding molecular diffusivities:

$$\frac{(k_l a)_{O2}}{(k_l a)_A} = \frac{D_{O2}}{D_A} \quad (10.33)$$

This is based on the assumption that the mass transfer coefficient for a certain component is proportional to its diffusion coefficient. This assumption is reasonable if the film theory can be applied (see Section 10.1.1). Often the volumetric mass transfer coefficient for oxygen is known, and equation (10.33) can then be used to find $k_l a$ for other components, e.g., carbon dioxide. A list of diffusion coefficients for different solutes in dilute aqueous solutions is given in Table 10.10.

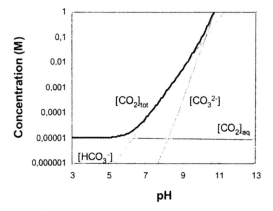

Figure 10.9. Equilibrium concentrations of dissolved CO_2, HCO_3^- and CO_3^{2-}. $[CO_2]_{tot}$ is the total dissolved concentration of all forms.

Mass Transfer

Table 10.10 Diffusion coefficients for different solutes in dilute aqueous solutions at 25 °C [a]

Component	D_A (10^{-9} m^2 s^{-1})	$D_A\eta/T$ (10^{-12} kg m s^{-2} K^{-1})
Oxygen	2.50	7.50
Carbon dioxide	1.96	5.88
Ammonia	2.00	6.00
Acetic acid	1.24	3.72
Methanol	1.60	4.80
Ethanol	1.28	3.84
Lactose	0.49	1.47
Glucose	0.69	2.07
Sucrose	0.56	1.68

[a]The ratio $(D_A\eta)/T$, which is approximately constant for a given solute-solution system (see Note 10.4) is also listed. T is the absolute temperature and η is the liquid (here water) viscosity. Data from Perry's Chemical Engineer's Handbook 6th edition, McGraw-Hill (1984).

Note. 10. 4. Estimating molecular diffusivities
Molecular diffusivities can be estimated by a number of different methods. The basis for many of these methods is the Stokes-Einstein equation. This equation is derived from considering the drag force on a large sphere, which moves in a solution (see e.g. Cussler, 1997). The diffusivity is found from the expression

$$D_A = \frac{k_B T}{6\pi\eta R_0} \tag{1}$$

where k_B is Boltzmann's constant (= 1.380662 10^{-23} kg m^2 s^{-2} K^{-1}), T is the absolute temperature, η is the viscosity of the solvent and R_0 is the radius of a spherical solute molecule. For non-spherical molecules, the equation can be modified in different ways, e.g. for ellipsoidal molecules

$$D_A = k_B T \Bigg/ \left(6\pi\eta \frac{(a^2-b^2)^{1/2}}{\ln\left(\frac{a+(a^2-b^2)^{1/2}}{b}\right)} \right) \tag{2}$$

where a and b are the length of the major axes of the ellipsoid. Since the Stoke-Einstein equation builds on the assumption that the diffusing compound is large relative the solvent compounds, it is mainly used to predict diffusivities for large molecules such as proteins. It may also be used the other way around, i.e. to predict the size of protein molecule from a measured diffusivity.

For prediction of diffusivities for smaller solute molecules A, diffusing in solvent B, the Wilke-Chang correlation (Eq. 3), which has a similar structure as the Stoke-Einstein equation, may instead be used (Reid et al., 1977).

$$D_{AB} = 7.4 \cdot 10^{-8} \frac{(M_B \phi)^{\frac{1}{2}} T}{v_A^{0.6} \eta_B}$$ (Note: Units cm²s⁻¹) (3)

where ϕ is the so-called association factor for the solute (=2.26 for water), M_B is the molar mass of solvent, v_A is the molar volume of A at its normal boiling point (Units: cm³/mol), and η_B is the viscosity of the solvent (Units: cP, where 1 cP = 10^{-3} kg m⁻¹ s⁻¹). This correlation can be used for liquid solutes, e.g. ethanol or glycerol diffusing in a solvent, but is of no use for predicting diffusivities of dissolved solid compounds, such as sugars. Both Eq. (1) and (3) predict that D_A varies inversely proportional to the viscosity of the solvent and proportional to the absolute temperature.

10.2. Mass Transfer To and Into Solid Particles

The mass transfer of substrates from the bulk liquid to the cells is normally rapid compared to the cellular consumption rates for a suspended cell culture (see e.g. Example 10.6). However, as is the case for many enzymatic processes, it is also for some microbial processes advantageous to apply *immobilized cells*. The cells are in such cases attached to the surface of a solid particle or are entrapped within a matrix structure, e.g. an alginate gel, forming either spherical pellets or immobilized films. Mass transfer from the bulk liquid will now occur to larger particles. Since a larger particle size gives a decreased specific surface area (cf. Eq. 10.17), the maximum obtainable volumetric mass transfer rate will be lower than for suspended cells. Furthermore, the rate of *internal mass transfer processes*, i.e. the diffusion of substrates and products within the pellet, becomes an important factor to consider. The transport processes will interact with the microbial conversion processes, and the overall conversion rates may, in fact, often be determined by the physical transport processes.

In this section we will therefore take a closer look at mass transfer processes in connection with the application of microorganisms present in a more or less solid matrix. This is also of interest in connection with the application of filamentous fungi, which have the ability to form pellets consisting of many individual hyphal elements. Our discussion of the mass transfer processes is divided into two parts: External mass transfer and intraparticle diffusion.

10.2.1. External Mass Transfer

External mass transfer carries the reacting species from the bulk liquid to the pellet surface. The treatment of mass transfer to the pellet is completely analogous to the treatment of mass transfer from a bubble. The concept of a stagnant film (often called the *Nernst diffusion layer* in the biochemistry literature) is thus used also for this transport process. The flux to the pellet surface is given by

$$J_A = k_s (c_A - c_{A,s})$$ (10.34)

$c_{A,s}$ is the concentration of A at the more or less well defined interface between liquid and e.g. a gel

Mass Transfer

surface or a pellet surface; c_A is the concentration in the bulk liquid. k_s is the mass transfer coefficient, which is a function of the physical properties of the considered system and of the liquid flow rate around the pellet. For pellets in suspension the value of k_s can be estimated using correlations of the type given by Eq. 10.28.

A certain concentration difference between the bulk phase and the pellet surface is obviously necessary to obtain a mass transfer of substrate to the pellet, and the reaction will thus take place at a slightly lower concentration than the bulk phase concentration. The question is now how much lower than the bulk concentration the surface concentration will be, and how that in turn will affect the conversion rate. Consider a system where cells are present only at the surface of a pellet. To emphasize that A is a substrate, its concentration is denoted s_A in the bulk liquid phase and $s_{A,s}$ at the pellet surface. The mass balance for the transport and reaction of the substrate at the pellet surface is given by

$$\frac{ds_{A,s}}{dt} = k_s a(s_A - s_{A,s}) + q_A \tag{10.35}$$

where $-q_A$ is the volumetric substrate consumption rate. At steady state and assuming Monod kinetics we get

$$k_s a(s_A - s_{A,s}) = Y_{xs} \mu_{max} \frac{s_{A,s}}{s_{A,s} + K_s} x \tag{10.36}$$

Note that the biomass concentration should be given with respect to the same volume as a, e.g. g (L medium)$^{-1}$.

Equation 10.36 can be rendered dimensionless by the following substitutions:

$$S_A = \frac{s_{A,s}}{s_A}; \qquad \alpha = \frac{K_s}{s_A}; \qquad Da = \frac{Y_{xs} \mu_{max} x}{k_s a s_A} \tag{10.37}$$

giving

$$1 - S_A = Da \frac{S_A}{S_A + \alpha} \tag{10.38}$$

Da is called the *Damköhler number*. The solution of the algebraic Eq. (10.38) is

$$S_A = -\frac{\beta}{2} + \sqrt{\frac{\beta^2}{4} + \alpha} \tag{10.39}$$

where

$$\beta = \text{Da} + \alpha - 1 \qquad (10.40)$$

The *Damköhler number* can, like the dimensionless groups of Table 10.6, be interpreted and given a physical meaning. It is the ratio between the maximum rate of substrate consumption by the reaction and its maximum mass transfer rate. Thus, if Da << 1 the maximum mass transfer rate is much larger than the maximum rate of reaction (low mass transfer resistance). S_A is in this case ≈ 1 and the conversion proceeds at the same rate as in the bulk phase. When Da >> 1, on the other hand, the mass transfer is the rate-limiting step and the reaction rate is significantly lower than at bulk phase conditions. These cases can be referred to as a *reaction-limited regime* and a *mass transfer-limited regime*, respectively.

Note 10.5. Effectiveness factors for film transport
Following the tradition in heterogeneous catalysis, the effect of external mass transfer on the overall conversion can be quantified by means of the *effectiveness factor for film transport*, η_f, which is defined as the ratio of the observed reaction rate and the reaction rate that would have occured without external mass transfer resistance. Thus with Monod kinetics

$$\eta_f = \frac{S_A/(S_A + \alpha)}{1/(1+\alpha)} \qquad (1)$$

Consequently, $\eta_f < 1$, and in general the effect of increasing mass transfer resistance is a reduction of the effectiveness factor.
For Da approaching zero (very slow reaction compared with the mass transfer process), we find from Eq. (10.39) that S_A must approach unity. For the reaction-limited regime we therefore have $\eta_f = 1$, and the observed reaction rate is given by

$$-q_A = Y_{xs}\mu_{max}\frac{S_A}{S_A + K_s}x \qquad (2)$$

i.e., the bulk liquid concentrations can be directly applied in the kinetic expression of the surface reaction. For very large values of Da, S_A will be close to zero, and from Eq. (10.39)

$$\frac{1}{\text{Da}} = \frac{S_A}{S_A + \alpha} \qquad (3)$$

which, inserted in Eq. (1) gives

$$\eta_f = \frac{(1+\alpha)}{\text{Da}} \qquad (4)$$

Here the observed reaction kinetics is given by

$$-q_A = k_s a s_A \qquad (5)$$

Mass Transfer

i.e., first order in the bulk substrate concentration and with rate constant k_s. Here the kinetics is totally independent of the intrinsic kinetic parameters μ_{max} and K_s.

Note 10.6. Finding values of the mass transfer coefficient k_s.
The same basic principles as previously described for finding mass transfer correlations for bubbles apply also for the mass transfer between a solid particle and a liquid. The difference is primarily that the assumption of a rigid particle surface is more credible. In Table 10.7 only the correlations for immobile interfaces are thus of interest. For suspended pellets the linear velocity of the pellets relative to the continuous phase, i.e. the surrounding medium, derives from the density difference. The velocity can therefore be described by Stokes law

$$u_b = \frac{g(\rho_p - \rho_l)d_p^2}{18\eta} \tag{1}$$

Inserting the expression for u_b Eq. (1) into the definition of the Reynolds number, we get:

$$\text{Re} = \frac{g(\rho_p - \rho_l)d_p^2}{18\eta} \frac{d_p \rho_l}{\eta} = \frac{Gr}{18} \tag{2}$$

The Grashof number can thus be applied in the correlations instead of the Reynolds number. However, the coefficients multiplying the Reynolds number will be different. For e.g. the third correlation in Table 10.7, valid for Re < 1, Pe < 10^4 we get:

$$Sh = \left(4 + 1.21\left(\frac{Gr}{18}\right)^{\frac{2}{3}} Sc^{\frac{2}{3}}\right)^{\frac{1}{2}} = \left(4 + 0.18 Gr^{\frac{2}{3}} Sc^{\frac{2}{3}}\right)^{\frac{1}{2}} \tag{3}$$

For a pellet with a diameter of 2 mm suspended in water, with a density difference of 20 kg m^{-3} between the liquid and the pellet, we find

$$Gr = \frac{(2 \cdot 10^{-3})^3 9.81 \cdot 997 \cdot 20}{(10^{-3})^2} = 1564 \tag{4}$$

The viscosity of water has been used in the calculation above. We see that

$$\frac{Gr}{18} > 1 \tag{5}$$

i.e. the validity range of correlation (3) is exceeded and another correlation should be used. The top correlation from Table 10.7 appears applicable since the validity range should be 36 < Gr < 23400. We get

$$Sh = 2 + 0.13 Gr^{\frac{1}{2}} Sc^{\frac{1}{3}} \tag{6}$$

Suppose the mass transfer coefficient for glucose is needed, and that the solution is very dilute. We have $D_{glucose} = 0.69\ 10^{-9}$ m^2 s^{-1} (at 25 °C). Assuming a viscosity and density of 10^{-3} kg m^{-1} s^{-1} and $\rho = 1000$ kg m^{-3} (close to water) we get

$$Sc = \frac{10^{-3}}{0.69 \cdot 10^{-9} \cdot 1000} = 1449 \tag{7}$$

Inserted into Eq. 6, we get $Sh \approx 60$ and

$$k_s = \frac{60 \cdot 0.69 \cdot 10^{-9}}{2 \cdot 10^{-3}} = 2 \cdot 10^{-5} \text{ m s}^{-1} \tag{8}$$

For a medium containing filamentous fungi, the viscosity is normally much higher than 10^{-3} kg m^{-1} s^{-1}, and for these systems the Grashof number may be smaller than 18. The above described correlations hold only for suspended pellets. For, e.g. packed beds, other correlations need to be used (see, e.g., Moo-Young and Blanch, 1981).

10.2.2. Intraparticle Diffusion

When microbial cells (or enzymes) are present within a pellet, the substrates have to be transported not only to the pellet, but also into the pellets, and metabolic products formed by the cells have to be transported out of the pellet. The convective transport inside a pellet is normally small, and the transport can therefore be well described by diffusion only. For diffusive transport to occur, a concentration gradient is needed, i.e. the concentration of a substrate inside the pellet must be smaller than at the surface for any transport to take place. The cells placed close to the center of a pellet will therefore not be exposed to the same substrate concentration as its more fortunate relatives located close to the pellet surface. If the substrate concentration difference is too large, starvation for carbon source, or anaerobic conditions, may occur in the center[1]. Again, it is important to analyze and compare the rate of reaction and the rate of the transport processes. The transient mass balance for substrate A can be written

$$\frac{\partial s_A}{\partial t} = D_{A,\text{eff}} \nabla^2 s_A + q_A^{'} \tag{10.41}$$

where $D_{A,\text{eff}}$ is the so-called *effective diffusivity*, ∇^2 is the Laplacian operator, and $q_A^{'}$ is the volumetric reaction rate. (Note that the volume is here the pellet volume).

---

[1] The fate of the cell at the pellet center can be compared to that of an unfortunate dinner guest placed at the very end of a long table. In the worst-case scenario, the plates may well be emptied before they reach the end of the table, particularly if all the dinner guests are very hungry.

Mass Transfer

It should be stressed that one cannot use the molecular diffusivity directly in Eq. (10.41) for two reasons; a) Diffusion only takes place in the void volume of the pellet, and b) The length needed for transport is increased due to the twisted path of the pores relative to the radial coordinate. This is called *tortuosity*. Both these factors tend to give an effective diffusivity significantly lower than the molecular diffusivity.

The void volume of the pellet can be expressed as:

$$V_{void} = V\varepsilon \tag{10.42}$$

The void volume is a function of the biomass concentration in the pellet, and of the matrix material used to immobilize the cell system. Thus if the biomass grows within the pellet, it takes up an increasing fraction of the void volume, resulting in a decreasing mass transfer. The effect of a longer diffusion pathlength can schematically be described by an overall tortousity factor, τ_t, defined as the ratio between the length of the straight path and the twisted pores. The tortuosity factor is usually in the range 1.5-5. The effective diffusivity is thus obtained as:

$$D_{A,eff} = \varepsilon \frac{D_A}{\tau_t} \tag{10.43}$$

The effective diffusivity in a pellet is therefore lower than the free diffusivity by at least a factor of 2.

In the case of large diffusing molecules, such as proteins, the pores may have diameters of the same order of magnitude as that of the diffusing species. In such a case, so-called hindered diffusion takes place and further corrections are needed in calculating the effective diffusivity (see e.g. Blanch and Clark, 1997, for further discussion).

At steady-state conditions, we have

$$D_{A,eff} \nabla^2 s_A + q_A' = 0 \tag{10.44}$$

Eq. 10.44 is in general difficult to solve analytically, but can rather easily be solved numerically provided of course, that an adequate expression for q_A is known. In such a case the concentration profile of the substrate in the pellet can be found. The analytical solutions for the simple cases of zeroth and first order kinetics are presented in Note 10.7. These cases are of interest, since the Monod kinetic expression gives an effective reaction order between zero and one.

Note 10.7. Finding the concentration profile for a zeroth and a first order reaction.
For very simple kinetic expressions and highly symmetric geometries, it is possible to find an analytical solution to Eq. (10.44). We will here consider the simple case of a spherical particle, with radius R_p for which Eq. (10.44) can be written:

$$\frac{D_{A,eff}}{\xi^2}\frac{d}{d\xi}\left(\xi^2\frac{ds_A}{d\xi}\right)+q_A^{'}=0 \tag{1}$$

where the radial distance from the center is denoted ξ. The boundary condition at the surface of the particle is:

$$s_A = s_{A,s} \text{ for } \xi = R_p \tag{2}$$

The symmetry of the solution requires

$$\frac{ds_A}{d\xi}=0 \text{ for } \xi = 0 \tag{3}$$

Eqs. (1)-(3) provide sufficient information for solving the concentration profile in the pellet. For a zeroth order reaction, we have

$$q_A^{'} = -k_0 \tag{4}$$

Integrating Eq. (1) and applying the boundary condition given by Eq (3) we get

$$\frac{ds_A}{d\xi}=\frac{k_0\xi}{3D_{A,eff}} \tag{5}$$

Eqs. (5) and (2) give

$$s_A = s_{A,s} + \frac{k_0}{6D_{A,eff}}\left(\xi^2 - R_p^2\right) \tag{6}$$

Eq. (6) thus gives the concentration of A at any given position in the pellet. Since there is no concentration dependence on the volumetric reaction rate for reaction order zero, one may get the impression that the diffusion will have no effect at all. However, one should remember the restriction $s_A > 0$, which requires that

$$s_{A,s} > \frac{k_0 R_p^2}{6D_{A,eff}} \tag{7}$$

If this requirement is not satisfied, the rate of diffusion will not be sufficient to provide substrate for the reaction to proceed in the central part of the pellet, and thus $q_A = 0$ for

$$\xi \leq \sqrt{R_p^2 - \frac{6D_{A,eff}}{k_0}s_{A,s}} \tag{8}$$

For a first order reaction, Eq (4) is replaced by Eq (9).

$$q_A^{\cdot} = -k_1 s_A \tag{9}$$

and we get

$$\frac{D_{A,eff}}{\xi^2} \frac{d}{d\xi}\left(\xi^2 \frac{ds_A}{d\xi}\right) - k_1 s_A = 0 \tag{10}$$

Eq. (10) can be rearranged into

$$\frac{d}{d\xi}\left(\xi^2 \frac{ds_A}{d\xi}\right) - \frac{\xi^2 k_1 s_A}{D_{A,eff}} = 0 \tag{11}$$

which can be solved by making the substitution $y = s_A \xi$ giving

$$\frac{d^2 y}{d\xi^2} - \frac{k_1}{D_{A,eff}} y = 0 \tag{12}$$

The general solution to Eq. (12) is given by

$$y = C_1 \sinh\left(\xi \sqrt{\frac{k_1}{D_{A,eff}}}\right) + C_2 \cosh\left(\xi \sqrt{\frac{k_1}{D_{A,eff}}}\right) \tag{13}$$

With the appropriate boundary conditions, the specific solution becomes

$$s_A = \frac{s_{A,s} R_p}{\xi} \frac{\sinh\left(\xi \sqrt{\frac{k_1}{D_{A,eff}}}\right)}{\sinh\left(R_p \sqrt{\frac{k_1}{D_{A,eff}}}\right)} \tag{14}$$

The calculated concentration profiles give an indication of the relation between reaction rate and diffusion rate. If the concentration difference between center and surface of the particle is small, the maximum diffusion rate is large compared to the reaction rate. However, if there is a large concentration difference between the center and the surface, the maximum reaction rate is larger than the diffusion rate. The concentration profiles suggest if starvation for substrate is likely to occur within the pellet.

The observed volumetric reaction rate, $q_{A,obs}$ is obtained by integration of the reaction rate over the pellet volume and division by the pellet volume i.e.

$$q'_{A,obs} = \frac{\int_0^{R_p} 4\pi\xi^2 q'_A d\xi}{4\pi R_p^3/3} \qquad (10.45)$$

The ratio between the observed volumetric reaction rate and that obtained in the absence of any mass transfer resistance is also called an effectiveness factor. This particular effectiveness factor is now an *internal effectiveness* factor, η_{eff}, to distinguish it from the previously defined film effectiveness factor, η_f.

$$\eta_{eff} = \frac{q'_{A,obs}}{q'_A(s_{A,s})} \qquad (10.46)$$

For a spherical geometry, Eq.10.44 may be transformed into the dimensionless form given by Eq. (10.47)

$$\frac{1}{\chi^2}\frac{d}{d\chi}\left(\chi^2 \frac{dS_A}{d\chi}\right) + \frac{R_p^2 q'_A(s_A = s_{A,s})}{D_{A,eff} s_{A,s}} \frac{q'_A}{q'_A(s_A = s_{A,s})} = 0 \qquad (10.47)$$

by introducing

$$\chi = \frac{\xi}{R_p}; \; S_A = \frac{s_A}{s_{A,s}} \qquad (10.48)$$

The dimensionless *Thiele modulus*, Φ, is defined by

$$\Phi = \frac{R_p q'_A(s_A = s_{A,s})}{3\sqrt{D_{A,eff} s_{A,s} q'_A(s_A = s_{A,s})}} \qquad (10.49)$$

For a first order reaction, i.e. $-q_A = k_1 s_A$, the Thiele modulus is given by

$$\Phi = \frac{R_p}{3}\sqrt{\frac{k_1}{D_{A,eff}}} \qquad (10.50)$$

It can be shown (see e.g. Fogler, 1999) that the internal effectiveness factor for a first order reaction in a sphere is given by:

Mass Transfer

$$\eta_{eff} = \frac{1}{\Phi}\left(\frac{1}{\tanh(3\Phi)} - \frac{1}{3\Phi}\right) \tag{10.51}$$

Like many other dimensionless parameters, also the Thiele modulus has a physical significance. We see that

$$\Phi^2 = \frac{R_p^2 k_1}{3D_{A,eff}} \leftrightarrow \Phi^2 = \frac{R_p^3 k_1 s_{A,s}}{3R_p^2 D_{A,eff}(s_{A,s}-0)/R_p} \tag{10.52}$$

i.e. Φ^2 can (loosely) be interpreted as the ratio between the reaction rate at the surface and a maximum "diffusion rate". A high value of the Thiele modulus therefore suggests a fast reaction compared to the diffusional transport, and the effectiveness factor will in such a case be low. For other geometries, e.g. a slab-formed solid of half-thickness L, it can be shown that

$$\eta_{eff} = \frac{\tanh\Phi}{\Phi} \tag{10.53}$$

where the Thiele modulus is defined by

$$\Phi = L\sqrt{\frac{k_1}{D_{A,eff}}} \tag{10.54}$$

As pointed out previously, it is for most kinetic expressions not possible to find an analytical expression for the concentration profile or the effectiveness factor. Instead Eq. (10.44) must be solved by numerical methods (e.g. by finite element, finite difference or collocation methods, see Villadsen and Michelsen, 1978) to give the concentration gradient in the pellet, and from that calculate the observed reaction rate by a quadrature. Aris (1975) demonstrated that Eq. (10.53) gives a satisfactory approximation for the effectiveness factor for any reasonable kinetics and any particle shape if the characteristic length Λ is defined as the ratio between pellet volume and pellet exterior surface area, V/A, and provided that the following *generalized Thiele modulus* is used

$$\Phi_{gen} = \frac{-q'_A(s_A = s_{A,s})\Lambda}{\sqrt{2D_{A,eff}\int_0^{s_{A,s}} -q'_A(s_A)ds_A}} \tag{10.55}$$

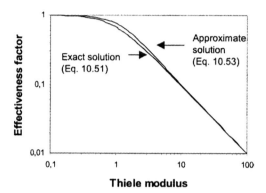

Figure 10.10 Plot of the effectiveness factor for a first order process. The effectiveness factor has been calculated both exactly, using Eq. (10.51) and the Thiele modulus for a sphere (Eq. 10.50), and approximately using Eq. (10.53) and the generalized Thiele modulus given by Eq. (10.55) with $\Lambda = R_p/3$.

As seen from Fig 10.10, the deviation between the exact and approximate solution is small for a first order reaction, with maximum deviation in the intermediate range (see also Example 10.7).

Example 10.7. Effectiveness factor for a pellet with Monod kinetics

We will now consider a pellet with biomass immobilized uniformly thorughout a pellet to a concentration of x kg m^{-3}. The uptake rate of glucose is assumed to follow Monod kinetics, i.e.

$$-q_s' = Y_{xs}\mu_{max}\frac{s}{s+K_s}x = k\frac{S}{S+\alpha} \tag{1}$$

where

$$S = \frac{s}{s_{surface}}, \quad k = Y_{xs}\mu_{max}x \quad \text{and} \quad \alpha = \frac{K_s}{s_{surface}} \tag{2}$$

The relevant data is summarized in Table 10.11. For a porosity of 0.5 and an assumed tortuosity factor of 1.5, we find the effective diffusion coefficient for glucose in the pellet by using Eq. (10.43):

$$D_{eff} = 6\cdot 10^{-10}\cdot\frac{0.5}{1.5} = 2\cdot 10^{-10} \text{ m}^2\text{ s}^{-1} \tag{3}$$

Table 10.11. Parameters for the pellet system considered.

Parameter	Value
k	0.50 kg m^{-3} (pellet) h^{-1}
α	0.2
$S_{surface}$	0.10 kg m^{-3}
D_S	6 10^{-10} m^2 s^{-1}
R_p	2 10^{-3} m
ε	0.5

We do not have an analytical expression for the effectiveness factor for a sphere with a reaction obeying Monod kinetics. However, Eq.(10.44) may be solved numerically, using a suitable numerical method.

With a centered finite difference method, using 200 grid points, the concentration profile given in Fig. 10.11 is obtained.

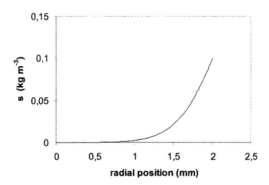

Fig. 10.11. The substrate concentration as a function of radial position in the pellet.

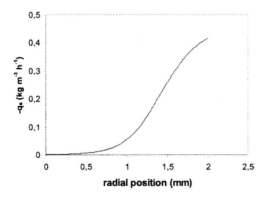

Fig. 10.12. The substrate consumption rate ($-q_s$) plotted versus the radial position in the pellet.

The volumetric substrate consumption rate ($-q_s$) is shown in Fig. 10.12. As could be expected from the concentration profile, the rate is not constant throughout the pellet, and the effectiveness factor is clearly smaller than one. The effectiveness factor is calculated numerically from Eq. (4) and Eq. (5).

$$q_{s,obs}' = \frac{\int_0^{R_p} 4\pi\xi^2 q_s' d\xi}{4\pi R_p^3 / 3} \qquad (4)$$

$$\eta_{eff} = \frac{q_{s,obs}'}{q_s'(s_s)} \qquad (5)$$

The numerically obtained value is $\eta_{eff} = 0.63$.

Alternatively, an approximate solution may be derived using the generalized Thiele modulus (Eq. 10.55). In this case, we get

$$\Phi_{gen} = \frac{kR_p}{3(1+\alpha)\sqrt{2D_{eff} s_s k(1-\alpha \ln(1+1/\alpha))}} \qquad (6)$$

or with inserted values

$$\Phi_{gen} = 1.29 \qquad (7)$$

The approximate effectiveness factor calculated by Eq. (10.53) is 0.66, which is a satisfactory approximation to the value obtained by numerically solving the complete equations.

Until now we have considered external and internal mass transport separately. However, in general the substrate must first traverse the external film or boundary layer and subsequently diffuse into the pellet where the reaction occurs. The mass balance of Eq. (10.44) of course still holds, but the outer boundary condition is now written

$$-D_{A,eff} \frac{ds_A}{d\xi}\bigg|_{\xi=R_p} = k_s(s_A(\xi=R_p) - s_{A,l}) \qquad (10.56)$$

This boundary condition (sometimes called a Neumann boundary condition) basically states that there can be no accumulation of substrate at the pellet interface.

The *overall effectiveness factor*, η_{tot}, is defined as the observed reaction rate divided by the reaction rate that would occur at bulk liquid condition e.g.

$$\eta_{tot} = \frac{q_{A,obs}'}{q_A(s_{A,l})} \qquad (10.57)$$

Mass Transfer

With slab geometry and first-order kinetics, the overall effectiveness factor can be calculated by

$$\frac{1}{\eta_{tot}} = \frac{\Phi}{\tanh(\Phi)} + \frac{\Phi^2}{Bi} \leftrightarrow \frac{1}{\eta_{tot}} = \frac{1}{\eta_{eff}} + \frac{\Phi^2}{Bi} \tag{10.58}$$

where Bi is the so-called *Biot number for mass transport* which expresses the ratio between the characteristic film transport rate and the characteristic intraparticle diffusion rate. For slab geometry

$$Bi = \frac{k_s L}{D_{A,eff}} \tag{10.59}$$

Eq. (10.58) may approximately be applied to other geometries as well (by defining $\Lambda = V/A$ as described for the generalized Thiele modulus). For a first order reaction we have

$$\frac{\Phi^2}{Bi} = \frac{\Lambda k_1}{k_s} \tag{10.60}$$

Clearly, this term is related to the external mass transfer, and it will only be of importance if the rate of reaction is large compared to the mass transfer rate. In the absence of intraparticle mass transfer resistance in a porous particle there is no reason to consider an effect of external resistance.

The mathematical treatment of both external and internal mass transfer is described extensively in chemical engineering textbooks due to their importance in traditional catalysis (see e.g. Levenspiel, 1999 or Fogler, 1999). For the purpose of immobilized cells, however, a detailed quantitative analysis of effectiveness factors serves little purpose. The kinetic expressions are often derived for suspended cell cultures and may therefore not be accurate, or even adequate, for immobilized cells. It is, furthermore, very likely that major changes in the metabolism sets in if oxygen depletion occurs inside the pellet (see also discussion in Section 11.4). The main reason for still taking the trouble of analyzing mass transfer processes into pellets is to make sure that the pellet size is small enough to avoid severe mass transfer limitations. For that purpose approximate kinetic expressions will do the trick.

PROBLEMS

Problem 10.1 Determination of $k_l a$ in a pilot plant bioreactor

The aim of a study of penicillin fermentation by Christensen (1992) and by Pedersen (1992) was to set up models for the microbial and reactor dynamics. A major purpose of this study was to see how mass transfer affects the overall performance of the process. A series of experiments was therefore designed for this purpose. The bioreactor can be assumed to be ideal, i.e., there are no concentration gradients in the medium. It is assumed that process air has the composition 20.95% O_2, 0.030 % CO_2, and 79.02% N_2. The total volume of the bioreactor was V = 15 L. The equilibrium concentration of O_2 in solutions of sulphite or sulphate is given by Eq. 1.

$$c_o^* = 5.909 \cdot 10^{-6} \exp\left(\frac{1602.1}{T} - \frac{0.9407 c_s}{1 + 0.1933 c_s}\right) \text{ M atm}^{-1} \qquad (1)$$

where c_s is the molarity of the salt solution and T is in K (Linek and Vacek, 1981). All experiments were carried out at 25 °C, with 1 atm head space pressure and constant stirrer speed.

a. *Determination of $k_l a$ by the sulphite method.*
The bioreactor was filled with 10 L of water, and sulphite was added to give a concentration of 0.26 mol L^{-1} at $t = 0$. The gas was turned on, and liquid samples were analyzed for sulphite concentration. The experimental results are shown in the table below. During the experiment the dissolved oxygen concentration can be assumed to be zero because of fast reaction. Calculate $k_l a$ in the bioreactor and discuss the sulphite method.

t (min)	Sulphite concentration (mM)
0	260
3	241
9	209
15	178
22	146
28	112
34	80

b. *Determination of $k_l a$ by the direct method.*
We now want to determine $k_l a$ during a fermentation experiment, and we therefore implement an exhaust-gas analyzer for measuring O_2 and CO_2. The O_2 analyzer is based on paramagnetic resonance, and the CO_s analyzer is based on infrared detection. Since the analyzers measure the partial pressures, it is essential to dry the gas by letting it pass through a column containing Drierite (anhydrous $CaSO_4$). During penicillin fermentation, the gas-flow rate is 1 vvm (volume per volume per minute), and the working volume is 10 L. At a certain time the O_2 content in the exhaust gas is 20.63%, and the dissolved O_2 concentration is measured to be constant at 69% of the equilibrium concentration, which is about 90% of that in water. The gas constant is R = 0.082045 L atm mol^{-1} K^{-1}. Calculate $k_l a$ during the fermentation, and discuss the difference from the value determined by the sulphite method.

Problem 10.2. Mass transfer during fermentations with mammalian cells

Mammalian cells are shear-sensitive, and aeration is often carried out by diffusion of O_2 across the wall of silicone tubing placed inside the bioreactor. The inlet gas (often pure O_2) normally has a higher pressure than the head space pressure, which ensures a pressure drop across the wall of the silicone tubing. $k_l a$ for this system depends on the pressure difference across the silicone tube wall, the temperature, and the porosity of the silicone tubing. When very little O_2 leaves the silicone tube through the tube wall, the O_2 concentration is almost constant throughout the tube. It is the aim of this exercise to quantify these effects in bioreactor with a volume of $V = 1$ L.

a. Show that:

$$k_l = \frac{D_{o,\text{eff}}}{\delta_t} \frac{p_g - H_o c_o}{RT(c_o^* - c_o)} \quad (1)$$

$D_{o,g}$ is the diffusion coefficient of O_2 in the inlet gas, pg is the concentration of O_2 in the inlet gas, c_o^* is the equilibrium concentration of O_2 in the liquid, and δ_t is the wall thickness of the silicone tubing.

b. Plot k_l versus in a system where $\delta = 0.35$ mm, the head space pressure is 1 atm, and the pressure in the inlet gas is 2 atm. The temperature is 30 °C. $D_o = 0.2 \cdot 10^{-4}$ m^2 s^{-1} and c_o^* (1 atm, 30 °C) = 1.16 mmoles L^{-1}.

c. State k_l as a function of the inlet gas pressure and the temperature, when the head space pressure is 1 atm.

d. For a given length l and inner diameter d_i, of the tube, calculate the effect of the wall thickness on the volumetric surface area a. The porosity of the tubing is ε.

e. For $l = 1$ m, $d_i = 1$ mm and $\varepsilon = 0.40$, calculate $k_l a$ for a bioreactor in which the inlet gas is pure O_2 with a pressure of 2 atm and a temperature of 30 °C. The head space pressure is kept constant at 1 atm, and the dissolved O_2 concentration is measured to be 0.6 mmol L^{-1}.

f. From experiments at different agitation speeds, you observe that $k_l a$ increases with the agitation speed. How do you explain this observation?

Problem 10.3. Mass transfer in a pilot plant bioreactor

Data for determination of $k_l a$ by the ^{85}Kr method are given in the following table:

v_g (vvm)	P_g (W)	$k_l a$ (h-1)
1.0	3000	951
0.2	3000	688
0.1	3000	598
0.05	3000	525
1.0	5000	1358
1.0	1000	440
1.0	500	268

A pilot plant bioreactor with a height-to-diameter ratio of 3:1 is equipped with two standard Rushton turbines. The bioreactor is filled with 1 m^3 of sterile medium. The aeration rate is 1 vvm, and with a total power dissipation of 3000 W the degree of filling is 69.5% and the gas holdup is measured to be 10%. The temperature is 30 °C, and the head space pressure is 1 atm.

a. Calculate the superficial gas velocity u_s, the mean bubble rise velocity u_b, and the average residence time of gas bubbles t_b.

b. Under the assumption that the medium is coalescing, calculate $k_l a$ using the correlation of Eq. (10.27) with k = 0.026, $\alpha = 0.5$, and $\beta = 0.4$. From measurements in the exhaust gas and the dissolved oxygen concentration during a fermentation experiment, you suspect that the calculated $k_l a$ value is too low. You therefore decide to examine the system more carefully. Using the ^{85}Kr method to determine $k_l a$ you observe the effect of variations in v_g and P_g when the bioreactor is filled with sterile medium. The results of the study are listed in Table 1.
Estimate the parameters for the correlation in Eq. (10.27) using these data. How do you explain that the measured $k_l a$ is significantly higher than the value calculated from the literature correlation?

c. With the estimated $k_l a$ value, find the largest possible oxygen consumption rate when $v_g = 1$ vvm and $P_g = 3000$ W. During penicillin fermentations it is found that the dissolved oxygen concentration should be above 30% of the saturation value if the penicillin production is not to be inhibited. Assume that the oxygen uptake rate can be described by Monod kinetics with $K_o = 0.015$ mmoles L^{-1} and $\mu_{max} = 0.65$ mmoles (kg DW)$^{-1}$ s^{-1}. Calculate the maximum biomass concentration that can be sustained without inhibition of the penicillin production.

Problem 10.4. Mass transport into pellets

The concentration profile for a solute A, which reacts as it is transported by molecular diffusion in a spherical pellet is given as the solution to Eq. (1)

$$\frac{1}{\chi^2}\frac{d}{d\chi}\left(\chi^2\frac{dS_A}{d\chi}\right) = 9\Phi^2 \frac{q_A'(s_A)}{q_A'(s_A = s_{A,s})} \tag{1}$$

where the dimensionless concentration, S_A, and the dimensionless radial coordinate, χ, are given by

$$S_A = \frac{s_A}{s_{A,s}} \quad \text{and} \quad \chi = \frac{\xi}{R} \tag{2}$$

and the Thiele modulus, Φ, is given by Eq. (3)

$$\Phi = \frac{q(s_{A,s})R^2}{9D_{eff}s_{A,s}} \tag{3}$$

$-q_A$ is the volumetric consumption rate (unit: mol m^{-3} (pellet) s^{-1}), R is the pellet radius, and D_{eff} is the effective diffusion coefficient in the pellet for the species being considered. c_b is the bulk concentration of the species. For zeroth-order kinetics the pellet size that gives a zero concentration precisely at the center of the pellet is given by (see Note 10.7)

$$R_{crit} = \sqrt{\frac{6D_{eff}s_{A,s}}{-q_A'(s_{A,s})}} \tag{4}$$

We now consider oxygen diffusion into a pellet of *Penicillium chrysogenum*. For this microorganism, the oxygen uptake rate can be approximated by

$$-q_o(s_o) = k\frac{s_o}{s_o + K_o}x \tag{5}$$

with $k = 4.0$ mmol O$_2$ (g DW)$^{-1}$ h^{-1} and $K_o = 58$ μmol O$_2$ L^{-1}. The biomass concentration in the pellet is approximately 99 kg DW m^{-3} of pellet (this corresponds to a void volume of around 70% and a water content of 67% in the cells) and the concentration of oxygen in the bulk medium is 17.4 μmole L^{-1}. Calculate the critical pellet radius when it is assumed that Eq. (4) can be used. Discuss the application of Eq. (4). Calculate the effectiveness factor using the generalized Thiele modulus (Eq. 10.55) as a function of the pellet radius (25 μm < R < 500 μm).

Mass Transfer

Problem 10.5. Measuring k_la by the hydrogen peroxide method

As described in Note 10.3, one method to measure k_la is based on the consumption of hydrogen peroxide in a catalase containing medium (see reaction below).

$$2H_2O_2 \xrightarrow{catalase} 2H_2O + O_2$$

After addition of a known amount of catalase to the reactor, a continuous feed of H_2O_2 is applied to the reactor. Initially, hydrogen peroxide will accumulate in the reactor medium, but a steady-state will soon be established at which the rate of decomposition of hydrogen peroxide equals the addition rate of hydrogen peroxide. By measuring the dissolved oxygen concentration, as well as the inlet and outlet oxygen mole fractions in the gas, it is possible to calculate the value of k_la.

In the original work by Hickman (1988), the values given in the table below, where measured dissolved oxygen tension in liquid phase and oxygen concentrations in inlet and exit gas streams are given, can be found.

Hydrogen peroxide addition rate 10^5 (mol s^{-1})	Dissolved oxygen concentration (% of saturation)	Measured exit oxygen concentration (mol%)	Stirrer rate (rpm)
1.6	111.5	21.14	500
3.2	123.8	21.36	500
6.4	147.2	21.78	500
12.8	195.8	22.6	500
3.2	110.6	21.35	900
6.4	121.7	21.77	900
12.8	143.4	22.60	900

(Note that the dissolved oxygen concentration is higher than 100%, since oxygen is generated in the liquid and transferred from the liquid phase to the gas phase.)

The following additional information is given:
Reactor volume: $V = 0.005$ m^3
Gas flow rate: $v_{gas} = 9.2 \cdot 10^{-5}$ m^3 s^{-1} (i.e. approx. 1 vvm)
Inlet oxygen concentration: 20.96%

a) Calculate the value of k_la for the two stirrer rates given in the Table 1. Assume that T = 20 °C. The saturation concentration of oxygen (corresponding to 100% DOT) is 0.289 mol m^{-3} under these conditions.

b) Assume that the gas analyzer has an absolute error of 0.1% of the measured value. How big is (in the worst case) the error in the calculated k_la-value? Under what conditions do you expect to get the most accurate measurements?

REFERENCES

Aris, R. (1975). *The Mathematical Theory of Diffusion and Reaction in Permeable Catalysts. Volume 1. The Theory of the Steady State*. Clarendon Press, Oxford.

Bailey, J. E. and Ollis, D. F. (1986). *Biochemical Engineering Fundamentals, 2nd ed*. McGraw-Hill, New York.

Bhavaraju, S. M., Russell, T. W. F., and Blanch, H. W. (1978). The design of gas sparged devices for viscous liquid systems. *AIChE. J.* 24:454-465.

Bird, R. B., Stewart, W. E., and Lightfoot, E. N. (2002). *Transport Phenomena, 2nd ed*. John Wiley and Sons, New York.

Blanch, H. W., and Clark, D. S. (1997). Biochemical Engineering. Marcel Dekker, New York

Brian, P. L. T., and Hales, H. B. (1969). Effect of transpiration and changing diameter on heat and mass transfer to spheres. *AIChE J.* 15:419-425

Calderbank, P. H., and Moo-Young, M. M. (1961). The continuous phase heat and mass transfer to spheres. *Chem. Eng. Sci.* 16:39-54

Christensen, L. H. (1992). Modelling of the Penicillin Fermentation. Ph.D. thesis, Technical University of Denmark, Lyngby.

Craig, V. S. J., Ninham, B. W., and Pashley, R. M. (1993). Effect of electrolytes on bubble coalescence. *Nature* 364:317-319.

Cussler, E. L. (1997). *Diffusion – Mass transfer in fluid systems, 2nd ed*. Cambridge University Press, Cambridge, U.K.

Danckwerts, P. V. (1970). *Gas Liquid Reactions*. McGraw-Hill, New York.

Fogler, H. S. (1999) *Elements of Chemical Reaction Engineering, 3rd ed*., Prentice Hall, Upper Saddle River, New Jersey

Froessling, N. (1938). Über die Verdunstung fallender Tropfen. *Gerlands Beitr. Geophys.* 32:170-216

Hickman, A. D. (1988). Gas-liquid oxygen transfer. A novel experimental technique with results for mass transfer in aerated agitated vessels. Proc. 6th Eur. Conf. Mixing, Pavia, 369-374

Geankoplis, C. J. (1993). *Transport processes and unit operations, 3rd ed*., Prentice Hall, Englewood Cliffs, New Jersey

Jones, R. P. and Greenfield, P. F. (1982) Effect of carbon dioxide on yeast growth and fermentation. *Enz. Microb. Technol.* 4:210-223

Keitel, G. and Onken, U. (1982). The effect of solutes on bubble size in air-water dispersions. *Chem. Eng. Commun.* 17:85-98.

Lacoursiere, A., Thompson, B.G., Kole, M. M., Ward, D, and Ferson, D.F. (1986) Effects of carbon dioxide concentration on anaerobic fermentations of *Escherichia* coli. *Appl. Microbiol. Biotechnol.* 23:404-406

Lee, Y. H. and Meyrick, D. L. (1970). Gas-liquid interfacial areas in salt solutions in an agitated tank. *Trans. Inst. Chem. Eng.* 48, T37-T45.

Lehrer, I. H. (1971). Gas hold-up and interfacial area in sparged vessels. *Ind Eng. Chem. Des. Dev.* 10: 37-40.

Levenspiel, O. (1999). *Chemical Reaction Engineering, 3rd ed*., John Wiley and Sons, New York

Linek, V. and Vacek, V. (1981). Chemical engineering use of catalyzed sulphite oxidation kinetics for the determination of mass transfer characteristics of gas liquid contractors. *Chem. Eng. Sci.* 36:1747-1768.

Linek, V., Vacek, V., and Benes, P. (1987). A critical review and experimental verification of the correct use of the dynamic method for the determination of oxygen transfer in aerated agitated vessels to water, electrolyte solutions and viscous liquids. *Chem. Eng. J.* 34:11-34.

Nernst, W. (1904) Theorie der Reaktionsgeschwindigkeit in heterogenen Systemen. *Zeitschrift für Physikalische Chemie* 47:52-55

McIntyre, M., and McNeill, B. (1997). Dissolved carbon dioxide effects on morphology, growth, and citrate production in *Aspergillus niger* A60. *Enz. Microb. Technol.* 20:135-142

Moo-Young, M. and Blanch, H. W. (1981). Design of biochemical reactors. Mass transfer criteria for simple and complex systems. *Adv. Biochem. Eng.* 19:1-69.

Pedersen, A. G. (1992). Characterization and Modelling of Bioreactors. Ph.D. thesis, Technical University of Denmark, Lyngby, Denmark.

Pedersen, A. G. (1997). k_La characterization of industrial fermentors. Proc. 4th Int. Conf. Bioreactor Bioprocess Fluid Dynamics, BHRG, 263-276.

Pedersen, A. G., Andersen, H., Nielsen, J., and Villadsen, J. (1994). A novel technique based on ^{85}Kr for quantification of gas liquid mass transfer in bioreactors. *Chem. Eng. Sci.* 49:803-810.

Popovic, M., Niebelschutz, H., and Reuss, M. (1979). Oxygen solubilities in fermentation fluids. *Eur. J. Appl. Microb. Biotechnol.* 8:1-15.

Reid, R. C., Prausnitz, J. M., and Sherwood, T. K. (1977). The properties of gases and liquids, 3rd ed. McGraw Hill, New York

Rowe, P. N, Claxton, K. T., and Lewis, J. B. (1965). Heat and mass transfer from a single sphere in an extensive flowing fluid. *Trans. Inst. Chem. Eng.* 43:14-31

Tse, K., Martin, T., McFarlane, C. M., and Nienow, A. W. (1998). Visualisation of bubble coalescence in a coalescence cell, a stirred tank and a bubble column. *Chem. Eng. Sci.* 53:4031-4036

van't Riet, K. (1979). Review of measuring methods and results in non-viscous gas liquid mass transfer in stirred vessels. *Ind. Eng. Chem. Process Dev.* 18:357-364.

Schugerl, K. (1981). Oxygen transfer into highly viscous media. *Adv. Biochem. Eng.* 19, 71-174.
Villadsen, J. and Michelsen, M. L. (1978). *Solution of Differential Equation Models by Polynomial Approximation.* Prentice-Hall, Englewood Cliffs, New Jersey.
Whitman, W. G. (1923). A preliminary experimental confirmation of the two-film theory of gas absorption. *Chem. Metal. Eng.* **29**:146-148.

11

Scale-up of bioprocesses

The previous chapters in this book have concerned stoichiometric, thermodynamic and kinetic analysis of the production microorganisms, as well as the operation of small-scale bioreactors. This is a necessary basis for exploitation of microorganisms in fermentation processes. The ultimate goal for process development is, however, the realization of a large-scale production. Scale-up is a very difficult task. Certainly, you would like to follow the advice given by H. Baekeland, the inventor of Bakelit, in 1916: *"Commit your blunders on a small scale and make your profits on a large scale"*.

One might regard scale-up as more an art than a science (Humphrey, 1998). This should not be understood in the sense that good engineering judgement is not helpful, on the contrary. However, many different aspects need to be taken into account (Leib et al., 2001), and the final scale-up will necessarily be a delicate compromise between inherently conflicting desirable characteristics. Certainly, this is a task for the experienced engineer. The purpose of this chapter is to give an understanding of the fundamental problems that arise when a process is scaled-up, and to provide some useful tools for analysis of critical scale-up factors.

11.1 Scale-up Phenomena

Production strains are normally first selected in the laboratory, under conditions different from the conditions in the production scale. Subsequently, the strain is tested in a number of bioreactors of increasing scale, and the final process verification is carried out in a pilot plant (reactor scale 50-3000 L). A considerable scale-up of the process can be said to take place already in the lab. The scale of operation changes significantly when a culture is taken from a petri-dish via a shake flask culture to a small-scale bioreactor. The environmental conditions for growth on a petri-dish are very different than those for growth in an E-flask culture, which in turn are rather different from those in a small scale aerated bioreactor. Normally, however, the term scale-up is used for the step from small scale to production scale. Using the definition by Bisio and Kabel (1985) scale-up can be defined as: "The successful startup and operation of a commercial unit whose design and operation procedures are in part based upon experimentation and demonstration at a smaller scale of operation." Scale-up does not only involve pure engineering considerations, but certainly also economic considerations. For instance, a new medium formulation may be found

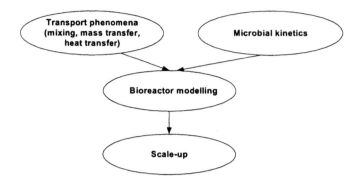

Figure 11.1. Schematic representation of analysis steps in scale-up

necessary since the costs for the medium used in the lab may be prohibitive for a large-scale operation. These additional economic restraints should be kept in mind, although we will restrict the discussion to the engineering aspects of scale-up.

The phenomena that need to be taken into account during scale-up of a fermentation process can be divided into physical processes (transport phenomena) and metabolic processes (microbial kinetics). These need to be combined in order to properly model the reactor and allow scale-up as schematically shown in Fig 11.1. The physical processes are typically described by classical mechanical or chemical engineering, and there are mathematical models of varying complexity available to describe these phenomena. The metabolic phenomena, on the other hand, are not *per se* scale-dependent. However, as a consequence of scale dependent transport phenomena, the local environment surrounding the cell will be different in a large-scale bioreactor than in a small-scale, typically well-mixed reactor. This changed environment may in turn cause metabolic changes. The consequences of these are sometimes difficult to predict, mainly due to insufficient experimental data. An important task during scale-up is therefore to identify potential gaps in knowledge, and take proper actions to acquire the missing information. If no uncertainties can be accepted, there is always the option of scaling-up by simply multiplying lab-scale production units. This, however, is expensive, and scale-up will therefore normally involve a substantial increase in reactor size.

11.2 Bioreactors

11.2.1. Basic requirements and reactor types

In the bioreactor a high product yield, a high productivity and a high reproducibility of the desired fermentation process should be achieved. These overall tasks can be decomposed into a number of subfunctions (Kossen, 1985, Lidén, 2001), and such a list of basic tasks is given in

Table 11.1. Basic tasks of the bioreactor

Function	Comment
Containment	Ensurance of sterility
Introduction of gaseous reactants	Typically oxygen, but also in some cases the carbon source e.g. methane
Introduction of liquid reactants	Typically the carbon source is a liquid sugar solution
Removal of gaseous products	Carbon dioxide is the most common gaseous product to be removed
Control of the physical environment	Temperature and pH normally need to be controlled. Shear rates may need to be limited
Suspension	Cells and particulate matter need to be suspended
Dispersion	Mixing of two-phase systems

Table 11.1. A number of different reactor types have been developed to fulfill these tasks. A rich source of inspiration for the design of novel bioreactors, is the great variety of natural bioreactors. As examples of these can be mentioned ponds, calf stomachs or termite guts (Cooney, 1983, Brune 1998).

Most bioreactors fall into one of the following categories; unstirred vessels, stirred vessels, bubble columns, airlift reactors, membrane reactors, fluidized beds, or packed beds. The reactor types differ primarily with respect to mode of agitation and aeration. In mechanically agitated reactors, mixing is obtained via internal stirrers, whereas in pneumatically agitated reactors the flow is achieved by the aeration only. In loop reactors part of the liquid is continuously withdrawn by means of a pump to an external circulation loop. Aeration, substrate addition and heat transfer may all be placed in the external loop.

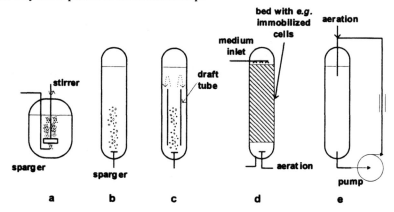

Fig. 11.2. Schematic representation of some reactor types. a) stirred tank reactor, b) bubble column, c) airlift reactor, d) trickle bed reactor, e) loop reactor with external circulation and heat exchanger.

11.2.2. The stirred tank bioreactor

The most important type of bioreactor in use in industry today is still the stirred tank bioreactor (Fig. 11.3). We will therefore focus on this reactor type in the rest of this chapter.

The reactor is typically cylindrical, with a ratio of reactor diameter to height – the so-called aspect ratio – between 1:2 and 1:5. The construction material is usually Type 316L stainless steel. The reactor has a round-bottom to facilitate cleaning and sterilization in place, and to avoid stagnant zones during operation. The reactor is equipped with a stirrer, consisting of a shaft upon which one or more impellers are placed. The stirrer shaft may be either top-mounted or bottom-mounted (most common). It is essential that the shaft enters the reactor in such a way that contamination is prevented. Typically, a double mechanical seal is used. The bioreactor may also be equipped with a mechanical foam breaker, which could be mounted on the stirrer shaft. A simple foam breaker could look like an extra, narrow impeller, but, located above the liquid surface level and with a larger diameter.

Fig. 11.3. An industrial scale bioreactor for production of enzymes. Picture courtesy of Novozymes.

Fig. 11.4. The inside of a stirred tank bioreactor. The impeller shaft with two six-bladed Rushton turbines, baffles and internal cooling coils can be seen in the picture. Picture courtesy of Novozymes.

Cooling (or heating) can take place through the reactor wall or by the use of internal coils. Alternatively, liquid may be pumped out of the reactor and cooled in an external heat exchanger. For small-scale reactors (below a few m^3) it is normally sufficient with wall cooling, whereas for large reactors it may be necessary to use also internal cooling coils. The reactor is often equipped with baffles (see Fig. 11.4). Typically 4 equally spaced baffles are used. The purpose of the baffles is to break the vortex that would otherwise form in the reactor, and which would decrease the mixing efficiency. The baffle width is normally 1/12 to 1/10 of the tank diameter.

There are several different types of impellers available (see Fig. 11.5). Their characteristics vary with respect to flow pattern, capacity for suspension and capacity for dispersion. Two main types of impellers can be distinguished; axial flow impellers (e.g. propellers) and radial flow impellers (e.g. flat blade turbine impellers). The design with six equally spaced blades mounted on a disk is often called a Rushton turbine, in honor of one of the pioneers in the field. The impeller diameter is typically around 1/3 of the reactor diameter for Rushton turbines, whereas fluid foil impellers may have a diameter exceeding half of the tank diameter. The characteristics of some impeller types are given in Table 11.2.

Aeration takes place by introducing air (or possibly oxygen) via a sparger, which is located below the lowest of the impellers. The sparger may consist of a single open tube, or a ring with fine orifices. The ring sparger typically has a diameter slightly smaller than the impeller.

Table 11.2. Impeller characteristics

Characteristics	Propeller	Disk turbine
Flow direction	axial	radial
Gassing	less suitable	highly suitable
Dispersing	less suitable	highly suitable
Suspending	highly suitable	less suitable
Blending	highly suitable	suitable

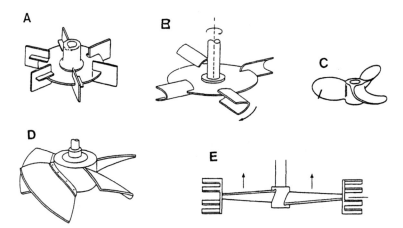

Fig. 11.5. Impellers used in stirred tank bioreactors. (a) Six-bladed Rushton turbine, (b) four-bladed Scaba agitator, (c) marine impeller (or propeller), (d) five-bladed Prochem Maxflo T, (e) Intermig. Adapted from Nienow 1990 and McDonough (1992)

The ratio between the volumetric air flow rate, v_g, and the cross sectional area of the reactor is called the superficial gas velocity, u_s. The superficial gas velocity should not be too large, to ensure an efficient dispersion and utilization of the gas. At too high superficial gas velocities, the impeller becomes fully surrounded by gas, and the dispersion capacity falls dramatically. This phenomenon is called flooding.

11.3. Physical Processes of Importance for Scale-up

A number of factors of a physical nature change with the scale of the reactor. In terms of a rigorous mathematical treatment, these problems are addressed by solving the governing equations of motion for the fluid in the reactor. This is the field of computational fluid dynamics (CFD). The mathematical treatment of flow phenomena is rather complex, and considerable fundamental problems relating to turbulence theory and multi-phase flow still remain to be solved. We will therefore in the following mainly resort to simplified models to illustrate the essential changes that occur with respect to mixing, power consumption, heat transfer, mass transfer and flow patterns in a stirred tank reactor as the scale changes.

11.3.1. Mixing

With *mixing* is understood *the process of achieving uniformity*. Mixing processes can be divided according to the number of phases involved in the mixing process; i.e. single phase liquid

Scale-up of Bioprocesses

mixing, liquid-liquid mixing, gas-liquid mixing, solids-liquid mixing and three-phase mixing (see Nienow et al., 1997). A bioreactor normally contains three phases, and therefore true uniformity cannot be achieved at the microscale. However, at very high stirring rates, the concentration within the liquid and gas phases will be approximately constant throughout the reactor volume. In the ideal tank reactor discussed in Chapter 9 complete mixing is assumed to take place instantaneously, i.e. any medium component added to the reactor is assumed to be homogeneously distributed in the system immediately. This assumption works reasonably well for small-scale (1-2 L) intensly stirred bioreactors, with mixing times in the order of 1 s, since most biological processes are relatively slow processes at a temperature of 30 - 40 °C. However, in large-scale systems, the time for achieving homogeneity in the reactor can be in the order of minutes and may no longer be neglected. Mixing depends on the scale of homogeneity considered. *Macromixing* refers to mixing on a conveniently observable scale, whereas *micromixing* refers to mixing on the molecular scale. On a macro-scale, mixing is achieved by bulk flow convection, i.e. the distribution caused by the main flow pattern in the reactor. Furthermore, in a turbulent flow field which is predominantly the case for stirred bioreactors, the mixing caused by turbulent eddies is highly important for mixing down to the Kolmogorov size (see Note 10.1.). The final micromixing below this size is achieved by molecular diffusion. However, since the length-scale over which this final mixing occurs depends on the turbulent eddy size, macromixing indirectly affects also micromixing (see Note 11.3).

One way of quantifying mixing is to make a pulse addition of a tracer, e.g. a dye, into the reactor, and then monitor the gradual return of homogeneity. The degree of mixing (or degree of homogeneity), m, is used to quantitatively describe such experiments. The degree of mixing is defined by

$$m = \frac{s(t) - s_0}{s_\infty - s_0} \tag{11.1}$$

where $s(t)$ is the concentration of the tracer (at a measuring point) at time t, s_0 is the initial concentration and s_∞ is the concentration for $t \to \infty$ where m will approach 1. The *mixing time*, t_m, is defined as the time needed to obtain a value of m larger than a specific threshold value. This value is in principle arbitrary. Commonly used values are e.g. 0.95 or 0.99. Here we will define t_m as the time needed to achieve a value of m equal to 0.632 (= $1-e^{-1}$). Several investigations have shown that mixing can be approximately regarded as a first order process. With the definition used here, t_m will therefore be the inverse of a first order rate constant for the mixing process, making it convenient to compare the rate of mixing with the rate of mass transfer, or the rate of reaction. The conversion of one definition to another is shown in Note 11.1. For a one-phase stirred tank system with baffles, the mixing time has been found to be approximately inversely proportional to the stirrer rate, N (unit s^{-1})[1], i.e.

$$t_m \propto \frac{1}{N} \tag{11.2}$$

---

[1] The stirrer rate is often given in the unit rpm = rotations per minute

The proportionality constant depends to a large extent on the impeller type. Another characteristic that depends on the impeller type is the *pumping capacity*, v_{pump}, which is defined as the volume of liquid that gets expelled from the impeller per unit time (m³ s⁻¹). It is to be expected that

$$v_{pump} \propto d_s^3 N \tag{11.3}$$

where d_s is the stirrer diameter. The proportionality constant is called *the flow number*, N_f, and it depends on the impeller type and on the viscosity of the medium. For a Rushton turbine and a low-viscosity medium, it has been found that $N_f = 0.72$. The *circulation time*, t_c, is defined as

$$t_c = \frac{V}{v_{pump}} \tag{11.4}$$

It is realized from Eqs. (11.2) - (11.4) that t_m is proportional to t_c for fully turbulent flow in a baffled tank reactor. As a first approximation, mixing can be assumed to have occured within 4 t_c.

Note 11.1: Mixing as a first-order process
If the mixing process is approximated by a first-order process, the deviation of the concentration of a tracer at a time t, $s(t)$, from its final value, s_∞, is given by

$$\frac{d[s_\infty - s(t)]}{dt} = -k[s_\infty - s(t)]; \; s(t=0) = s_0 \tag{1}$$

for which the solution is

$$\frac{s_\infty - s(t)}{s_\infty - s(t)} = e^{-kt} \Leftrightarrow 1 - m = e^{-kt} \tag{2}$$

Rearrangement of Eq. (2) gives

$$t_m = -\frac{1}{k}\ln(1-m) \tag{3}$$

which shows that for $m = 1 - 1/e = 0.632$, we have $t_m = 1/k$. Furthermore, by using Eq. (3) we can calculate the mixing time for any value of m^* when the mixing time is known for m:

$$t_m^* = t_m \frac{\ln(1-m^*)}{\ln(1-m)} \tag{4}$$

For example, for m* = 0.99 (and m = 0.632) we get

$$t_m^* = \left(\frac{\ln(0.01)}{-1}\right) t_m = 4.6 t_m \tag{5}$$

Note 11.2: Methods for characterizing mixing

Many different characterization methods have been applied to describe the flow pattern and mixing in bioreactor systems. The simplest methods are based on tracer techniques in which a tracer is added to the bioreactor and its concentration is measured as a function of time. A short description will be given of these methods, which are easily implemented and therefore recommendable for initial studies of mixing phenomena. For detailed studies of the mixing, characterization of the entire flow field is necessary as discussed in section 11.3.5. Many tracer techniques have been successfully applied to determine the mixing time in a bioreactor:

1. *Conductivity methods*, based on electrolytes as tracer
2. *pH methods*, based on acids or bases as tracer
3. *Coloration methods*, in which dyes are used as tracer
4. *Heat pulse methods*, in which a warm liquid is used as a tracer
5. *Isotope methods*, based on radioactive isotopes as tracer

All these methods are easy to implement, but unfortunately each has one or more drawbacks in connection with measurement in fermentation media. The choice of method should therefore be based on an evaluation of the drawbacks (see below) when the method is applied in the actual system.

The *conductivity method* is inexpensive and easily implemented, but it has the disadvantage that most fermentation media are themselves good conductors, and the sensitivity is therefore poor.

The *pH method* is very easily implemented since most bioreactors are equipped with pH electrodes. By measuring the change in pH after addition of base (or acid), the mixing time can be determined. Pulses of bases (or acids) can normally be added to a bioreactor without seriously affecting the fermentation process, and the method can therefore be applied under real process conditions. However, microbial activity may influence the results since many microorganisms produce acids as metabolic products, and it is therefore important that the mixing time is much smaller than the characteristic time for acid production. The major disadvantage of the pH method is that most fermentation media have a high buffer capacity, and large pulses are therefore required in order to obtain good sensitivity.

The *coloration method* is based on measurement of an inert dye, e.g. a fluorophore such as NADH, riboflavin or coumarin. There are commercially available fluorescence probes that can be inserted through standard ports in most bioreactors, and it is therefore possible to quantify the mixing time under real process conditions. However, for many fermentation media the background fluorescence is high and the sensitivity may therefore be poor. Instead of coloring the medium it is also possible to decolorize the medium. This can be done by for instance lowering pH by adding acid in an alkaline phenolphtalein solution. This is very useful for detecting stagnant zones in a reactor, since these zones will remain colored (Stein, 1992).

The *heat pulse method* is based on following the spread of warm water by high precision temperature sensors. One problem with this method is to achieve a local distribution of heat.

The *isotope method* is based on addition of radioactive isotopes and measurement of the radioactivity using scintilation counters. An advantage of this method is that the sensor can be placed outside the bioreactor, and it can therefore easily be fitted. With the right choice of isotope a good sensitivity can be obtained for most systems and there will be no traceable influence on the microbial activity from the radiation. A disadvantage of the method is, however, the concern caused in production facilities when radioactive isotopes are added to the medium.

There have been relatively few studies concerning mixing in aerated systems compared to the case of one-phase systems. At high aeration rates, mixing in the liquid phase is significantly improved. However, the positive effects on mixing take place after flooding of the impeller. This is not a desirable situation, since the dispersion of gas drastically decreases at that point giving poor gas-liquid mass transfer. At moderate aeration rates, below flooding, aeration appears to only slightly improve mixing in the liquid phase (Nienow, 1997; Vrabel et al 2000).

Note 11.3. Micromixing and macromixing.
The turbulent mixing contributes to mixing down to the Kolmogorov size of the eddies. As discussed in Note 10.1 this size is given by

$$l_{min} = \frac{\eta^{\frac{3}{4}}}{\rho_l^{\frac{1}{2}}} \left(\frac{P_g}{V_l} \right)^{-\frac{1}{4}} \quad (1)$$

Molecular diffusion is the mechanism for mixing below this length scale. The diffusion time is highly dependent on the length over which diffusion occurs. The time required for diffusion, t_D, into a sphere of radius l can be obtained from solving Fick's second law, which gives:

$$t_D = \frac{l^2 \beta}{D_i} \quad (2)$$

The value of β depends on the stated requirements for completion of mixing. The requirement that all molecules within the sphere should have been exchanged with the molecules in the surrounding, gives $\beta = 0.5$ (Bourne, 1997). For a power input of 1 W L^{-1}, a viscosity of 1 mPa s (typical of water), and a density of 1000 kg m^{-3} we get l_{min} = 32 µm. To get an estimate of the time for the final diffusive micromixing of e.g. glucose, we can assume that this occurs in a sphere with the radius equal to half l_{min}. The time for micromixing of glucose (D = 0.7 10^{-9} m^2 s^{-1} in water at 25 °C) is thus obtained from Eq. (2) to t_D = 0.18 s. This time is normally negligible compared to t_m, which shows that the final molecular diffusion is a rapid process compared to the mixing processes by convective transport and turbulent mixing in most stirred tank reactors.

The measurement of mixing times by tracer experiments is useful. It is, however, important to keep in mind the following inherent limitations in these experiments:
1. Mixing is a local property, which depends on the chosen location for measuring the tracer concentration. A more strict definition should therefore be based on the averaging over many sensors in the reactor (Lundén et al, 1995). For practical reasons, however, this is seldom done.
2. The mixing time is dependent on the point of addition of tracer. A badly chosen point of injection may give a significantly higher value of t_m than a well-chosen injection point.
3. Mixing is a function on the scale of magnification, and the size of the sensor will decide what level of mixing is studied (cf. Note 11. 3).

11.3.2. Power consumption

The mixing process requires energy. The power input, P (W), to a stirred tank bioreactor consists of two parts; the power for stirring and the compression power for aeration. The first term is completely dominating for stirred tank reactors, whereas the compression power is dominant in bubble columns and airlift reactors. Additional energy for liquid circulation may also be supplied

Scale-up of Bioprocesses

by external pumps in e.g. airlift fermentors. The power input for stirring is a major cost for aerobic bioprocesses, and it depends on several factors. Most important are the stirrer rate, N, the impeller diameter, d_s, the density of the fluid, ρ_l, and the viscosity of the fluid, η. By dimensional analysis (see e.g. Geankoplis, 1993), it is possible to combine these factors into dimensionless numbers, and the power consumption for stirring is normally expressed as a function of the dimensionless *power number*, N_p (sometimes also called the Newton number) defined as

$$N_p = \frac{P}{\rho_l N^3 d_s^5} \tag{11.5}$$

The power number can normally be expressed as a function of the Reynolds number, Re_s, defined by Eq. (11.6).

$$Re_s = \frac{\rho_l N d_s^2}{\eta} \tag{11.6}$$

The variation of N_p with Re_s is shown schematically in Fig. 11.6. The flow in a stirred tank reactor is fully turbulent if Re_s defined in Eq. (11.6) is larger than 10^4. This is true in most cases

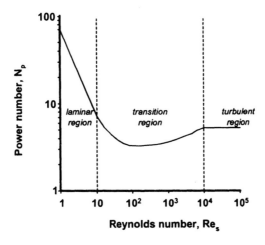

Fig 11.6. Graph showing schematically the variation of the power number, N_p, with Reynolds number, Re_s, for a bioreactor equipped with one six-bladed Rushton turbine impeller and baffles

Table 11.3. Power number at fully turbulent flow for various impeller designs in a bioreactor system equipped with baffles.

Impeller type	N_p	Reference
Rushton	5.20	Nienow (1990)
Intermig	0.35	Nienow (1990)
Prochem	1.00	Nienow (1990)
Marine impeller	0.35	Schügerl (1991)

for stirred tank bioreactors. In a baffled reactor, only the inertial forces are important for a fully turbulent flow, and N_p has a constant value independent of Re_s. The power number will, however, depend on the impeller type (see Table 11.3). In the laminar flow regime, $Re_s < 10$, viscous forces dominate and the power dissipation will depend on Re_s. For laminar flow it has been shown (Rushton et al, 1950a) that

$$N_p \propto \frac{1}{Re_s} \qquad (11.7)$$

In between fully laminar flow and fully turbulent flow, there is a transition regime, in which both inertial and viscous forces need to be considered (Rushton et al, 1950b). The functional relation is more complex in this case. For a known power number, the power consumption can obviously be calculated from:

$$P = N_p \rho_l N^3 d_s^5 \qquad (11.8)$$

Influence of multiple impellers
Equation (11.8) holds for single-impeller systems. In practice, several impellers are normally put on the same shaft. For systems with multiple impellers, the situation is more complicated (see e.g., Nienow and Lilly, 1979), but as a first approximation the power input can be calculated by just multiplying the power consumption for a single impeller with the number of impellers. This is an overestimation, since the power consumption increases slightly less than proportional to the number of impellers.

Influence of aeration
The power consumption for stirring is affected by aeration. The dispersion of gas in the liquid causes a decrease in liquid density. Furthermore, when gas is sparged to a tank, gas bubbles are drawn to regions of low pressure. This results in the formation of gas-filled areas (called *cavities*) behind the stirrer blades. The formation of these cavities depends on the ratio between the volumetric gas flow rate and the pump capacity. This ratio is often expressed as a dimensionless group, the so-called *aeration number* N_A:

$$N_A = \frac{v_g}{Nd_s^3} = \frac{v_g}{v_{pump}} N_f \qquad (11.9)$$

where N_f is the flow number, (Eq. 11.3).

Scale-up of Bioprocesses

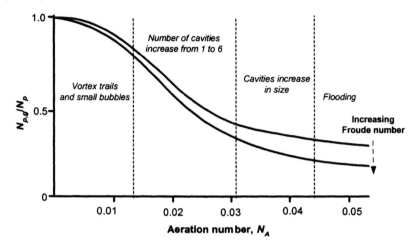

Fig. 11.7. The ratio $N_{p,g}/N_p$ shown (qualitatively) as a function of the aeration number for a six-bladed Rushton turbine. The Froude number is given in Eq. (11.28). (Adapted from Ekato, Handbook of mixing technology (www.EKATO.com)).

The cavities will grow with increasing N_A until eventually the impeller becomes fully immersed in gas, a condition called flooding. The power number $N_{p,g}$ gradually falls with increasing N_A in a manner shown by Fig. 11.7. Typically, a fluid foil impeller will have a significantly smaller drop in the ratio $N_{p,g}/N_p$ with increasing aeration rate than a Rushton turbine (see Problem 10.2).

An approximate empirical correlation describing the relation between aerated, P_g, and unaerated, P, power consumption is (Michel and Miller, 1962):

$$P_g = k \left(\frac{P^2 N d_s^3}{v_g^{0.56}} \right)^{0.45} \quad (11.10)$$

where the value of k depends on the impeller type.

Power input in bubble columns

The power input by the aeration is normally small in stirred tank reactors. In a bubble column, on the other hand, all mixing energy is supplied by the aeration. The compression power input for aeration in a bubble column operating at moderate superficial velocities can be expressed as (Chisti and Moo-Young, 2001)

$$P_{comp} = g\rho_l u_s V \quad (11.11)$$

where g is the gravitational acceleration (9.82 m s^{-2}).

Example 11.1. Power input to a laboratory bioreactor
In a study of yeast physiology, a 1-L (liquid volume, 800 mL) laboratory bioreactor is used. It is equipped with one Rushton turbine with $d_s = 47$ mm. The density and the viscosity of the medium containing yeast cells are approximately the same as for water, i.e. 997 kg m^{-3} and 10^{-3} kg m s^{-1}, respectively. With a stirring speed of 1200 rpm, the stirrer Reynolds number is therefore (at 25 °C):

$$\text{Re}_s = 44047 \tag{1}$$

This is well within the turbulent regime, and a power number of $N_p = 5.2$ can be used. The power input is therefore

$$P = 5.2 \cdot 997 \text{ kg m}^{-3} (20 \text{ s}^{-1})^3 (47 \cdot 10^{-3} \text{ m})^5 = 9.51 \text{ W} \tag{2}$$

For gassed conditions, we assume that the power input is 50% of the power input at ungassed conditions, and therefore the power input per volume of medium is

$$\frac{P_g}{V} = 5.95 \text{ W L}^{-1} \tag{3}$$

This is a typical figure for laboratory bioreactors. When the bioreactor is used for aerobic yeast fermentation, it is sparged with air at 1 vvm, i.e., 800 mL min^{-1}. The tank diameter is 106 mm, and the superficial gas velocity is therefore

$$u_s = \frac{13.33 \cdot 10^{-6}}{\pi \cdot 0.106^2 / 4} = 1.51 \cdot 10^{-3} \text{ m s}^{-1} \tag{4}$$

Assuming that the medium is noncoalescing, we can calculate the volumetric mass transfer coefficient for the bioreactor using Eq. (10.27) and the parameter values listed in Table 10.4 (try also to calculate $k_l a$ using the parameter sets specified for noncoalescing media in Table 10.4):

$$k_l a = 0.0018 \, (1.51 \cdot 10^{-3} \text{ m s}^{-1})^{0.3} (5.95 \cdot 10^3 \text{ W m}^{-3})^{0.7} = 0.113 \text{ s}^{-1} \tag{5}$$

In Example 10.1 it was found that the oxygen requirement of a rapidly respiring yeast culture is approximately 79.3 mmoles of O_2 L^{-1} h^{-1}. With the calculated volumetric mass transfer coefficient above, we therefore find that under steady-state conditions the dissolved oxygen concentration in the medium is

$$s_0 = s_o^* + \frac{q_0}{k_l a} = 0.26 \text{ mmoles L}^{-1} - 79.3 \text{ mmoles L}^{-1} \text{ h}^{-1}/(406.8 \text{ h}^{-1}) = 0.065 \text{ mmoles L}^{-1} \tag{6}$$

which corresponds to 25% of the saturation value. This will normally not result in oxygen limitation. However, if the dissolved oxygen concentration is to be maintained at a higher level it is necessary to modify the bioreactor setup. This is probably done most easily by mounting an additional Rushton turbine on the impeller shaft. Thereby the power input increases by a factor of approximately 2. This leads to an increase of approximately 62% in the $k_l a$ value, and the dissolved oxygen concentration can be maintained at around 55% of the saturation value.

Example 11.2 Maintaining mixing times at different scales.

A process in the lab-scale reactor described in Example 11.1 is to be scaled up to a 1 m^3 reactor, with the same aspect ratio and the same impeller diameter to tank diameter ratio. It is desirable to maintain the same mixing time in the large-scale reactor as in the lab-scale reactor. Since the mixing time is inversely proportional to the stirrer rate, N, (cf. Eq 11.2) the stirrer rate in the large reactor need to be the same as in the small scale, i.e. N= 20 s^{-1}.

The impeller diameter in the 1 m^3 reactor is 0.047 10 = 0.47 m. For the large scale reactor we get

$$\text{Re}_s = \frac{997 \cdot 20 \cdot (0.47)^2}{10^{-3}} = 4.4 \; 10^6 \tag{1}$$

This is clearly in the turbulent regime, and the power number can be considered constant, Np = 5.2. The power consumption can be calculated to

$$P = 5.2 \; 997 \text{ kg m}^{-3} \; (20 \text{ s}^{-1})^3 \; (0.47 \text{ m})^5 = 951 \text{ kW} \tag{2}$$

If we assume that the power consumption falls with 50% at gassed conditions, we get

$$\frac{P_g}{V} = 595 \text{ kW m}^{-3} \tag{3}$$

This power consumption is far too high to be acceptable for a large-scale bioreactor. As a rule of thumb, the power consumption should be about 1-5 kW m^{-3} for large-scale bioreactors. The important conclusion from this calculation is that *it is not possible to maintain the same mixing time in a large-scale bioreactor as in a well-mixed laboratory scale reactor if the scale-up is made using geometrically similar reactors*. If instead the same power consumption per volume is maintained during scale-up, the mixing time can be shown from Eq. (11.2) and (11.8) to increase approximately according to

$$\frac{t_{m,l}}{t_{m,s}} = \left(\frac{d_{s,l}}{d_{s,s}}\right)^{2/3} \tag{4}$$

where $t_{m,l}$ and $t_{m,s}$ are the mixing times for the large scale reactor and small scale reactors, and $d_{s,l}$ and $d_{s,s}$ are the respective stirrer diameters. In the current example, it can be estimated that the mixing time in the 1 m^3 reactor is approximately 5 times that in the 1 L reactor if the power consumption of 5.95 kW m^{-3} is maintained.

11.3.3. Heat transfer

Biological reactions are very temperature sensitive and a careful temperature control is therefore absolutely essential. Heat is generated by agitation of the medium and – as discussed in Chapter 4 – by the large heat of reaction, which often accompanies bioreactions. Hence cooling is required, and removal of heat is done by heat exchange.

Heat transfer correlations

We will first consider heat transfer through the reactor wall. The rate of heat transfer, Q_{HE}, (W) is proportional to surface area, A_{HE}, and the temperature difference, ΔT, across the surface over which transfer occurs as described by Eq (11.12)

$$Q_{HE} = U_{HE} A_{HE} \Delta T \quad (11.12)$$

In the chemical engineering literature, the proportionality constant, U_{HE}, (W m^{-2} K^{-1}) is called the overall heat transfer coefficient. The overall heat transfer can be decomposed into heat transfer through a boundary layer on the inside of the wall, through the wall itself and through a boundary layer on the outside of the wall. It can be shown that the overall heat transfer coefficient can be calculated from:

$$\frac{1}{U_{HE}} = \frac{1}{\alpha_i} + \frac{d_w}{\lambda_w} + \frac{1}{\alpha_o} \quad (11.13)$$

where α_i is the heat transfer coefficient on the inside (W m^{-2} K^{-1}), α_o is the heat transfer coefficient on the outside, d_w is the wall thickness and λ_w (W m^{-1} K^{-1}) is the heat conductivity of the wall material. In many cases, the first term of Eq. (11.13) will be the largest, i.e. the heat transfer rate will be largely determined by the heat transfer coefficient on the inside. However, stainless steel has a rather low heat conductivity (<20 W m^{-1} K^{-1}), which is several times lower than that of steel. The resistance for conduction through the wall (the second term on right hand side of Eq. 11.13) must be taken into account when $d_w > 5$ mm. Fouling of the wall, i.e. deposition of a biofilm on the wall, will give a further decreased value of U_{HE}. Fouling not only results in a thickening of the wall - the thermal conductivity of the deposited film is typically significantly lower than that of the wall material.

The values of the internal and external heat transfer coefficient depend on the flow rate and material properties such as viscosity, η, heat conductivity, λ_l, and heat capacity, C_p (J kg^{-1} K^{-1}), of the fluid. Correlations are normally based on dimensionless numbers, which give *the Nusselt number, Nu,* as a function of the Reynolds number, Re_s, (cf. Eq. 11.6) and *the Prandtl number, Pr.* In addition, a correction is sometimes made for viscosity changes that may occur close to the reactor wall due to temperature effects or lower shear rate (see further 11.3.5.). One correlation given for a stirred baffled tank of standard geometry ($d_s/d_t = 1/3$) is (Chilton et al., 1944)

$$Nu = 0.36 \, Re_s^{0.66} \, Pr^{0.33} \left(\frac{\eta_l}{\eta_{wall}}\right)^{0.14} \quad (11.14)$$

where η_{wall} is the viscosity close to the wall, and η_l is the bulk viscosity. (For a low viscosity fluid like water the last factor will be 1.) The dimensionless numbers are defined as

$$Nu = \frac{\alpha_i d_t}{\lambda_l} \tag{11.15}$$

$$\Pr = \frac{C_p \eta_l}{\lambda_l} \tag{11.16}$$

As size increases, cooling often becomes a larger problem than oxygen transfer. This is due to the fact that the ratio between surface area and volume is inversely proportional to the length scale of the reactor, if scale up is made with a conserved aspect ratio (cf Table 11.4). Since the metabolic heat production grows linearly with volume, it is necessary to increase the heat transfer capacity either by increasing the temperature difference across the wall, or by increasing the heat transfer coefficient. The cooling water is normally available at around 10-15 °C, and this temperature cannot be easily or cheaply lowered. The internal heat transfer coefficient, α_i, can be increased by increasing the stirrer rate, as seen from Eq (11.14). However, α_i is only a weak function of the stirring power, P. From Eq (11.14) and (11.8) it can be estimated that $\alpha_i \propto P^{0.15}$. A doubling of the power input thus only increases the heat transfer by approximately 10%. Furthermore, the increased power dissipation due to mixing also need to be transferred. In practice, it is therefore necessary to complement the external wall cooling with internal coils above a certain reactor size (see Example 11.3).

A correlation for calculation of heat transfer coefficients for internal coils will necessarily be more complicated, since the flow pattern in the tank is influenced by the added internal structures. An approximate correlation is given by Oldshue and Gretton (1954):

$$\frac{\alpha_i d_{coil}}{\lambda_l} = 0.17 \, \mathrm{Re}_s^{0.67} \, \Pr^{0.37} \left(\frac{d_s}{d_t}\right)^{0.1} \left(\frac{d_{coil}}{d_t}\right)^{0.5} \tag{11.17}$$

where the left hand side of Eq (11.17) is in fact a Nusselt number based on the diameter of the cooling tubes, d_{coil}. A word of caution is that internal coils may influence the flow pattern in a non-desirable way, which could give stagnant zones close to the reactor wall for non-Newtonian media (Kelly and Humphrey, 1998). Furthermore, the desire to obtain a good mixing may be in conflict with the desire to obtain a good heat transfer. Axial flow impellers, e.g. fluid foil impeller, which are good for mixing, may be less suitable for good heat transfer to internal coils (Oldshue, 1989).

Table 11.4. Ratio of total outer surface area to volume for geometrically similar cylindrical reactors (h_t:d_t = 1)

Reactor volume (m³)	Surface area/ volume (m⁻¹)
0.001	55.5
0.1	11.9
1	5.5
50	1.5
100	1.2

Example 11.3. Estimation of typical cooling requirements.
We will here consider approximate cooling requirements for two stirred tank reactors; one of size 100 L, and the other of size 50 m^3. Both reactors should be able to handle an aerobic process with a cell density of 20 g DW L^{-1} at a specific growth rate of 0.2 h^{-1}. The overall reaction can be regarded as simply production of biomass according to

$$-CH_2O - Y_{so}O_2 - Y_{sn}NH_3 + Y_{sx}CH_{1.8}O_{0.5}N_{0.2} + Y_{sc}CO_2 + Y_{sw}H_2O \quad (1)$$

The two major sources of heat generation in the bioreactor are the metabolic heat evolution and the dissipated stirring power. The cooling capacity of the reactor should thus exceed the sum of these two heat sources. The metabolic heat evolution for an aerobic process will depend on the cell density, x, and the specific oxygen uptake rate, r$_O$. As shown in Chapter 4, a reasonable approximation of the metabolic heat evolution, Q_{met}, can be obtained by the simple relation

$$\frac{Q_{met}}{V} = 460 \cdot 10^3 r_o x \quad (2)$$

where (if maintenance oxygen requirements are neglected)

$$r_o = Y_{xo}\mu \quad (3)$$

For a value of Y_{sx} = 0.56 C-mol/C-mol in Eq. (1), we get Y_{xo} = 0.735 mol/C-mol. By combining Eqs. (2) and (3) we thus get

$$\frac{Q_{met}}{V} = \frac{460 \cdot 10^3 \cdot 0.735 \cdot 0.2 \cdot 0.813 \cdot 1000}{3600} = 15270 \text{ W m}^{-3} \quad (4)$$

A typical value for the stirring power input is 1 kW m^{-3}. The total cooling requirement is thus approximately 16 kW m^{-3} for both reactors, or from Eq. (11.12)

$$U_{HE}\Delta T = \frac{Q_t/V}{A_{HE}/V} \quad (5)$$

By assuming that 50% of the total surface area from Table 11.3 is available for cooling, we get the following requirements for the product of the heat transfer coefficient and the temperature difference across the wall:

$U_{HE}\Delta T > 2566$ W m^{-2} for V = 0.1 m^3 and $U_{HE}\Delta T > 20360$ W m^{-2} for V = 50 m^3

With an assumed maximum value of ΔT of about 20 K, this gives the requirement
$U_{HE} > 1067$ W m^{-2} K^{-1} for the larger reactor. Internal cooling coils may be considered to ensure a sufficient cooling in this case.

11.3.4. Scale-up related effects on mass transfer

The fundamentals of mass transfer were covered in Chapter 10. Here we will only point out some specific scale dependent mass transfer phenomena, which deserve special attention. The concern over oxygen transfer was certainly the first scale-up problem faced by the early pioneers trying to scale-up the production of penicillin in the 1940s (Humphrey, 1998). Initially, it was a problem to give even a crude quantitative description of the transfer rate of oxygen from the gas phase to the liquid. Systematic studies resulted in very useful correlations of the kind discussed in Chapter 10, i.e. that the overall value of $k_l a$ depended on the specific power input, P/V, and the superficial gas velocity, u_s by equations of the type given by Eq. (11.18)

$$k_l a = k u_s^\alpha \left(\frac{P}{V_l}\right)^\beta \tag{11.18}$$

Although very useful, these correlations should be used with care when estimating mass transfer at widely different scales.

Superficial gas velocity
The correlations of the kind given by Eq. (11.18) are only valid as long as the impeller is not flooded, i.e. totally immersed in gas. A perhaps not immediately obvious consequence of scale-up is that the ratio between aeration rate and liquid volume may need to be changed. The aeration rate is often specified in *vvm*, or *volume of gas per volume of liquid per minute* (e.g. m^3 gas m^{-3} liquid min^{-1}). As seen in Table 11.5, the superficial gas velocity, u_s, increases as $V^{1/3}$ if the same vvm-value is maintained. If the superficial gas velocity exceeds the bubble rise velocity, u_b, flooding will definitely occur. For a broth viscosity about 1 mPa s, u_b can be estimated to be about 0.2 m s^{-1}. Already at a superficial gas velocity approaching 25-50% of the bubble rise velocity, flooding is likely to occur. Thus, the superficial gas velocity should be kept below 0.05 m s^{-1} to safely avoid flooding for water-like broths. However, u_b depends strongly on viscosity, and the allowed superficial gas velocity may therefore be an order of magnitude lower for highly viscous media.

Surface aeration
Oxygen transfer to the liquid takes place not only via dispersed bubbles, but also via the surface. In small scale reactors (< 50 l), the contribution of this surface oxygen transfer may be quite

Table 11.5. Superficial velocity as a function of scale for an aeration rate of 1 vvm in a cylindrical stirred tank reactor with $h/d_t = 1$.

Reactor volume (L)	Superficial gas velocity (m s^{-1})
10	0.006
100	0.013
1000	0.028
10000	0.063

significant, whereas for larger reactors the contribution can be neglected. In the type of correlation given by Eq (11.18), this can be seen as a decreasing value of the exponents α and β with increasing reactor size.

Non-constant power dissipation
Another factor that needs to be considered is that the power dissipation in the reactor is spatially inhomogeneous. There is a much higher power dissipation, up to a factor 100, close to the impeller. This gives a difference in bubble size distribution, with smaller bubbles close to the impeller and bigger bubbles close to the wall. This is particularly true for a coalescing medium. Again, a lower k_la than predicted from correlations based on experiments made in smaller scale bioreactors may result.

Gradients of other compounds than oxygen
As stated in section 11.3.1, it is not possible to maintain the same mixing time in a large-scale bioreactor as in a lab-scale bioreactor. Mass transfer in the bulk liquid phase can no longer be assumed to take place instantaneously, i.e. the assumption of an ideal tank reactor is no longer valid. The increased circulation times, t_c, will cause concentration gradients for any component, which is added during operation. Besides for oxygen, this typically applies when base is added for pH control (see problem 11.4). In fed-batch or continuous processes, gradients will occur also for medium components, e.g. the carbon source added.

11.3.5. Rheology of fermentation broths

The flow pattern in a stirred tank reactor is determined by the reactor and impeller geometry, the power input to the system and the properties of the fluid. Before defining the fundamental equations governing flow, we will introduce some useful rheological concepts.

Rheology is the study of flow and deformation of matter. A *fluid* is strictly defined a material phase, which cannot sustain *shear stress, i.e.* forces acting in the tangential direction of the surface. (The forces acting in the normal direction of the surface are the pressure forces.) A fluid will respond to shear stress by continuously deforming, i.e. by flowing. The *shear rate*, $\dot{\gamma}$, (s^{-1}) in a two-dimensional flow is defined as

$$\dot{\gamma} = \frac{du_y}{dz} \quad (11.19)$$

where z is the direction perpendicular to the direction of the flow. Isaac Newton was the first to propose a relation between shear stress and the shear rate in 1687, when he stated that "the resistance (*sic* the shear stress) which arises from the lack of slipperiness originating in a fluid, other things being equal, is proportional to the velocity by which parts of the fluid are being separated from each other (*sic* the shear rate)". The *dynamic viscosity*, η, (kg m s^{-1}) of a fluid is the fundamental property which relates the *shear stress*, τ, (N m^{-2}) to the shear rate. For the one-dimensional case described in Eq. (11.19) we get:

$$\tau = -\eta\dot{\gamma} \qquad (11.20)$$

We will in the following simply call η "viscosity", but it is important not to confuse the dynamic with the *kinematic viscosity*. The latter is the ratio between the dynamic viscosity and the density of a fluid. The kinematic viscosity has the unit ($m^2\ s^{-1}$).

Note 11.4. Shear stress as a tensor property
In many textbooks outside the field of fluid mechanics, the concept of stress is mathematically abused much in the same way as done above. Stress should properly be treated as a tensor property (see e.g. Mase, 1970). The state of stress in a material phase is fully described by the stress tensor, **T** (N m^{-2}). The stress tensor includes both normal stress (i.e. pressure) and shear stress and can be written:

$$\mathbf{T} = -p\mathbf{I} - \boldsymbol{\tau} \qquad (1)$$

where p is the pressure (N m^{-2}), **I** is the identity matrix and $\boldsymbol{\tau}$ is *the shear stress tensor* (N m^{-2}). The stress acting on a surface at any point, characterized by a normal vector, **n**, can be found from the stress tensor by scalar multiplication according to:

$$\mathbf{t_n} = \mathbf{n} \cdot \mathbf{T} \leftrightarrow \mathbf{t_n} = -\mathbf{n} \cdot (p\mathbf{I} + \boldsymbol{\tau}) \qquad (2)$$

where $\mathbf{t_n}$ is the state of stress. Note that $\mathbf{t_n}$ has the same unit as **T**, i.e. N m^{-2}, but is a vector instead of a tensor. For a Newtonian incompressible fluid, the shear stress tensor can be written:

$$\boldsymbol{\tau} = -\eta(\nabla \boldsymbol{u} + \nabla \boldsymbol{u}^T) \qquad (3)$$

where $\nabla\mathbf{u}$ is the velocity gradient (s^{-1}), which can be said to be the tensor counterpart of the previously defined shear rate, $\dot{\gamma}$.

For *Newtonian fluids* the viscosity is independent of the shear rate - *i.e.* it is constant - whereas for *non-Newtonian fluids* it is a function of the shear rate. The viscosities of fluids vary over a wide range, as seen in Table 11.6. Water, and many fermentation broths containing yeasts or bacteria, can be considered to be Newtonian fluids. However, fermentations involving filamentous fungi, or fermentations in which polymers are excreted, will often exhibit non-Newtonian behaviour.

There are several different kinds of non-Newtonian behavior. The most important type is the s*hear rate dependent* viscosity. A fluid for which the viscosity decreases with increasing shear rate is called a *pseudoplastic fluid*. The opposite, *i.e.* a fluid with an increasing viscosity with increasing shear rate is called a *dilatant fluid*. Polymer solutions, or solutions containing filamentous fungi, are often pseudoplastic, whereas e.g. rice straw in water or whipped cream exhibit dilatant behavior. Certain materials do not flow until a treshold shear stress is exceeded. This is true for e.g. damp clay or cheese. Although they are not in a strict sense fluids, they behave like fluids once the yields stress is exceeded. These materials are called Bingham plastics or Bingham fluids.

Table 11.6. Viscosity of some fluids (Adapted from Johnson, 1999).

Fluid	Temperature (°C)	η (kg m^{-1} s^{-1})
Water	0	1.793 10^{-3}
	21	9.84 10^{-4}
	100	5.59 10^{-4}
Ethanol	20	1.20 10^{-3}
Glycerol	60	0.98
Sucrose solution (20 wt%)	21	1.916 10^{-3}
Sucrose solution (60 wt%)	21	6.02 10^{-2}
Olive oil	30	8.40 10^{-2}
Molasses	21	≈6.6

A typical functional relationship for shear rate dependent viscosity is the so-called *power law model* (or *Ostwald de Waele model)*:

$$\eta = K|\dot{\gamma}|^{n-1} \qquad (11.21)$$

K is the *consistency index* and *n* is the *power law index*. In accordance with previous definitions, it can be seen that if $n > 1$, the fluid is dilatant, and if $n < 1$, it is pseudoplastic. For a Newtonian fluid $n = 1$.

For Bingham fluids, the shear stress is given by

$$\tau = \tau_0 - \eta\dot{\gamma} \qquad (11.22)$$

where τ_0 is the yield stress. A material which exhibits both a yield stress and pseudoplasticity is often described by an empirical model named after Casson:

$$|\tau|^{0.5} = |\tau_0|^{0.5} + K_c|\dot{\gamma}|^{0.5} \qquad (11.23)$$

The different kinds of shear-rate dependent rheological behaviors are schematically shown in Fig. 11.8.

The viscosity may also change with time. A *decrease of viscosity with time* is called *thixotropy*, whereas *an increase of viscosity with time* is call *rheopexy*. Thixotropy, as well as pseudoplastisticy, is often observed in polymer solutions. It is caused by irreversible breaking of bonds between long macromolecules, or breakdown of hyphae. Rheopectic behavior is rather unusal, but clotting blood can be named as one example.

Scale-up of Bioprocesses

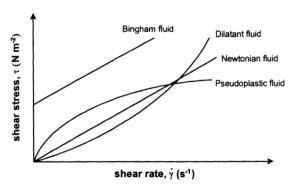

Fig. 11.8 Shear stress as a function of shear rate for different fluids.

A medium containing unicellular microorganisms is normally Newtonian, and the viscosity is close to that of pure water (except for the case of very high biomass concentrations). If the microorganisms produce extracellular polysaccharides, however, there is a significant effect on the rheology of the medium. The change in viscosity can be ascribed to the formation of the polymers, and the effect by the cells themselves is negligible. The rheology of these media can therefore normally be described with a rheological model derived from a pure polymer solution (often the power law model). In fermentations with filamentous microorganisms, the medium gradually becomes very viscous. The reason for this non-Newtonian behavior is the formation of a mycelial network. The rheological properties depend on whether free mycelia or agglomerates of hyphal elements, in the form of pellets, are present, but the power law model can be used to describe the rheology of both morphological forms (Pedersen et al., 1993). Normally, both the degree of shear thinning and the viscosity are higher in a medium containing a mycelium than in a medium where pellets are formed (at the same biomass concentration), i.e., the power law index is smaller and the consistency index is higher for media with mycelia than for media containing pellets. Other models, e.g., the Casson model, have also been used to describe the rheology of fermentation media containing filamentous microorganisms, but for reasonably high shear rates in a bioreactor ($\dot{\gamma} > 20$ s^{-1}) it is not possible to distinguish between the power law model and the Casson model (Roels et al., 1974).

In Fig. 11.9, the power law parameters are shown as functions of the biomass concentration in fermentations with *P. chrysogenum* (pellet morphology). It is observed that the power law index is approximately constant n ≈ 0.45, whereas the consistency index increases with the biomass concentration.

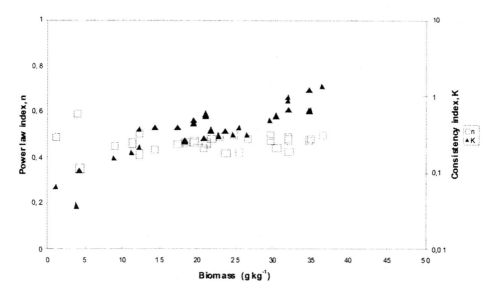

Figure 11. 9. Power law constants as functions of the biomass concentration of *P. chrysogenum* (pellet morphology). The data are taken from Pedersen *et al.* (1993).

In a stirred bioreactor there are large variations in the shear rate throughout the tank, with a high shear rate found close to the impeller and a low shear rate close to the walls. It is therefore not possible to specify a value of the viscosity for a medium with non-Newtonian properties. The *average shear rate*, however, is approximately proportional to the stirrer rate, N, i.e.

$$\dot{\gamma}_{av} = kN \qquad (11.24)$$

where k is a characteristic empirical constant for the system being. k is reported to be in the range 4-13, depending on the system, and with a standard Rushton turbine impeller $k = 10$ may be used (Nienow and Elson, 1988). The *maximum shear r*ate, on the other hand, is approximately proportional to the impeller tip speed, i.e.

$$\dot{\gamma}_{max} = k' N d_s \qquad (11.25)$$

The stirrer rate is typically lower in a large-scale reactor than in a small scale. This means that the average shear rate is lower in the large-scale reactor. The tip speed is, however, normally higher, which means that the maximum shear rate is higher in the large-scale reactor. For a pseudoplastic fluid, the decreasing shear rate when moving towards the reactor wall will give an increased viscosity. In extreme situations an inner core of medium is agitated while a shell of liquid at the wall does not move at all.

11.3.6. Flow in stirred tank reactors

All physical processes depend on the flow pattern and energy dissipation in the reactor. These quantities are described by partial non-linear differential equations obtained from conservation of mass and momentum. The flow pattern is fully described if the velocity, which is a vector property $\mathbf{u} = \mathbf{u}(x,y,z)$, and the pressure which is a scalar property, $p = p(x,y,z)$, is known at each point in the reactor. The numercial treatment to determine the flow field is the subject of computational fluid dynamics, CFD, which will not be described here (see e.g Kuipers and van Swaaij, 1998 for an introduction). However, it is useful to understand the basis for these calculations and their limitations. The fundamental relation between the pressure, velocity and the material properties viscosity and density, is given by the Navier-Stokes equation. This equation states that the time rate of change of momentum in a fluid volume is the sum of pressure forces, viscous forces and gravity forces acting on that volume. For a one-phase incompressible flow, i.e. when $\nabla \mathbf{u} = 0$, the equation can be compactly written using vector notation as:

$$\rho_l \frac{D\mathbf{u}}{Dt} = -\nabla p + \eta_l \nabla^2 \mathbf{u} + \rho_l \mathbf{g} \tag{11.25}$$

where p is the pressure, $\mathbf{g}$ is the gravity vector, ∇ is the gradient operator, and ∇^2 is the Laplacian operator. (For translation of these operators into the correct component equations for the respective coordinate system see e.g. Bird et al., 2002). The left hand side of the equation, the so-called material derivative, can be expanded into:

$$\frac{D\mathbf{u}}{Dt} = \frac{\partial \mathbf{u}}{\partial t} + \mathbf{u} \cdot \nabla \mathbf{u} \tag{11.26}$$

The Navier-Stokes equation can be transformed into dimensionless form, in which the familiar Reynolds number and *Froude number, Fr*, occur. The choice of characteristic length depends on the geometry of the system studied. For a stirred tank reactor, the impeller diameter is often chosen. The dimensionless form of the Navier-Stokes equation reads:

$$\frac{D\mathbf{u}^*}{Dt} = -(\nabla p)^* + \frac{1}{\text{Re}_s} \nabla^2 \mathbf{u}^* + \frac{1}{Fr} \mathbf{g}^* \tag{11.27}$$

where

$$\mathbf{u}^* = \frac{\mathbf{u}}{Nd_s}; \; \mathbf{g}^* = \frac{\mathbf{g}}{g}; \; Fr = \frac{d_s N^2}{g} \tag{11.28}$$

From Eq. (11.27), the Reynolds number is seen to be a parameter in the governing flow equation, which further justifies its appearance in simplified correlations for flow related phenomena. In baffled reactors, the influence of the Froude number on the flow pattern is normally small. For flow in stirred tank bioreactors, two important complications arise for obtaining a numerical

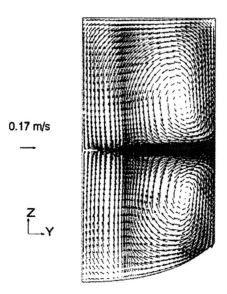

Fig. 11.10. Calculated velocity profiles in a round-bottom stirred tank reactor equipped with a Rushton turbine. (from Lundén (1994)).

solution of the flow field: 1.) The flow is normally turbulent. 2). The flow is not a one-phase flow, since both gas and liquid are present. This makes it necessary to introduce modelling assumptions concerning both turbulence and two-phase flow before a numerical flow pattern can be calculated (See Note 11.5). Simulated flow patterns in bioreactors are thus to be seen only as approximations. Even so these simulations can be highly illustrative for showing flow patterns, and thereby help to identify possible scale-up problems (see e.g. Kelly and Humphrey, 1998). An example of a flow pattern obtained in a CFD simulation is shown in Fig. 11.10.

Note. 11.5. Modelling of turbulent flow

The turbulence of the flow in stirred tank reactors does not invalidate Eq. (11.25), and time resolved simulations of the turbulent flow have indeed been made (Revstedt et al., 1998). However, to resolve the turbulent flow, which contains highly irregular flow patterns down to the Kolmogorov size, a grid size of the same length scale is needed, which is computationally very expensive. In practice, the fine details in the turbulent flow are therefore not resolved, but are instead treated by a simplified model, in which the time-averaged flow pattern is shown. The velocity is in these simulations divided into a time-averaged velocity and a fluctuating (turbulent) velocity according to:

$$\mathbf{u} = \bar{\mathbf{u}} + \mathbf{u}' \qquad (1)$$

where

$$\bar{\mathbf{u}} = \int_{t}^{t+\Delta t} \mathbf{u} \, dt \tag{2}$$

The time-averaged Navier-Stokes equation can thus be written:

$$\rho_l \frac{D\bar{\mathbf{u}}}{Dt} = -\nabla p + \eta \nabla^2 \bar{\mathbf{u}} + \rho_l \mathbf{g} - \rho_l \nabla \cdot \overline{\left(\mathbf{u}'\mathbf{u}'\right)} \tag{3}$$

As seen from comparing Eq (3) with Eq. (11.25), the time-averaged equation looks very similiar to the original equation with the exception of one additional term. This additional term, called the Reynolds stress tensor, is due to the turbulent velocity fluctuations. A very often used model is the so-called k-ε model, in which it is assumed that the Reynolds stress tensor can be represented by a term involving a *turbulent viscosity*.

$$-\rho_l \nabla \cdot \overline{\left(\mathbf{u}'\mathbf{u}'\right)} = \eta_t \nabla^2 \bar{\mathbf{u}} \tag{4}$$

The turbulent viscosity, η_t, is assumed to be a function of the velocity fluctuations, k, (m s^{-1}), and the energy dissipation, ε, (m^2 s^{-3}) according to the following expression

$$\eta_t = c_\eta \rho \frac{k^2}{\varepsilon} \tag{5}$$

with

$$k = \frac{1}{2}\left|\mathbf{u}'\right| \text{ and } \varepsilon = \frac{P}{\rho V} \tag{6}$$

c_η is a universal constant. Note that the turbulent viscosity is not related to the normal (laminar) viscosity of the fluid, but is entirely a function of the flow field. The presence of both gas and liquid leads to the need of modelling both phases. This can be done by solving the flow equations for each phase separately (e.g. Morud and Hjertager, 1996), or by considering the fluid to be pseudo-homogeneous at low aeration rates.

Simplified flow models
Solving the fundamental flow equations for a two-phase flow requires quite substantial computing power as well as rather elaborate modeling. An alternative is therefore to use a simpler so-called *compartment model*. These models often give a useful qualitative description of the main flow or mass transfer phenomena. In these models, the reactor is divided into regions (compartments) based on the qualitatively known bulk flow properties in the reactor. The compartments can be modeled as ideally mixed (i.e. small tanks) or fully segregated (i.e. plug flow reactors). The area surrounding the impeller can for instance be modeled as a well-mixed region, whereas the bottom region may be regarded as a poorly mixed region (see also example 11.5). Further refinement is possible by introducing more compartments to give a better resemblance to known flow patterns as shown in Fig 11.11.

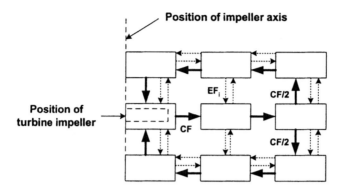

Fig. 11.11. Representation of an axi-symmetric compartment model of a stirred tank reactor equipped with one turbine impeller giving a radial flow pattern. Each compartment is assumed to be ideally mixed. The picture shows the right half of the reactor cross section. The thick arrows represent the main circulation pattern (flow rate = CF along the central axis, which is split into two flows at the reactor wall), and the dotted arrows represent the exchange flows between the compartments. The exchange flows, EF_i, need to be modeled for each individual compartment.

Use of multiple impellers on the same shaft tends to give rise to horizontally divided flow patterns. Each such area can be modeled as a compartment, or a set of compartments as shown in Fig. 11.10. To describe the overall reactor behavior with respect to e.g. gas-liquid mass transfer or mixing, the compartments need to be connected via exchange flows (see e.g. Vrábel et al., 2000). The necessary information for making these models are thus; the number of compartments, a flow model for each compartment, and a model for the exchange flows between the compartments. The exchange flow can be of two different kinds; the exchange caused by turbulence and exchange caused by aeration-induced flow. It is possible to vary the level of detail by changing the number of compartments used. The major weakness of these models is probably the difficulty in getting an estimate of the exchange flows between compartments that would allow extrapolations to different scales.

Example.11.4. A two-compartment model for oxygen transfer in a large-scale bioreactor.
An illustrative example of the usefulness of a simple compartment model is given by Oosterhuis and Kossen (1983). These authors were concerned with the validity of using Eq. (10.27) for calculation of oxygen transfer rates for widely different scales using the same parameter values. A bioreactor equipped with two Rushton turbines was used in the study, and the basic data for the reactor is given in Table 11.7.

Scale-up of Bioprocesses

Table 11.7. Reactor properties of the modeled reactor

Property	Notation	Value
Stirrer:vessel diameter	d_s/d_t	0.32
Number of impellers	n	2
Stirrer rate	N	2.6 s^{-1}
Baffle diameter/vessel diameter	d_{baffle}/d_t	0.09
Number of baffles		4
Liquid volume	V_l	varying - up to 25 m^3

The microbial (oxygen consuming) reaction in the reactor is not given in the paper, but it is described by Monod type kinetics. The dissolved oxygen concentration was measured in the reactor by a movable oxygen electrode, which allowed measurement at different axial positions. Furthermore, the overall oxygen transfer rate was calculated. The measurement showed (as could be expected) that the oxygen concentration was not constant throughout the reactor space. Futhermore, the oxygen transfer rate was overestimated using a single correlation for $k_l a$ of the type given by Eq. 1 or 2.

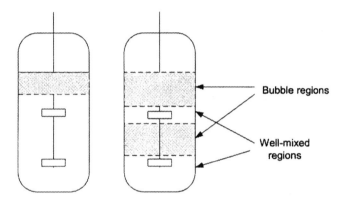

Coalescing case   Non-coalescing case

Figure 11.12. Schematic representation of compartments in the model by Oosterhuis and Kossen, 1983

To better model the oxygen transfer rate, the reactor was divided into one or two well mixed-regions close to the impellers and one or two bubble regions (i.e. regions which behave like bubble columns) one located in the top part of the reactor and the other located in the region between the turbines as shown in Fig. 11.12. For a coalescing medium, i.e. for the case where the time for coalescence was lower than the circulation time, only one well-mixed region and one bubble region was used. For the case of a non-coalescing medium, however, the time for coalescence to occur was expected to be larger than the circulation time, and two mixed regions and two bubble regions were used. The main reason for making this compartment division was that the oxygen transfer rate could be expected to be higher in the well-mixed region close to the impellers and somewhat lower in the bubble region, and two different correlations should be used for the gas-liquid mass transfer to the well-mixed and bubble regions respectively (Fig. 11.13).

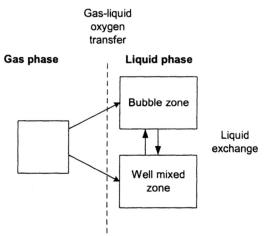

Figure 11.13. Schematic representation of oxygen transport processes.

The correlations for $k_l a$ used for the well-mixed region (subscript m) were

$$k_l a_m = 2.6 \cdot 10^{-2} u_s^{0.5} \left(\frac{P_g}{V}\right)^{0.4} \quad (1)$$

for the coalescing case and

$$k_l a_m = 2 \cdot 10^{-3} u_s^{0.2} \left(\frac{P_g}{V}\right)^{0.7} \quad (2)$$

for the non-coalescing case, respectively. The $k_l a$ value in the bubble region (subscript b) was given by

$$k_l a_b = 0.32 u_s^{0.7} \quad (3)$$

The top bubble region was assumed to start just above the top impeller in both cases. For the non-coalescing case, the volume of the well mixed region, V_m, was assumed to be

$$V_m = n d_w d_t^2 \frac{\pi}{4} \quad (4)$$

where n was the number of stirrers (=2), d_w was the impeller blade width and d_t was the tank diameter. The liquid flow between the compartments was estimated from the pumping capacity, v_{pump}, which was given by:

$$v_{pump} = 0.2 N_A d_s^3 N \quad (5)$$

Scale-up of Bioprocesses

where N_A is the aeration number defined by Eq. (11.9). To make a balance for the gas phase, the gas-hold ups in the different compartments are needed. This was estimated from the power input in the impeller region and from the superficial gas velocity in the well-mixed region according to

$$(1-\varepsilon_m) = 0.13 \left(\frac{P_G}{V_l}\right)^{0.3} u_s^{0.67} \quad (6)$$

$$(1-\varepsilon_b) = 0.6 u_s^{0.7} \quad (7)$$

Steady-state conditions requires

$$k_l a_m (c_O^* - c_{O,m}) + \frac{v_{pump}}{V_m}(c_{O,b} - c_{O,m}) + q_{O,m} = 0 \quad (8)$$

for the well mixed regions and

$$k_l a_b (c_O^* - c_{O,b}) + \frac{v_{pump}}{V_b}(c_{O,m} - c_{O,b}) + q_{O,b} = 0 \quad (9)$$

for the bubble regions.

The oxygen concentration predicted by the compartment model in the non-coalescing case as well as the measured oxygen concentration are shown in Fig. 11.14.

Obviously, this simple compartment model will not be able to predict the observed gradually changing oxygen concentration in the axial direction. However, qualitatively, the increased dissolved oxygen concentration close to the impellers is indeed predicted. With respect to the overall oxygen transfer rate, a better agreement was found using the compartment model than the application of either Eq. 1 or 2 for the entire reactor volume (Fig. 11.15).

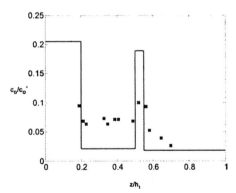

Figure 11.14. Calculated oxygen concentration for coalescing case compared to measured oxygen concentration as a function of axial position in the reactor (data from Oosterhuis and Kossen, 1983).

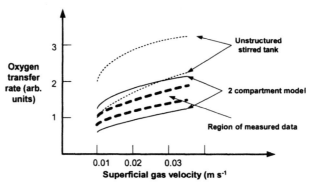

Figure 11.15. Comparison of predicted oxygen transfer rates for different models as a function of u_s. The graph shows predictions for an unstructured model (i.e. a stirred tank) based on Eq. 2 (non-coalescing - top dotted curve) and Eq. 1 (coalescing liquid – bottom dotted curve) as well as the previously described 2 compartment model (non-coalescing case top curve and coalescing case bottom solid curve). Experimentally determined values of q_O^i for h_t/d_t in the range 1.75 to 1.35 were between the thick broken lines.

This example shows that a better accuracy of predicted oxygen transfer rates may be obtained using a physically motivated compartment model of low complexity in combination with standard correlations for $k_l a$.

11.4. Metabolic Processes Affected by Scale-up

We have so far only discussed the physical processes that are affected by scale-up. The next question is: How will these changes affect the microbial kinetics (or physiology) in the large-scale process? Our prime concern is the concentration gradients that may occur due to poor mixing in large-scale reactors, but also effects by shear stress may need to be considered. In large-scale aerated reactors there will almost certainly be gradients present, both with respect to oxygen, and, in case of fed-batch or continuous operation, most likely also with respect to the limiting substrate. In the presence of gradients, the overall average volumetric reaction rate vector, $\mathbf{q}_{av}$, is therefore an integral property according to Eq. (11.29).

$$\mathbf{q}_{av} = \frac{1}{V} \iiint_V \mathbf{q}(c(x,y,z)) dV \qquad (11.29)$$

Thus, the entire concentration field need to be known, as well as the kinetic expression giving the direct concentration effects on volumetric rates, as discussed in chapter 7. Major changes of the overall metabolism may occur as a function of substrate or oxygen concentration, e.g. in organisms exhibiting overflow metabolism or anaerobic metabolism. This is true for both the industrially important organisms *S. cerevisiae* and *E. coli* (see Fig 2.6 and Example 7.3). Ethanol formation in *S. cerevisiae* will rapidly occur if oxygen is depleted. The intracellular response in NADH levels to a high glucose concentration has been found to occur within a few seconds.

Microbial kinetics is, furthermore, not only determined by the local concentrations, but also by the time history of the organisms. Cells will experience a changing environment as they circulate in a large-scale reactor. This periodic change in extracellular conditions may trigger regulatory phenomena, *e.g.* gene expression turn-on or turn-off (see Fig. 2.2). It is for example well known that both oxygen and glucose trigger several regulatory responses. The response time will be determined by the rate of mRNA polymerization and the rate of translation (Konz *et al.*, 1998). In a study by Schweder *et al.* 1999, seven different mRNA levels in *E. coli* were studied in a scaled-down system. It was found that the transcription of several genes, such as the *proU* gene, which is involved in osmoregulation, responded within 15 s of exposure to high glucose concentration. The rate of protein synthesis is probably somewhat lower. The peptide elongation rate has been estimated to be between 13-16 amino acids per second in *E. coli* (Einsele et al., 1978, see also Note 7.6). For this reason, a single "dip" in oxygen concentration of short duration does not necessarily result in a change in the overall protein synthesis pattern.. However, the effects of repeated depletions, as occuring for a cell circulating in a large-scale bioreactor, are indeed difficult to predict and need to be studied experimentally.

It is clearly expensive to do the experiments in the actual large-scale bioreactor, and *scale-down experiments* are therefore made. Concentration gradients can be created by making step-change experiments in a laboratory scale stirred tank reactor, or by connecting two small reactors. Step-change experiments in one reactor can be used to study dynamic effects of a sudden increase in glucose concentration or aerobic/anaerobic transitions. In a sense one can say that an experimental "compartment system" is used to mimic the large-scale reactor. A compilation of such experimental studies is given by Lidén (2001). A very rapid mixing of glucose can be obtained in a laboratory reactor. However, the oxygen transfer rate is not sufficiently high to enable studies of fast dynamics related to aerobic/anaerobic transitions. A combination of a stirred tank reactor and a plug flow reactor may in this case be a better option (George et al., 1993).

The *a priori* expectation is probably that gradients will have a negative effect on process performance. This has also been reported for *e.g.* biomass yield. However, not all results show a decreased performance in large-scale. The leavening capacity of Baker's yeast was found to be higher in both a scaled-down model system and a large-scale process compared to an ideally mixed system (George et al., 1998). Furthermore, the stability of a heterologous protein was found to increase in a non-ideally mixed system (Bylund et al, 2000).

Also shear rates may be a concern for shear sensitive organisms. This normally does not apply to yeast or bacteria, whereas mammalian cells or plant cells are shear sensitive (Tanaka, 1981). The situation is complicated by the fact that the average shear rate normally decreases during scale-up, whereas the maximum shear rate normally increases. Again, an "oscillating environment" with respect to shear will be sensed by the organisms during circulation.

Another factor to keep in mind is that the requirement for culture stability increases. The actual number of generations in the large-scale process depends on the inoculation density. However, even if a high inoculation density is used, the number of generation from the stock culture to the final harvesting increases rather much, and possible genetic instabilities of the production strain will therefore be more pronounced.

A final, unpleasant, surprise experienced in many large-scale processes is foaming. Foaming is caused by surface-active components that are excreted or released through cell breakage. Foaming is normally manageable in lab-scale experiments, but in a large-scale reactor unexpected foaming can become a major problem.

11.5. Scale-up in Practice

At this point, we may ask ourselves, what is a suitable practical approach to scale up? Suppose that a very successful lab-scale or pilot scale process has been developed. Could we not just maintain the same process conditions and scale up the process? From the previous sections in this chapter it should be clear that this, unfortunately, is physically impossible. This is further illustrated in Table 11.8, which is based on a classical table published by Oldshue (1966).

Four different parameters were maintained constant during the scale-up in Table 11.8, i.e. the specific power input, the stirrer rate (which corresponds to maintaining the mixing time), the tip speed (which gives approximately the same maximum shear rate) and the Reynolds number (which is suggested by the dimensionless Navier-Stokes equation). It is clear that in fact most of the other parameter values, except the one chosen constant, change during scale-up.

Four in principle different approaches to scale-up in practice can be distinguished (Kossen and Oosterhuis, 1985):

1. Fundamental methods
2. Semifundamental methods
3. Dimensional analysis
4. Rules of thumb

Table 11.8. Effect of scale-up on charateristic properties when scaling up from a 100 L to 12.5 m³ reactor. (adapted from Oldshue, 1966)

Property	Pilot scale (100 L)	Plant scale (12.5 m³)			
P	1	125	3125	25	0.2
P/V	1	1	25	0.2	0.0016
N	1	0.34	1	0.2	0.04
d_s	1	5	5	5	5
v_{pump}	1	42.5	125	25	5
t_c	1	0.34	1	0.2	0.04
Nd_s	1	1.7	5	1	0.2
Re_s	1	8.5	25	5	1

Scale-up of Bioprocesses

Table 11.9. Characteristic times for important processes in fermentations

Process	Variable
Circulation	t_c
Mixing	$t_m \approx 4\, t_c$
Gas flow	$t_{gas} = (1 - \varepsilon)V/\upsilon_g$
Gas-liquid mass transfer	$t_{otr} = 1/k_l a$
Biomass growth	$t_{bio} = 1/\mu$
Substrate consumption	$t_{sc} = s/q_s$
Substrate addition	$t_{sa} = Vs/\upsilon_f s_f$

The *fundamental method* is based on actually solving the governing equations of flow(see section 11.3.5) in combination with the reaction kinetics. In a strict sense this is not even theoretically possible, since in particular the microbial kinetics is not fully known (see section 11.4). Assumptions concerning both the kinetics and flow modeling (see Note 11.5) are thus necessary.

In the *semifundamental methods*, simpler flow models are applied, e.g., the compartment models described in Section 11.3.5. When such a model is combined with a suitable kinetic model, a reasonably precise description of the process may be obtained. However, despite the extensive simplification of the problem when one moves from the fundamental models to the semifundamental models, the complexity of the model is still substantial if e.g. a structured kinetic model is applied.

Regimen analysis is based on a comparison of characteristic times for the different mechanisms involved in the overall fermentation process. The characteristic time for a certain process, which is modeled as a first-order process, is defined as the reciprocal of the rate constant. For processes, which are not first order, the characteristic time can be calculated as the ratio between capacity (e.g., the volumetric content of the considered species) and the flow (e.g., the volumetric consumption rate of the species). Table 11.9 lists the most important characteristic times in fermentation processes. By analyzing the characteristic times, it is possible to identify potential problems such as oxygen limitation in the large-scale process. The regimen analysis may also suggest further scale-down studies, such as repeated exposure to substrate depletion.

Example 11.5. Regimen analysis of penicillin fermentation
In the literature, one finds many applications of time-scale analysis to identify the limiting regime in a given process (see Sweere et al., 1987 for a review). Here we will consider the penicillin fermentation for which Pedersen (1992) found the characteristic times listed in Table 11.10. The characteristic times are given for the two different phases of a typical fed-batch fermentation (i.e., the growth phase and the production phase), carried out in a 41-L pilot plant bioreactor. The smallest characteristic time is certainly t_{mix} (which was experimentally determined using isotope techniques). This indicates that no mixing problems arise during the fed-batch fermentation in the pilot plant bioreactor. Thus there will not be any concentration gradients for the substrates (glucose and oxygen). This conclusion will, however, definitely be different in a large production vessel, where the mixing time may approach 20-50 s. The characteristic mixing time for mixing in the gas, t_{gas}, is high, indicating that this mechanism is slow. The characteristic times for glucose addition, t_{sa}, and glucose consumption, t_{sc}, are of the same order of magnitude. This is quite obvious, since it

Table 11.10. Characteristic times during a penicillin fermentation carried out in a 41-L pilot plant bioreactor.

Characteristic time	Growth phase (s)	Production phase (s)
t_m	1.5	1.5
t_{gas}	60.3	60.3
t_{otr}	22.2	22.2
t_{sa}	95.6	126.7
t_{oc}	11.5	12.4
t_{sc}	189.3	102.2

was intended to keep the glucose concentration at a constant level by controlling the feed addition. Also, the characteristic times for oxygen supply t_{otr}, and oxygen consumption, t_{oc}, are of the same order of magnitude. This indicates that oxygen limitations may occur if the oxygen supply for some reason fails or the oxygen requirements increase (e.g., due to an increasing feed addition of glucose). Thus the gas liquid mass transfer seems to be the limiting regime for this process, and this conclusion would be even more pronounced in a production-scale vessel.

It is understandable that *rules of thumb* (i.e. practical "one-liners") are much desired by the plant engineer. Although all parameters cannot be maintained constant (cf. Table 11.8), there has been a widespread belief that one could (and perhaps even should!) choose one to be maintained constant during scale-up. Most often the specific stirrer power input, or the $k_l a$ value, has been suggested for this purpose. However, there are strong reasons to caution against these simple scale-up criteria. There is very little theoretical support that a successful scale-up will result from such simplistic procedure (Oldshue, 1997). In fact, it is more than likely that in the successfully scaled-up process no parameter values will be exactly the same as in the small-scale or pilot-scale reactor. In fact, not only may the parameter values change, a completely different reactor design may also be preferable (see Example 11.6). Fig. 11.16 is offered as a final help for the difficult task of scale-up. If the qualitative procedure of Figure 11.16 is followed one may hope to avoid most of the pitfalls associated with scale-up.

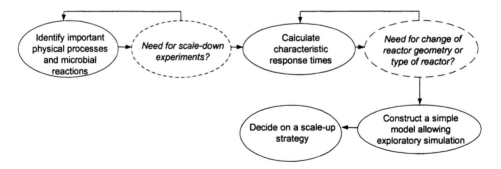

Figure 11.16. Important principal steps during scale-up.

Example 11.6. Loop-reactor design for production of single-cell protein
In a production plant close to Trondheim, Norway, about 9000 ton/year of protein is produced from methane. The methane is obtained from natural gas in the North Sea and the methanotrophic microorganism *Methylococcus capsulatus* is used as production organism. The overall reaction stoichiometry for production of biomass from methane is approximately given by

$$-CH_4 - 1.52O_2 - 0.09NH_3 + 0.546CH_{1.8}O_{0.5}N_{0.2} + 0.544CO_2 + 1.72H_2O \quad (1)$$

The protein content in the biomass is quite high (70%), which gives a protein yield of about 0.5 kg (kg methane)$^{-1}$. The reaction takes place at atmospheric conditions and at a temperature of 45 °C.
The bioreactor design for this process is rather demanding for several reasons. The solubility of methane (and oxygen) is low in water. The saturation concentrations are approximately 1 mM for both compounds at process conditions (see Table 10.1 and 10.8). The oxygen yield is, furthermore, high since a very reduced substrate is used. This in turn gives a high heat yield. The heat of reaction is approximately given by (see Section 4.2)

$$-\Delta H = \frac{1.52 \cdot 460 \cdot 10^3}{0.546} = 1.28 \cdot 10^6 \text{ J (C-mol biomass)}^{-1} = 52 \text{ kJ (g biomass)}^{-1} \quad (2)$$

At a biomass production rate of one ton per hour, the heat effect is thus approximately 14.4 MW. To obtain a stoichiometric gas phase composition according to Eq. 1 using air at atmospheric pressure, the partial pressures of CH$_4$ and O$_2$ should be 0.116 and 0.177 atm, respectively. The volumetric mass transfer coefficient, $k_l a$, obtained in a mechanically stirred bioreactor can be approximated to about 600 h^{-1}. We can therefore estimate the required reactor volume based on the mass transfer capacity (oxygen will be the limiting component) from Eq.3.

$$k_l a c_o^* V \geq Y_{xo} q_x V \quad (3)$$

Assuming that c_o is ≈ 0 and that $c_o^* = 0.177$ mol m^{-3}, the volume required for production of 1000 kg biomass h^{-1}, is

$$V \geq \left(\frac{1}{600 \cdot 0.177}\right)\left(\frac{1.52 \cdot 10^6}{0.456 \cdot 24.6}\right) = 1276 \text{ m}^3 \quad (4)$$

The reactor volume in the actual process is, however, only 300 m^3. Instead of a stirred tank reactor, a loop reactor (see Fig 11.2), with two long horizontal tubes and two short vertical tubes is used. The liquid is circulated at a linear flow rate of about 1 m s^{-1}. Furthermore, pure oxygen is used instead of air. (The plant is located very close to a major methanol production facility, which makes it easy to obtain pure oxygen in sufficient quantities.) The gas containing methane and oxygen (partial pressures 0.397 and 0.603 atm, respectively) is injected in the liquid flowing downwards in one of the vertical tubes. The tube is equipped with regularly spaced static mixers, giving a high volumetric mass transfer rate ($k_l a > 900$ h^{-1}). Produced CO$_2$ and the remainder of O$_2$ and CH$_4$ is separated from the liquid phase in the top, horizontal part of the loop. The considerable cooling requirement is met by the use of external, very efficient, heat exchangers. Part of the liquid is pumped out of the main loop to these external heat exchangers at several places along the loop.

PROBLEMS

Problem 11.1. Scale-up without maintaining geometrical similarity.

In example 11.2, the scale-up of a stirred tank bioreactor from 1 L to 1 m^3 was considered. In the example it was found that the mixing time increased by a factor of 5, if the same geometry was maintained during scale up. Suppose now instead that the diameter of the 1 m^3 reactor is chosen to 0.187 m (i.e. the reactor geometry is changed), but the requirement of a constant P/V (= 5.95 kW m^{-3}) is maintained.

a) How much larger is the mixing time in the large-scale bioreactor now?
b) What value of P/V would give the same mixing time in the large-scale reactor as the small-scale reactor?

Problem 11.2. Exchanging impellers

One drawback of the traditional Rushton impeller is that the ratio between aerated and unaerated power consumption falls rapidly with an increased aeration rate. The hydrofoil impeller, typically has a lower power number, but the aerated power consumption falls less rapidly with increasing aeration rate. For stirring speeds close to 300 rpm the ratio between aerated and unaerated power consumption is given in the figure for a Rushton turbine and a Prochem impeller.

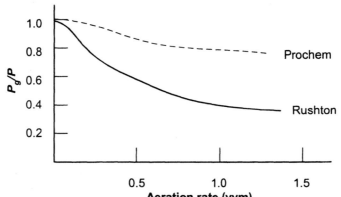

You are replacing a Rushton turbine in a 100 m^3 reactor (d_t = 3 m, d_s = 1 m) by a Prochem impeller. Assume that the stirring rate should be maintained the same.

Scale-up of Bioprocesses

a) What diameter of the Prochem impeller will give the same unaerated power consumption as the Rushton turbine? The values of N_p for the Rushton turbine and the Prochem impeller are 5.2 and 1.0, respectively.

b) How much higher (approximately) will the k_la-value be for the Prochem impeller at an aeration rate of 0.5 vvm under the conditions in problem a, (i.e. the same unaerated power consumption is achieved)? The k_la value for the same specific power input has been found to be identical for the two impeller types. Assume a non-coalescing medium.

Problem 11.3. Design of a pilot plant bioreactor

In connection with the purchase of a new pilot plant bioreactor (41 L) to be used for penicillin fermentation, it is desired to examine whether one of the manufacturer's standard-design bioreactors (equipped with Rushton turbines) can be used. The dimensions of this bioreactor are specified in the table below.

Aspect ratio, d_t/d_s	3
Tank diameter, d_t	0.267 m
Stirrer diameter, d_s	0.089 m
Number of impellers*	3
Maximum stirring speed, N_{max}	600 rpm

*The impellers are six-bladed Rushton turbines.

a. Show that with a non-Newtonian medium, for which the rheology is described by a power law expression (Eq. 11.21), Re_s is given by Eq. (1)

$$\mathrm{Re}_s = \frac{\rho_l N^{2-n} d_s^2}{K k^{n-1}} \quad (1)$$

where $k = 10$ is the constant used for calculation of the average shear rate in Eq. (11.24). Plot Re_s versus the stirring speed for a medium containing respectively 0, 20, and 40 g L^{-1} biomass (of the fungus *P. chrysogenum*). The power law parameters can be taken from Fig. 11.10. Discuss the results.

b. Determine the power number with $N = 600$ rpm for the three cases considered in (a), using Fig. 11.10, and calculate the power input per unit volume when the bioreactor contains 25 L of medium. Calculate the average viscosity in the bioreactor.

c. For non-Newtonian media and the 41 L bioreactor, the following correlation was found:

$$k_l a = 0.226 \cdot 10^{-3} u_s^{0.4} \left(\frac{P_g}{V}\right)^{0.6} \eta^{-0.7} \quad (2)$$

Can the dissolved oxygen concentration be maintained above 30% in the bioreactor (which for some strains is a critical level for penicillin production) when the oxygen requirement for a *rapidly growing* culture of *P. chrysogenum* is r_o = 2.3 mmoles of O_2 (g DW)$^{-1}$ h^{-1} ? Discuss how the bioreactor can be modified to satisfy the oxygen requirement.

d. You decide to examine the effect of increasing the stirrer diameter. Start with d_s/d_t = 0.4. Can the critical level of dissolved oxygen concentration be maintained with this diameter ratio?

Problem 11.4. Scaled-down experiment

The pH-value is often controlled using only a single pH electrode and a single point of addition of base or acid, also in large-scale bioreactors. The pH-electrode is typically located in a well-mixed region, and addition of base or acid is typically made at the liquid surface. Since concentrated solutions are used, pH-gradients in large-scale reactors are likely to be present.

Amanullah et al. (2001) used of a scaled-down system to study effects of pH gradients. The scaled-down system consisted of a 2 L standard stirred tank reactor, with a working volume of 1 L, equipped with two Rushton turbines (d_s/d_t = 0.33). To the reactor was connected a piece of tubing (L = 2.75 m, d_i = 4.8 mm), through which liquid from the reactor was pumped. The pH-value in the reactor was measured and pH control was achieved by adding base to a small mixing bulb (volume about 1.5 ml) located just before the tubing (see figure below).

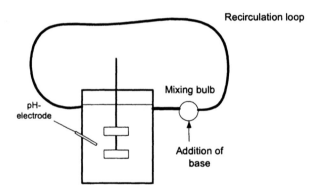

This system was used to experimentally simulate a three compartment reactor model, with a direct feed zone (bulb), a poorly mixed zone (tubing part), and a well-mixed zone (the reactor).

a) Discuss what residence time in the tubing that should be chosen to simulate a large-scale reactor (100 m^3). The same residence time in the tube can be achieved by different ratios of tube volume, V_{tube} and recirculation flow, v_{rec}. Discuss how the values of V_{tube} and v_{rec} should be chosen.

In their study, Amanullah et al. (2001) used a strain of *Bacillus subtilis*. This organism produces acetoin and 2,3 butanediol under oxygen limited conditions. The formation of acetoin and butanediol is described by:

2 pyruvate → CO_2 + acetolactate   *(acetolactate synthase)*

Acetolactate → CO_2 + Acetoin   *(acetolactate decarboxylase)*

Acetoin + NADH + H^+ ↔ 2,3 butanediol   *(butanediol dehydrogenase)*

At pH-values higher than 6.5 also acetate is formed.

b) Compare the residence time in the tube to the characteristic times for oxygen consumption and substrate consumption. Assume that the dissolved oxygen concentration in the reactor is 10% of DOT, and that the system is to be operated as a chemostat with a glucose concentration of 10 mg L^{-1} and a biomass concentration of 4 g L^{-1}.
What are your conclusions?

c) The experiments by Amanullah et al, were in fact made as batch cultivations, with oxygen limited conditions also in the stirred tank reactor. A control experiment, in which base addition was made in the reactor instead of the loop was also made. The following yields were found

Residence time in loop (s)	Yield of acetic acid (g g^{-1})	Sum of yields of acetoin and butanediol (g g^{-1})	Biomass yield (g g^{-1})
(no loop)	0	0.30	0.49
30	0.002	0.31	0.52
60	0.023	0.29	0.51
120	0.076	0.23	0.50
240	0.083	0.22	0.50
120*	0	0.30	0.50

*pH control made in the stirred tank reactor instead of in the loop.

Discuss the results.

REFERENCES

Amanullah, A., McFarlane, C. M., Emery, A. N., and Nienow, A. W. (2001). A scale-down model to simulate spatial pH variations in large-scale bioreactors. *Biotechnol. Bioeng.* 73:390-399
Bird, R. B., Stewart, W. E., and Lightfoot, E. N. (2002). Transport phenomena, 2nd ed., John Wiley&Sons, New York
Bisio, A., and Kabel, R. L. (1985) Scale-up of Chemical Processes – Conversion of Laboratory Scale Tests to Successful Commercial Size Design. John Wiley & Sons, New York
Blanch, H. W., and Clark, D. S. (1997). Biochemical Engineering, Marcel Dekker, New York
Bourne, J. R (1997). Mixing in single-phase chemical reactors, in Mixing in the Process Industries, 2nd edition, Butterworth Heinemann,
Brune, A. (1998). Termite guts: the world's smallest bioreactors. *Trends Biotechnol.* 16:16-21
Bylund, F., Castan, A., Mikkola, R., Veide, A., and Larsson, G. (2000) Influence of scale-up on the quality of recombinant human growth hormone. *Biotechnol. Bioeng.* 69:119-128
Chisti, Y., Moo-Young, M. (2001) Bioreactor design, in: *Basic Biotechnology* (C. Ratledge and B. Kristiansen, eds), Cambridge University Press, Cambridge
Chilton, T. H., Drew, B., Jebens, R. H. (1944). Heat transfer coefficients in agitated vessels. *Ind. Eng. Chem.* 36:510-516
Cooney, C. L. (1983). Bioreactors: Design and operation. *Science* 19:728-740
Einsele, A., Ristroph, D. L., and Humphrey, A. E. (1978) Mixing times and glucose uptake measured with a fluorometer. *Biotechnol. Bioeng.* 20:1487-1492
Geankoplis, C. J. (1993). Transport processes and unit operations, 3rd ed., Prentice Hall, Englewood Cliffs, New Jersey
George, S., Larsson, G., Enfors, S.-O. (1993). A scale-down two-compartment reactor with controlled substrate oscillations. Metabolic response of *Saccharomyces cerevisiae*. *Bioproc. Eng.* 9:249-257

George, S., Larsson G., Olsson, K., and Enfors, S.-E. (1998) Comparison of the Baker's yeast process performance in laboratory and production scale. *Bioproc. Eng.* **18**:135-142

Humphrey, A. (1998) Shake flask to fermentor: What have we learned? *Biotechnol. Prog.* **14**: 3-7

Johnson, A. T. (1999) Biological process engineering, John Wiley & Sons, New York

Kelly, W. J., and Humphrey, A. E. (1998). Computational fluid dynamics model for predicting flow of viscous fluids in a large fermentor with hydrofoil flow impellers and internal cooling coils, *Biotechnol. Prog.* **14**:248-258

Kossen, N. W. F. (1985). Bioreactors: Consolidation and innovation. In: Proc. 3rd European Congress on Biotechnology, vol 4, pp. 257-279, VCH Verlag

Kossen, N. W. F., and Oosterhuis, N. M. G. (1985). Modelling and scale-up of bioreactors, in Biotechnology, 2nd ed., (Rehm, H.J. and Reed, G. eds.) VCH-Verlag, Weinheim, Germany

Konz, J. O.; King, J.; Cooney, C. L. (1998). Effects of oxygen on recombinant protein production. *Biotechnol. Prog.* **14**:393-409

Kuipers, J. A. M., and van Swaaij, W. P. M. (1998) Computational fluid dynamics applied to chemical reaction engineering, *Adv. Chem. Eng.* **19**:227-328

Leib, T. M., Pereira, C. J., and Villadsen, J. (2001) Bioreactors. a chemical engineering perspective. *Chem. Eng. Sci.* **56**:5485-5497

Lidén, G. (2001) Understanding the bioreactor. *Bioproc. Biosys. Eng.* **24**:273-279

Lundén, M. (1994) A computational investigation of transport phenomena in stirred reactors and catalyst particles. Ph.D. thesis, Chalmers University of Technology, Gothenburg, Sweden

Lundén, M., Stenberg, O., Andersson, B. (1995). Evaluation of a method for measuring mixing time using numerical simulation and experimental data. *Chem. Eng. Commun.* **139**: 115-136

Mase, G. E. (1970) Continuum mechanics. Schaum outline series, McGraw Hill, New York.

McDonough, R. J. (1992). Mixing for the process industries. Van Nostrand-Reinhold, New York

Michel, B. J., Miller, S. A. (1962) Power requirements of gas-liquid agitated systems. *AIChE J.* **8**: 262-266

Nienow, A. W. (1990). Agitators for mycelial fermentations. *Trends in Biotechnol.* **8**:224-233

Nienow, A. W. (1997). On impeller circulation and mixing effectiveness in the turbulent flow regime, Heilical coil heat transfer in mixing vessels. *Chem. Eng. Sci.* **15**:2557-2565

Nienow, A. W. and Elson, T. P. (1988). Aspects of mixing in rheologically complex fluids. *Chem. Eng. Res. Des.* **66**:5-15

Nienow, A. W., Harnby, N., and Edwards, M. F. (1997). Introduction to mixing problems, in *Mixing in the Process Industries*, 2nd edition, Butterworth Heinemann,

Nienow, A. W., and Lilly, M. D. (1979). Power drawn by multiple impellers in sparged vessels. *Biotechnol. Bioeng.* **21**:2341-2345

Oldshue, J. (1966). Fermentation mixing scale-up techniques. *Biotechnol. Bioeng.* **8**:3-24

Oldshue, J. Y. (1989). Fluid Mixing in 1989. *Chem. Eng. Prog.* 85(5):33-42

Oldshue, J. Y. (1997) Heed this advice in scaling up mixers, *Chem. Eng. Prog.* **93**:70-73

Oldshue, J. and Gretton, A. T. (1954) Heilical coil heat transfer in mixing vessels *Chem. Eng. Prog.* **50**: 615-621

Oosterhuis, N. M. G., and Kossen, N. W. F. (1983) Oxygen transfer in a production scale bioreactor. *Chem. Eng. Res. Dev.* **61**:308-312

Pedersen, A. G., Bundgård, M., Nielsen, J., Villadsen, J. and Hassager, O. (1993). Rheological characterization of media containing *Penicillium chrysogenum*. *Biotechnol. Bioeng.* **34**:1393-1397

Revstedt, J., Fuchs, L., and Trägårdh, C. (1998). Large eddy simulations of the turbulent flow in a stirred reactor. *Chem. Eng. Sci.* **53**:4041-4053

Roels, J. A., van den Berg, J., and Voncken, R. M. (1974). The rheology of mycelial broths. *Biotechnol. Bioeng.* **16**:181-208

Rushton, J. H., Costich, E. W. and Everett, H. J. (1950a). Power characteristics of mixing impellers – Part I. *Chem. Eng. Prog.* **46**:395-404

Rushton, J. H., Costich, E. W. and Everett, H. J. (1950b). Power characteristics of mixing impellers – Part II. *Chem. Eng. Prog.* **46**:467-476

Schügerl, K. (1991). Bioreaction Engineering, vol 2, John Wiley&Sons, Chichester, U. K.

Schweder, T., Kruger, E., Xu, B., Jurgen, B., Blomsten, G., Enfors, S.-E., and Hecker, M. (1999). Monitoring of genes that respond to process-related stress in large-scale bioprocesses. *Biotechnol. Bioeng.* **65**:151-159

Stein, W. A. (1992). Mixing times in bubble columns and agitated vessels, *Int. Chem. Eng.* **32**: 449-474

Sweere, A. P. J., Luyben, K. Ch. A. M., and Kossen, N. W. F. (1987). Regimen analysis and scale-down: tools to investigate the performance of bioreactors. *Enz. Microb. Technol.* **9**:386-398

Tanaka, H. (1981). Technological problems in cultivation of plant cells at high density. *Biotechnol. Bioeng.* **23**:1203-1218

Vrabel, P., van der Lans, R. G. J.M., Luyben, K. Ch. A. M., Boon, L., Nienow, A. L. (2000) Mixing in large-scale vessels stirred with multipel radial or radial and axial up-pumping impellers: modelling and measurements. *Chem. Eng. Sci.* **55**:5881-5896

Westerterp, K. R., van Swaaij, W. P. M., and Beenackers, A. C. M. (1984) Chemical Reactor Design and Operation, J. Wiley & Sons, New York.

Index

Abiotic phase 4
Acetic acid 11-12
  as substrate for PHB production 91
  degree of reduction 62
  diffusion coefficient in aqueous solutions 455
  heat of combustion 104
  heterofermentative fermentation 148-149
  permeability coefficient 17
  transport of 11-12, 16-17
Acetone
  acetone butanol fermentation 181-184
  degree of reduction 62
  heat of combustion 104
AcetylCoA
  in fermentation pathways 29
  in TCA cycle 27
  requirement for synthesis of an *E. coli* cell 25
Activation energy
  for *E. coli* growth 262-263
  for *K. pneumoniae* growth 262-263
  for maintenance processes 263
Active transport: see Cell transport mechanisms
Adenosine nucleotides 31-33, 99-100, 109-110
  ADP 25
    phosphorylation of 27-28, 99, 109-110
    regulatory role of 31
  AMP 31-33
    regulatory role of 31
  ATP 12, 18-20, 22-28, 30-34
    consumption in futile cycles 18-20, 126, 128
    energy charge 31-34
    formation of 25-27, 30-31, 33-34, 110
    in ammonia assimilation 155
    in maintenance models 125-128, 175
    in maintenance requirement 127
    in transport processes 18-20, 22-24, 109-111, 128
    maintenance requirement 127
    net yield
      in EMP pathway 27
      in the oxidative phosphorylation 27, 109-115, 129
      in TCA cycle 27, 63
    pseudo steady state 19, 33, 112, 124, 133
    requirement
      for synthesis of an *E. coli* cell 32
      for growth of *L. cremoris* 129-131
    steady state concentration 32, 100
    substrate level phosphorylation 26-27, 63
    turnover time 31, 124-125
    yield on biomass (Y_{XATP}) 127, 132
      in *B. Clausii* 136-138
      in aerobic *S. cerevisiae*
      fermentation 140-141
ADP: see Adenosine nucleotides
Aeration 481

Aeration number 488
  influence on power input 488
  in pilot plant bioreactors 515
  surface aeration 495
  using a nozzle 479
Aerobacter aerogenes
  ash content and elemental composition 56
  biomass yield on ATP 127
  biomass yield on glucose and glycerol 255
  growth on glycerol 249, 254
  maintenance coefficients 254
  saturation constant for glucose 247
Allosteric enzymes 201, 204
Amino acids
  composition of and degree of reduction of 62
  in defined media 129-132
  transport of 12, 16, 129
  uptake of amino acids 12, 177, 290
Ammonia
  as nitrogen source 64-65
  heat of combustion 104
  production of glutamate by GDH or by the GS-GOGAT system 155-156
  thermodynamic properties 115
  uptake of 59, 274
AMP: see Adenosine nucleotides
Anabolic reduction charge 34
Anaplerotic pathways 34, 154, 156
Antibiotics 11, 35-36, 40
Antifoam agents
  effect on coalescence 435
Apex: see Filamentous microorganism
Apical cell: see Hyphae
Apical compartment: see Filamentous microorganism
Arrhenius plot 262
Aspect ratio 480
Aspergillus awamori
  biomass yield on glucose 254
  maintenance coefficients 254
  morphologically structured model 303
Aspergillus nidulans
  biomass yield on glucose 254
  duplication cycle 301
  maintenance coefficients 254
Aspergillus niger 10, 453
  ash content and elemental composition of 56
  citric acid production by 71
ATP: see Adenosine nucleotides
Attractor (in solution of differential eq.) 391
Axial diffusion 379
Axial dispersion 379
Azotobacter 24, 92
  sugar transport in 24

Bacillus megatarium
  biomass yield on glycerol 254
  maintenance coefficients 254
Bacillus subtilis 290
  single cell model 290
  used in scale down experiment 516
Baker's yeast: see *Saccharomyces cerevisiae*
Balanced growth 6, 30-32, 119-120
Bingham fluids 498
Biomass
  ash content and composition of 55-57
  degree of reduction 61, 63-66
  heat of combustion 103-105, 108
Bioreactor 6-7
  airlift 479
  auxostat (definition) 49
  chemostat (definition) 49
  continuous stirred tank 47-53, 352-359
  design 2-4, 7-9, 45, 339-340, 438-439, 512-515, 517-518
  ideal 7, 50-51, 237, 339-340
  industrial 7, 43-45, 478-482
  loop reactor 50, 414-415, 479, 513, 517
  natural flow bubble column 479
  operation modes 342
  productostat (definition) 49
  stirred tank reactor 7, 47, 340-342, 480-482
  tubular reactor 7, 372-379
  turbidostat (definition) 49
Biot number 469
Biotic phase 4
Black box model
  for growth kinetics 245-253
  for stoichiometry 57-60, 73-74
    error analysis 77-88
  limitation of 59-60, 252
Blackman model: see Growth rate equations
Bottleneck 207
  in aerobic growth 256
  in aerobic growth of *S. cerevisiae* 67, 256-261
  model 256, 274
Branching: see Filamentous microorganism branching
Brewer's yeast: see *S. cerevisiae*
Briggs-Haldane model for enzymes 191-192
Bubble
  break up 432
  coalescence 431, 434-435, 505
  column 435, 438, 479, 489
  diameter 431-439, 443, 471, 474, 505-506
    mean Sauter diameter 431, 434-435
  formation 323, 431-432, 435
  maximum stable diameter 432-433, 435, 437
  rise velocity 471, 495
Budding index 298, 328
Budding yeast: see Yeast budding
Butanol 30, 129
  degree of reduction 62
  fermentation 62, 181-183
  heat of combustion 104
Candida lipolytica
  citric acid fermentation 156-159
Candida utilis 56
  ash content and elemental composition 56

biomass yield on glucose 111, 254
Carbohydrate 156, 158, 275, 277
  carbon and energy source 11
  content in *E. coli* 32
  content in *S. cerevisiae* 55
  elemental composition 55
  in a flux model 156-158
  storage in yeast 294
Carbon balance 54, 58, 62-66, 68-73
Carbon dioxide 11
  as carbon source 11, 107
  as electron acceptor 72
  diffusion coefficient in dilute aqueous solutions 454
  gas liquid mass transfer 453-454
  gas phase resistance 427
  Henry's constant 426
  permeability coefficient in membranes 17
  uptake 18, 30, 68, 83, 88, 149, 158-159, 175, 256-260, 295, 425
Carbon labeling
  ^{13}C labeling 163-172
Casson equation for description of fermentation media 498
Catabolic reduction charge 34
Catabolic repression 13
  glucose repression of the oxidative system 256
  lac operon 279
Catabolism 12, 24-30
  in lactic acid bacteria 148-150, 180
Cell composition 15, 30-32, 54-57, 294, 298
  elemental, for different microorganisms (table) 56
  macromolecular
    S. cerevisiae 55
Cell cycle
  E. coli 290, 323-325
  in filamentous fungi 330-335
  in models for oscillations 298-299
  in single cell models 289-290, 315
  S. cerevisiae 290, 294-297
Cell differentiation: see Filamentous microorganism
Cell membrane 12, 15-24, 112-113, 115, 126, 160
  active transport 16, 18, 23-24, 115, 195
  facilitated diffusion 16, 20-22
  free diffusion 16, 18, 264
Cell size
  critical 294, 297-298, 337
  distributions in *S. cerevisiae* 326-329
  distributions in *S. pombe* 319
  in morphologically structured models 294
  in yeast oscillations 297-298
Cell transport mechanisms
  active transport 16
    antiport 24
    symport 23-24
    uniport 24
    primary 22
    secondary 22
  facilitated transport 16, 20-22
  of organic acids 18
  of phenoxyacetic acid 19
  passive transport 16, 109
  product excretion 13
  PTS in *E. coli* 24

Index

uncoupling agents 18-22, 264
Cell wall 13, 15, 268, 301, 332
Cephalosporins 10, 35-36, 39
Cephalosporin C 35
Chemostat: see Bioreactor
Chi-square distribution 85
  use in error identification 85-88
Chromosome 13, 15, 288-289, 309
  duplication in yeast 294
Citric acid
  degree of reduction 62
  heat of combustion 104
  production by *A. niger* 71-72
  production by *C. lipolytica* 156-159
Coexistence of several organisms 397-402
  of predator and prey 404-409
Colonial mutants 294
Compartment 29, 61, 160, 239
Compartment models
  for metabolism and growth 160, 239, 265-271, 276-277, 284
  for mixing 503-504, 508, 511
Competition
  between organisms using the same substrate 397-401
  between prey and predator 404
  in enzyme kinetics 195-204
Consistency analysis 67-71
Consistency index 498-499
Continuous Stirred Tank Reactor (CSTR): see Bioreactor
Contois model: see Growth rate equations
Cooperativity 201-204
Corn steep liquor 40
  oxygen solubility in 448
Correlated measurements 83-85, 107, 159, 199-201, 334
C. glutamicum
  lysine production 76, 164, 180
Coulter counter 319-321
Crabtree effect 30, 171, 256-260
Cultivation
  batch 246-249, 341-352
  fed-batch 367-371
  plug flow reactor 372-379
  stirred tank, continuous
    general mass balance 342
    mixed microbial population 397-409
    steady state 47-53, 341, 352-364
      with biomass recirculation 359-364
      with gas phase substrate 364-366
    transient 380-409
Culture parameters 261
Culture variables 261
Cybernetic model 274-278
  matching law model 277
  of *K. oxytoca* 278
  variables 265, 276

Damköhler number 458
Daughter cell: see Yeast budding
Degrees of freedom 73, 81, 85-86, 123, 135, 142-143, 444
Deoxyribonucleic acid: see DNA
Diauxic growth 13, 251, 274
  modeling 274, 277-278, 281
  of *S. cerevisiae* 13

Diffusion 16-18
  facilitated 16, 20-22
  free (molecular passive) 16
  into pellets 424, 456, 463
  of oxygen 16, 423-424, 430, 452, 454-455
  of phenoxyacetic acid 19
Diffusion coefficient 21-22, 379, 428-430, 452
  in dimensionless groups 443
  in lipid membrane 16-18
  in pellets effective 460
  for solutes in dilute aqueous solutions 454
  relation to volumetric mass transfer coefficient 430
Dilution rate 7, 49-50
  critical 50, 67-71
  maximum 249, 273, 329, 354, 359
Dimensionless groups related to mass transfer 442-444
  aeration number 479, 488-489, 504, 507, 514-515
  Biot number 469
  Damköhler number 458
  Froude number 437, 442, 489, 501
  Grashof number 443, 445, 459-460
  Peclet number 443, 445
  Reynolds number 437, 442-443
  Schmidt number 443, 445
  Sherwood number 443, 445-447
Dispersion model 379
DNA
  content in *E. coli* 32, 39, 266, 288, 290
  content in *S. cerevisiae* 39, 55, 290, 321
  elemental composition 55
  industrial production of 10
  measurements of 32, 159, 266, 295, 321
  replication 339, 281-283
  technology 1-2, 39, 43
Duplication cycle: see Filamentous microorganism

Effectiveness factor
  for reaction and film transport 458
  for reaction and solid phase transport 464-468
Electron acceptor 72
  anaerobic growth of *S. cerevisiae* 129
Electron transport 23, 28, 110, 129
Elemental composition 55-56, 62, 136
  matrix 73, 81, 145
Elemental mass balance 68, 73-74
Embden-Meyerhof-Parnas pathway 26
Energy charge 31
Enthalpy
  changes of, table 104
  free 95, 97
  of combustion 102
Error diagnosis 86-88
  chi-square distribution 85
  redundancy 86-87
    matrix 86-87
  test function 86-88
  variance covariance matrix 87
E. coli 4, 14, 29-30
  ash content and composition 56
  ATP and NADPH requirement for biosynthesis 32, 133
  ATP requirement for maintenance, biomass yield 127
  diauxic growth 251, 274, 278-284

electron transport 129, 133
facilitated diffusion 20
fermentative metabolism 24, 29-30, 127, 175
  growth of 20, 39-40, 174-177, 251, 453
    growth model by linear programming 174-177
    single cell model for growth 278-284, 286-288
    temperature influence on 263
    true yield coefficient and maintenance 254
  lactose uptake 24, 274, 279
  precursor needed to synthesize 25
  recombinant 37, 39-40, 279, 286, 288
    population balance 323
  RNA composition 32
    at different growth rates 266
  saturation constant for growth on glucose 247
  substrate uptake via PTS system 23-24
Ethanol 10-13
  as metabolic product 29, 58-59, 64-65, 68-71, 92, 138-148, 179
  biomass yield on 111, 256-260
  diffusion coefficient in aqueous solutions 455
  degree of reduction 62
  heat of combustion 102-104
  influence on growth of $S.$ $cerevisiae$ 58, 136, 256
  in oscillating yeast cultures 297
  maximum yield on glucose 65, 176-177, 259
  permeability coefficient 17
  uptake 10, 12-13, 30, 88-89, 175-177, 256-260, 295
Exponential growth 246, 265, 334, 352, 369
  of filamentous fungi 330
Exponential phase: see Growth phases

FAD/FADH$_2$ 27-28, 61, 110
Facilitated diffusion: see Diffusion
Fat(s): see also lipids
  composition of an $E.$ $coli$ cell 32
  elemental composition of 55
    neutral fat 55
    phospholipids 55
Fatty acids 32
  as antifoam agents 435
Fermentations
  batch: see Cultivation, batch
  continuous: see Cultivation, plug flow
  fed-batch: see Cultivation, fed-batch
  stirred tank: see Cultivation, stirred tank
  for single cell protein (SCP) 65, 89, 365, 414, 513
  $A.$ $teichomyceticus$ 416
  $A.$ $aerogenes$ 56
  $A.$ $niger$ 56, 453, 474
  $C.$ $acetobutylicum$ anaerobic 181-184
  $C.$ $glutamicum$ aerobic 74, 164, 180
  $L.$ $cremoris$ anaerobic 129-132
  $M.$ $capsulatus$ aerobic 65
  $M.$ $vaccae$ aerobic 89
  $P.$ $chrysogenum$ 56, 186-187, 313, 334-335, 499, 518
  $P.$ $pantotrophus$ 91
  $S.$ $cerevisiae$ 59, 136, 313
    aerobic 67-71
    anaerobic 186-187
    on ethanol 136
Fermentors: see Bioreactor

Filamentous microorganism 4
  $Acremonium$ $chrysogenum$ 36
    morphological structured model 302-303
  $A.$ $nidulans$ 254
    duplication cycle 301
  $A.$ $niger$ 10, 56, 71-72, 93, 254, 453
    citric acid production 71
    apex 300-301, 304
  $A.$ $awamori$ 254, 302-303
  branching 12, 302-303, 331, 333
  compartment
    apical 300-305
    hyphal 300-305
    subapical 300-304
  conidiophore 302
  differentiation 291, 302-303
  duplication cycle 301
  fragmentation
    breakage function 331-333
    of bubbles 432
    partitioning function 331-333
    specific rate of fragmentation 331-332, 334
  $G.$ $candidum$
    branching 302
    morphology 302, 305-306
  imperfect fungi 302
    life cycle 302
  medium rheology 500, 515
  model
    intracellular structured model 291-292, 294-295, 298
    morphologically structured model 291-292, 302-303, 305
    population model 319-320, 330-335
  $P.$ $chrysogenum$ 19, 38-39, 56, 90, 132, 334-335, 499-500, 515-516
    penicillin production 19, 38-39, 90, 516
    population model of 334
    uptake of phenoxyacetic acid 19
  septum 300-302
  spontaneous mutants 293-294
  sugar uptake 24, 204
  tip extension 300-301, 305, 331, 333
Flavin adenine dinucleotide: see FAD/FADH$_2$
Flow cytometry 321-322, 328
Flux control: see Metabolic control analysis
Foam 435, 480
Fragmentation: see Filamentous microorganism
Free energy: see Gibbs free energy
Fractional enrichment 163-169
Froude number 437, 442, 489, 501
Functional genomics 2-3, 8, 14
Futile cycle 20, 126

Gas hold up 474
Gene dosage 288
Generalized degree of reduction balance 66
Generalized Thiele modulus: see Thiele modulus, generalized
Genetic instability
  modeling plasmid instability 292-293, 323-326, 402-403, 416
  recombinant $E.$ $coli$ 292-293
Gibbs free energy 11-12, 23-25, 27, 96-97, 100-101, 264
  of combustion 102, 106
  thermodynamic efficiency 114

Index

Glutamic acid/glutamate 10, 62-63, 92, 155, 181
  annual sales 10
  degree of reduction 62
Glycogen
  ATP and NADPH requirement for biosynthesis 32
  content in *E. coli* 32, 290
  in *S. cerevisiae* 57
  elemental composition 57
  content in *S. cerevisiae* during oscillations 297
Glycolysis 25-27
  analysis for different organisms 148-149, 178, 181,184-186
Glyoxylate cycle 28, 154-158
Grashof number 443, 459-460
Gross measurements error 86-88
Growth phases
  diauxic: see Diauxic growth
  exponential 246
    of budding yeast 294, 337
    of *L. cremoris* 130
  lag phase 268-274
Growth rate limiting compound/substrate 245-246, 252
  multiple limiting substrates 251
  saturation constant values for sugars 247
Growth rate equations
  balanced growth 6, 30-32, 111, 119-120, 367
  cybernetic model 274-278
  pH effects 261-264
  population models 238-239, 290, 295, 314-318, 323-326
    based on
      cell number 239, 315, 319, 326
      mass fractions 315
    single cell models 289-294
  structured 238-239, 251-253, 265-274, 278, 281, 283-295
    morphologically 239, 290-295, 298-299,302-305, 310-311
  temperature effects 261-264
  unstructured 6, 238-240, 245-253
    Blackman 250
    Contois 250
    for growth on multiple substrates 250-251
    logistic 250
    Monod 246-249, 342-345, 353-355
      with maintenance 254-255, 348-350, 357, 358
    Moser 250
    Teissier 250
    with product inhibition 252, 345, 357-358
    with substrate inhibition 252, 345, 357-358
Growth rate specific: see Specific growth rate

Heat balance: see Enthalpy
Heat of combustion 102
  of biomass 104, 108
  of compounds 102-104
  correlation to degree of reduction 105
Heat generation 95, 109, 494
  in aerobic processes 106
  in anaerobic processes 106-109
Henry's law 52, 108
Henry's constant 426
Heterofermentative metabolism: see Metabolism
Heterogeneous microbial culture: see Mixed cultures
Homofermentative metabolism: see Metabolism
Hyphae 300-301
  apex 300, 304
  apical cell 303, 332
  apical compartment 300, 302-305
  branching 302-303, 331, 333
  differentiation 302-303
  duplication cycle 301
  extension zone 301
  fragmentation 331-333, 338
  growth of 300, 302-305, 314, 331-333, 338
  hyphal compartment 300, 302-305
  life cycle 302
  modeling 305
  septum 300, 302
  subapical compartment 300, 302-304
  tip extension 300, 305, 331, 333
Hyphal growth unit 300, 304-306

Immobilized cells 323, 359, 456, 466-468
  mass transfer 456, 469
  population balances 323
Impeller: see Stirring
Imperfect fungi: see Filamentous microorganism
Inducer 279-282, 286, 308-309, 367
Induction 155
  of lac operon 283, 285, 288, 309
Inhibition 15
  of enzymatic reactions 189, 196-200, 204, 207-213, 233
  on growth by
    a product 252, 453
    high biomass concentration 250
    substrate concentration 252
    lactic acid 307
Interfacial area
  specific 427, 430-431, 433, 435-436
  of a cell 447
Interfacial film 427, 435
Interfacial saturation concentration 51-53, 364-366, 379, 427, 447, 451-453
Isotopomers 166-167

Jacobian matrix 389
  eigenvalues
    and stability of steady state 390
    calculation of 389
    of (definition) 389

Kinetics: see Growth rate equations
Krebs cycle: see TCA cycle

Lactic acid/lactate
  as a primary metabolite 37
  catabolic reactions 25, 124, 129-132, 148-150
  fermentative metabolism 29-30, 127
  formula and degree of reduction 62
  heat of combustion 104
  in corn steep liquor 40
  inhibitory effect 307-308
  production 10, 40, 363-364, 410-411
  transport process 16, 364
Lactic acid bacteria 4, 10, 124, 127, 129
  fermentative metabolism of 29
  heterofermentative metabolism 148

maintenance 127, 131, 348
  two compartment model for 271-274, 353
Lactobacillus casei 127
Lactococcus cremoris: see also *Lactococcus lactis*
  requirement of ATP for growth 30, 129
  RNA content 267
  structured model 271-274
Lactococcus lactis shift in metabolic product 396
Lac operon 280
Lag phase: see Growth rate equations
Lipid: see also Fat(s) 16-18, 156
  in *E. coli* 24-25, 32, 127-129, 133, 174-177
  in metabolic flux analysis 153, 156-159, 174-177
  lipopolysaccharide 32
    in *E. coli* 32
  neutral fat 55
  transport across lipid membrane 16, 18, 20
Logistic law 250
Lotka Volterra equation for prey predator interaction 404
Luedeking and Piret equation
  lactic acid production by *Lactococcus delbruekii* 130

Macromixing: see Mixing, macromixing
Macromolecule 11
  elemental composition 55
Macroscopic morphology: see Morphology
Maintenance 18, 20, 69, 122, 125-126, 128, 130-138, 175, 308
  influence on cell recirculation reactor 361-363, 414
  kinetics in batch culture 348-350, 416
  kinetics in continuous culture 254-256, 356-359, 391, 416
Mass transfer 7, 16, 67, 339-340, 423-430
  gas liquid mass transfer 52, 88-89, 364-366, 415, 511-513
    of other components than oxygen 453-456
    of oxygen 448-453, 473
    volumetric mass transfer coefficient (definition) 427
    scale up 495-496
  mass transfer into solid particles 456
    external 456-460
    intraparticle diffusion 460-469, 472
  regimen analysis of the penicillin fermentation 511
Matching law model: see Cybernetic model
Mean residence time 429
Mean Sauter diameter 431, 434-435
Medium 4, 6-7, 11
  coalescence 435, 451, 505
  complex 32, 40-41, 49, 127, 129-131, 148, 191, 204, 236
  containing filamentous microorganisms 499
  defined 32, 41, 236
  methods for characterization of mixing 485
  rheology 496-500, 515
  stirrer design 439, 515
  viscous 432, 439, 443, 452, 499
    mixing 452
    resistance to bubble formation 432
Membrane: see Cell membrane
Messenger RNA: see RNA
Metabolic control analysis 5, 189-190, 207-233
  branched pathway 221
  concentration control coefficient 213
  elasticity coefficient 214
  flux control connectivity theorem 214

flux control (sensitivity) coefficient 211-212
  flux control summation theorem 212
Metabolic engineering 1-4, 42-45, 151, 170, 184, 221-222, 478
Metabolic flux analysis 5, 44, 120-122, 151, 153
  citric acid fermentation 156
  heterofermentative metabolism of lactic acid bacteria 148-151, 180
  of aerobic *S. cerevisiae* 138-142
  of anaerobic *S. cerevisiae* 143-148, 179
  of solvent fermentation 181-184
  propane 1,3-diol fermentation 184-186
  using linear programming 171-178
Metabolism 5, 11, 35
  citric acid fermentation 67
  of *Clostridium acetobutylicum* 181, 187
  heterofermentative 148
  secondary 35-36
Metabolite
  growth with metabolite formation 25, 123-124
  primary 29
  secondary 12
Methane 62, 415, 479
  aerobic growth of *Methylococcus capsulatus* 65-66, 365, 414-416
  as product from CO_2 and H_2
  as substrate for production of single cell protein 412, 414-416, 513
  from *Methanobacterium thermoautotrophicum* 107
  heat of combustion 104
  Henry's constant for 426
  reduction of carbon dioxide 65, 72, 107-109
Michaelis Menten equation 190-194
  variants of 195-201
Micromixing: see Mixing, micromixing
Microscopic morphology: see Morphology
Mitochondria: see Phosphorylation oxidative
Mitosis in cell cycle for budding yeast 294
Mixed acid fermentation 29, 148-150, 181, 396
Mixed cultures
  as a result of infection 390-391
  competition between prey and predator 404
  competition between two microorganisms 397-401
  reversion of a desired mutant 402
Mixed substrate: see also Diauxic growth
  cybernetic model for growth on 277, 313-314
Mixing 4, 339-340, 370, 433, 481-486, 504, 508-511, 516-518
  different stirrer design for 482
  in regimen analysis of the penicillin fermentation 511
  macro mixing 483
  micromixing 483, 486
  mixing time in (ideal) bioreactors 339-340, 483, 491, 496
  visualized by computational dynamics (CFD) 500
Monod equation/model 246
  estimation of parameters 236-237, 306
  and external mass transfer 456-458
  with inhibition by substrate or product 345-348, 354, 356-358, 389-391, 393
  model for growth of filamentous fungi 291, 294, 302-303
  model for growth of *S. cerevisiae* 256-261
  Monod kinetics including maintenance 254-255
  structured models 251-253, 268-269, 276, 284-285, 303
Morphology 7, 291, 300-306, 499-500

Index

growth of filamentous microorganisms 302
morphologically structured models 291, 302, 303-306
population balance for hyphal elements of filamentous fungi 330-335
Mother cell: see Yeast budding
mRNA: see RNA, mRNA
Mutant: see Mutation
Mutation
  colonial mutants 294
  competition between two microorganisms 390, 399-404, 416
  description of spontaneous occurrence of mutants 294
  in obtaining control coefficients 222-223
  improved penicillin production 42-43
Mycelium 300-301
  breakage of mycelium 331
  rheological properties of medium containing 499

NADH, NAD^+, NADPH, $NADP^+$: see Nicotinamide adenine dinucleotide
Networks of zones models 503-504
Nicotinamide adenine dinucleotide
  NADH
    conversion to NADPH 61, 92, 117
    conversion to NAD^+ in respiration 27-28, 109-111
    degree of reduction 61
    determination of, using a cyclic enzyme assay 205-206
    formation in biomass production 132-133
    role in glycerol production 144
  NADH and $FADH_2$
    in TCA cycle 28
    in respiration 27, 110
  NADPH
    conversion to NADH 61, 92, 117
    lumping with NADH 61, 137, 138-148
    production in PP pathway 26, 133
    requirement in biomass production 132-133
    requirement in *E. coli* 32
    requirement in lysine production 180
Newtonian fluid 496-497
Non Newtonian fluid 497
  Bingham fluid 498
  Casson fluid 498
  power law model for viscosity of 498, 500
  Pseudoplastic fluid 498-499
Non-observability: see Observability
Nuclear magnetic resonance (NMR) 166-167
Nusselt number 492

Observability 150
Operator 79, 279-282, 284, 308-309, 460
  in a model for diauxic growth 281, 284
  of the lac operon 280
Oscillations 48, 295-299
  criteria for oscillations in a chemostat with one microorganism 390, 392
  in a chemostat with a mixed microbial population 405-406, 416-417
  of predator prey interactions 405-406, 416-417
  of yeast cultures 295, 297-298
Over-determined system: see Error diagnosis
Oxidative phosphorylation: see Phosphorylation, oxidative

Oxygen
  consumption 54, 128-129, 133
    aerobic growth 64-66
    citric acid production by *Aspergillus niger* 71-72
    E. coli 176-177
    propane 1,3-diol 184-185
    penicillin production by *Penicillium chrysogenum* 90, 472
    single cell protein production 65-66, 89-90, 414-415
    S. cerevisiae 67-71, 138-140, 370-371
  determination of oxygen in biomass 88
  diffusion coefficient 452, 454-455, 471-472
  diffusion into a pellet of *Penicillium chrysogenum* 472
  dissolved oxygen concentration 88-89, 425, 448-453, 470-473
    fermentations with mammalian cells 470
    laboratory bioreactor 428, 449
    pilot plant bioreactor 471
    production of single cell protein 425
    spontaneous oscillations of yeast cultures 295
  Henry's constant 426
  in compartment model for *S. cerevisiae* 508-509
  maintenance requirement 128, 138, 175
  operational P/O ratio 128, 133-135, 137-138
  regimen analysis of the penicillin fermentation 511
  requirement
    different organisms 254
    yeast culture 425, 490
  respiratory chain 27, 136, 142
  specific uptake rate 18, 68, 88, 175-177, 202, 256-260, 449
  transport 16, 18, 26, 109-110, 423-424, 506
    into a single cell 447
    into pellets 472
  volumetric mass transfer coefficient 365, 371 447, 471, 513
    empirical correlations 430, 439
    in a laboratory bioreactor 439, 449, 490
    measurement 449-453

Parasite: see Parasitism
Parasitism 401
Partitioning function: see Filamentous microorganism,
Peclet number 443, 445
Penicillin
  critical dissolved oxygen concentration 474, 516
  design of pilot plant bioreactor for production of 512
  determination of k_la in a bioreactor for producing 469-472
  increase in productivity 41-43
  production of 9, 19, 35-36, 38-39, 41-43, 48, 53, 90
  yield coefficients for NADH/NADPH 133
Penicillium chrysogenum: see Filamentous microorganism, *P. chrysogenum*
Pentose phosphate pathway/PP pathway 25
Permeability coefficient 17-19
pH
  auxostat 49
  difference over cell membrane 113
  gradient in the proton motive force 112-113
  influence on
    transport of organic acids 18-20
    growth kinetics 261-264
    citric acid fermentation 71-72
    lactate fermentation 18, 130, 411
    CO_2 uptake 453-454

thermodynamic reactions 103-104
   uptake of NH_3 59-60
method for determination of mixing time 485, 516-517
Phase plane plot 395, 405-408
Phosphorylation
   oxidative 27, 109-110, 129
   substrate level 127, 129
Phosphotransferase system (PTS) 16
   uptake of glucose in *E. coli* 24
Plasma membrane: see Cell membrane
Plasmid
   copy number, and stability of 288-290, 292, 325-330, 401-402, 416
   flow cytometry for measuring plasmid DNA 321
   in age distribution model for *S. cerevisiae* 326
   loss of plasmid 324, 329
   modeling of recombinant *E. coli* 288
      Cornell model 289-290
      population balance 323
Plug flow reactor: see Bioreactor, tubular
P/O ratio
   definition 28
   operational 111, 114, 127-128, 132-133, 135-142, 185
Polyhydroxybutyrate (PHB) 57, 91
Population balance 315-318
   age distribution model of *S. cerevisiae* 326
   general form 316
   hyphal elements of filamentous fungi 330
   plasmid content in recombinant *E. coli* 323
   size distribution of *Schizosaccharomyces pombe* 320
Power input/dissipation/consumption
   correlations involving power input 432-434, 436, 438, 440, 486-489
      in bubble size determination 436-437
      in k_la determination 438-439, 515
      in mixing 486-489
Power law
   model 498-499
   index 498-499
Power number 487-491
   for different stirrer designs 488
Prandtl number 493
Precursor 11-12, 25, 27, 30
   for synthesis of cell material 120
   in antibiotics 35, 39, 90
Procaryotes 301
   sugar transport by PTS system 23
   fermentative pathways for (Fig. 2.6 A, B) 29
   proton transport of oxidative phosphorylation in 112
   transhydrogenase taken from Azeobachter 92
Product formation stoichiometry: see Stoichiometry
Product yield: see Yield coefficient
Promoter
   in a model for diauxic growth 285
   in inducible plasmids 279-280
   of the lac operon 280
Protein 2-3, 10
   as a biomass component 32, 55, 266, 268, 513
   as a foam stabilizing compound 435
   catabolite activator protein (CAP) 282
   Cornell model 290
   degradation rate 128, 268, 286, 288

   denaturation of 96, 262
   elemental composition 55, 136
   in *E. coli* 14, 24, 30, 32, 39, 262, 266, 279-281, 286, 288
   in metabolic flux analysis 156
   in ribosomes 266
   in *S. cerevisiae* 10, 14, 37, 39, 55, 57, 260-261
   in structured model 251, 268-269, 281, 289-291
   in uptake of ions 24
   measurement by flow cytometry 321
   number of molecules per cell 281
   produced by protein synthesizing system (PSS) 31, 287-288
   recombinant 10, 37, 39, 43, 260-261, 286-289, 326
   repressor 279-282, 286, 288-289
   single cell (SCP) 10, 53, 57, 65, 365
      bioreactor design 415, 513
      use of methane for production of 65, 513
      use of methanol for production of 65, 513
      use of ethane for production of 89
   synthesis model 268-269, 281, 286-288
   ribosomal 266
   transport by 14-15, 22, 24
Proton transport 18, 20, 24, 113
Pseudosteady state 5
   intracellular metabolites 15, 100, 119-124, 134, 142, 151-152, 155-156, 205
   mass transfer processes 16, 423-425, 448-449, 453, 456, 460, 511
   of ATP 12, 22-28, 30-34, 62-63, 99-100, 109-112, 161, 175-177
   pathway intermediates 100-101, 148-149, 163, 201, 212-215
Pseudoplasticity 498

Quasistationarity: see Pseudosteady state

Reaction limited regime 458
Recirculation of biomass
   in a stirred tank reactor 359-364
   in a plug flow reactor 372, 418-420
Recombinant: see Protein, recombinant
Redundancy: see Error diagnosis
Regimen analysis 511, 518
Repression/repressor 286, 288-289
   in *E. coli* model 308-309
   model for diauxic growth 281, 284
   of lac operon 280
Respiratory chain: see Phosphorylation, oxidative
Respiratory quotient (RQ)
   aerobic growth with NH_3 as N-source 64
   definition 54
   for growth of *S. cerevisiae*
      with ethanol formation 70-71
      without ethanol formation 69
Response coefficient (in MCA) 228
Reynolds number 442-443, 445, 501, 510
   for stirring 442, 487, 490
   of gas stream at the orifice of a sparger 437
Rheology of fermentation media 499, 515, 518
Ribonucleic acid: see RNA
Ribosome 267, 287, 313
   in protein synthesis model 290
   in structured models 290, 313
RNA (mRNA, rRNA, tRNA)

in *E. coli* 32
in lactic bacteria 266
mRNA 15
rRNA 266-267

Saccharomyces cerevisiae: see Yeast
Saturation constant
 in allosteric enzymes 201
 in Michaelis Menten kinetics 191
 in Monod model 246
 oxygen uptake by *S. cerevisiae* 449
 values for sugars in different microorganisms 247
Scale up 2-3, 7, 477-478, 482, 510-512
 effects on
  metabolic processes 508-509
  mixing time 339, 379, 491, 496, 509-510, 514, 517-518
  mass transfer 495-496
  reactor to surface area 493
Schmidt number 443
Secondary metabolite 12, 35-37
 influence of morphology 312, 334, 338
 Luedeking and Piret model for production of 130, 187
 penicillin 9, 19-20, 35-36, 38-39, 41-43
Septum: see Filamentous microorganism
Shear stress
 definition of
  for bubbles 432
  for liquid flow 496-498
 on hyphae 331
Shear rate
 definition 496
 effect on hyphae 331-332
Sherwood number 443
 mass transfer into a single cell 445
Single cell protein: see Protein, single cell
Solubility
 of carbon dioxide 16, 425, 427
 of oxygen 16, 425, 427, 448, 450-451, 513
 partitioning coefficient 16
Specific growth rate
 definition 53
 general calculation of 242-245
 in black box model 245-246, 252
 influence of temperature and pH 261-263
Statistical theory of turbulence 432-433
Stability
 asymptotic 389
 conditional 399
 of a microbial strain 294, 416
Stirred tank reactor: see Bioreactor
Stirring
 influence on k_la 439-442, 490
 influence on mixing 483-488
 power input 439, 441-442, 486-488, 490, 493-494, 515
Stochastic model 238, 281
Stoichiometry
 biochemical reaction networks
  aerobic processes 132-142
  in simple networks 142-150
 black box model 57-60, 73-77
 morphologically structured model 291-292
 pathway reactions 63

stoichiometric matrix 122-123
Stokes law 459
Storage carbohydrates: see Glycogen and Trehalose
Subapical compartment: see Filamentous microorganism
Substrate level phosphorylation: see Phosphorylation, substrate level
Superficial gas velocity
 definition of 436
 upper limit of 482
 use in correlations 436, 438, 441-442, 498
Synchronous culture 295, 312
Systems biology 1-4, 7-8
TCA cycle 27-28, 132-133, 171, 178, 185
 anaplerotic pathways 34, 154, 156
 energetics of aerobic processes 132
 in metabolic flux analysis of citric acid fermentation 155
Teissier model: see Growth rate equations, unstructured
Temperature
 influence on growth kinetics 261-263
 influence on oxygen solubility 448
Thermodynamic efficiency 28, 109, 111, 114-115, 142
Thiele modulus 464-466
 generalized 468-469
Tip extension: see Filamentous microorganism
Transport process
 active transport 22-24, 114-115
 facilitated diffusion 20-22
 free diffusion 16-20
 See also: Mass transfer and Cell membrane, active transport
Trehalose
 in *S. cerevisiae* 57
Tricarboxylic acid cycle: see TCA cycle
Turbulence
 isotropic turbulence 433-434
 statistical theory of 432-433
 turbulence models 482, 504
Turnover
 of ATP 31, 34, 100, 124-126, 128
 of macromolecules 126, 128, 269
 of mRNA 126, 128, 285-286
 of NADH and NADPH 34, 124-125

Uncorrelated measurements 82-83
Unstructured model 238-240
 for growth 245-253
 alternatives to the Monod model 251
 black box model 245-246, 252-253
 Monod model 246-249

Variance covariance matrix: see Error diagnosis
Viscosity
 for fluids 496-498
 definition 443, 459, 486
 in dimensionless groups 443
 in medium containing filamentous fungi 460
 influence on volumetric mass transfer 439
 resistance to bubble formation 431-432
Volumetric mass transfer coefficient
 definition 427

Wall growth
 filamentous fungi 300-301

influence on reactor design 418
Wash out
  dilution rate 354
    for a mixed microbial population 399-402
  design of cell recirculating system 362
Wastewater 9
  bioreactors for waste water treatment 340, 376
  removal of methane and carbon dioxide 425
  removal of nitrate 72

Xanthomonas campestris 10
  xanthan gum fermentation 10

Yeast
  baker's yeast production 370-371
  budding 294, 297-299, 313, 326, 328-330, 337-338
    age distribution model 298, 326, 329-330
    cell cycle 294, 297, 299, 330, 337
    index 298, 328
    morphological model 294, 298-299, 337
    oscillating cultures 294, 297-298
  Candida lipolytica 156-157, 186
    metabolic flux analysis 156
  Candida utilis 56, 111, 254
    elemental composition of 56
  down scaling of yeast fermentation 518
  oscillating cultures of 48, 294, 297-298
  Saccharomyces cerevisiae 4, 10, 172, 247, 254
    aerobic growth of 29, 51, 64, 67-71, 110, 425
    aerobic growth with ethanol formation 138-142
    anaerobic growth of 29, 45, 57-59, 63, 143-148
    ATP requirement for maintenance 138
    biomass yield on ATP 18, 127, 138
    consistency analysis of aerobic fermentation 67-71
    elemental composition of biomass 56, 127
    error diagnosis of fermentation 86-88
    glycerol formation in 116, 143
    growth on ethanol 13, 92, 138, 143
  Schizosaccharomyces pombe 319-320, 338
    population model 319-320, 338
Yield coefficient
  apparent (observed) and true 253
  definition 18, 54
Yield stress 498

Zymomonas mobilis 10, 65, 271
  ethanol production 65

LaVergne, TN USA
17 February 2011
216859LV00004B/5-18/A